Biodiversity, Functional Ecosystems and Sustainable Food Production

Charis M. Galanakis
Editor

Biodiversity, Functional Ecosystems and Sustainable Food Production

Springer

Editor
Charis M. Galanakis
Galanakis Laboratories
Chania, Greece

ISBN 978-3-031-07433-2 ISBN 978-3-031-07434-9 (eBook)
https://doi.org/10.1007/978-3-031-07434-9

This Springer imprint is published by the registered company Springer Nature Switzerland AG
The registered company address is: Gewerbestrasse 11, 6330 Cham, Switzerland

Preface

Over the last decades, different practices like the cultivation of a few high-yielding crop varieties on large scales, the application of heavy machinery and continued mechanization of agriculture, the removal of natural habitats, as well as the application of pesticides and synthetic have resulted in the simplification of agro-ecosystems at various spatial scales. Agriculture's intensification has resulted in a significant increase in food production, but at the same time, it has transformed landscapes. Indeed, there is a concern that declines in biodiversity affect the delivery of ecosystem services. Although they have increased food production, the above practices cannot be considered sustainable in long-term applications. Subsequently, the importance of diversity and diversification in increasing the resilience of food systems is becoming a key issue. Biodiversity and microbiome activities support processes across soils, plants, animals, the marine environment, and humans. The resilience of food systems can be achieved through better use of the plant, animal genetic, and microbial genetic resources. To this line, there is a need for a new reference.

Food Waste Recovery Group (www.foodwasterecovery.group) is a leading consulting network that has developed numerous initiatives such as workshops, e-courses, reports for governmental bodies and companies, a new open-access journal (*Discover Food*, Springer Nature), and more than 50 books in the broad fields of bioresources, environment, sustainability, food, and nutrition. The present book covers all the essential aspects of biodiversity, functional ecosystems, and sustainable food production. It covers ecosystems in terms of emerging risks to plant health, pest and disease control, crop and animal production, soil fertility, and productivity. It emphasizes the interconnection of ecosystem functions, food production, and quality with consumer health. It also exhibits diverse practices that contribute to sustainable food production at different levels.

The book consists of ten chapters. Chapter 1 provides a synthesis of published evidence of the complex and crucial relationships between elements of agro-biodiversity, climate change, and the food chain. It highlights the trends of changes in the components of agro-biodiversity, the factors enhancing such changes, and the points needed to be considered to maintain a sustainable way of food production for obtaining a stable food chain. In addition, the present status of studies relating to both agro-biodiversity and genetic resources is denoted. Chapter 2 revises climate change, its impacts on pests' behavior, and the spread of emerging pests. Since climatic factors such as temperature, relative humidity, solar radiation, precipitations, and carbon dioxide level significantly impact different organisms, climate change can lead to various challenges for organisms such as pests. Furthermore, climate change consequences can lead to changes in the abundance and geographic distribution of different pests. These changes are responsible for emerging pests commonly related to the global trade in agricultural products.

Chapter 3 discusses the future-proofing of plants against climate change, as the world must immediately act on it to end hunger and malnutrition. To ensure rapid and advanced agricultural development in a short period, precision farming practices and innovative breeding strategies need to follow, such as machine learning, deep learning, big data analysis, remote sensing, artificial intelligence, system biology study, genomic prediction, speed breeding, and haplotype breeding. These techniques can prove the future plants against climate variability with increased yield potential and improved resilience to achieve the goal of resilient climate agriculture.

Chapter 4 deals with the role of integrated pest management for sustainable food production, giving soybean as an example. Integrated pest management is based on the principle that some degree of plant injury is tolerable without requiring pest control. Moreover, the most environment-friendly pest control tools should be combined to have a longer-lasting pest solution. Those tools that include biological control and transgenic plants, among others, are discussed in detail, highlighting commercially available ones combining environmental, economic, and social benefits.

Over the years, the indiscriminate use of pesticides has caused several problems, including pest resistance and contamination of important global sources such as water, air, and soil. Therefore, plant-based pesticides can be an ecological alternative to synthetic pesticides to improve the efficiency of agricultural production and sustainably reduce the food crisis while protecting consumers' health. In addition, they are cheap, biodegradable, environmentally friendly, and act more specifically through multiple mechanisms of action. Chapter 5 presents in detail the use and activity of plant-derived pesticides.

Chapter 6 discusses antimicrobial use in animal food production. The push towards intensive livestock production systems to meet food demand has increased antimicrobial use in livestock metaphylactically and prophylactically as growth-promoting agents. However, it has become increasingly recognized that the widespread application of antimicrobials in food production contributes to the emergence and proliferation of antimicrobial-resistant species. The presence of clinically relevant multidrug-resistant species in food-producing animals may result in human cases of infectious disease. As a result, antibiotics in human medicine are being monitored in most developed countries. Nevertheless, antimicrobial use in livestock and food production is poorly scanned and assessed. Better monitoring, surveillance, and understanding of the consequences of the expressive use of antibiotic agents in veterinary medicine are needed to determine their potential impact on animals and humans health wise.

Chapter 7 discusses the functional properties of bee pollen of some environmental-friendly novel unit operations. In this context, fluidized bed-assisted cold drying, microwave, freezing, vacuum, infrared, and microwave-assisted vacuum drying are investigated as new environment-friendly unit operations. The chapter also discusses the protective properties of these unit operations on the raw material properties during processing, storage, and kitchen applications. In addition, many specific bioactive properties of bioactive pollen components, such as antimicrobial, antioxidant, and anti-carcinogenic properties, are discussed. Finally, the pollen production chain, sustainability in this life cycle, and the environmentally-friendly features of these new applications on sustainability are revealed.

Chapter 8 is divided into two parts. The first part provides critical insights to allow scientists to generate discoveries across microbiome applications for sustainable food systems. It offers a broad view of research of interest to early and experienced scientists and an understanding of the role of microbiomes as vital ecosystems and inter-relations among microbiomes across food chains. In the second part, the reuse of spent coffee grounds to increase the resilience of agro-food systems is described as an example of a successful application of a microbiome-related intervention.

Chapter 9 focuses on the role of dynamic value chains using underutilized biodiversity crops to improve food system resilience and deliver foods with good nutritional and healthy properties while ensuring a low environmental impact. Consumers pay attention to healthier food attributes and diets due to sustainable production processes. Other significant trends include increasing demand for less processed and regionally supplied food. To meet these demands, food production and processing need to evolve to preserve raw materials' "natural" character while ensuring sustainable, tasty, and, most importantly, healthy food. Likewise, it is vital to understand the influence of consumers' preferences in preserving the beneficial attributes of food up to the time of consumption.

Chapter 10 deals with new alternative protein and traditional protein sources of terrestrial origin for food and feed such as insects, plants (legumes and grasses), and by-products of crops. Protein supply is critical for animal feed and human consumption. Therefore, integrating a variety of alternative protein sources into existing products or processes should be explored to ensure more resilient supply chains, highlighting consumer preference by a clean labeling strategy and respective market opportunities.

Conclusively, the current book assists food producers and researchers working at the edge of food and environmental fields, agriculturalists and food scientists seeking to improve production system efficiency, and professionals active in the food supply chain from farm to fork. Likewise, university libraries and institutes worldwide could suggest this reference as ancillary reading in graduate and postgraduate courses dealing with agricultural and environmental science, sustainable food systems, and bioresources.

I want to thank and acknowledge, one by one, all authors of this book. Their acceptance of my invitation, and their dedication to the editorial guidelines and schedules are much recognized. I would also like to thank Arjun Narayanan, the book manager; Daniel Falatko, the acquisition editor; and the production team of Springer Nature for their assistance during the development of this book. Finally, I have a message for all the readers of this book. This collaborative effort ended up in a manuscript containing hundreds of thousands of words, and thus it may collect some errors. Criticism and constructive comments are welcome, so please do not hesitate to contact me to suggest changes if you have any objections.

Chania, Greece Charis M. Galanakis

Contents

Chapter 1
Agro-Biodiversity Across the Food Chain

Shamim Ahmed Kamal Uddin Khan, Md. Moshiur Rahman, and Md. Matiul Islam

Abstract Currently, we are at a point of unprecedented changes in the climate, which is affecting the food chain across the agro-biodiversity. The changes in species diversity in the food web indicate the changes in variety at any of the trophic levels. Adaptation with the environmentally changed conditions depends on the effective use of biological components of the local agro-ecosystem, which is also the focal point of sustainable approaches. Sustainable management of natural resources in the agro-biodiversity is essential for food and livelihood security of the living beings in an ecosystem. In this chapter, a synthesis of published evidence of the complex and crucial relationships between elements of agro-biodiversity, climate change, and the food chain is provided. A review of published articles highlights the status and trend of changes in the components of agro-biodiversity, the factors enhancing such changes, and the points needed to be considered to maintain a sustainable way of food production for obtaining a stable food chain. Finally, the present status of studies and researches relating both of the agro-biodiversity and genetic resources are identified. Nevertheless, despite the need for more knowledge of agro-biodiversity and the food chain, it is clear that more effective action would be taken.

Keywords Agro-biodiversity · Change in diversity · Affecting factors · Sustainability

S. A. K. Uddin Khan (✉) · Md. Matiul Islam
Khulna University, Khulna, Bangladesh
e-mail: samkuk_bd@ku.ac.bd; matiul_rubel@ku.ac.bd

Md. Moshiur Rahman
Khulna University, Khulna, Bangladesh

Research Scholar, University of California, Davis, Tracy, CA, USA
e-mail: moshiurku@fmrt.ku.ac.bd; momrahman@ucdavis.edu

C. M. Galanakis (ed.), *Biodiversity, Functional Ecosystems and Sustainable Food Production*, https://doi.org/10.1007/978-3-031-07434-9_1

Highlights

- The mass extinction of biodiversity in the ecosystems is associated with climate warming, increased nitrogen deposition, land-use change, biotic exchange, and an increased atmospheric CO_2 level.
- The status of preservation of crop germplasm is insufficient to the alarming pace of threats posed by climate change. The collection and preservation status is low for crop wild relatives, wild food plants, and neglected and underutilized crop species.
- About 75% of all crop genetic diversity has been lost since the previous century, primarily due to changes in the agricultural food system, which values uniformity.
- The productivity of a fishery remains high, especially in inland waters, as lower trophic-level species increase in abundance without larger predators.
- Introduction of alien species causing genetic erosion by the disappearance of traditional species through introgression.
- The knowledge on the effects of environmental stressors on biodiversity and the food chain is still far from our understandings as the studies, including multiple environmental stressors, are scarce.
- During the last two decades, studies on food-web ecology are being conducted, including smaller trophic entities consisting of only two to seven species which are unable to predict emergent patterns of food webs. A robust geographic bias in respect of sampling was observed, and the number of studies conducted within each biome is not proportional to the overall size of the biome.

1 Introduction

The concept of ecosystem differs in respect of economic, cultural, and societal needs where biological diversity is one of the central components. 'Biological diversity' refers to the variability among living organisms in terrestrial and aquatic ecosystems, also known as 'biodiversity.' It also includes the ecological complexes within and between species in an ecosystem (Secretariat of the Convention on Biological Diversity, 2004). Biodiversity has enormous impacts on agriculture and food productions. It embraces not only the domesticated crops and livestock but also other species of plants, animals, and micro-organisms. These components of biodiversity deliver a range of vital services such as building healthy soils, pollinating plants, purifying water, providing protection against extreme weather events, enabling ruminant animals to digest fibrous plant materials, etc. (FAO, 2019a). Ecological components also contribute to agricultural production as the tree or herbaceous crops are protecting the soil against erosion and creating a favorable microclimate for other components (e.g., soil-inhabiting flora and fauna) of the production systems (Balvanera et al., 2017).

The ecosystems used for agriculture and the production of food and non-food products are known as 'agroecosystems,' which comprise all biological resources and their biodiversity, including the physical environment and the practices managed by human beings (FAO, 2018a). The words 'agro diversity' and 'agro-biodiversity' are used interchangeably to indicate the biological diversity of lands used for agricultural purposes (Brookfield & Stocking, 1999). Agro-biodiversity, as defined by FAO (1999), is the combination of varieties and variability of different biological components related directly or indirectly to food and agriculture. These components include crops, livestock, forestry, and fisheries (considered as agricultural products) and comprise the diversity of their genetic resources. The variety of non-harvested species playing the supporting roles in production (soil microorganisms, predators, pollinators) and maintaining the wider environment to keep the diversity of agro-ecosystems (agricultural, pastoral, forest, and aquatic) are also included in the agro-biodiversity. However, Thrupp (1998) mentioned the agro-biodiversity as the vital sub-set of biodiversity, which results from natural selection processes maintained by the peoples related to agriculture over millennia. Sustainable management of biological resources in the agro-biodiversity is vital for the food and livelihood security of the living beings in an ecosystem (FAO & PAR, 2011). All species of a biological community form a 'food web' through trophic interactions between them (Banašek-Richter, 2004). Elton (1926) first acknowledged the scientific significance of food webs and used the term 'food chain' to express the relationships between the animals that feed on one another (Lawton, 1989) (Fig. 1.1).

A functional food web or chain is an expression of the relationship between biodiversity and the ecosystem through resource use complementarity (Norberg, 2000) and species abundances (Wootton, 1994). Ecosystem response to nutrient enrichment for crops through the trade-offs between competing plants and their resistance to herbivores (Chase et al., 2000). Species coexistence and the trophic position of species in the food web are inter-related. For instance, increasing consumer diversity may decrease producers' biomass through predator-mediated coexistence (Duffy et al., 2003). Similar effects of plant diversity have been recorded on soil-resource depletion (Symstad & Tilman, 2001). The changes in species diversity in the food web depend on the changes in variety at any of the trophic levels, which ultimately propagate to both higher and lower trophic levels of an agro-biodiversity (Dyer & Letourneau, 2002).

Agro-biodiversity plays a vital role as a source of food for humanity and also maintains a healthy environment for better living and sustainable development (Esquinas-Alcázar, 2005). Adaptation with the current environmentally changed conditions by reducing the risk exposures depends on the richness of agro-biodiversity and crop genetic diversity (Vigouroux et al., 2011). Increasing the quantity and the nutritional quality of food products through more effective use of the functions performed by the local agroecosystem's biological components is the focal point of sustainable approaches (FAO, 2019a). Sustainable agro-biodiversity ensures the better utilization of the components of biological diversity that does no harm to the biodiversity. It also maintains the present and future potentials of the

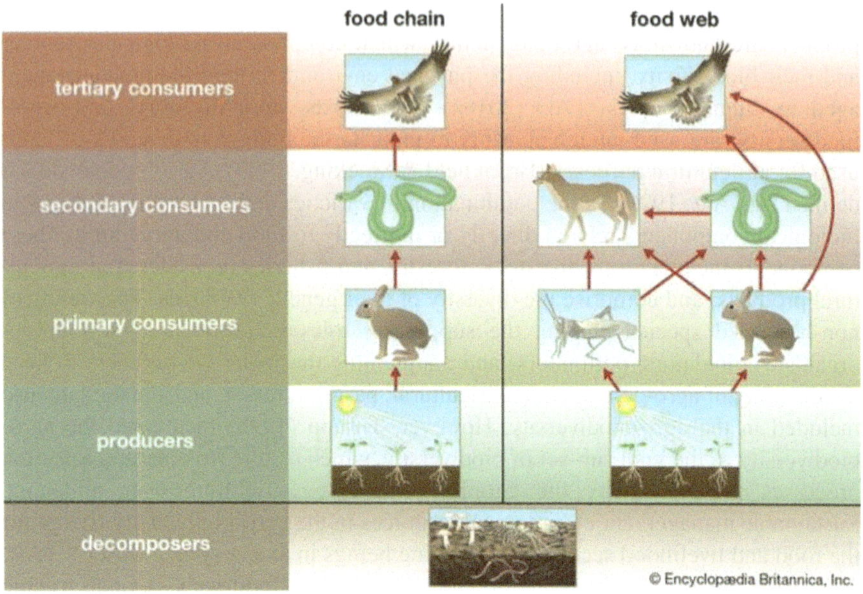

Fig. 1.1 A food chain shows how matter and energy from food are transferred from one organism to another. In a natural ecosystem, many food chains intertwine to form a complex food web (Encyclopedia Britannica, 2020)

biodiversity for better agricultural production (Secretariat of the Convention on Biological Diversity, 2004). Sustainable use of biological components and their conservation are interrelated. Zimmerer et al. (2019) highlighted four significant themes in the Anthropocene as ecology, governance, nutrition-health, and global change to explain the impact of ongoing complex human-environment interactions on agro-biodiversity. The sustainability of an agro-biodiversity depends on the conservation of neighboring or distant ecosystems that provide it with essential services (FAO, 2019a).

Sustainable use of agro-biodiversity is inevitable to ensure a 50% increase in food production (FAO, 2017) for a world population predicted to increase to almost 9.8 billion by 2050 (United Nations, 2017). The majority of the smallholder farmers and the 2.7 billion poor who live on less than two dollars a day depend on locally grown food for their living (Rapsominikis, 2015). Food security for these people falls at risk when agro-biodiversity faces threats due to changes in production systems. As a part of agricultural production systems in the 1990s, the developing countries adopted modern varieties of wheat, rice, and maize at around 90%, 70%, and 60%, respectively, which threatened the agro-biodiversity there (FAO, 2017). In the last couple of decades, the modern rice varieties leaped from 4% to 58% in Latin America and 12% to 67% in Asia (Dronamraju, 2008). FAO (2017) reported that more than 90% of crop varieties had been disappeared from farmers' fields in the past 100 years. Crop varieties and livestock breed are being lost annually at a rate of

2% and 5%, respectively (United Nations, 2019). The modern cultivation practices simplified the crop production system leading to a less resilient agro-ecosystem (Altieri et al., 2015). For example, over 50 pollinator species are at risk of extinction, and wild honeybee populations have dropped 25% since 1990 in the USA (Dronamraju, 2008). Maintenance of a sustainable agro-biodiversity through traditional crop production system is successful in restoring yields (Altieri, 1999). Extensive studies and researches warrant further attention on agro-biodiversity across the food chain in these circumstances. The overall objective of this chapter is to explore the status and trend of changes in the components of agro-biodiversity, the factors enhancing such changes, and the points needed to be considered to maintain a sustainable way of food production for obtaining a stable food chain.

2 Changing Biodiversity and Food Chain in the Global Aspect

Nature includes living organisms, their diversity, and interactions (among themselves and with their abiotic environment). Nature's contributions to people (NCP) may be either positive or negative to people's quality of life (IPBES, 2017; Díaz et al., 2018). The positive or negative contribution of nature depends on spatial, temporal, social, or cultural context (Saunders & Luck, 2016; Rasmussen et al., 2017). Though the positive contributions may be due to the expression of dominant climatic and socio-economic factors, which may become recessive in times, and the less powerful elements can take place (Cáceres et al., 2015). In this respect, it is essential to verify the trend of the changes in the factors governing the well beings, including the food web in the agro-biodiversity.

2.1 Demographic and Economic Trends

The demographic status and trends indirectly influence the changes in nature, which ultimately govern the NCP (Nature's contribution to people) and GQL (Good Quality of Life) at local, national, regional, and global levels. The World Bank (2017) has estimated a 2.5 times increase in world population from 1960 to 2016. The growing population has impacted the use of land through urbanization and the development of infrastructures and transportation networks (IPBES, 2016). By the mid of this century, all archetype scenarios show a significant increase in population size, which reduced by the end of the century except in regional rivalry (O'Neill et al., 2017; Samir & Wolfgang, 2017) (Fig. 1.2). The increasing economic globalization in recent decades has expedited the economic activities worldwide, which ultimately influenced the changes in the ecosystem, biodiversity, NCP, and GQL through various direct and indirect pathways (IPBES, 2016).

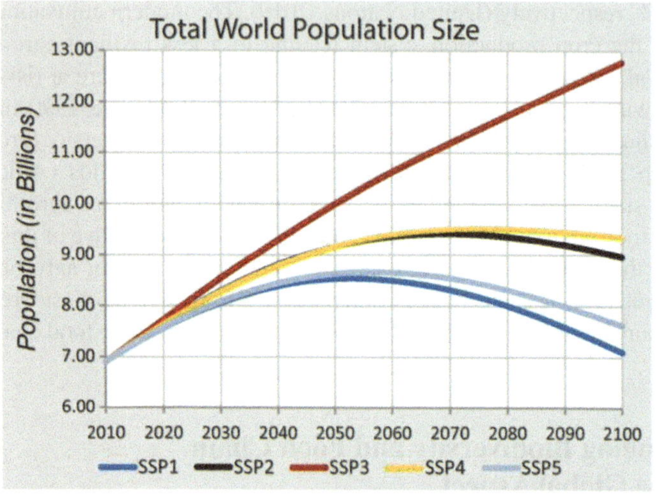

Fig. 1.2 Projected changes in world population according to the five Shared Socio-economic Pathways (SSP). (Adapted from Samir & Wolfgang, 2017). SSP1: Sustainability (Taking the Green Road- Low challenges to mitigation and adaptation), SSP2: Middle of the Road (Medium challenges to mitigation and adaptation), SSP3: Regional Rivalry (A Rocky Road- High challenges to mitigation and adaptation), SSP4: Inequality (A Road divided- Low challenges to mitigation and high challenges to transformation) and SSP5: Fossil-fueled Development (Taking the Highway- High challenges to comfort and common challenges to adaptation) (Riahi et al., 2017)

The worldwide economic development is the critical indirect anthropogenic driver of changes in nature, ecosystem, and biodiversity across all scales (global, regional, national and local). According to the World Bank (2017), the global GDP (at constant 2010 USD) increased about 7-times from 1960 to 2016. The current global economic trends have generated stresses on natural resources, the environment, and ecosystem functions (Schneider et al., 2011). Growing per capita GDP increased the demand for critical natural resources like food, water, and energy, adversely affecting the ecosystems and biodiversity through unsustainable patterns of production and consumption. These demands have exceeded the planet's biocapacity for more than 40 years. From an assessment by WWF (2016), it has been found that 1.6 Earths would be required to meet the human demands each year. Ecological Footprint also shows that the consumption patterns in high-income countries make stresses renewable resources, often at the expense of people and nature elsewhere in the world (WWF, 2016).

2.2 *Trends in Land Use Pattern*

Depending on the trade and the degree of globalization of future agricultural and forestry markets, all SSP (Shared Socio-economic pathways) scenarios show land-use changes due to agricultural and industrial demands for food, timber, and

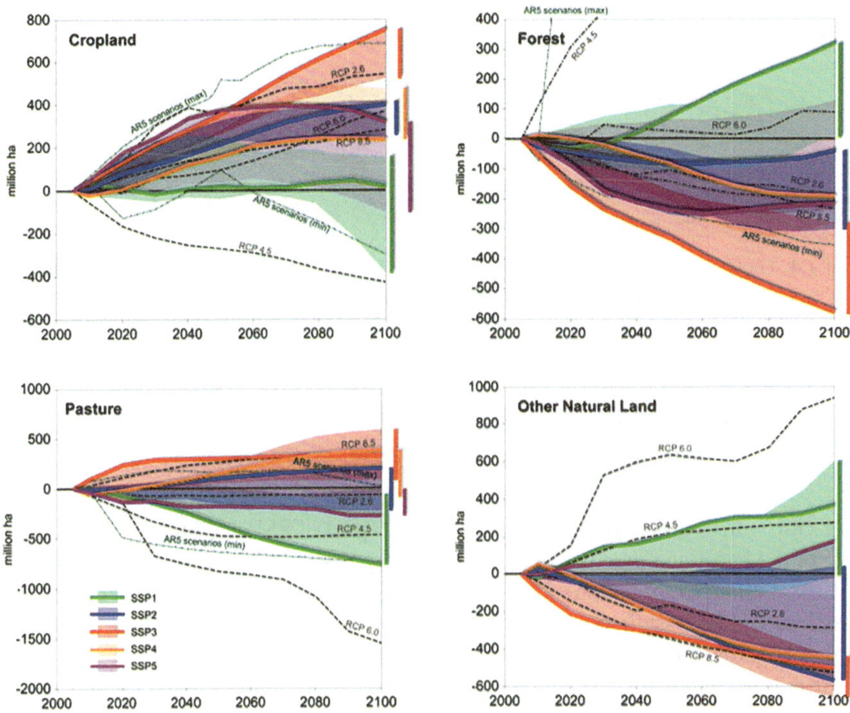

Fig. 1.3 Trends in land use patterns for different SSPs. The thick lines show the SSP marker baseline scenarios and colored areas range from other non-marker strategies. (Adapted from Riahi et al., 2017). Changes shown here are relative to the base year of 2010 = 0. The development of the RCPs (Representative Concentration Pathways) (van Vuuren et al., 2011) and the range of the IPCC AR5 (Intergovernmental Panel on Climate Change Fifth Assessment *Report*) scenarios (Clarke et al., 2014) have also been showing here. The croplands include energy crops, and other natural land consists of all land-categories beyond forests, pasture, cropland, and build-up areas

bioenergy (Popp et al., 2017). The SSP scenarios show that the total cultivated land can be expanded or contracted by hundreds of millions of hectares over this century (Fig. 1.3). Comparatively large pressure on the global land-use system has been reported for SSP3 featuring massive growth of population, relatively low agricultural productivity, and little emphasis on environmental protection. Such a land-use pattern in SSP3 leads to large-scale losses of forests and other natural lands due to an expansion of cropland and pasture land (Fig. 1.3). The SSP1 scenario features a sustainable land transformation with comparatively little pressure on land resources due to low population projections, healthy diets with limited food waste, and high agricultural productivity. All other SSP scenarios feature modest land-use changes with some expansion of overall cultivated lands (Riahi et al., 2017) (Fig. 1.3).

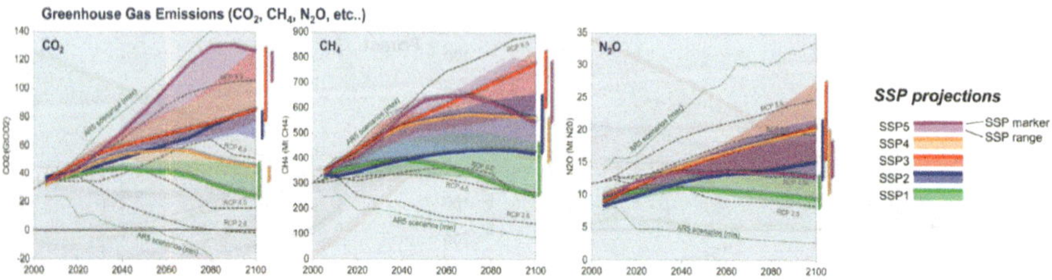

Fig. 1.4 Global emissions and the global average change in radiative forcing. (Adapted from Riahi et al., 2017). SSP baseline marker scenarios (and ranges of SSP non-marker baseline scenarios) are compared to the RCPs (van Vuuren et al., 2011) and the full range of the IPCC AR5 scenarios (Clarke et al., 2014)

2.3 Climate Change Scenario with Respect to Greenhouse Gases

The emission of greenhouse gases (GHG) in respect of SSP and broadly representative of the baseline range has been shown in Fig. 1.4. CO_2 emission is strongly correlated with the combustion of fossil fuels. Higher dependence on fossil fuels in the SSP3 and SSP5 baselines results in higher CO_2 emissions. On the other hand, low CO_2 emission in the SSP1 and SSP4 has been estimated due to lower dependency on fossil fuel and increased deployment of non-fossil energy sources. An intermediate emission of CO_2 was reported for the SSP2 baseline (Riahi et al., 2017). CH_4 (the second largest contributor to global warming) is emitted from non-energy sources like manure from livestock, rice cultivation, and enteric fermentation. In contrast, the energy-related sources, including the production and transport of coal, natural gas, and oil, contribute to its emission to a lesser extent. Changes in demographic status and food demand determine the future CH_4 emissions across the SSPs, which is depicted through higher CH_4 emission in the SSP3 baseline and lowest in SSP1. All other SSPs show intermediate levels of CH_4 where the combination of different energy and non-energy drivers leads to emissions in the long term. The major contributors to N_2O emission across all the SSPs are agricultural soils and fertilizer use. N_2O emissions are lowest in SSP1, featuring sustainable farming practices and low population. On the other hand, the highest emission of N_2O was reported for SSP3 and SSP4 baselines due to increased population and a higher rate of fertilizer use for more food production (Riahi et al., 2017).

2.4 Changes in Plant Genetic Resources

Food security, nutrition, and livelihood of the people depend on the plant genetic resources. The richness of plant genetic resources can allow crops to adapt to evolving environmental conditions and sustainable intensification of agricultural production. According to FAO (2020a), the plant genebank holdings in 103 countries, 17

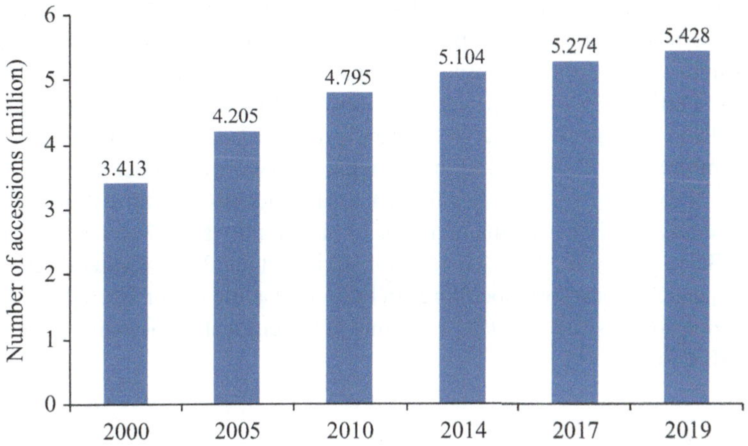

Fig. 1.5 A number of accessions of plant genetic resources secured in medium- or long-term conservation facilities in the world, 2000–2019 (FAO, 2020a)

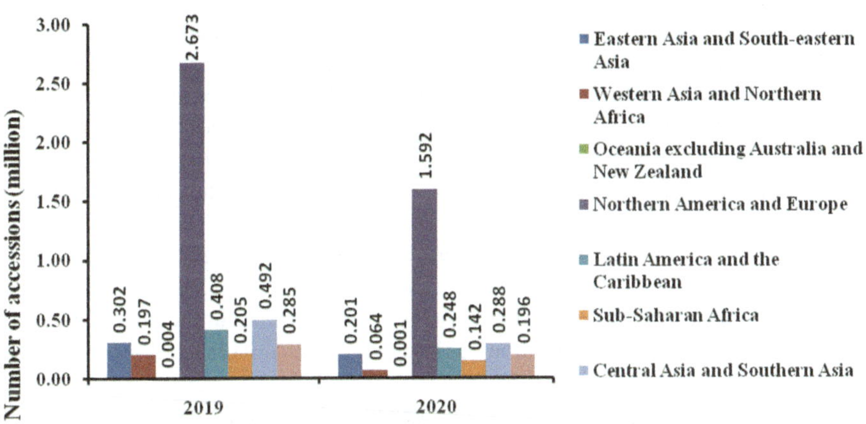

Fig. 1.6 A number of accessions of plant genetic resources secured in conservation facilities under medium- or long-term conditions by region in 2000 and 2019. (Adapted from FAO, 2020a)

regional and international research centers were estimated at 5.43 million accessions in 2019 (Fig. 1.5), which was about a 1% increase on the previous year. The gene bank holdings virtually increased worldwide between 2015 and 2019. Over the year, the conserved germplasm increased in 40 out of 59 countries and 7out of 12 international centers. However, a net decrease was also reported by FAO (2020a) (Fig. 1.6). A greater than 1% decrease in germplasm was recorded in six countries, out of which three are in Europe and one each in Western Asia (−1.7%), Eastern Africa (−10.7%), and South America (−11.4%).

Almost 96,000 germplasm samples from over 1700 species listed in the IUCN categories of primary global concern were conserved in 290 genebanks around the world by the end of December 2019. These samples include wild relatives of crops significant for global and local food security (FAO, 2020a). However, in recent years the worldwide response of preserving crop germplasm has been insufficient to the alarming pace of threats posed by climate change to crop and crop-associated diversity. Particularly for crop wild relatives, wild food plants, and neglected and underutilized crop species, the collection and preservation status is low (FAO, 2020a).

A combination of short-lived and perennial crops and timber and non-timber products developed over centuries in rural areas made diverse agricultural systems. Such diversity facilitates the maintenance of biodiversity and lowers nature's degradation (Balvanera et al., 2014; González-Esquivel et al., 2015; Kanter et al., 2018). Though pesticide-based monocultures in the Asia-Pacific region caused a reduction in genetic resources and about 70% decline in the cultivation of native plant varieties (IPBES, 2018).

2.5 *Changes in Forest Genetic Resources*

Precise monitoring of the status and trends of forest genetic resources as well as monitoring the quality of forest degradation, forest restoration, and species composition are difficult tasks (Newton et al., 2015). The number of extant tree species in the world has been estimated by (Beech et al., 2017) as about 60,000. However, FAO (2014) on The State of the World's Forest Genetic Resources (SoWFGR) listed nearly 8000 species of trees, shrubs, palms, and bamboo, of which about 2400 are actively managed for cultivation, and more than 700 species are now included in tree-breeding programs globally.

About 93% (3.75 billion ha) of the forest area globally is composed of naturally regenerating forests, and 7% (290 million ha) is planted (Figs. 1.7 and 1.8). The size of naturally regenerating forests has decreased since 1990, but the area of planted forests has increased by 123 million ha though the rate of increase slowed down in the last 10 years (FAO, 2020b).

Forest genetic resources are being threatened and eroded globally by converting forests to agriculture, unsustainable harvesting of trees for wood and non-wood products, grazing and browsing, climate change, forest fires, and invasive species (FAO, 2014). Forests cover about 30.6% of the world's land area, and the occupied space is continuously shrinking (FAO, 2018b). The extension of commercial and

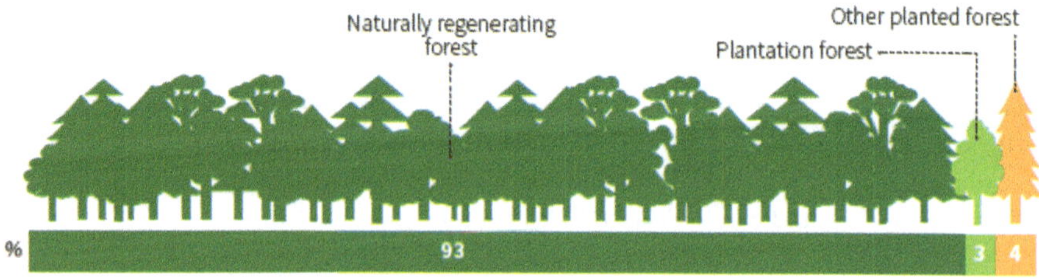

Fig. 1.7 Naturally regenerating versus planted forests, 2020 (% of global forest area) (FAO, 2020b)

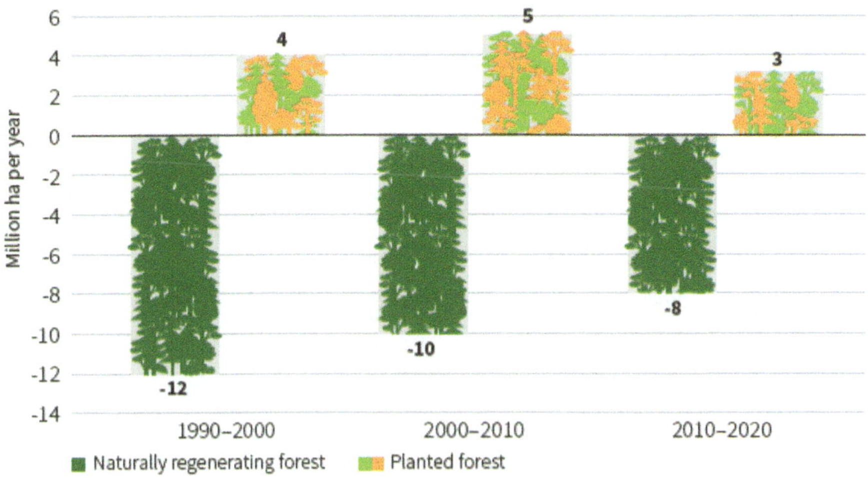

Fig. 1.8 Annual net change in the area of naturally regenerating and planted forest, by decade, 1990–2020 (FAO, 2020b)

large-scale agriculture accounted for 40% of forest loss from 2010 to 2015 (FAO, 2016a). On the other hand, smallholder farming accounted for 33% of the failure, urbanization, and infrastructure for 10% each, and mining causing 7% of forest loss (FAO, 2016a). The world has lost 178 million ha of forest since 1990, but the loss rate has slowed down. The quality of net forest loss over the period 1990–2020 has reduced due to a reduction in deforestation in some countries, increases in forest area in others through afforestation, and the natural expansion of forests. The rate of net forest loss declined from 7.8 million ha per year in the decade 1990–2000 to 5.2 million ha per year in 2000–2010 and 4.7 million ha per year in 2010–2020 (Fig. 1.9) (FAO, 2020b).

A reasonable amount of forest (420 million ha) has been lost worldwide through deforestation since 1990 with a declining loss rate. The annual rate of deforestation in the most recent years (2015–2020) has been estimated as 10 million ha, down from 12 million ha in 2010–2015 (Fig. 1.10) (FAO, 2020b).

However, the pattern of forest loss varies considerably from region to region. For example, from 2000 to 2010, the forest-area loss in Latin America, Africa, and Asia was 70%, 35%, and 40%, respectively, due to the transformation to commercial agriculture (Hosonuma et al., 2012).

2.6 *Trends in Animal Genetic Resources*

Accurately measuring global animal genetic resources is a challenging task due to the under-reporting of national inventories. About 94.5% of local livestock breeds have no information as to their conservation status. Only about 30 countries

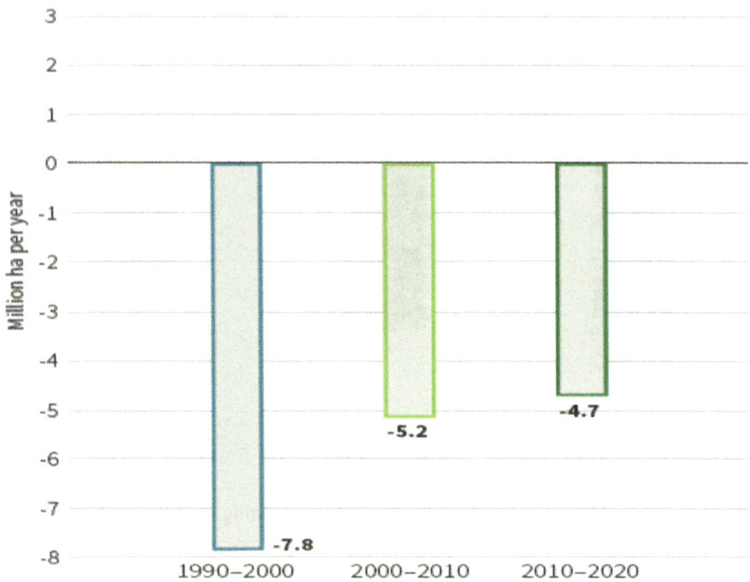

Fig. 1.9 Global annual forest area net change, by decade, 1990–2020 (FAO, 2020b)

Fig. 1.10 Annual rate of forest expansion and deforestation, 1990–2020 (FAO, 2020b)

(majority in Western Europe) report data on this indicator, and even this data is not regularly updated. More than half of these countries have not provided new data since 2015 (FAO, 2020a). Ongoing efforts to preserve animal genetic resources appear inadequate in the face of climate change and the rising demand for livestock products to achieve the SDG for food and agriculture. Between 2010 and 2019, the number of local breeds stored in genebanks increased from 10 to 101. However, it is a fraction of the approximately 7600 species reported globally and is still far from the targeted SDG for food and agriculture. By the end of 2019, 7643 local breeds were registered globally (including extinct ones), out of which 400 (5.2%) were reported with some genetic material stored, and 101 (1.3%) were reported with sufficient material stored to allow them to be reconstituted. This was a little progress compared to the preceding year when only 3.3% of local animal breeds had some material stored. Only 0.9% had enough material to allow the species to be reconstituted in case of extinction (FAO, 2020a).

An alarming proportion of local breeds are at risk of extinction as the animal genetic resources are not being adequately conserved in medium- and long-term conservation facilities. In 2019, most assessed local livestock breeds (2025 out of 2761) were determined to be at risk of extinction based on their population size, reproductive rates, and other biological characteristics (FAO, 2020a). This is marginally better than 1 year ago when 78% of assessed breeds were determined to be at risk of extinction. However, the results on the risk of extinction for local livestock breeds in different regions differ significantly. For example, 84%, 44%, and 71% of local species are considered at risk in Europe, South America, and South Africa, respectively (FAO, 2020a). For most local breeds around the world (4343), the risk status remains unknown due to a lack of data. Only 77 countries reported data in 2020 – seven more than the previous year (FAO, 2020a).

2.7 *Trends in Aquatic Genetic Resources*

Marine genetic resources (AqGR) for food and agriculture include both capture fisheries and aquaculture. The growing number of the human population shows an increase in fish consumption of approximately 1.2% annually up to 2030 when the total production of fish and fish products (capture plus aquaculture, excluding aquatic plants) will reach 201 million tonnes (FAO, 2019b). The global production from capture fisheries has plateaued at about 90–95 million tonnes per annum with little scope for additional production, and thereby, the targeted production goal needs to be achieved mainly from aquaculture (Fig. 1.11) (FAO, 2019b).

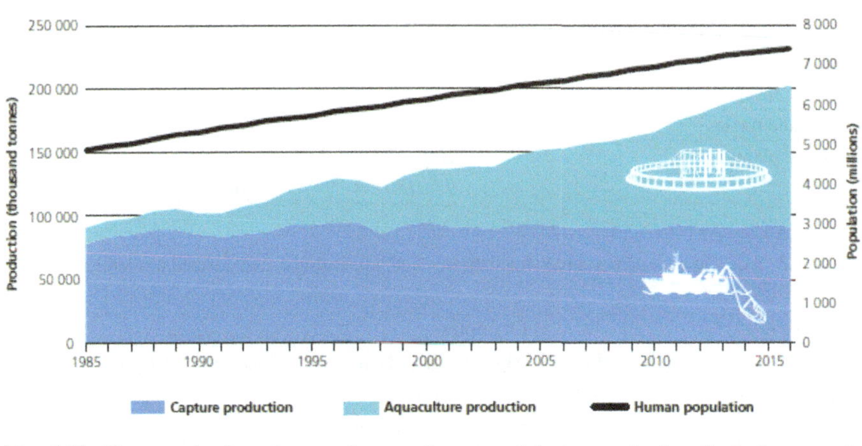

Fig. 1.11 The growth of total aquaculture and capture fisheries production (including aquatic plants) relative to human population growth (FAO, 2020a)

In recent decades, the relatively new food production system named aquaculture is spreading rapidly in the developing countries (FAO, 2019b). Over 1800 species, including fish, crustaceans, molluscs, echinoderms, coelenterates, and aquatic plants have been harvested globally in 2016 (FAO, 2018c) where the farmed marine species occupy comparatively a small portion. FAO recorded a total of 598 species as being cultivated around the world (Table 1.1). The total production from capture fisheries and aquaculture, including aquatic plants, was 202.2 million tonnes in 2016 (Table 1.2). The species diversity of AqGR for food and agriculture includes several phyla as reported by FAO (2018c) and listed in Table 1.3.

Table 1.1 Diversity of aquatic species identified in the wild and the number of farmed and fished species or species items and families represented in FAO production statistics, 2016 (FAO, 2019c)

Taxon	Wild species (marine)	Wild species (freshwater)	Number of farmed species	Number of farmed families	Number of captured species	Number of captured families
Finfish	18,768	12,834	344	80	1452	237
Molluscs	47,844	4998	95	27	151	37
Crustaceans	52,412	11,990	60	13	181	34
Other aquatic animals	[a]	[a]	15	10	26	13
Aquatic plants	12,128	2614	40	21	29	14
Total	*131,152*	*32,436*	*554*	*151*	*1839*	*335*

[a]These include echinoderms, coelenterates, and tunicates too numerous to list (many of which have no potential as food and are all marine species) and a few amphibians and reptiles

Table 1.2 World total capture fisheries and aquaculture production, 2016 (thousand tonnes, live weight) (FAO, 2019c)

Taxon	Capture fisheries	Aquaculture	Total
Finfish	77,267	54,091	131,359
Molluscs (edible)	6326	17,139	23,465
Molluscs (pearls and ornamental shells)	9	38	47
Crustaceans	6711	7862	14,573
Aquatic invertebrates	608	443	1051
Frogs and turtles	2	495	497
Aquatic plants	1091	30,139	31,230
Total	*92,015*	*110,208*	*202,223*

Table 1.3 Aquatic genetic resources for fisheries and aquaculture, categorized according to phyla (FAO, 2019c)

Phylum	Examples
Aquatic plants (multiple phyla)	Algae (seaweeds and microalgae), vascular plants
Phylum Chordata	Finfish, amphibians, reptiles
Phylum Mollusca	Bivalves (clams, mussels, oysters), gastropods (snails, abalone), cephalopods (octopus, squid)
Phylum Arthropoda	Crabs, shrimps, lobsters, cladocerans, brine shrimp
Phylum Cnidaria	Jellyfish, corals
Phylum Echinodermata	Sea urchins and sea cucumbers

More than 31,000 species of finfish, 52,000 species of aquatic molluscs, 64,000 species of marine crustaceans, and 14,000 species of aquatic plants are there in the world (WoRMS, 2018). In recent days, more species are being farmed than ever before, primarily as more marine fishes are being bred in captivity (FAO, 2016b). FAO (2018c) assessed the state of the world's marine fisheries by analyzing over 400 stocks of fish and reported that the percentage of overfished, maximally sustainably fished, and overfished in 2015 were 33.1%, 59.9%, and 7% respectively. It was also said that the share of fish stocks within biologically sustainable levels (maximally sustainably fished or underfished) declined from 90% in 1974 to 66.9% in 2015 (ibid). A decline in the number of top-level carnivores has been recorded from many marine and inland fisheries (Pauly, 1998), which is referred to as "fishing down the food web" and it reflects the state of overfishing (ibid.). However, in such cases, the productivity of a fishery remains high, especially in inland waters, as lower trophic-level species increase in abundance in the absence of larger predators (Welcomme, 1999).

Aquaculture reflects diversity than fisheries in terms of fish varieties and their environmental tolerances, food habits, feeding strategies, and culture systems. Only a few well-developed farmed types are there in aquaculture, which is highly similar to their wild relatives. Hence the wild relatives in aquaculture are essential fishery resources, but they are also exploited as sources of eggs and broodstock. To support capture fisheries, farmed aquatic species are often released back into the wild, and such interactions highlight the crucial linkages between capture fisheries and aquaculture. This interactivity between cultured species and their wild relatives might cause threats to AqGR through declining populations of important wild relatives on which aquaculture depends (FAO, 2019c).

3 Factors Affecting the Changes in Agro-Biodiversity and Food Chain

The biodiversity that includes crops, animals, and other interacting species of symbionts, pests, parasites, predators, and their competitors is known as agro-biodiversity (Boef, 2000). Such a subset of biodiversity is affected by several factors, which are also associated with the changing of an agro-ecosystem that ultimately affects the food chain. The followings are the major factors influencing the change in an agro-biodiversity:

3.1 Climate Change

Climate changing variables such as drought, heavy rainfall, elevated CO_2, and high temperature significantly affect crop production and quality, dynamics of pests and pathogens, and plant and soil biodiversity (Dwivedi et al., 2013). These climate-changing variables can also enhance the spread and severity of pathogens (Jeger & Pautasso, 2007), increase infestation of insects (Holton et al., 2003), and infection rate of diseases to cause severe losses in yields (Soriano et al., 2004). Parmesan (2006) demonstrated that the increase in global temperature alone could affect species abundance, their habitats, and interactions with their ecosystems. In a study, Lurgi et al. (2012) showed that global warming sufficiently generated changes in the size structure, distribution, and food web properties of the vertebrates in the Pyrenees. In addition to climate change, soil biodiversity vulnerability has escalated the threats on agro-ecosystems at a global scale (Dwivedi et al., 2013).

3.2 Nutrient Enrichment

Nitrogen and phosphorus are the essential nutrients that are very important for terrestrial as well as aquatic ecosystems (Tilman & Pacala, 1993; Leibold, 1999). Clark et al. (2013) identified nitrogen enrichment as a primary threat to any ecosystem and biodiversity because the more considerable increase of this nutrient can stimulate eutrophication, cause acidification, enhance secondary stressors (i.e., fire, drought, frost, or pests) and directly damage leaves by toxicity. Thus, the vast amount of nitrogen deposition can reduce plant biodiversity that may lead to a reduction in species diversity, loss of habitats, increase infestation of pest organisms, and change in soil microbial activities (McKinney & Lockwood, 1999; Treseder, 2004). Evidence showed that an excessive amount of N and P in a marine ecosystem could cause algae blooms, anoxic conditions, and ocean acidification, leading to kill fish, produce toxins, affect species diversity, disrupt food webs, etc. (Ngatia et al., 2019).

3.3 Homogenization of Agricultural Production Systems

The global agricultural production becomes more homogenized by producing only some very particular species to fulfill the consumers' demand (Thrupp, 2000). The monoculture of fishes and high-yielding varieties (HYV) of rice in India, Bangladesh, and the Philippines decreased rice varieties and fish species (Shiva, 1991; Hussein, 1994). The 'Green Revolution' movement also reduced native species diversity of agricultural items in Africa by producing modern crop varieties following monoculture rather than the polyculture of traditional species (Mann, 1990; IFOAM, 1994). Traditional diverse varieties of bananas, cacao, and cotton are virtually disappeared from South America because of the competitive production of selected high-value export species (Fowler & Mooney, 1990). Many varieties of fruits and vegetables in North America have become extinct because of the selective output of select varieties, while thousands of flax, wheat, oats, and rye vanished after HYVs were introduced in Europe (Harlan & Bennett, 1979; Thrupp, 2000).

3.4 Invasive Species

Many agricultural items are being introduced from one part of the world to another component causing the rehabilitation and establishment of the submitted items outside their natural grounds and in the new ecosystems. The trades, transport, travel, and tourism across the world, accidentally or intentionally, are spreading the invasive species globally. Paini et al. (2016) observed that 40 out of 124 countries are highly invaded by many insect or pathogen species. Some similar studies also reported that species invasion could alter biodiversity (Vilà et al., 2011) and the functionality of the ecosystem (Pejchar & Mooney, 2009). As a consequence of the introduction of non-native species to some areas (such as rabbits and cats in Australia, goats in St. Helena, and American mink in Great Britain), vulnerable ecosystems face threats of losing native species and thereby diminishing biodiversity (Wittenberg & Cock, 2001).

3.5 Genetic Erosion

The loss of genetic diversity in agriculture quantifies the genetic decline as a complete loss of crop varieties or alleles (Tsegaye & Berg, 2007; Willemen et al., 2007). As a result, a reduction in richness (i.e., a reduction in the total number of crop varieties or alleles) (Nabhan, 2007) or evenness (i.e., the diversity indices used in vegetation ecology and population genetics) (Ford-Lloyd, 2006) may arise in the agro-ecological systems. Such reductions indicate the erosion of genetic resources. A number of species become endangered and some disappear from agro-biodiversity through this genetic erosion (Hammer & Khoshbakht, 2005). On the other hand, new species are still being domesticated (Boches et al., 2006). Although introducing some closely related species can increase the diversity level of local genetic resources, the ultimate genetic erosion may be caused by the disappearance of traditional species through introgression (Ishikawa et al., 2006).

3.6 Destruction, Conversion, or Degradation of Agro-Ecosystems

Agricultural lands or areas rich in biodiversity are continuously transformed, degraded, or destroyed by urbanization (McDonald et al., 2013), industrialization (Richer, 2008), changes in land use policy (Huijun et al., 2002), alteration of cropping patterns (Mmom, 2009), pollution (Swaminathan, 2003) and natural disasters (Sudmeier-Rieux et al., 2006). Different studies have projected the significant impact of urban growth on the biodiversity of the eco-regions, protected areas across the world, and rare species (McDonald, 2008; McDonald et al., 2013). For example, the unprecedented industrialization and development of Qatar threatened the ecosystem services and biological diversity, particularly the mega projects for coastal development negatively affected marine species and threatened commercial fishing (Richer, 2008). The desertification and soil salination due to over-exploitation of water also affect the overall condition of biodiversity. Evidence showed that natural disasters negatively affect biodiversity by spreading invasive species, killing mass species, and destroying habitats (Sudmeier-Rieux et al., 2006).

3.7 *Unsustainable Management Practices and Pollutions in Agricultural Landscapes*

Unsustainable farming practices cause pollution of farming landscapes by using non recommended amounts and banned pesticides and fertilizers. The expansion of agricultural cultivation in forests, marginal and grazing lands, for example, in Nepal, causes the loss of genetic diversity, reduction/extinction of some plant and animal species, change of micro-environment, etc. (Upreti, 2000). Studies also showed that the expansion of agricultural farming reduced species number in tropical forests (Newbold et al., 2014), the land-use changes in Italy declined some Mediterranean species (Falcucci et al., 2007), and the unplanned livestock grazing in Australia was associated with losses of different plant diversity (Eldridge et al., 2015). Recent evidence showed the devastating effects of land-use changes on diverse insect populations essential to many ecosystems (Hallmann et al., 2017; Sánchez-Bayo & Wyckhuys, 2019). The unprecedented climate change and unsustainable shrimp farming increase salinity in the coastal areas (e.g., in Bangladesh), which causing severe anomalies or loss in agro-biodiversities of rice, vegetables, and fruit trees and threatening the production of agricultural commodities (Rahman et al., 2011). Both natural (e.g., climate change) and anthropogenic activities pollute the air, water, and soil (Karmakar et al., 2016; Mo et al., 2017), which ultimately affect agro-biodiversity. Isenring (2010) has placed many examples of how pesticides affect birds, mammals, beneficial insects, aquatic species, and plant communities. Applied fertilizers and pesticides also affect the soil microflora by influencing a range of soil functions and properties (Prashar & Shah, 2016).

3.8 *Population Growth and Socio-Economic Factors*

In recent decades, rapid population growth poses severe threats to our biodiversity by encroaching on the natural environments to fulfill increasing basic demands (e.g., food, house, medications, job, etc.) (Marques et al., 2019). For instance, rapid population growth, unplanned urbanization, and intense poverty cause severe pressures on habitats of wilds through deforestation, conversion of agricultural lands, and unsustainable resource exploitation for creating dwelling facilities, expanding municipalities, and producing foods for all (Upreti & Upreti, 2002). The complex combination of environment, socio-culture, and economic factors affect agro-biodiversity in a region (Rana et al., 2007; Paudel et al., 2012). For example, educated farmers tended to have more diverse yields because of better awareness than illiterate farmers (Winters et al., 2006). The changes in socio-economic and demographic characteristics result in higher opportunities for non-agricultural jobs, which create pressure in agricultural production systems by increasing the density of farming throughputs. The natural patterns of organic matter cycling in the agro-ecosystem can no longer be maintained for fossil energy inputs and monoculture (Giampietro, 1997). Thus, agrobiodiversity can be negatively influenced because of the changes in socio-economic and demographic characteristics (Kahane et al., 2013; Marques et al., 2019).

3.9 Overexploitation

Overexploitation is considered one of the most devastating factors that cause a significant decline in the species diversity of different animals (Rosser & Mainka, 2002). For example, bushmeat hunting (Ripple et al., 2016), wildlife poaching and trafficking (Gray et al., 2018), and commercial overfishing (Klautau et al., 2016) put enormous pressure on wild species to be declined. In a study, Rosser and Mainka (2002) showed that overhunting of different species of birds for food and body parts was ranked as the second most threatening factors following habitat loss. Thus, overexploitation, overharvesting, or overhunting of any species must be stopped to keep them protected and thereby ensure the biodiversity to be functioning well.

4 Sustainable Use of Agro-Biodiversity and Food Chain

Sustainability is a complex idea in respect of agro-biodiversity that includes economic, social, and environmental issues. Rotating crops to embrace diversity, planting cover crops, reducing or eliminating tillage, applying integrated pest management, integrating livestock and crops, adopting agroforestry practices etc. are considered as the sustainable approaches. Agro-biodiversity is fundamentally different from other approaches to sustainable development where innovations are based on the co-creation of knowledge, combining science with the traditional, practical, and local experience of producers (FAO, 2018d). The ways followed in agro-biodiversity to empower the producers and communities as critical agents for making a sustainable change have been discussed as follows-.

4.1 Managing an Environment Free from Pollution

Pollution is the introduction of contaminants into the natural environment that causes an adverse change (Merriam-Webster, 2020). Our surrounding environment is getting polluted due to human interventions, either industrially and agriculturally, or due to infrastructural developments. Energy conservation should be emphasized, and renewable resources should be encouraged to combat the adverse effects of global pollution and maximum protection of public health (Owusu & Asumadu-Sarkodie, 2016). Measures should be formulated and implemented to reduce pollution and protect surface water, groundwater, and drinking water. To reduce Reuse and recycling of solid and hazardous wastes should be promoted too to ensure maximum protection of public health and safety and preservation of the environment (Lumen Learning, 2020). Increasing concerns about global environmental change and food security have focused attention on the need for environmentally sustainable agriculture. One aspect of this 'sustainability' is environmental sustainability (Dicks et al., 2013), which includes efficient use of natural resources and does not degrade the ecological systems that underpin it or deplete natural capital stocks (Dobermann & Nelson, 2013).

4.2 Integrated Management of Land and Water

Water Resources Management is one of the pillars of sustainable ecosystems (Turton et al., 2007). It is related to various sectors like agriculture, industry, domestic water supply, and sanitation, hydropower, health, and environment. Agriculture is the largest water user worldwide but is also a significant polluter; large scale land-use changes impact the hydrology of river basins resulting in the downstream water availability and flood risks (ITT, 2019).

Integrated water resources management (IWRM) is a widely used framework, but water efficiency in agriculture does not figure prominently in many IWRM plans (Calder, 2005; Lenton, 2011). In response to the decreasing availability of water in many regions, a need for a better water management strategy to achieve water productivity in rain-fed agriculture is especially urgent (Duda, 2003). Effective water management is essential for sustainable agriculture, both to produce food crops and for the sustainable production of biofuels and energy. Concepts like "more crop per drop" need to be added in information technologies for controlling irrigation water flow (Rosegrant et al., 2009; Monaghan et al., 2013).

Soil conservation measures are essential for land conservation and rehabilitation (Jahangir et al., 2019). The participation of women in decision-making processes, as the managers of water and land resources, is also essential to ensure the sustainable use of land and water resources (UN Chronicle, 2020).

4.3 Integrated Management of Plant Nutrition and Pest

Integrated Plant Nutrient Management (IPNM) is promoted by improving the timing, dosing, and application method of nutrients and thus minimizing the potential impact of weeds (Chapagain & Gurung, 2010). IPNM is interpreted as a more holistic sense of "land husbandry," which provides tangible benefits in terms of higher yields and simultaneously conserves the soil resources (Mbuthia et al., 2015). The field-level management practices in IPNM would include the use of farmyard manure, natural and mineral fertilizers, soil amendments, crop residues and farm wastes, agroforestry, and tillage rules, green manures, cover crops, legumes, intercropping, crop rotations, fallows, irrigation, drainage, plus a variety of other agronomic, vegetative and structural measures designed to conserve both water and soil (FAO, 2019a). Better plant management is needed focusing on improved crop establishment at the beginning of the rains through protective ground cover to reduce splash erosion, enhancing infiltration, and biological activity. Combinations of the complementary crop, livestock, and land husbandry practices can maximize organic materials' additions and recycle farm wastes to maintain and enrich the soil with organic matter. Such combined approaches can also reduce rainfall impact, improve surface infiltration, and reduce the velocity of surface runoff (FAO, 2020c).

Integrated plant nutrient management also contributes to pest management (FAO, 2019a). Stressed crops are more susceptible to disease and pest attacks. Crops growing in poorly structured soil, under low or unbalanced nutrient conditions, or inadequate water supply or retention will be stressed (FAO, 2020c). The application of pesticides is a costly symptomatic approach to a syndrome that is better addressed by improving the ecological conditions and systems within which the crops are cultivated (FAO, 2020c). Such production efficiencies are gained through the integrated nutrient management practices promoting the combined use of mineral, organic, and biological resources that ensure ecosystem sustainability through nutrient cycling (MEA, 2005). Integrated Weed Management may also be facilitated by improving the timing, dosing, and application method of nutrients (Mortensen et al., 2012; FAO, 2019a).

4.4 Integrated Crop-Livestock Management

Mixed production system between crops and animals (also called agropastoral system) are being practiced mainly in developing countries in sub-Saharan Africa, South and Southeast Asia, and Latin America, where mixed rain-fed farming zones account for 25% of the total land areas, account for 42% of the total population, 37% of the total cattle and 36% of the total sheep and goats (FAO, 2020c). In the integrated crop-livestock systems, cover crops and crop residues provide feed to livestock, while plants capture nutrients from the livestock waste. The use of cover crops will not only provide economic benefits but can also be used to achieve multiple environmental benefits (FAO, 2019a). Potential economic benefits include reduced fertilizer cost for the cash crop, yield or profit increase from subsequent cash crop, and additional cost savings from supplemental hay. The cover crops planted after harvesting the main crop can reduce nitrogen (N) loss by absorbing residual fertilizer N (Tonitto et al., 2006).

Moreover, cover crops provide vegetal cover during critical periods, which can significantly reduce topsoil loss from croplands (De Baets et al., 2011). However, there are still some concerns regarding the role of hoof traffic from livestock that can adversely affect the near-surface soil conditions, soil health, and hydrological properties (Bell et al., 2011). The use of a diverse cover crop mixture such as radish, peas, oats, lentils, and sorghum can provide increased biomass on the soil surface that can alleviate the compaction impact under these integrated crop-livestock systems (Blanco-Canqui et al., 2015). Short-term grazing does not have adverse effects on the soil properties, including soil water retention, bulk density, soil organic carbon, and total nitrogen contents (Lu et al., 2015). The use of cover crops after a small grain harvest might be a good practice in this regard (Velazco-Bedoya et al., 2014).

The crop rotations in the crop-livestock systems would preferably also include crops that are beneficial from a nutrient point of view (e.g., legumes). A balanced nutrient level in the soil will contribute to sustainable crop and livestock production intensification by shifting from static nutrient balances to nutrient flows in cycles. Integrating crops with livestock provides benefits to the environment by creating a 'closed'-system for nutrients and to animal welfare by providing sufficient space for the animals to graze (FAO, 2020c).

4.5 *The Role of Organic Agriculture*

The US Department of Agriculture (USDA) defines organic farming as a production system without synthetic fertilizers, pesticides, growth regulators, and livestock feed additives (FiBL, 2019). More than 95% of the food we produce is directly or indirectly linked to soil (FAO, 2015). In order to achieve the zero hunger goal of SDG, it is imperative to take into account the health of the grounds. FAO (2011) estimated that about 25% of the soils suffer from high soil degradation causing threats to the environment and food security (Pimentel, 2006). The addition of organic matter and practicing crop rotation with diverse crops for a more extended period, including cover and catch crops, can contribute to the reduction of soil erosion and fertility decline. Organically managed fields contain higher doses of organic matter and facilitate an abundant presence of active soil microbial communities (Meemken & Qaim, 2018), which stabilize the soil aggregates to form a better soil structure (Nichols, 2015). This results in improved soil capacity to absorb and hold more water during rainfalls (IFOAM, 2012; Nichols, 2015), which provides greater resilience in agro-biodiversity during water scarcity or heavy precipitations (Scialabba & Müller-Lindenlauf, 2010). Moreover, the use of pesticides and mineral fertilizers is prohibited in organic farming, which also allows variability in crop rotation and enhances biodiversity in agro-ecosystems (Gabriel & Tscharntke, 2007).

Bengtsson et al. (2005) found that 30% more varieties of flora and fauna and 50% more individual plants are grown in organically managed lands. Though crop yields from organic cultivation are generally reduced as 19–25% (Meemken & Qaim, 2018), it can play a vital role in the long-term provision of food through keeping better soil quality, resulting in less farmland loss over time and a better climate-resilience (Scialabba & Müller-Lindenlauf, 2010). On the other hand, organically produced foods are healthier and more nutritious than the conventional counterpart containing higher levels of antioxidants, vitamin C, Omega 3 fatty acids, and the omega 3–6 ratios (Reganold & Wachter, 2016). From a review and meta-analysis, Barański et al. (2014) also reported higher concentrations of antioxidants and lower prevalence of cadmium (4 times less) and pesticide residues in organic products.

Reganold and Wachter (2016) suggested four key sustainability pillars as productivity, environmental impact, economic viability, and social wellbeing. The Fig. 1.12 illustrates that organic agriculture is more profitable and environmentally friendly, though it gives slightly lower yields compared to its conventional counterpart. Additionally, organic practices of crop cultivation provide equally or more nutritious foods with less (or no) pesticide residues in comparison to traditional farming. Reganold and Wachter (2016) also claimed that organic agricultural systems deliver more critical ecosystem services and social benefits. Organic agriculture also brings economic benefits to farmers. Most certified organic farmers from developing countries produce cash crops (e.g., coffee, tea, cocoa, tropical fruits) and

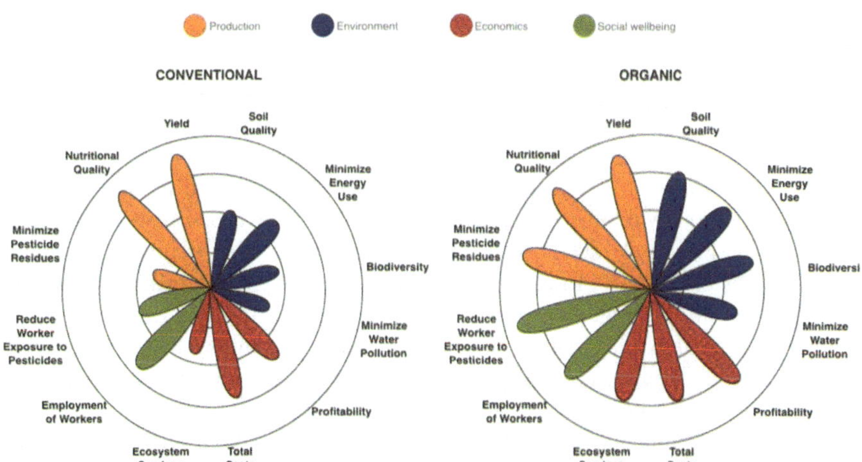

Fig. 1.12 A comparative illustration of organic and conventional systems of crop cultivation in respect of four areas of sustainability (Reganold & Wachter, 2016)

export them to developing countries, where they receive a significant price premium for their products (Lernoud & Willer, 2019).

4.6 Practicing Conservation Agriculture

Conservation agriculture is a farming system that maintains a permanent soil cover to assure its protection, avoids soil tillage, and cultivates a diverse range of plant species to improve soil conditions, reduce land degradation and increase water and nutrient use efficiency (FAO, 2016c). It is a crop management system based on three principles: (a) minimum soil disturbance, (b) permanent soil surface cover with crop residues and cover crops, and (c) crop rotations that include diverse species (FAO, 2016c; CIMMYT, 2020; Climate-ADAPT, 2020).

In Conservation Agriculture systems, pest and disease control are based on Integrated Pest Management (IPM) technologies. Conservation agriculture and organic farming maintain a balance between agriculture and resources, use crop rotation, and protect the soil's organic matter. However, the main difference between these two types of farming is that organic farmers use a plow or soil tillage. In contrast, farmers who practice conservation agriculture use natural principles and do not till the soil (CIMMYT, 2020). Organic farmers apply tillage to remove weeds without using inorganic fertilizers. While conservation agriculture and climate-smart agriculture are similar, though, their purposes are different. Conservation agriculture aims to sustainably intensify smallholder farming systems and positively affect the environment using natural processes. It helps farmers to adapt and increase profits despite climate risks (CIMMYT, 2020).

4.7 Diversifying Aquaculture Practices

Aquaculture is one of the world's fastest-growing agricultural production systems, which enhanced its overall global production from 125.6 million tonnes in 1996 to 178.5 million tonnes in 2018 (FAO, 2020d). It is one of the most diversified farming systems where many numbers of species (e.g., both aquatic and terrestrial: Sahoo & Singh, 2005; Frei et al., 2007), several approaches (e.g., from traditional to intensive: Bostock et al., 2010; Li & Liu, 2013) and different environments (e.g., across fresh to deep marine water, clean water to sewage or wastewater: Li & Liu, 2013; Kumar et al., 2015) are involved for the production of fish as well as other agro-products. Like other agricultural production systems, this sector also depends on several physicochemical (e.g., water and soil parameters), biological (e.g., food web), and environmental (e.g., climate change) factors (Boyd & Tucker, 2014). Therefore, to keep this sector sustainable, diversification of species, farming systems, and products could provide at least some assurances to the farmers and investors.

The species diversification in this sector generally depends on culture systems, breeding and feeding technologies, farming environments, and of course, their market demands (Harvey et al., 2017). Aquaculture diversification can happen in various ways such as introduction and farming of new species or strains (e.g., salmon in Chile and tilapia in Asian countries) and increasing the number of existing culture species (e.g., increased trout and salmon culture in Chile). Practicing polyculture or other integrated farming approaches (e.g., integrated farming of fish-shrimp-vegetable-chicken to increase post product diversification), generating hybrids from new or existing species (e.g., pacu hybrids in Brazil, Thai catfish in Asia) and producing genetically improved species (e.g., monosex tilapia, triploid oyster) can also bring diversification in aquaculture (Harvey et al., 2017). Thus, species diversification and different farming systems and technologies in a wide range of environments create enormous opportunities for farmers, investors, policy-makers, environmentalists, researchers, and even politicians to work with.

The boom of aquaculture farming with massive diversification around the world not only increases the overall production; it also creates jobs, alleviates poverty, develops economic growth, and changes socio-economic conditions of the farming communities. According to Beveridge et al. (1994), unplanned utilization of natural resources (i.e., land, water, food, broods, larvae, etc.), land and pond transformation process, and continuous wastes production are the major factors that may impact biodiversity. For examples, vast areas of mangroves and coastal lowlands have been converted into shrimp and other fish farms (Primavera, 1998), which are releasing different types of solid waste, chemicals, therapeutics, bacteria, pathogen, etc. to the environment (Schwitzguébel & Wang, 2007; De Silva, 2012). Introduction of new aquaculture species may increase alien or invasive species number (Welcomme, 1988), and catching wild larvae of aquaculture species cause over-exploitation or mortality of other non-target species (Islam & Haque, 2004). Thus, unplanned development, illegal deforestation, untreated water effluents, different contaminants, new species introduction, etc. may alter the structural and functional food webs (Islam & Haque, 2004; Abery et al., 2005; De Silva, 2012). These activities can also affect other living organisms in the ecosystems by increasing pathogenicity, eutrophication and competitions between the invasive and local species (Bartley & Casal, 1998; Flegel, 2006; Jensen et al., 2010).

4.8 Genetic Improvement of Plant, Animal, Forest, and Aquatic Resources

Genetic materials refer to the materials of plant, animal, microbial or other origins containing functional units of heredity (Biodiversity. fi, 2016). The fundamental objective of genetic resource conservation is the maintenance of broad genetic diversity within each of the species (i.e., intra-specific genetic diversity) with a known or potential value to ensure availability for exploitation by present and future generations (Khanna & Singh, 1991). Genetic diversity serves as a way for populations to adapt to changing environments and determines the potential fitness of a people and their extent of persistence (Dyke, 2008).

To explain the present status of crop genetic resources, Krug (2018) claimed that about 75% of all crop genetic diversity had been lost since the previous century, primarily due to changes in the agricultural food system, which values uniformity. Of the remaining 25%, one third is expected to become extinct by 2050. However, agricultural genetic diversity is imperative to provide a robust food security system capable of adapting to pest and environmental stressors (Krug, 2018).

The global food supply is increasingly under threat from climate change, world population growth, and the introduction or range expansion of disease and insects (FAO, 2017). To adequately store genetic diversity, the germplasm collections must be well maintained, and backup supplies are created to ensure survival in case of a natural disaster or political unrest (Krug, 2018).

4.9 Use of Micro-Organisms in Agro-Industries

Unraveling the biota black box using modern molecular methods helps find new suites of beneficial microorganisms that can help improve agricultural production worldwide (Gupta, 2012). Current farming practices mainly rely on mineral fertilizers' high inputs to increased yields and involve applications of chemical pesticides to protect crops from diseases and pests. The use of microbial inoculants in agriculture would be an attractive eco-friendly alternative to further the application of mineral fertilizers and chemical pesticides (Patil & Solanki, 2016). The beneficial influences of microorganisms on plant growth include nitrogen fixation, acquisition and uptake of major nutrients, promotion of shoot and root growth, disease control or suppression, and improved soil structure (de Souza et al., 2015). Some of the commonly promoted and used beneficial microorganisms in agriculture worldwide include *Rhizobia*, *Mycorrhizae*, *Azospirillum*, *Bacillus*, *Pseudomonas*, *Trichoderma*, *Streptomyces* species, and many more (Gupta, 2012). A continued exploration of the natural biodiversity of soil microorganisms and the manipulation of microbial interactions in the rhizosphere of crops represents a prerequisite step to develop more efficient microbial inoculants (Jacoby et al., 2017).

5 Status of Studies and Research on Agro-Biodiversity and Food Chain

The so-called sixth wave of mass extinction of biodiversity in the ecosystems (Barnosky et al., 2011) is associated with climate warming, increased nitrogen deposition, land-use change, biotic exchange, and an increased atmospheric CO_2

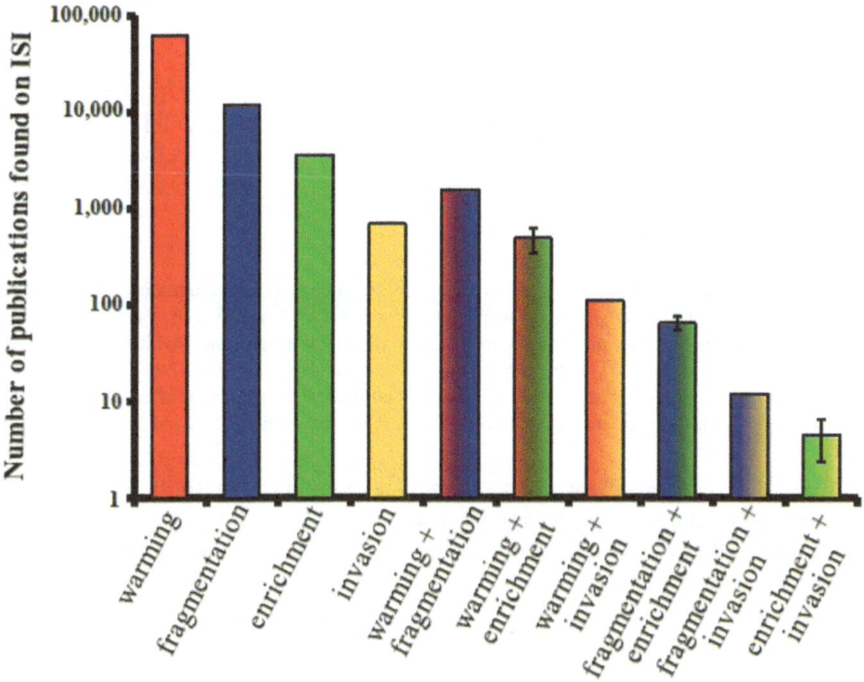

Fig. 1.13 A number of publications on different drivers of global change were published between 2000 and December 2014, as reviewed by Schwarzmüller (2015). Numbers are based on a search in the ISI Web of the knowledge database. Error bars result from an altered order of search terms

level (Dirzo et al., 2014). A reduction in suitable habitat for the wilds (Eklöf et al., 2012) or the range shift of species leading to the invasion of other ecosystems (Lurgi et al., 2012) might be the result of the combined effect of the drivers of the environmental changes. However, increased nutrient availability via enrichment might satisfy the increased metabolic demands of a species due to warming (Binzer et al., 2012). The knowledge about the mechanism of the effect of environmental stressors on biodiversity is still far from our understandings as multiple environmental stressors are still scarce in most of the studies (O'Gorman et al., 2012) (Fig. 1.13).

During the last two decades, food-web ecologists followed "community modules" (Holt, 1997) or "network motifs" (Stouffer & Bascompte, 2010). Smaller trophic entities or simple representations of entire food webs consisting of two to seven species were considered in these categories (Otto et al., 2007; Stouffer & Bascompte, 2010). These trophic entities cannot predict emergent patterns of food webs (Cohen et al., 2009).

Climate change is not occurring evenly across the globe. Large shifts in temperature and precipitation are being observed in the Arctic regions (IPCC, 2014). Similarly, such changes in climatic factors are not equally affecting the biomes and communities having adaptability in different environmental conditions. Marino

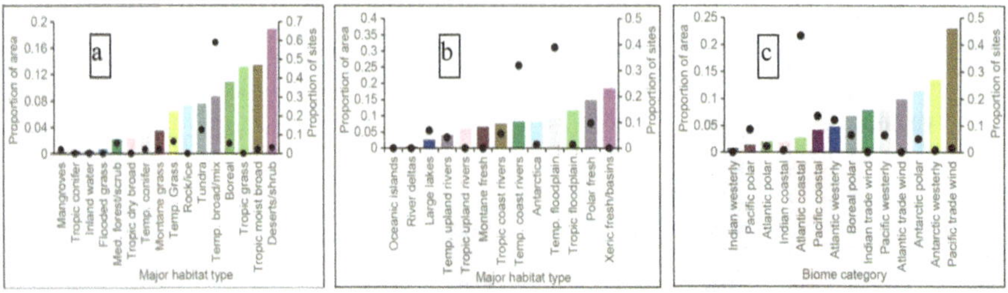

Fig. 1.14 Global maps of studies as reported by Cameron et al. (2019) from a review ($n = 308$) conducted on climate change impacts on species interactions for (**a**) terrestrial systems; (**b**) freshwater systems; and (**c**) marine systems about terrestrial/ freshwater major habitat types (Olson et al., 2001; Abell et al., 2011) and ocean biomes (Flanders Marine Institute, 2009). Graphs show the proportion of the total area (colored bars; y-axis on the left) and the proportion of study sites (black points; second y-axis on the right) concerning the major habitat types and biomes (x-axis)

et al. (2018) reported that warming strengthens top-down control in food webs in colder regions and reverses in warmer areas.

A balanced global picture of trophic interactions can be drawn from reasonably collected representative samples of food webs. However, studies on food webs across major geographic/climatic zones are very scarce as the sampling is generally extremely spatially patchy (Martin et al., 2012; Sotomayor & Lortie, 2015; Bellard & Jeschke, 2016). From a review of literature focusing on climate change effects on food webs, Cameron et al. (2019) observed a robust geographic bias in respect of sampling. They reported that most of the studies occurred in the USA and Europe. They also noted that the number of studies conducted within each biome is not proportional to the overall size of the biome. For example, only 3.5% and 3% of the reviews were completed in the tropics and xeric shrublands. However, these regions account for ~40% and ~19% of the Earth's surface and support the majority of the world's species. None of the studies were conducted in fresh xeric water and closed basin ecosystems, and only 1% of reviews were shown in the Pacific Trade wind biome. However, they comprise 19% and 23% of the global freshwater biomes, respectively. In contrast, temperate broadleaf and mixed forests, gentle coastal rivers, temperate floodplains, and the Atlantic coastal regions are substantially oversampled relative to their spatial extent (Fig. 1.14a–c).

In respect of the food web, assessing the impact of biotic interactions between trophic levels are more challenging than examining the response of a single species or general biodiversity changes (McCann, 2007; Tylianakis et al., 2008). Hence, information on multi-trophic interactions on the food web is also scarce, as described by Cameron et al. (2019). Almost half of the total articles reviewed by them included only two trophic levels where studies in terrestrial systems examined more than two trophic levels (58%), slightly more than lessons in marine (49%) or freshwater (45%) systems. These understudied regions should be included in future studies to update our understanding of climate change impacts on the food web across various biomes under current and future changed climatic conditions.

6 Conclusion

Food chains or webs in many ecosystems become highly vulnerable because of climate change, land degradation and other anthropogenic activities (IPCC, 2018). The extinction of biodiversity in the ecosystems is happening with climate warming, increased nitrogen deposition, land-use change, biotic exchange, and an increased atmospheric CO_2 level (Dirzo et al., 2014). Changes in demographic status and food demand are determining the emission of GHG (e.g., CH_4 and N_2O) across the SSPs (Shared Socioeconomic Pathways) (Riahi et al., 2017). Due to the combined effect of the drivers of the environmental changes, suitable habitats for the wilds have been reduced (Eklöf et al., 2012), alien species are invading the ecosystems (Lurgi et al., 2012) or the wilds are being attacked by newly emerging diseases (Harvell et al., 2009). Changes in climatic variables are needed to be minimized for a sustainable agro-biodiversity and food chain (Dwivedi et al., 2013). Sustainability in agro-biodiversity may be attained through crop diversity, planting cover crops, reducing or eliminating tillage, applying integrated pest management, integrating livestock and crops, adopting agroforestry practices, etc. (Oberč & Arroyo Schnell, 2020). Integrated Plant Nutrient Management (IPNM) would also be considered as a more holistic sense of "land husbandry" (Roy & Nabhan, 2001). Organic farming would be another option for attaining a sustained agro-biodiversity (IFOAM, 2012; Eyhorn et al., 2019). Ensuring sustainability in the food chain and food web also necessitates assessing the impact of biotic interactions between trophic levels (Cameron et al., 2019). However, the knowledge about the mechanism of effects of environmental stressors on biodiversity and the food chain is still far from our understandings as the studies including multiple environmental stressors are scarce (O'Gorman et al., 2012). These aspects should be included in the future studies to update our understanding of the relationship between climate change and the food web across a range of biomes.

References

Abell, R., Thieme, M., Ricketts, T. H., et al. (2011). Concordance of freshwater and terrestrial biodiversity. *Conservation Letters, 4*, 127–136. https://doi.org/10.1111/j.1755-263X.2010.00153.x

Abery, N. W., Sukadi, F., Budhiman, A. A., et al. (2005). Fisheries and cage culture of three reservoirs in west Java, Indonesia; A case study of ambitious development and resulting interactions. *Fisheries Management and Ecology, 12*, 315–330. https://doi.org/10.1111/j.1365-2400.2005.00455.x

Altieri, M. A. (1999). Applying agroecology to enhance the productivity of peasant farming systems in Latin America. *Environment, Development and Sustainability, 1*, 197–217. https://doi.org/10.1023/A:1010078923050

Altieri, M. A., Nicholls, C. I., Henao, A., & Lana, M. A. (2015). Agroecology and the design of climate change-resilient farming systems. *Agronomy for Sustainable Development, 35*, 869–890. https://doi.org/10.1007/s13593-015-0285-2

Balvanera, P., Siddique, I., Dee, L., et al. (2014). Linking biodiversity and ecosystem services: Current uncertainties and the necessary next steps. *Bioscience, 64*, 49–57. https://doi.org/10.1093/biosci/bit003

Balvanera, P., Quijas, S., Karp, D. S., et al. (2017). Ecosystem services. In *The GEO handbook on biodiversity observation networks* (pp. 39–78). Springer International Publishing.

Banašek-Richter, C. (2004). *Quantitative descriptors and their perspectives for food-web ecology* (Dissertation). Universite De Neuchatel.

Barański, M., Średnicka-Tober, D., Volakakis, N., et al. (2014). Higher antioxidant and lower cadmium concentrations and lower incidence of pesticide residues in organically grown crops: A systematic literature review and meta-analyses. *The British Journal of Nutrition, 112*, 794–811. https://doi.org/10.1017/S0007114514001366

Barnosky, A. D., Matzke, N., Tomiya, S., et al. (2011). Has the Earth's sixth mass extinction already arrived? *Nature, 471*, 51–57. https://doi.org/10.1038/nature09678

Bartley, D., & Casal, C. V. (1998, September 3–5). Impacts of introductions on the conservation and sustainable use of biodiversity. In *International conference on sustainable use of aquatic biodiversity: Data, tools and collaboration*, Lisbon.

Beech, E., Rivers, M., Oldfield, S., & Smith, P. P. (2017). GlobalTreeSearch: The first complete global database of tree species and country distributions. *Journal of Sustainable Forestry, 36*, 454–489. https://doi.org/10.1080/10549811.2017.1310049

Bell, L. W., Kirkegaard, J. A., Swan, A., et al. (2011). Impacts of soil damage by grazing livestock on crop productivity. *Soil and Tillage Research, 113*, 19–29. https://doi.org/10.1016/j.still.2011.02.003

Bellard, C., & Jeschke, J. M. (2016). A spatial mismatch between invader impacts and research publications. *Conservation Biology, 30*, 230–232. https://doi.org/10.1111/cobi.12611

Bengtsson, J., Ahnström, J., & Weibull, A.-C. (2005). The effects of organic agriculture on biodiversity and abundance: A meta-analysis. *Journal of Applied Ecology, 42*, 261–269. https://doi.org/10.1111/j.1365-2664.2005.01005.x

Beveridge, M. C. M., Ross, L. G., & Kelly, L. A. (1994). Aquaculture and biodiversity. *Ambio, 23*, 497–502.

Binzer, A., Guill, C., Brose, U., & Rall, B. C. (2012). The dynamics of food chains under climate change and nutrient enrichment. *Philosophical Transactions of the Royal Society B, 367*, 2935–2944. https://doi.org/10.1098/rstb.2012.0230

Biodiversity.fi. (2016). *What are genetic resources? Genetic resources and legislation*. United Nations Decade on Biodiversity. Available from https://www.biodiversity.fi/geneticresources/genetic-resources/what-are-genetic-resources. Accessed 25 Nov 2020.

Blanco-Canqui, H., Shaver, T. M., Lindquist, J. L., et al. (2015). Cover crops and ecosystem services: Insights from studies in temperate soils. *Agronomy Journal, 107*, 2449–2474. https://doi.org/10.2134/agronj15.0086

Boches, P., Bassil, N. V., & Rowland, L. (2006). Genetic diversity in the highbush blueberry evaluated with microsatellite markers. *Journal of the American Society for Horticultural Science, 131*, 674–686. https://doi.org/10.21273/JASHS.131.5.674

Boef, W. S. (2000). *Learning about institutional frameworks that support farmer management of agrobiodiversity: Tales of the unpredictables* (Ph.D. thesis). Wageningen University.

Bostock, J., McAndrew, B., Richards, R., et al. (2010). Aquaculture: Global status and trends. *Philosophical Transactions of the Royal Society B, 365*, 2897–2912. https://doi.org/10.1098/rstb.2010.0170

Boyd, C. E., & Tucker, C. S. (2014). *Handbook for aquaculture water quality* (pp. 1–438). Craftmaster Printers, Inc.

Brookfield, H., & Stocking, M. (1999). Agrodiversity: Definition, description and design. *Global Environmental Change, 9*, 77–80. https://doi.org/10.1016/S0959-3780(99)00004-7

Cáceres, D. M., Tapella, E., Quétier, F., & Díaz, S. (2015). The social value of biodiversity and ecosystem services from the perspectives of different social actors. *Ecology and Society, 20*, art62. https://doi.org/10.5751/ES-07297-200162

Calder, I. R. (2005). Integrated land and water resources management. In *Encyclopedia of hydrological sciences*. John Wiley & Sons, Ltd.

Cameron, E. K., Sundqvist, M. K., Keith, S. A., et al. (2019). Uneven global distribution of food web studies under climate change. *Ecosphere, 10*, e02645. https://doi.org/10.1002/ecs2.2645

Chapagain, T., & Gurung, G. B. (2010). Effects of integrated plant nutrient management (IPNM) practices on the sustainability of maize-based farming systems in Nepal. *The Journal of Agricultural Science, 2*, 1–7. https://doi.org/10.5539/jas.v2n3p26

Chase, J. M., Leibold, M. A., Downing, A. L., & Shurin, J. B. (2000). The effects of productivity, herbivory, and plant species turnover in grassland food webs. *Ecology, 81*, 2485–2497. https://doi.org/10.1890/0012-9658(2000)081[2485:TEOPHA]2.0.CO;2

CIMMYT. (2020). *What is conservation agriculture?* International Maize and Wheat Improvement Center (CIMMYT), Carretera México-Veracruz, Km. 45, El Batán, 56237 Texcoco, MÉXICO. Available from https://www.cimmyt.org/news/what-is-conservation-agriculture/. Accessed 25 Nov.

Clark, C. M., Bai, Y., Bowman, W. D., et al. (2013). Nitrogen deposition and terrestrial biodiversity. In S. A. Levin (Ed.), *Encyclopedia of biodiversity* (Vol. 5, 2nd ed., pp. 519–536). Academic Press.

Clarke, L., Jiang, K., Akimoto, K., et al. (2014). Assessing transformation pathway. In O. Edenhofer, et al. (Eds.), *Climate change 2014: Mitigation of climate change* (Working group III to the Fifth assessment report of the Intergovernmental Panel on Climate Change). Cambridge University Press.

Climate ADAPT. (2020). *Conservation agriculture. Structural and physical: Ecosystem-based adaptation options, Climate change adaptation in the agriculture sector in Europe* (EEA report No 4/2019). Available from https://climate-adapt.eea.europa.eu/metadata/adaptation-options/cons

Cohen, J. E., Schittler, D. N., Raffaelli, D. G., & Reuman, D. C. (2009). Food webs are more than the sum of their tritrophic parts. *Proceedings of the National Academy of Sciences, 106*, 22335–22340. https://doi.org/10.1073/pnas.0910582106

De Baets, S., Poesen, J., Meersmans, J., & Serlet, L. (2011). Cover crops and their erosion-reducing effects during concentrated flow erosion. *Catena, 85*, 237–244. https://doi.org/10.1016/j.catena.2011.01.009

De Silva, S. S. (2012). Aquaculture: A newly emergent food production sector—And perspectives of its impacts on biodiversity and conservation. *Biodiversity and Conservation, 21*, 3187–3220. https://doi.org/10.1007/s10531-012-0360-9

de Souza, R., Ambrosini, A., & Passaglia, L. M. P. (2015). Plant growth-promoting bacteria as inoculants in agricultural soils. *Genetics and Molecular Biology, 38*, 401–419. https://doi.org/10.1590/S1415-475738420150053

Díaz, S., Pascual, U., Stenseke, M., et al. (2018). Assessing nature's contributions to people. *Science (80-), 359*, 270–272. https://doi.org/10.1126/science.aap8826

Dicks, L., Bardgett, R., Bell, J., et al. (2013). What do we need to know to enhance the environmental sustainability of agricultural production? A prioritisation of knowledge needs for the UK food system. *Sustainability, 5*, 3095–3115. https://doi.org/10.3390/su5073095

Dirzo, R., Young, H. S., Galetti, M., et al. (2014). Defaunation in the anthropocene. *Science (80-), 345*, 401–406. https://doi.org/10.1126/science.1251817

Dobermann, A., & Nelson, R. (2013). *Opportunities and solutions for sustainable food production*. Sustainable Development Solutions Network. Available from https://resources.unsdsn.org/opportunities-and-solutions-for-sustainable-food-production. Accessed 20 Nov 2020.

Dronamraju, K. (2008). *Emerging consequences of biotechnology* (pp. 1–490). World Scientific. Foundation for Genetic Research, USA. https://doi.org/10.1142/6632.

Duda, A. M. (2003). Integrated management of land and water resources based on a collective approach to fragmented international conventions. *Philosophical Transactions of the Royal Society B, 358*, 2051–2062. https://doi.org/10.1098/rstb.2003.1410

Duffy, J. E., Richardson, J. P., & Elizabeth, A. C. (2003). Grazer diversity effects on ecosystem functioning in seagrass beds. *Ecology Letters, 6*, 637–645. https://doi.org/10.1046/j.1461-0248.2003.00474.x
Dwivedi, S., Sahrawat, K., Upadhyaya, H., & Ortiz, R. (2013). Chapter 1: Food, nutrition and agrobiodiversity under global climate change. In D. L. Sparks (Ed.), *Advances in agronomy* (1st ed., pp. 1–128). Academic Press.
Dyer, L. A., & Letourneau, D. (2002). Top-down and bottom-up diversity cascades in detrital vs. living food webs. *Ecology Letters, 6*, 60–68. https://doi.org/10.1046/j.1461-0248.2003.00398.x
Dyke, V. (2008). Genetic diversity – Understanding conservation at genetic levels. In *Conservation biology* (pp. 153–184). Springer Netherlands.
Eklöf, A., Kaneryd, L., & Münger, P. (2012). Climate change in metacommunities: Dispersal gives double-sided effects on persistence. *Philosophical Transactions of the Royal Society B, 367*, 2945–2954. https://doi.org/10.1098/rstb.2012.0234
Eldridge, D. J., Poore, A. G. B., Ruiz-Colmenero, M., et al. (2015). Ecosystem structure, function and composition in rangelands are negatively affected by livestock grazing. *Ecological Applications, 26*, 1273–1283. https://doi.org/10.1890/15-1234.1
Elton, C. S. (1926). *Animal ecology* (pp. 1–296). University of Chicago Press.
Encyclopedia Britannica. (2020). *Community*. Available from https://www.britannica.com/science/community-biology. Accessed 15 Nov 2020.
Esquinas-Alcázar, J. (2005). Protecting crop genetic diversity for food security: Political, ethical and technical challenges. *Nature Reviews Genetics, 6*, 946–953. https://doi.org/10.1038/nrg1729
Eyhorn, F., Muller, A., Reganold, J. P., et al. (2019). Sustainability in global agriculture driven by organic farming. *Nature Sustainability, 2*, 253–255. https://doi.org/10.1038/s41893-019-0266-6
Falcucci, A., Maiorano, L., & Boitani, L. (2007). Changes in land-use/land-cover patterns in Italy and their implications for biodiversity conservation. *Landscape Ecology, 22*, 617–631. https://doi.org/10.1007/s10980-006-9056-4
FAO. (1999, September). *Agricultural biodiversity, multifunctional character of agriculture and land conference* (Background paper 1). Maastricht.
FAO. (2011). *The state of the world's land and water resources for food and agriculture*. Available from http://www.fao.org/3/i1688e/i1688e.pdf. Accessed 25 Nov 2020.
FAO. (2014). *The state of the world's forest genetic resources*. Available from http://www.fao.org/3/a-i3825e.pdf. Accessed 15 Nov 2020.
FAO. (2015). *Healthy soils are the basis for healthy food production*. Available from http://www.fao.org/3/a-i4405e.pdf. Accessed 25 Nov 2020.
FAO. (2016a). *State of the world's forests 2016. Forests and agriculture: Land-use challenges and opportunities*. Available from http://www.fao.org/3/a-i5588e.pdf. Accessed 15 Nov 2020.
FAO. (2016b). Report of the 32nd session of the committee on fisheries. Fortieth session. *Conference*, Rome, 11–15 July 2016. Available from http://www.fao.org/3/a-mr484e.pdf. Accessed 11 Dec 2020.
FAO. (2016c). *Conservation agriculture*. Food and Agriculture Organization of the United Nations. Available from http://www.fao.org/3/a-i6169e.pdf. Accessed 25 Nov 2020.
FAO. (2017). *The future of food and agriculture – Trends and challenges*. Available from http://www.fao.org/3/a-i6583e.pdf. Accessed 10 Dec 2020.
FAO. (2018a). *Agrobiodiversity: A training manual for farmer groups in East Africa*. Available from http://www.fao.org/3/I9307EN/i9307en.pdf. Accessed 8 Sept 2020.
FAO. (2018b). *The state of the world's forests 2018. Forest pathways to sustainable development*. FAO. Available from http://www.fao.org/3/ca0188en/ca0188en.pdf. Accessed 10 Dec 2020.
FAO. (2018c). *The state of world fisheries and aquaculture 2018 - Meeting the sustainable development goals*. Licence: CC BY-NC-SA 3.0 IG. Available from http://www.fao.org/3/I9540EN/i9540en.pdf. Accessed 10 Dec 2020.
FAO. (2018d). *Guiding the transition to sustainable food and agricultural systems*. Food and Agriculture Organization of the United Nations. Available from http://www.fao.org/documents/card/en/c/I9037EN/. Accessed 25 Nov 2020.

FAO. (2019a). The state of the world's biodiversity for food and agriculture. In J. Bélanger & D. Pilling (Eds.), *FAO commission on genetic resources for food and agriculture assessments* (p. 572). Available from http://www.fao.org/3/CA3129EN/CA3129EN.pdf
FAO. (2019b). *The state of the world's aquatic genetic resources for food and agriculture*. FAO commission on genetic resources for food and agriculture assessments. Available from http://www.fao.org/3/ca5256en/CA5256EN.pdf. Accessed 10 Dec 2020.
FAO. (2019c). *Fishery and aquaculture statistics. Global production by production source 1950–2017* (FishstatJ). FAO Fisheries and Aquaculture Department (online). Available from www.fao.org/fishery/statistics/software/fishstatj/en. Accessed 25 Nov 2020.
FAO. (2020a). *Tracking progress on food and agriculture-related SDG indicators 2020- A report on the indicators under FAO custodianship*. Available from http://www.fao.org/sdg-progress-report/en/#chapeau. Accessed 25 Oct 2020.
FAO. (2020b). *Global forest resources assessment– Key findings*. Available from http://www.fao.org/forest-resources-assessment/2020. Accessed 10 Dec 2020.
FAO. (2020c). *How to practice conservation agriculture?* FAO. Available from http://www.fao.org/agriculture/crops/thematic-sitemap/theme/spi/scpi-home/managing-ecosystems/conservation-agriculture/ca-how/en/. Accessed 25 Nov 2020.
FAO. (2020d). *The state of world fisheries and aquaculture 2020*. Available from http://www.fao.org/3/ca9229en/ca9229en.pdf. Accessed 25 Nov 2020.
FAO, & PAR. (2011). Biodiversity for food and agriculture contributing to food security and sustainability in a changing world outcomes of an expert workshop held by FAO and PAR (the Platform on Agrobiodiversity Research) from 14–16 April 2010 in Rome.
FiBL. (2019). *The world of organic agriculture: Statistics and emerging trends*. Available from http://www.fao.org/agroecology/database/detail/en/c/1262695/. Accessed 25 Nov 2020.
Flanders Marine Institute. (2009). *Longhurst provinces*. Available via http://www.marineregions.org. Accessed 15 Nov 2020.
Flegel, T. W. (2006). The special danger of viral pathogens in shrimp translocated for aquaculture. *ScienceAsia, 32*, 215. https://doi.org/10.2306/scienceasia1513-1874.2006.32.215
Ford-Lloyd, B. V. (2006). Realistic population and molecular genetic approaches togeneticassessment. In B. V. Ford-Lloyd, S. R. Dias, & E. Bettencourt (Eds.), *Genetic erosion and pollution assessment methodologies*. Proceedings of PGR forum workshop 5, Terceira Island.
Fowler, C., & Mooney, P. (1990). *Shattering: Food, politics, and the loss of genetic diversity* (p. 104). University of Arizona Press.
Frei, M., Razzak, M. A., Hossain, M. M., et al. (2007). Performance of common carp, Cyprinus carpio L. and Nile tilapia, Oreochromis niloticus (L.) in integrated rice–fish culture in Bangladesh. *Aquaculture, 262*, 250–259. https://doi.org/10.1016/j.aquaculture.2006.11.019
Gabriel, D., & Tscharntke, T. (2007). Insect pollinated plants benefit from organic farming. *Agriculture, Ecosystems and Environment, 118*, 43–48. https://doi.org/10.1016/j.agee.2006.04.005
Giampietro, M. (1997). Socioeconomic constraints to farming with biodiversity. *Agriculture, Ecosystems and Environment, 62*, 145–167. https://doi.org/10.1016/S0167-8809(96)01137-1
González-Esquivel, C. E., Gavito, M. E., Astier, M., et al. (2015). Ecosystem service trade-offs, perceived drivers, and sustainability in contrasting agroecosystems in central Mexico. *Ecology and Society, 20*, art38. https://doi.org/10.5751/ES-06875-200138
Gray, T. N. E., Hughes, A. C., Laurance, W. F., et al. (2018). The wildlife snaring crisis: An insidious and pervasive threat to biodiversity in Southeast Asia. *Biodiversity and Conservation, 27*, 1031–1037. https://doi.org/10.1007/s10531-017-1450-5
Gupta, V. V. (2012). Beneficial microorganisms for sustainable agriculture. *Microbiology Australia, 33*, 113. https://doi.org/10.1071/MA12113
Hallmann, C. A., Sorg, M., Jongejans, E., et al. (2017). More than 75 percent decline over 27 years in total flying insect biomass in protected areas. *PLoS One, 12*, e0185809. https://doi.org/10.1371/journal.pone.0185809

Hammer, K., & Khoshbakht, K. (2005). Towards a 'red list' for crop plant species. *Genetic Resources and Crop Evolution, 52*, 249–265. https://doi.org/10.1007/s10722-004-7550-6
Harlan, J., & Bennett, E. (1979). *Seeds of the earth: A private or public resource?* (p. 12). Canadian Council for International Cooperation.
Harvell, D., Altizer, S., Cattadori, I. M., et al. (2009). Climate change and wildlife diseases: When does the host matter the most? *Ecology, 90*, 912–920. https://doi.org/10.1890/08-0616.1
Harvey, B., Soto, D., Carolsfeld, J., et al. (2017). *Planning for aquaculture diversification: The importance of climate change and other drivers*. FAO Technical Workshop, 23–25 June 2016, FAO Rome. FAO Fisheries and Aquaculture Proceedings No. 47, p. 166.
Holt, R. D. (1997). From metapopulation dynamics to community structure: Some consequences of spatial heterogeneity. In I. Hanski & M. E. Gilpin (Eds.), *Metapopulation biology* (pp. 149–164). Academic Press.
Holton, M. K., Lindroth, R. L., & Nordheim, E. V. (2003). Foliar quality influences tree-herbivore-parasitoid interactions: Effects of elevated CO_2, O_3, and plant genotype. *Oecologia, 137*, 233–244. https://doi.org/10.1007/s00442-003-1351-z
Hosonuma, N., Herold, M., De Sy, V., et al. (2012). An assessment of deforestation and forest degradation drivers in developing countries. *Environmental Research Letters, 7*, 044009. https://doi.org/10.1088/1748-9326/7/4/044009
Huijun, G., Padoch, C., Coffey, K., et al. (2002). Economic development, land use and biodiversity change in the tropical mountains of Xishuangbanna, Yunnan, Southwest China. *Environmental Science & Policy, 5*, 471–479. https://doi.org/10.1016/S1462-9011(02)00093-X
Hussein, M. (1994, January). *Regional focus news: Bangladesh. Ecology and farming: Global monitor* (p. 20). International Federation of Organic Agricultural Movements (IFOAM).
IFOAM. (1994, January). *Biodiversity: Crop resources at risk in Africa. Ecology and farming: Global monitor* (p. 5). International Federation of Organic Agricultural Movements (IFOAM).
IFOAM. (2012). Organic agriculture – A strategy for climate change adaptation. Available from https://www.ifoameu.org/sites/default/files/page, Accessed 15 Nov 2020.
IPBES. (2016). The methodological assessment report on scenarios and models of biodiversity and ecosystem services. In S. Ferrier, K. N. Ninan, P. Leadley, R. Alkemade, L. A. Acosta, H. R. Akçakaya, L. Brotons, W. W. L. Cheung, V. Christensen, K. A. Harhash, J. Kabubo-Mariara, C. Lundquist, et al. (Eds.). https://doi.org/10.5281/zenodo.3235428. Accessed 10 Dec 2020.
IPBES. (2017). *IPBES-5 plenary. Report of the plenary of the intergovernmental science-policy platform on biodiversity and ecosystem services on the work of its fifth session*. Available from https://enb.iisd.org/ipbes/ipbes5/. http://enb.iisd.org/ipbes/ipbes5/. Accessed 11 Dec 2020.
IPBES. (2018). *Summary for policymakers (SPM) of the global assessment report*. Intergovernmental Science-Policy Platform on Biodiversity and Ecosystem Services (IPBES). A media released on Monday, 6 May 2019. Available from bit.ly/IPBESReport. Accessed 11 Dec 2020.
IPCC. (2014). *Climate change 2014: Impacts, adaptation, and vulnerability. Part B: Regional aspects*. Contribution of working group II to the fifth assessment report of the intergovernmental panel on climate change. Cambridge University Press.
IPCC. (2018). *Global warming of 1.5°C. An IPCC special report on the impacts of global warming of 1.5°C above pre-industrial levels and related global greenhouse gas emission*. Available from https://www.ipcc.ch/site/assets/uploads/sites/2/2019/05/SR15_Citation.pdf. Accessed 23 Nov 2020.
Isenring, R. (2010). *Pesticides and the loss of biodiversity: How intensive pesticide use affects wildlife populations and species diversity* (pp. 2–17). Pesticide Action Network Europe.
Ishikawa, R., Yamanaka, S., Fukuta, Y., et al. (2006). Genetic erosion from modern varieties into traditional upland rice cultivars (Oryza sativa L.) in northern Thailand. *Genetic Resources and Crop Evolution, 53*, 245–252. https://doi.org/10.1007/s10722-004-6132-y

Islam, M. S., & Haque, M. (2004). The mangrove-based coastal and nearshore fisheries of Bangladesh: Ecology, exploitation and management. *Reviews in Fish Biology and Fisheries, 14*, 153–180. https://doi.org/10.1007/s11160-004-3769-8

ITT. (2019). *Integrated land and water resources management*. Available from https://www.tt.th--koeln.de/research/chairs-researchgroups/ribbe/detailed-description-of-the-chair-integrated-land-and-water-resources-management/. Accessed 25 Nov 2020.

Jacoby, R., Peukert, M., Succurro, A., et al. (2017). The role of soil microorganisms in plant mineral nutrition—Current knowledge and future directions. *Frontiers in Plant Science, 8*, 1617. https://doi.org/10.3389/fpls.2017.01617

Jahangir, M., Jahan, I., & Mumu, N. (2019). Management of soil resources for sustainable development under a changing climate. *Journal of Environmental Sciences & Natural Resources, 11*, 159–170. https://doi.org/10.3329/jesnr.v11i1-2.43383

Jeger, M. J., & Pautasso, M. (2007). Plant disease and global change - The importance of long-term data sets. *The New Phytologist, 177*, 8–11. https://doi.org/10.1111/j.1469-8137.2007.02312.x

Jensen, Ø., Dempster, T., Thorstad, E., et al. (2010). Escapes of fishes from Norwegian sea-cage aquaculture: Causes, consequences and prevention. *Aquaculture Environment Interactions, 1*, 71–83. https://doi.org/10.3354/aei00008

Kahane, R., Hodgkin, T., Jaenicke, H., et al. (2013). Agrobiodiversity for food security, health and income. *Agronomy for Sustainable Development, 33*, 671–693. https://doi.org/10.1007/s13593-013-0147-8

Kanter, D. R., Musumba, M., Wood, S. L. R., et al. (2018). Evaluating agricultural trade-offs in the age of sustainable development. *Agricultural Systems, 163*, 73–88. https://doi.org/10.1016/j.agsy.2016.09.010

Karmakar, R., Das, I., Dutta, D., & Rakshit, A. (2016). Potential effects of climate change on soil properties: A review. *Science International, 4*, 51–73. https://doi.org/10.17311/sciintl.2016.51.73

Khanna, P. P., & Singh, N. (1991). *Conservation of plant genetic resources*. Available from https://www.bioversityinternational.org/fileadmin/bioversity/publications/Web_version/174/ch13.htm#bm1-Exsitu%20conservation. Accessed 25 Nov 2020.

Klautau, A. G. C. D. M., Cordeiro, A. P. B., Cintra, I. H. A., et al. (2016). Impacted biodiversity by industrial piramutaba fishing in the Amazon river mouth. *Boletim do Instituto de Pesca, 42*, 102–111. https://doi.org/10.20950/1678-2305.2016v42n1p102

Krug, C. J. (2018). *Importance of genetic biodiversity in agriculture*. Available from https://medium.com/thenextnorm/importance-of-genetic-diversity-in-agriculture-b9f88f5fda55. Accessed 25 Nov 2020.

Kumar, D., Chaturvedi, M. K. M., Sharma, S. K., & Asolekar, S. R. (2015). Sewage-fed aquaculture: A sustainable approach for wastewater treatment and reuse. *Environmental Monitoring and Assessment, 187*, 656. https://doi.org/10.1007/s10661-015-4883-x

Lawton, J. H. (1989). Food webs. In J. M. Cherrett (Ed.), *Ecological concepts* (pp. 43–78). Blackwell Scientific.

Leibold, M. A. (1999). Biodiversity and nutrient enrichment in pond plankton communities. *Evolutionary Ecology Research, 1*, 73–95.

Lenton, R. (2011). Integrated water resources management. In *Treatise on water science* (pp. 9–21). Elsevier.

Lernoud, J., & Willer, H. (2019). *Organic agriculture worldwide 2017: Current statistics*. Available from https://orgprints.org/33355/5/lernoud-willer-2019-global-stats.pdf. Accessed 13 Nov 2020.

Li, D., & Liu, S. (2013). Remote monitoring of water quality for intensive fish culture. In *Smart sensors for real-time water quality monitoring* (pp. 217–238). Springer.

Lu, X., Yan, Y., Sun, J., et al. (2015). Short-term grazing exclusion has no impact on soil properties and nutrients of degraded alpine grassland in Tibet, China. *Solid Earth, 6*, 1195–1205. https://doi.org/10.5194/se-6-1195-2015

Lumen Learning. (2020). *Environmental biology*. Available from https://courses.lumenlearning.com/suny-monroe-environmentalbiology/chapter/15-2-waste-management-strategies/. Accessed 8 Dec 2020.

Lurgi, M., López, B. C., & Montoya, J. M. (2012). Climate change impacts on body size and food web structure on mountain ecosystems. *Philosophical Transactions of the Royal Society B, 367*, 3050–3057. https://doi.org/10.1098/rstb.2012.0239

Mann, R. D. (1990). Time running out: The urgent need for tree planting in Africa. *Ecologist, 20*, 48–53.

Marino, N. D. A. C., Romero, G. Q., & Farjalla, V. F. (2018). Geographical and experimental contexts modulate the effect of warming on top-down control: A meta-analysis. *Ecology Letters, 21*, 455–466. https://doi.org/10.1111/ele.12913

Marques, A., Martins, I. S., Kastner, T., et al. (2019). Increasing impacts of land use on biodiversity and carbon sequestration driven by population and economic growth. *Nature Ecology & Evolution, 3*, 628–637. https://doi.org/10.1038/s41559-019-0824-3

Martin, L. J., Blossey, B., & Ellis, E. (2012). Mapping where ecologists work: Biases in the global distribution of terrestrial ecological observations. *Frontiers in Ecology and the Environment, 10*, 195–201. https://doi.org/10.1890/110154

Mbuthia, L. W., Acosta-Martínez, V., DeBruyn, J., et al. (2015). Long term tillage, cover crop, and fertilization effects on microbial community structure, activity: Implications for soil quality. *Soil Biology and Biochemistry, 89*, 24–34. https://doi.org/10.1016/j.soilbio.2015.06.016

McCann, K. (2007). Protecting biostructure. *Nature, 446*, 29–29. https://doi.org/10.1038/446029a

McDonald, R. I. (2008). Global urbanization: Can ecologists identify a sustainable way forward? *Frontiers in Ecology and the Environment, 6*, 99–104. https://doi.org/10.1890/070038

McDonald, R. I., Marcotullio, P. J., & Güneralp, B. (2013). Urbanization and global trends in biodiversity and ecosystem services. In *Urbanization, biodiversity and ecosystem services: Challenges and opportunities* (pp. 31–52). Springer Netherlands.

McKinney, M. L., & Lockwood, J. L. (1999). Biotic homogenization: A few winners replacing many losers in the next mass extinction. *Trends in Ecology & Evolution, 14*, 450–453. https://doi.org/10.1016/S0169-5347(99)01679-1

MEA. (2005). *Millennium ccosystem assessment: Ecosystems and human well-being (current state and trends)*. Available from https://agris.fao.org/agris-search/search.do?recordID=XF2006408644. Accessed 25 Nov 2020.

Meemken, E.-M., & Qaim, M. (2018). Organic agriculture, food security, and theenvironment. *Annual Review of Resource Economics, 10*, 39–63. https://doi.org/10.1146/annurev-resource-100517-023252

Merriam-Webster. (2020). *Definition from the Merriam-Webster Online dictionary*. Available from https://www.merriam-webster.com/dictionary/pollution. Accessed on 8 Dec 2020.

Mmom, P. C. (2009). Impact of poverty and changing cropping systems on agro-crop diversity in the upper Niger Delta, Nigeria. In *IAIA09 conference proceedings', Impact assessment and human well-being 29th annual conference of the international association for impact assessment*, pp. 16–22.

Mo, X.-G., Hu, S., Lin, Z.-H., et al. (2017). Impacts of climate change on agricultural water resources and adaptation on the North China Plain. *Advances in Climate Change Research, 8*, 93–98. https://doi.org/10.1016/j.accre.2017.05.007

Monaghan, J. M., Daccache, A., Vickers, L. H., et al. (2013). More 'crop per drop': Constraints and opportunities for precision irrigation in European agriculture. *Journal of the Science of Food and Agriculture, 93*, 977–980. https://doi.org/10.1002/jsfa.6051

Mortensen, D. A., Egan, J. F., Maxwell, B. D., et al. (2012). Navigating a critical juncture for sustainable weed management. *Bioscience, 62*, 75–84. https://doi.org/10.1525/bio.2012.62.1.12

Nabhan, G. P. (2007). Agrobiodiversity change in a saharan desert oasis, 1919–2006: Historic shifts in tasiwit (Berber) and Bedouin crop inventories of Siwa, Egypt. *Economic Botany, 61*, 31–43. https://doi.org/10.1663/0013-0001(2007)61[31:ACIASD]2.0.CO;2

Newbold, T., Hudson, L. N., Phillips, H. R. P., et al. (2014). A global model of the response of tropical and sub-tropical forest biodiversity to anthropogenic pressures. *Proceedings of the Royal Society B: Biological Sciences, 281*, 20141371. https://doi.org/10.1098/rspb.2014.1371

Newton, A., Oldfield, S., Rivers, M., et al. (2015). Towards a global tree assessment. *Oryx, 49*, 410–415. https://doi.org/10.1017/S0030605315000137

Ngatia, L., Grace, J. M., III, Moriasi, D., & Taylor, R. (2019). Nitrogen and phosphorus eutrophication in marine ecosystems. In *Monitoring of marine pollution* (pp. 1–17). IntechOpen.

Nichols, R. (2015). *A hedge against drought: Why healthy soil is "Water in the bank"*. USDA. Available from https://www.usda.gov/media/blog/2015/05/12/hedge-against-drought-why-healthy-soilwater-bank. Accessed 3 Nov 2020.

Norberg, J. (2000). Resource-niche complementarity and autotrophic compensation determines ecosystem-level responses to increased cladoceran species richness. *Oecologia, 122*, 264–272. https://doi.org/10.1007/PL00008855

O'Gorman, E. J., Fitch, J. E., & Crowe, T. P. (2012). Multiple anthropogenic stressors and the structural properties of food webs. *Ecology, 93*, 441–448. https://doi.org/10.1890/11-0982.1

O'Neill, B. C., Kriegler, E., Ebi, K. L., et al. (2017). The roads ahead: Narratives for shared socioeconomic pathways describing world futures in the 21st century. *Global Environmental Change, 42*, 169–180. https://doi.org/10.1016/j.gloenvcha.2015.01.004

Oberč, B. P., & Arroyo Schnell, A. (2020). In IUCN, International Union for Conservation of Nature (Ed.), *Approaches to sustainable agriculture: Exploring the pathways towards the future of farming*.

Olson, D. M., Dinerstein, E., Wikramanayake, E. D., et al. (2001). Terrestrial ecoregions of the world: A new map of life on earth: A new global map of terrestrial ecoregions provides an innovative tool for conserving biodiversity. *Bioscience, 51*, 933–938. https://doi.org/10.1641/0006-3568(2001)051[0933:TEOTWA]2.0.CO;2

Otto, S. B., Rall, B. C., & Brose, U. (2007). Allometric degree distributions facilitate food-web stability. *Nature, 450*, 1226–1229. https://doi.org/10.1038/nature06359

Owusu, P. A., & Asumadu-Sarkodie, S. (2016). A review of renewable energy sources, sustainability issues and climate change mitigation. *Cogent Engineering, 3*. https://doi.org/10.1080/23311916.2016.1167990

Paini, D. R., Sheppard, A. W., Cook, D. C., et al. (2016). Global threat to agriculture from invasive species. *Proceedings of the National Academy of Sciences, 113*, 7575–7579. https://doi.org/10.1073/pnas.1602205113

Parmesan, C. (2006). Ecological and evolutionary responses to recent climate change. *Annual Review of Ecology, Evolution, and Systematics, 37*, 637–669. https://doi.org/10.1146/annurev.ecolsys.37.091305.110100

Patil, H. J., & Solanki, M. K. (2016). Microbial inoculant: Modern era of fertilizers and pesticides. In *Microbial inoculants in sustainable agricultural productivity* (pp. 319–343). Springer India.

Paudel, B., Rana, R. B., Sthapit, B., et al. (2012). *Determinants of agriculture biodiversity in Western Terai landscape complex of Nepal* (MPRA paper No. 58181). Munich Personal RePEc Archive (MPRA). Available from MPRA. https://mpra.ub.uni-muenchen.de/58181/. Accessed 10 Dec 2020.

Pauly, D. (1998). Fishing down marine food webs. *Science (80-), 279*, 860–863. https://doi.org/10.1126/science.279.5352.860

Pejchar, L., & Mooney, H. A. (2009). Invasive species, ecosystem services and human well-being. *Trends in Ecology & Evolution, 24*, 497–504. https://doi.org/10.1016/j.tree.2009.03.016

Pimentel, D. (2006). Soil erosion: A food and environmental threat. *Environment, Development and Sustainability, 8*, 119–137. https://doi.org/10.1007/s10668-005-1262-8

Popp, A., Calvin, K., Fujimori, S., et al. (2017). Land-use futures in the shared socio-economic pathways. *Global Environmental Change, 42*, 331–345. https://doi.org/10.1016/j.gloenvcha.2016.10.002

Prashar, P., & Shah, S. (2016). Impact of fertilizers and pesticides on soil microflora in agriculture. In E. Lichtfouse (Ed.), *Sustainable agriculture aeviews* (Vol. 19, pp. 331–361). Springer.

Primavera, J. (1998). Tropical shrimp farming and its sustainability. In S. S. De Silva (Ed.), *Tropical mariculture* (pp. 257–290). Academic Press.

Rahman, M. H., Lund, T., & Bryceson, I. (2011). Salinity impacts on agro-biodiversity in three coastal, rural villages of Bangladesh. *Ocean and Coastal Management, 54*, 455–468. https://doi.org/10.1016/j.ocecoaman.2011.03.003

Rana, R. B., Garforth, C., Sthapit, B., & Jarvis, D. (2007). Influence of socio-economic and cultural factors in rice varietal diversity management on-farm in Nepal. *Agriculture and Human Values, 24*, 461–472. https://doi.org/10.1007/s10460-007-9082-0

Rapsominikis, G. (2015). *The economic lives of smallholder farmers: An analysis based on household data from nine countries* (p. 48). Food and Agriculture Organization of the United Nations.

Rasmussen, L. V., Kirchhoff, C. J., & Lemos, M. C. (2017). Adaptation by stealth: Climate information use in the Great Lakes region across scales. *Climatic Change, 140*, 451–465. https://doi.org/10.1007/s10584-016-1857-0

Reganold, J. P., & Wachter, J. M. (2016). Organic agriculture in the twenty-first century. *Nature Plants, 2*, 15221. https://doi.org/10.1038/nplants.2015.221

Riahi, K., van Vuuren, D. P., Kriegler, E., et al. (2017). The shared socioeconomic pathways and their energy, land use, and greenhouse gas emissions implications: An overview. *Global Environmental Change, 42*, 153–168. https://doi.org/10.1016/j.gloenvcha.2016.05.009

Richer, R. (2008). *Conservation in Qatar: Impacts of increasing industrialization* (CIRS occasional paper No. 2 (2008)). Available via SSRN. https://ssrn.com/abstract=3447013. Accessed 10 Dec 2020.

Ripple, W. J., Abernethy, K., Betts, M. G., et al. (2016). Bushmeat hunting and extinction risk to the world's mammals. *Royal Society Open Science, 3*, 160498. https://doi.org/10.1098/rsos.160498

Rosegrant, M. W., Ringler, C., & Zhu, T. (2009). Water for agriculture: Maintaining food security under growing scarcity. *Annual Review of Environment and Resources, 34*, 205–222. https://doi.org/10.1146/annurev.environ.030308.090351

Rosser, A. M., & Mainka, S. A. (2002). Overexploitation and species extinctions. *Conservation Biology, 16*, 584–586. https://doi.org/10.1046/j.1523-1739.2002.01635.x

Roy, R. N., & Nabhan H. (2001). *Soil and nutrient management in Sub-saharan Africa in support of the soil fertility initiative- proceedings of the expert consultation, Lusaka, Zambia, 6–9 Dec 1999*. Available from http://www.fao.org/tempref/agl/agll/docs/misc31.pdf. Accessed 10 Dec 2020.

Sahoo, U. K., & Singh, S. L. (2005). Integrated fish-pig and fish-poultry farming in East Kalcho, Saiha district of Mizoram, North-East India: An economic analysis. *International Journal of Agriculture and Forestry, 5*, 281–286. https://doi.org/10.5923/j.ijaf.20150505.03

Samir, K. C., & Wolfgang, L. (2017). The human core of the shared socioeconomic pathways: Population scenarios by age, sex and level of education for all countries to 2100. *Global Environmental Change, 42*, 181–192. https://doi.org/10.1016/j.gloenvcha.2014.06.004

Sánchez-Bayo, F., & Wyckhuys, K. A. G. (2019). Worldwide decline of the entomofauna: A review of its drivers. *Biological Conservation, 232*, 8–27. https://doi.org/10.1016/j.biocon.2019.01.020

Saunders, M. E., & Luck, G. W. (2016). Limitations of the ecosystem services versus disservices dichotomy. *Conservation Biology, 30*, 1363–1365. https://doi.org/10.1111/cobi.12740

Schneider, U. A., Havlík, P., Schmid, E., et al. (2011). Impacts of population growth, economic development, and technical change on global food production and consumption. *Agricultural Systems, 104*, 204–215. https://doi.org/10.1016/j.agsy.2010.11.003

Schwarzmüller, F. A. (2015). *Global change effects on the stability of food-web motifs* (Dissertation). Georg-August-Universität Göttingen.

Schwitzguébel, J.-P., & Wang, H. (2007). Environmental impact of aquaculture and countermeasures to aquaculture pollution in China. *Environmental Science and Pollution Research, 14*, 452–462. https://doi.org/10.1065/espr2007.05.426

Scialabba, N. E.-H., & Müller-Lindenlauf, M. (2010). Organic agriculture and climate change. *Renewable Agriculture and Food Systems, 25*, 158–169. https://doi.org/10.1017/S1742170510000116

Secretariat of the Convention on Biological Diversity. (2004). *The ecosystem approach* (CBD guidelines, p. 50). Secretariat of the Convention on Biological Diversity.

Shiva, V. (1991). The green revolution in the Punjab. *Ecologist, 21*, 57–60.

Soriano, I. R., Riley, I. T., Potter, M. J., & Bowers, W. S. (2004). Phytoecdysteroids: a novel defense against plant-parasitic nematodes. *Journal of Chemical Ecology, 30*, 1885–1899. https://doi.org/10.1023/B:JOEC.0000045584.56515.11

Sotomayor, D. A., & Lortie, C. J. (2015). Indirect interactions in terrestrial plant communities: Emerging patterns and research gaps. *Ecosphere, 6*, art103. https://doi.org/10.1890/ES14-00117.1

Stouffer, D. B., & Bascompte, J. (2010). Understanding food-web persistence from local to global scales. *Ecology Letters, 13*, 154–161. https://doi.org/10.1111/j.1461-0248.2009.01407.x

Sudmeier-Rieux, K., Masundire, H., Rizvi, A., & Rietbergen, S. (2006). *Ecosystems, livelihoods and disasters: An integrated approach to disaster risk management, Ecosystem*. IUCN.

Swaminathan, M. (2003). Bio-diversity: An effective safety net against environmental pollution. *Environmental Pollution, 126*, 287–291. https://doi.org/10.1016/S0269-7491(03)00241-0

Symstad, A. J., & Tilman, D. (2001). Diversity loss, recruitment limitation, and ecosystem functioning: Lessons learned from a removal experiment. *Oikos, 92*, 424–435. https://doi.org/10.1034/j.1600-0706.2001.920304.x

The World Bank. (2017). *The World Bank database*. Available from http://www.worldbank.org. Accessed 10 Dec 2020.

Thrupp, L. A. (1998). The central role of agricultural biodiversity: Trends and challenges. In *Conservation and sustainable use of agricultural biodiversity*. CIP-UPWARD in partnership with GTZ, IDRC, IPGRI and SEARICE.

Thrupp, L. A. (2000). Linking agricultural biodiversity and food security: The valuable role of agrobiodiversity for sustainable agriculture. *International Affairs, 76*, 265–281. https://doi.org/10.1111/1468-2346.00133

Tilman, D., & Pacala, S. (1993). The maintenance of species richness in plant communities. In *Species diversity in ecological communities* (pp. 13–25). University of Chicago Press.

Tonitto, C., David, M. B., & Drinkwater, L. E. (2006). Replacing bare fallows with cover crops in fertilizer-intensive cropping systems: A meta-analysis of crop yield and N dynamics. *Agriculture, Ecosystems and Environment, 112*, 58–72. https://doi.org/10.1016/j.agee.2005.07.003

Treseder, K. K. (2004). A meta-analysis of mycorrhizal responses to nitrogen, phosphorus, and atmospheric CO_2 in field studies. *The New Phytologist, 164*, 347–355. https://doi.org/10.1111/j.1469-8137.2004.01159.x

Tsegaye, B., & Berg, T. (2007). Genetic erosion of Ethiopian tetraploid wheat landraces in Eastern Shewa, Central Ethiopia. *Genetic Resources and Crop Evolution, 54*, 715–726. https://doi.org/10.1007/s10722-006-0016-2

Turton, A. R., Hattingh, J., Claassen, M., et al. (2007). Towards a model for ecosystem governance: An integrated water resource management example. In *Governance as a trialogue: Government-Society-Science in Transition* (pp. 1–28). Springer.

Tylianakis, J. M., Didham, R. K., Bascompte, J., & Wardle, D. A. (2008). Global change and species interactions in terrestrial ecosystems. *Ecology Letters, 11*, 1351–1363. https://doi.org/10.1111/j.1461-0248.2008.01250.x

UN Chronicle. (2020). *Women and agricultural water resource management*. United Nations. Available from https://www.un.org/en/chronicle/article/women-and-agricultural-water-resource-management. Accessed 8 Dec 2020.

United Nations. (2017). *World urbanization prospects: The 2017 revision* (ESA/P/WP/248). UN, Department of Economic and Social Affairs, Population Division.

United Nations. (2019). Available from https://www.un.org/sustainabledevelopment. Accessed Sept 2020.

Upreti, B. R. (2000). The effects of changing land use systems in agricultural biodiversity: Experiences and lessons from Nepal. In J. Xu (Ed.), *Links between the culture and biodiversity* (pp. 327–337). Yunnan Science and Technology.

Upreti, B. R., & Upreti, Y. G. (2002). Factors leading to agro-biodiversity loss in developing countries: The case of Nepal. *Biodiversity and Conservation, 11*, 1607–1621. https://doi.org/10.1023/A:1016862200156

van Vuuren, D. P., Edmonds, J., Kainuma, M., et al. (2011). The representative concentration pathways: An overview. *Climatic Change, 109*, 5–31. https://doi.org/10.1007/s10584-011-0148-z

Velazco-Bedoya, D. M., Camargo, A. S. N., Yu, A. S. O., & Nascimento, P. T. D. S. (2014). Integrated crop-livestock systems: Technology strategy adoption, management and firm competences. In *International Association for Management of Technology Conference*, Washington, DC, Vol. 23. https://doi.org/10.1016/j.livsci.2017.03.010

Vigouroux, Y., Barnaud, A., Scarcelli, N., & Thuillet, A.-C. (2011). Biodiversity, evolution and adaptation of cultivated crops. *Comptes Rendus Biologies, 334*, 450–457. https://doi.org/10.1016/j.crvi.2011.03.003

Vilà, M., Espinar, J. L., Hejda, M., et al. (2011). Ecological impacts of invasive alien plants: A meta-analysis of their effects on species, communities and ecosystems. *Ecology Letters, 14*, 702–708. https://doi.org/10.1111/j.1461-0248.2011.01628.x

Welcomme, R. L. (1988). *International introductions of inland aquatic species* (FAO fisheries technical paper, pp. 294–318).

Welcomme, R. L. (1999). A review of a model for qualitative evaluation of exploitation levels in multi-species fisheries. *Fisheries Management and Ecology, 6*, 1–19. https://doi.org/10.1046/j.1365-2400.1999.00137.x

Willemen, L., Scheldeman, X., Soto Cabellos, V., et al. (2007). Spatial patterns of diversity and genetic erosion of traditional cassava (Manihot esculenta Crantz) in the Peruvian Amazon: An evaluation of socio-economic and environmental indicators. *Genetic Resources and Crop Evolution, 54*, 1599–1612. https://doi.org/10.1007/s10722-006-9172-7

Winters, P., Cavatassi, R., & Lipper, L. (2006). *Sowing the seeds of social relation: The role of social capital in crop diversity* (ESA working paper No. 06–16). Food and agriculture organization.

Wittenberg, R., & Cock, M. J. W. (2001). *Invasive alien species: A toolkit of best prevention and management practices* (Vol. xvii, p. 228). CAB International.

Wootton, J. T. (1994). The nature and consequences of indirect effects in ecological communities. *Annual Review of Ecology and Systematics, 25*, 443–466. https://doi.org/10.1146/annurev.es.25.110194.002303

WoRMS. (2018). World Register of Marine Species. Available from http://www.marinespecies.org. Accessed 8 Nov 2020.

WWF. (2016). *Living planet report 2016. Risk and resilience in a new era.* WWW International. Available from https://www.worldwildlife.org/pages/living-planet-report-2016. Accessed 10 Dec 2020.

Zimmerer, K. S., de Haan, S., Jones, A. D., et al. (2019). The biodiversity of food and agriculture (Agrobiodiversity) in the anthropocene: Research advances and conceptual framework. *Anthropocene, 25*, 100192. https://doi.org/10.1016/j.ancene.2019.100192

Chapter 2
Emerging Risks to Plant Health

Homa Hosseinzadeh-Bandbafha, Mohammadali Kiehbadroudinezhad, Majid Khanali, and Afrooz Taghizadehghasab

Abstract Changes in world climate caused by human activities, especially fossil fuel combustion, are gradually increasing, and its intensity is expanding day by day. Since the climatic factors such as temperature, relative humidity, solar radiation, precipitations, carbon dioxide level, etc. have significant impacts on different organisms, climate change can lead to various challenges for organisms such as pests. More specifically, climate change is a multifaceted challenge that can affect Pest's dynamics and behavior. Besides, climate change consequences can lead to changes in the abundance and geographic distribution of different pests. These changes are responsible for the emergence of emerging pests that are commonly thought to be related to the global trade in agricultural products. For various reasons, such as the lack of a natural enemy, it is impossible to control emerging pests by biological methods, while many pests are resistant to pesticides. Due to this fact, awareness of climate change's effects on emerging pests' emergence is now essential. This chapter reviews climate change and its impacts on pests' behavior and the spread of emerging pests.

Keywords Climate change · Global warming · Pest's behavior · Emerging pests

H. Hosseinzadeh-Bandbafha (✉) · M. Khanali
Department of Mechanical Engineering of Agricultural Machinery, Faculty of Agricultural Engineering and Technology, College of Agriculture and Natural Resources, University of Tehran, Karaj, Iran
e-mail: homa.hosseinzadeh@ut.ac.ir

M. Kiehbadroudinezhad
Division of Engineering, Saint Mary's University, Halifax, NS, Canada

A. Taghizadehghasab
Department of Soil Science, College of Agriculture, Isfahan University of Technology, Isfahan, Iran

C. M. Galanakis (ed.), *Biodiversity, Functional Ecosystems and Sustainable Food Production*, https://doi.org/10.1007/978-3-031-07434-9_2

1 Introduction

Currently, pests are the most critical biological limitations in crop production. As one of the crucial components of crop ecosystems, biodiversity could partly overcome pests by biocontrol, i.e., pests' control by their natural enemies (Seppelt et al., 2020). However, researchers believe that the occurrence of pests in farms could lead to reducing yields of agricultural products, up to 40% (Oerke et al., 2012). Despite pests' role in threatening food products, because of the world's growing population and the increase in living standards, local and global food security is currently vital (Bommarco et al., 2018). To support food availability and security, pesticide application in crop production has been successful in pest control.

However, the long-term utilization of pesticides has led to resistance of more than 500 species of pests into pesticides since 1945 (Liang et al., 2016). In addition to biological factors, climate change as a critical factor is responsible for pest resistance to pesticides (Bajwa et al., 2020). Various evidence indicates that climate change is one of the most critical environmental challenges that lead to the emergence of different ecosystem quality issues, such as reducing biodiversity (Lepetz et al., 2009). For example, altered precipitation patterns caused by climate change could create a favorable environment for Pest's attacks (Roos et al., 2011) and advances in phenology (Fontaine et al., 2009). It is interesting to note that competition or predation as interspecific relationships between organisms could be changed owing to climate variability (Wang et al., 2009). Consequently, climate change could be useful in altering the geographical distributions and outbreak frequencies of pests.

More specifically, all organisms, such as plants and insects for their growth and development, are dependent on climatic factors. Plant phenology is directly influenced by temperature, relative humidity, photoperiod, carbon dioxide, and precipitation (Bale et al., 2002; García de Cortázar-Atauri et al., 2010; Caffarra & Donnelly, 2011; Bregaglio et al., 2013). In addition to the direct relation with climate factors, habitat structure, food quality, length of the growing season, overwintering, and oviposition of pests is related to climatic factors (Moreau et al., 2008; Reineke & Thiéry, 2016). According to this fact, previous studies have emphasized that climate change is effective in phenological stages of pests population dynamics, limits of development, adaptation, etc. (Caffarra et al., 2012), and interactions between plants and pests (Castex et al., 2018). More explicitly, since mobile organisms with shorter generation times, i.e., pests, could move and escape from adverse conditions than sessile species, i.e., plants, insects could evolve adaptations faster. Accordingly, a warmer climate favor insects over plants (Kurz et al., 2008).

In general, climate change is defined as a change in the climate situation that persists for a lengthy period (decades or longer) that these changes are identified through changes in the mean/properties of climate using statistical tests (Nkomwa et al., 2014). However, climate change does not limit to an increase in temperature and subsequent phenomena such as precipitation patterns. Still, also it is a multifaceted challenge that is with increased carbon dioxide and other greenhouses that could affect pest dynamics and behavior (Jactel et al., 2019). For example, aphids,

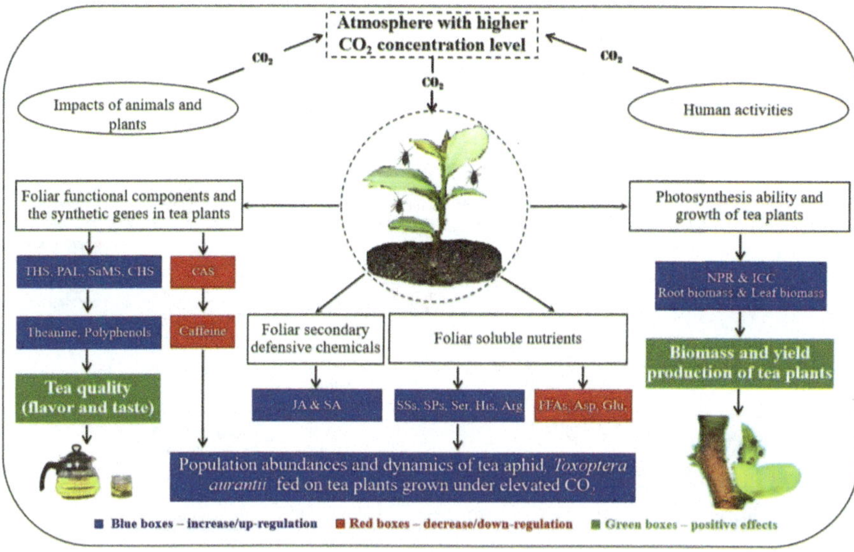

Fig. 2.1 A schematic model of tea aphid and *Toxoptera aurantii* interactions under elevated carbon dioxide (Li et al., 2019). (Copyright © 2019 The Authors. Published by Elsevier Masson SAS.; License Number: 4924790521174. Arg arginine acid, Asp aspartic acid, CHS chalcone synthase, CAS caffeine synthase, FFAs free fatty acids, FAAs free amino acids, Glu glutamic acid, His histidine acid, ICC intercellular carbon dioxide concentration, JA jasmonic acid, NPR net photosynthetic rate, PAL phenylalanine ammonia-lyase, SaMS s'-adenosine methionine synthetase, SA salicylic acid, SSs: soluble sugars, SPs soluble proteins, Ser serineacid, THS theanine synthetase)

as one of the most harmful pests to many crop plants globally (Züst & Agrawal, 2016), usually benefit from elevated carbon dioxide (Johnson & Züst, 2018). Also, increasing the concentration of carbon dioxide in the atmosphere leads to more vegetative growth that subsequently creates higher relative humidity as a favorable environmental factor to pests' growth and development (Gani & Ghosh, 2018). Interestingly, Li et al. (2019) provided a schematic model of tea aphid and *Toxoptera aurantii* interactions under the elevated carbon dioxide that is illustrated in Fig. 2.1.

On the other hand, pesticides are also exposed to environmental influences caused by climate change (Fig. 2.2), which causes a change in the dosage of the pesticides and, thus, a change in the Pest's behavior and more Pest's resistant.

In detail, elevated temperatures, high moisture content, and direct exposure to sunlight as the main factor of climate strongly effect on degradation and volatilization of pesticides (Johnson et al., 1995; Otieno et al., 2013). Interestingly, concentrations of pesticides are strongly affected by degradation and volatilization (Zhang et al., 2006; Noyes et al., 2009). Accordingly, when the volatilization and degradation of pesticides are risen due to climate change, pesticides are reduced. Also, the rainfall pattern, i.e., timing and intensity of rainfall that is strongly affected by climate, influences pesticide's efficiency and persistence (Rosenzweig et al., 2001;

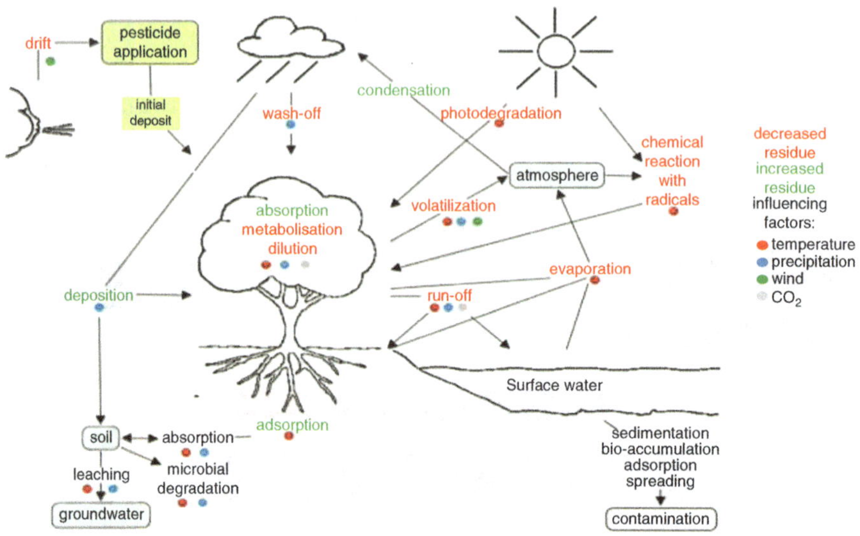

Fig. 2.2 The schematics of the environmental factors and climate change's role in applied pesticides (Delcour et al., 2015). (With permission from Elsevier. Copyright© 2014; License Number: 4924800523684)

Bailey, 2004). Therefore, altering the climate and afterward, rainfall patterns could reduce pesticide's efficiency and persistence.

Moreover, temperature and light through chemical alteration could affect pesticide persistence (Rosenzweig et al., 2001). Finally, in turn, items listed above could lead to more pesticide consumption and, subsequently, pest resistance. In other words, pesticide application patterns should be changed in response to climate change.

It should be noted that the critical role and the adverse effects of pesticides on climate change have been proven (Claver et al., 2006). As a result, it could be said that there is a close relationship between climate change and pests. Not only, this interaction does not end with the increase in Pest's growth and development as well as pest resistance to pesticides; but also it is responsible for the spread of *emerging pests* in different areas (Miraglia et al., 2009).

Generally, "emerging pests"are defined as species (known or new) whose occurrence or geographical distribution is spreading (Athanassiou & Rumbos, 2018). In general, (i) the spread of insect vectors, (ii) accidental entry of pests to new areas that often are linked to trade, (iii) modification of cultural practice, e.g., optimal consumption of pesticides, and (iv) environmental changes such as climate change are possible reasons for the occurrence of emerging pests (Suffert & Ward, 2014). Management of various emerging pests is faced with different challenges and issues such as lack or very few promising natural enemies, broad host range and polyphagy in nature, surviving even in weeds, recognizing only after symptoms, resistance to pesticides, etc. (Rai et al., 2014).

Table 2.1 A category for "emerging pest/emerging risk" in stored foods (Stejskal et al., 2015)

Emerging pests/emerging risks	
Categories	Causes
Category I. New (emerging) species	Invasive species spreading in new geographic area Newly described species Newly discovered native or cryptogenic species in given area Newly increased population density/incidence/frequency of infestation
Category II. New levels of population density, changed pest spectra, a percentage of store infestation in the geographical area	Newly elevated population densities change pest status and newly emerged key species
Category III. New (emerging) biotypes	Emergence and spread of new pest biotypes (resistance, virulence, food preference)
Category IV. New adverse effects and risks	Newly discovered and previously unrecognized pest effects (allergy, contamination)
Category V. Newly endangered and packaged commodities	Newly colonized commodities, plants, materials (either native or newly introduced pests). New issues are also packages. They either increase the risk of infestation (not preventing invasion/penetration of pests) or provide shelters and enable the geographical spread of pests and their introduction into new territories

Emerging pests not only cause significant damage to crops during their growing season, but they also threaten post-harvest crops (stored food). Stejskal et al. (2015) presented an overview of primary categories and causes of "emerging pest/emerging risk" in stored foods shown in Table 2.1.

Although it is believed that global agriculture confronts with emerging pests due to the exchange of uncontrolled plant material (Boukhris-Bouhachem, 2017) but, climate change is also involved in this phenomenon. Temperature changes result in altering the geographical range of pest and pest migration to other areas, and subsequently provide conditions for the emergence and spread of emerging pests (Reddy, 2013). More specifically, climate analysis shows that increasing winter temperature leads to an earlier start to the flying of pests (Rosenzweig et al., 2002). In addition to long-distance movements in different pests, species could be altered due to wind patterns (Miraglia et al., 2009). This chapter will first note about climate change and then will focus on its role in spreading some of the most critical emerging pests.

2 Climate Change

Currently, climate change is one of the global issues which affects the economic (Alagidede et al., 2016), energy (Li et al., 2012), and most importantly, the environment (Masse et al., 2006). The scientific research has extensively documented that climate change is originated from the accumulation of greenhouse gases of anthropogenic in the atmosphere (Krause & Farina, 2016). Carbon dioxide, methane, nitrous oxide, hydrofluorocarbons, and sulfur hexafluoride are known as greenhouse gases based on the Kyoto protocol (Oberthür & Ott, 1999). More specifically, carbon dioxide, methane, and nitrous oxide have been dedicated to themselves for more than half of the total greenhouse gases (Oberthür & Ott, 1999). It should be noted that methane and nitrous oxide are roughly 28 and 256 times more potent than carbon dioxide, respectively, owing to their 100-year global warming potential (Gorsky et al., 2019).

In 2017, greenhouse gases' total concentration, including cooling aerosols, is estimated at 454 ppm carbon dioxide equivalents. Then, 2016 had a growth of more than 4 ppm while compared with 10 years ago, had a rise of more than 37 ppm. It is estimated that if greenhouse gas emissions are not limited, the global average temperature will be increased 2 °C above preindustrial levels (Vetter et al., 2017). This increase in temperature could eventually lead to droughts and heatwaves; moreover, it could increase precipitation and storms and floods risk and the ensuing problems (Agovino et al., 2019). Warm temperature leads to altering pests' growth and development rate, including insects, weeds, and diseases (Macfadyen et al., 2018). However, invertebrate pests, through three ways listed in Table 2.2, respond to warm temperatures.

These responses lead to altering pests' dynamics and behavior that more details are provided in Sect. 3.

3 Pests and Climate Change

Although agriculture accounts for about 5.0–5.8 Gt carbon dioxide equivalents or, in better words, 10–12% of greenhouse gas emissions (Tubiello et al., 2013; Maraseni & Qu, 2016), it is mostly affected by threats of climate change due to it is a climate-sensitive sector (Sabbaghi et al., 2020). One of the most critical challenges of agriculture in the face of climate change is pests' behavior. It has been documented that climatic barriers hampered the movement of species by commercial networks and their establishment in new areas in the past (Robinet & Roques, 2010); however, climate change can lift barriers and causes the proliferation and spread of pests (Walther et al., 2002; Grünig et al., 2020). Climate change could lead to positive and negative effects on various pests (Sharma & Dhillon, 2020). Positive effects lead to pests' growth and development, and adverse effects lead to pests' migration to other areas and emerging pests in different areas.

As previously mentioned, climate change has various consequences, such as rising temperatures, rising carbon dioxide, changing rainfall patterns, etc. These consequences, in turn, play a role in the growth and development of pests. For example, feeding and hosting patterns of pests and their oviposition are significantly sensitive to carbon dioxide (Castex et al., 2018). This sensitivity is in line with the fact that carbon dioxide by altering the plants' physiology and morphology leads to growth and development; this altering could subsequently affect pests' feeding behavior

Table 2.2 An overview of the responses of pests to climate change (Macfadyen et al., 2018)

Response group	Species	Details	Citation
Shifting distributions	*Penthaleus* spp. (blue oat mites)	Distributions of the three *Penthaleus* species in Australia are correlated with different climatic variables, suitable climate space likely to decrease in the future. Cryptic species respond differently	Hill et al. (2012)
	Leptinotarsa decemlineata, Colorado potato beetle, *Ostrinia nubilalis*, European corn borer	The models suggest a widening of the area of suitable habitat for both pests in central Europe	Kocmankova et al. (2011)
	Diabrotica virgifera virgifera, western corn rootworm	The models showed a northward advancement of the upper physiological limit in the Northern hemisphere, which might lead to increased outbreaks at higher latitudes	Aragón and Lobo (2012)
	12 pest fruit fly species (Tephritidae)	Results from distribution models revealed general patterns of poleward movement for the group. For individual species, distribution shifts also appear to be eastward, and at finer scales, varying amounts of species turnover was apparent. These changes in response across different scales present regional management challenges for these species under climate change	Hill et al. (2016)
Altering phenology	*Leptinotarsa decemlineata*, Colorado potato beetle, *Ostrinia nubilalis*, European corn borer	Models suggest an increase in the number of generations per year. Area of arable land affected by a third-generation per season of L. *decemlineata* in 2050 is c. 45% higher, and by the second generation of *O. nubilalis* is nearly 61% higher, compared to present levels	Kocmankova et al. (2011)
	Cydia pomonella, Codling Moth	Under future conditions of increased temperatures (2045–2074) in Switzerland, the risk of an additional third generation will increase from 0–2% to 100%, and there will be a two-week shift in earlier overwintering adult flight. The shifts in phenology and voltinism will require a change to plant protection strategies	Stoeckli et al. (2012)
	13 agriculturally important pest insect species	Degree-day models were used to predict the voltinism of 13 agronomically important pests in California, USA. Under future climate change, all species are likely to see an increase in voltinism per year, with different climate change models contributing variance across results	Ziter et al. (2012)
	Halotydeus destructor (redlegged earth mite)	Models suggest that the temperature cues for post-diapause egg hatch have evolved markedly between the western Australian "Mediterranean" environment (20.5 °C) and the southeastern Australian (16 °C) more temperate environment	McDonald et al. (2015)

(continued)

Table 2.2 (continued)

Response group	Species	Details	Citation
Adjusting to persist *in situ* (phenotypic plasticity or adaptation)	*Halotydeus destructor* (redlegged earth mite)	Species distribution models indicate that invasive populations of *H. destructor* in Australia have undergone a recent range shift into hotter and drier inland environments since establishing a stable distribution in the 1960s. Experiments measuring physiological traits reported greater thermal tolerance in Australian populations than South African (native)	Hill et al. (2013)
	Zaprionus indianus (African fig fly)	Invasive populations in India display latitudinal clines indicative of rapid adaptive shifts. Traits included in studies were desiccation and starvation tolerance of adults, body weight, wing length, thorax length, and the number of ovarioles	Gibert et al. (2016)
	Aedes albopictus (Asian tiger mosquito)	Invasive populations in the USA from Japan demonstrated rapid adaptive evolution (in 20 years) of the photoperiodic response during invasion and range expansion into higher latitudes. Change in photoperiodism has been an essential adaptation to climatic variation across the invasive range	Urbanski et al. (2012)

and diet quality of herbivore pests (Ryan et al., 2014). Interestingly, pests such as *Lobesia botrana*, can increase their eggs laying rate when carbon dioxide concentration is a higher level (Guerenstein & Hildebrand, 2008). Also, it was reported that an increase in carbon dioxide concentration is responsible for the rising of food consumption and metabolism of the *Helicoverpa amigera* larvae (Akbar et al., 2016). Besides, as previously noted, sucking insects such as aphids could also be spared in the presence of high concentrations of carbon dioxide (Newman, 2004; Ryan et al., 2014; Johnson & Züst, 2018). It should be noted that diseases such as *Phytopthora infestans*, *Pyricularia oryzae*, and *Rhizoctonia solani* are also affected by rising concentrations of carbon dioxide (Gautam et al., 2013).

Like high concentrations of carbon dioxide, high temperatures could also alter pests' growth and development and their fecundity and mortality (Khaliq et al., 2014). Generally, temperature could alter the development rate of pests, i.e., changing in life cycle duration of pests, population density, genetic composition, the extent of host plant exploitation, voltinism, size, and local and geographical distribution (Bale et al., 2002). Researchers believe that warmer weather could result in accelerating the growth and development rates of many pests. Warmer weather leads to accumulating earlier and faster heat and subsequently increases the number of generations a pest in 1 year named voltinism (Honek, 2013; Castex et al., 2018). For example, *Delia radicum (L.)* becomes active a month earlier, when the temperature becomes 3 °C warmer; moreover, increases of 5–10 °C in temperature could complete the four generations each year (Collier et al., 1991). Meisner et al. (2014) claimed that when temperatures are high, the development rate of and subsequently, the number of generations of aphids is risen, leading to an increase in their populations.

It is also documented that higher temperatures in winter lead to reduce the mortality rates of pests (Harrington et al., 2001) that, in turn, increase the distribution of pests (Battisti et al., 2005). Higher temperatures in spring and winter result in favorable overwintering of pupae and lengthen growth season of pests (Bale et al., 2002). It is reported that during the warm winter season, *Herlicoverpa amigera* could easily overwinter (Reddy, 2013). More specifically, the researchers indicate that milder winters could increase the survival of more pests species at higher latitudes (Bale et al., 2002; Bebber et al., 2013).

High temperatures could also short the Pest's life cycle and increase pest populations (Van Dyck et al., 2015). As a sample, Table 2.3 shows the role of temperature on the life cycle duration of looper *Hyposidra talaca* (Roy et al., 2019).

The distribution of pests in latitude and altitude may change because of increased voltinism and faster development rates (Svobodová et al., 2014a, b). The fossil evidence demonstrates that many pests preferred that have tracked climate change rather than adapted to it (Coope, 1978). Based on Porter et al. (1991), the increase of 1 and 3 °C in temperature has led to northward shifts in the distribution of the *Ostrinia nubilalis* to 1220 km, while in almost all areas generate a different generation. Parmesan and Yohe (2003) predicted that some pests would shift by 6.1 km to the north per decade because of increased temperatures.

In line with that, Yan et al. (2017) investigated species richness for the world's current climate. They indicated that southeast Asia, western and southern India, eastern USA, Mexico, and southwest China have higher species richness (Fig. 2.3). They predicted that in areas with higher altitudes, e.g., the western USA, western China, southeast Brazil, the northeast part of South Africa, and northwest Ethiopia, will face increases in species pests (Fig. 2.4). More specifically, due to an increase in the temperatures, the migration rate of the butterflies and lepidopteran moths is increasing (Sparks et al., 2007).

Table 2.3 The life cycle duration of tea looper *Hyposidra talaca* in different temperatures[a] (Roy et al., 2019)

	Temperature (°C)			
	10 °C	20 °C	30 °C	35 °C
Stages (Instars)	Days (mean ± SD)			
Eggs–1st	6.6 ± 2.5[a]	6 ± 2.64[a]	5.3 ± 3.21[a]	4 ± 1.73[a]
1st–2nd	5 ± 1.0[a]	4.6 ± 1.15	4.6 ± 1.15[a]	3.7 ± 0.57[a]
2nd–3rd	5 ± 1.0[a]	4.6 ± 1.15[a]	4.7 ± 1.14[a]	4 ± 1[a]
3rd–4th	5.6 ± 2.08[a]	5.3 ± 2.30[a]	4.6 ± 2.88[a]	3.3 ± 0.58[a]
4th–5th	5 ± 1.0[a]	4.7 ± 2.08[a]	4.5 ± 1.15[a]	4.3 ± 0.60[a]
5th–pupa	4.6 ± 2.00[a]	5 ± 1.73[a]	4.3 ± 2.30[a]	3.6 ± 1.15[a]
Pupa–adult	7 ± 2.64[a]	5.6 ± 2.88[a]	5.3 ± 3.21[a]	4.3 ± 1.14[a]
Total developmental period	37.6 ± 12.22[a]	35.8 ± 13.93[b]	33.33 ± 15.04[c]	26.5 ± 6.77[d]

[a]Different letters in the column indicates a significant difference between the means @ $p \leq 0.05$

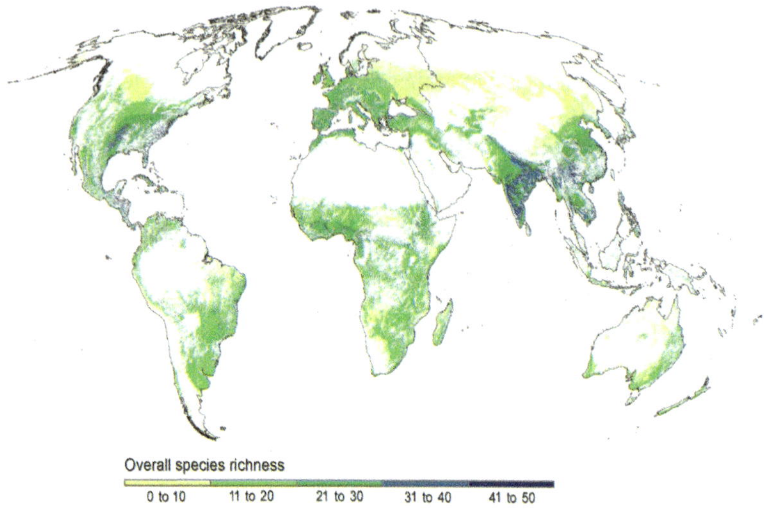

Fig. 2.3 The species richness of pests according to the current climate (Yan et al., 2017). (With permission from Elsevier. Copyright© 2017; License Number: 4924840647433)

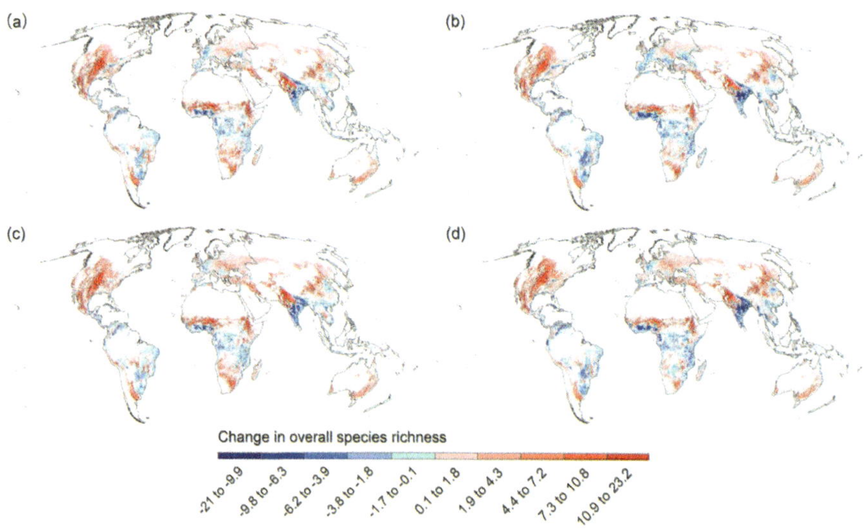

Fig. 2.4 Predicted changes in the species richness of Pest according to future climatic. (**a**) 2050-RCP2.6; (**b**) 2050-RCP4.5; (**c**) 2070-RCP2.6; (**d**) 2070-RCP4.5 (RCP: Representative Concentration Pathway) (Yan et al., 2017). (With permission from Elsevier. Copyright© 2017; License Number: 4924840807079)

Also, the time of the flight thresholds, pests such as aphids (Woiwod & Harrington, 1994; Fleming & Volney, 1995; Zhou et al., 1995) and moths (Woiwod, 1997) are advanced in warmer weather that leads to an increase in the possibility of early immigration (Bale et al., 2002). It should be pointed out photoperiod induced by the altering in distribution range cases quick diapause induction and reduction of the metabolic action period (Nagarkatti et al., 2003; Bale & Hayward, 2010; Stoeckli et al., 2012; Reineke & Thiéry, 2016).

Changes in the rainfall pattern, followed by an increase in relative humidity or drought, could alter Pest's behavior. More specifically, according to de Sassi and Tylianakis (2012) and Ortega-Lopez et al. (2014), higher relative humidity and drought could respectively result in the duration of early development and shorten the lifespan of eggs and larvae and increase their mortality. It may seem impressive, but drought could increase the pests population growth and reproduction rate (Adamo et al., 2012). For example, the abundance of *A. tubercular* has negatively correlated with relative humidity (del Pino et al., 2020); thus, drought could lead to an increase in its publication (Popov et al., 2006). The feeding behavior and some female insects'fecundity are affected by drought (Bajwa et al., 2020). For instance, Gutbrodt et al. (2011) reported that *Spodoptera frugiperda* Smith prefers to feed on drought-stressed plants. Furthermore, the phenology of host–natural enemy populations affected by drought and, as a result, leads to disorder in biological control (Romo & Tylianakis, 2013).

Climate change leads to a change in plant's nutritional quality, e.g., C/N ratio that causes allure the chewing, biting, and sucking pests such as caterpillars, beetles, thrips, insects plant bugs, and scale insects (Caulfield & Bunce, 1994; Roth & Lindroth, 1995; Coviella & Trumble, 1999; Musser & Shelton, 2005). Also, drought results in more appealing of the plant for pest (Sherrard et al., 2009). In better words, drought results in a decrease in secondary defense compounds. Simultaneously, items that are favorable for pest growth and development are, i.e., soluble proteins, free amino acids, and carbohydrates are increased (Gutbrodt et al., 2011).

Finally, it can be said that climate change plays a significant role in changing pests' behavior, which in turn will lead to emerging pests, followed by changes in distribution, migration, and changes in growth and expansion rate.

4 Emerging Pests

In this section, some emerging pests in different regions of the world have been introduced and discussed:

4.1 Aromia bungii

One of the emerging pests observed in Europe, especially Germany and Italy, is *Aromia bungii,* named red-necked longhorn beetle (Russo et al., 2020). Depending on the climate, the life cycle of *Aromia bungii* is reported 2–4 years (Gui-ping,

2005; Ma et al., 2007). This Pest that overwinters as larvae, and then it feeds the outer sapwood, cambium, and nutrient-rich phloem. In this step from the life cycle, larvae are latent within the subcortical tissues; thus, their control with pesticides is difficult. To face this challenge, Xu et al. (2017) suggested that pheromone could be a crucial tool to eradicate *Aromia bungii* from regions of the world that it has already invaded as Japan. Mature larvae bore the xylem in order to form a pupation chamber, and finally, in midsummer, emerges as an adult pest (Xu et al., 2017). This Pest native to East Asia (China, Taiwan, Korea, Mongolia, Vietnam, and eastern Russia) as a severe threat is identified to stone fruit trees (Kano et al., 2014; Li et al., 2016; Xu et al., 2017), due to *Aromia bungii* targets healthy host plants (Gressitt, 1942).

Based on Russo et al. (2020), due to the severe threat to stone fruit trees, this Pest is added to the priority pest list as well as the EPPO A1 list of quarantine species in Europe owing to its more serious social, environmental, economic impacts. They reported that the genetic variability of red-necked longhorn beetle populations is not well known in the area of origin. More specifically, the single haplotype that is found in Italy dose match neither the haplotype found in China nor Germany. They reported that due to this fact could not identify the accurate region of origin of studied specimens in Europe. Russo et al. (2020) emphasized that insect life history, such as longevity, fertility, developmental period, and fecundity and flight and behavior, are affected by abiotic factors, including temperature. In particular, the temperature is a critical factor in embryonic development and, subsequently, the egg developmental times of longhorn beetles.

4.2 Drosophila suzukii

Another emerging pest in Europe, as well as north America (reported in 2008) that is usually recognized as the spotted winged Drosophila (*Drosophila suzukii*), is native to Asia (Calabria et al., 2012). This Pest is recognized as a highly polyphagous invasive pest that, through larval infestations of berry and stone fruits, lead to economic damage in Europe and North America (Walsh et al., 2011; Cini et al., 2012). This Pest penetrates and lays eggs in unripe fruit, unlike other fruit flies that attract to decaying overripe fruit (Audsley et al., 2015). According to the available estimation, *Drosophila suzukii* could lead to a decrease of 20% in crop yield to preferred hosts, including blueberries, blackberries, cherries, raspberries, and strawberries that are estimated at more than $5 million (Langille et al., 2017).

Previous researches on this Pest showed that it could not tolerate and survive extended periods of cold such as in regions such as Michigan, Washington, eastern Oregon, and Canada (Dalton et al., 2011). However, this Pest is now established in those areas (Burrack et al., 2012; Isaacs et al., 2010; Jakobs et al., 2015) and has expanded from southern California to British Columbia, Canada (Asplen et al., 2015). Climatic change and warmer temperatures in southern areas lead to pests' unfavorable conditions; thus, northern regions have become favorable areas for *Drosophila suzukii* population growth (Langille et al., 2017).

4.3 Balaustium medicagoense and Bryobia sp.

Arthur et al. (2008) introduced *Balaustium medicagoense and Bryobia sp.* as emerging pests in southern Australia that attack crops and pastures and led to significant damage in the past decade. They reported that *Balaustium medicagoense and Bryobia sp.* in southern Australia have a high natural tolerance instead of conventional pesticides that these properties caucuses that their control is difficult. In another study by Arthur et al. (2010), they reported that these pests naturally endure current pesticides; thus, it is essential to develop alternative control strategies.

In another study, Arthur et al. (2011) reviewed the distribution, abundance, and life cycle of the *Bryobia* spp. and *Balaustium medicagoense* in Australia. They cited that *Balaustium medicagoense* that is previously introduced from South Africa is observed in the Mediterranean climate areas in southern Australia from autumn to spring, i.e., throughout the winter growing periods (Halliday, 2001; Hoffmann et al., 2008). This Pest is unusual due to its feeding behavior; in better words, it is regarded as a crop pest (Micic et al., 2008) and identified as a beneficial predator (James, 1995; James et al., 1995; Halliday & Paull, 2004). Arthur et al. (2011) argued that climate change and management practices lead to developing invertebrate pest complexes in Australia. They also reported that regions such as King Island and Kangaroo Island have a favorable climate for *Balaustium medicagoense* that these conditions may extend further into southern regions. They concluded that *Bryobia spp.* and *Balaustium medicagoense* can convert the serious pests due to high widespread distributions, levels of abundance, and an immense range of activities.

Another study also showed that the Australian grains industry, due to climate change and evolving management practices, deals with emerging pests (Hoffmann et al., 2008). Interestingly, based on Hoffmann et al. (2008) from the mid-1990s, in western Australia, vegetable weevils, aphids, and armyworms are observed while the population of *Balaustium* mites, snails, red-legged earth mites, blue oat mites, pasture cockchafer, and lucerne flea are developed. On the other hand, eastern Australia faced with the development of Balaustium mites, Bryobia mites, lucerne flea, and blue oat mites and reduced armyworms and pea weevils from the early 1980s to 2006–07. They argued that dry climate that intensifies by climate change, lead to alter the migration patterns of pests e.g., decrease the build-up of migratory pests from inland Australia.

4.4 Neomaskellia andropogoni

Nikpay (2017) introduced *Neomaskellia andropogoni* as an emerging pest in the sugarcane fields of Iran. They reported that the increase of relative humidity leads to an increase in population pests until 21 October, and then because of a decrease in temperature, the population pest is decreased. It should be noted that the activity of this Pest starts in late August, and in late September until late October population of Pest reaches its peak. The total population of this Pest is parthenogenesis (male Pest rarely seen in the population). Nikpay (2017) stated that emersion of this Pest previously reported in Pakistan (Inayatullah, 1984), India (Mann & Singh, 2003), and West Malaysia and Sri Lanka (Mound & Halsey, 1978). Concerning this Pest, it must be said that its nymphs suck the phloem sap, and afterward, excretes the honeydew that is a medium for *Capnodium sp.* and other fungi that disrupt the photosynthesis (Pandya, 2005).

4.5 Nysius cymoides

Scaccini and Furlan (2019a) investigated the status and management of *Nysius cymoides* that attacks multiple crop species in Europe and the Middle East as an emerging pest. They reported that this Pest, similar to other species of the same genus, due to its polyphagy and outbreak during warm and dry periods (from 26 to 36 °C), could become a significant pest. They reported that previous literature emphasized outbreaks following warm for *Nysius cymoides* (Farrell & Stufkens, 1993) and *Nysius raphanus* (Demirel, 2007) as conventional pest species in New Zealand and North America, respectively. They also claimed that temperature could be a significant factor in flight dispersal of *Nysius huttoni* (Wei, 2014) and *Nysius vinitor* (Moradi Vajargah & Parry, 2017) and subsequently be effective on outbreaks of *Nysius spp.* Through their movement and migration. For example, it is reported that during a warm and dry summer, *Nysius cymoides* attacks soybean as a second crop in northern Italy (Scaccini & Furlan, 2019b).

4.6 Steneotarsonemus spinki

Mutthuraju et al. (2014) reported that *Steneotarsonemus spinki* Smiley from the *Tarsonemidae* family is observed in West Bengal and Gujarat as an emerging pest and cause significant loss in the yield of rice fields. This Pest leads to infest leaf sheath of rice and subsequently results in brown discoloration. Also, pests' infestation on the panicle could lead to chaffy grains, ill-filled, or discoloration of filled grains. Mutthuraju et al. (2014) cited that humidity between 83.8% and 89.5%, and temperatures range 25.5–27.50 °C, lead to developing this pest (Miranda Cabrera et al., 2003). As a result, climate change could be significant in the growth and development rate of this Pest. In another study, Saha et al. (2016) also introduced *Steneotarsonemus spinki* Smiley as a severe emerging pest of rice crop in India and West Bengal. They reported that this pest population is dependent on climate; in particular, it is negatively correlated with minimum temperature and rainfall and positively correlated with maximum temperature and sunshine hours.

4.7 Rastrococcus iceryoides

Recently, Halder et al. (2019), for the first time, record the emergence of mealybug, *Rastrococcus iceryoides,* in Arecanut orchards of West Bengal, India. *Rastrococcus iceryoides* was previously observed in neotropical and afrotropical oriental regions (countries such as Pakistan, Srilanka, Singapore, Congo, Malaysia, French Guiana, Thailand, Ghana, Benin, Vietnam, Gabon, Senegal, Bangladesh, Togo, Bhutan, Philippines, and some regions in India. They observed *Rastrococcus iceryoides* attack the first young leaves of terminal shoots, then fruit and inflorescence faced with Pest. Finally, an increase in the pest population leads to pests movement on the peduncle of the nuts. They reported that this emerging Pest leads to 30–40% damage to fruit. Watson and Mifsud (2017) reported that the initial establishment of foreign pests that originated from tropical and subtropical countries could be facilitated by climate change.

4.8 Deanolis albizonalis

In many southeast Asia regions, *Deanolis albizonalis* is recognized as a monophagous pest on mango (Gibb et al., 2007). Larvae of this Pest that bore tip or narrow apex of the fruit result in severe fruit (Reddy et al., 2018). In an interesting study, Singh and Kundan (2014) reported the widespread distribution of *Deanolis albizonalis* from India to Australia. They declared that *Deanolis albizonalis* as an emerging pest leads to 42% damage to mango in India's east coast, especially in Andhra Pradesh, Bihar, Orissa, and West Bengal. While in India's major mango belts, i.e., south, north, and west, this Pest is not observed. It should be noted that climate change leads to shift in mango phenology which has a critical role in the complexity of pest challenges (Reddy et al., 2018).

4.9 Cacosceles newmannii

Another emerging pest of sugarcane in South Africa is *Cacosceles newmannii* (Coleoptera: Cerambycidae) (Way et al., 2017). The sugarcane stalk is attacked by larvae of this Pest, leading to a significant sugar production reduction. Way et al. (2017) reported that in 2015, *Cacosceles newmannii* larvae found on sugarcane in the KwaZulu-Natal of South Africa. *Cacosceles newmannii* larvae are mostly observed in the stools and stalk bases of sugarcane. Javal et al. (2019) also emphasized the severe economic impacts worldwide due to Cerambycids' strongly adverse impacts on sugarcane. They gave an example that in Thailand, populations of *Dorysthenes buqueti* (Coleoptera: Cerambycidae) increased tenfold within a year and has severely attacked the sugarcane crop (Pliansinchai et al., 2007).

4.10 Tetranychus evansi

The *Tetranychus evansi*, for the first time in 1952, was observed in north-east Brazil (Silva, 1954). However, the status of *Tetranychus evansi* altered in the past decade. They moved to the Mediterranean basin as well as different parts of Africa. Its behavior in Africa was too threatening, so that in west Africa (Duverney & Ngueye-Ndiaye, 2005) and southeast Africa (Sibanda et al., 2000) led to yield losses to 90% of tomatoes in the dry season. Boubou et al. (2011) believed that climate change and human activities play a critical role in distributing *Tetranychus evansi* in Africa. They cited that invasion of *Tetranychus evansi* to new origins is because it prefers a broad range of temperatures (Bonato, 1999).

4.11 Acronicta rumicis

It is reported that 90% of produced apple in India is grown in the hills of Uttarakhand Himachal Pradesh and Jammu and Kashmir (Indian Horticulture Database, 2014). However, these gardens are severely threatened by various pests (Chundawat, 1993; Singh et al., 2010). In line with that, Dar et al. (2019) studied *Acronicta rumicis* as an emerging pest in Kashmir Valley's apple gardens. They reported that larvae of the moth *Acronicta rumicis* lead to significant harm to new apple plantation. This Pest that is widely distributed throughout the world, such as Europe; Eurasia; Greece; Spain; Virginia; North America; Mexico; USA; Argentina; Iran; Campbellpur; Azerbaijan; Japan, and China, in Kashmir lead to 31.07% incidence and 14.03% severity. Based on Dar et al. (2019), this Pest attacks the terminal shoots and leads to undesired apical dominance and subsequently disrupts the new apple plantation. Franzén and Johannesson (2007) reported that all butterfly species could benefit from climate change and global warming, but it could deteriorate the habitat of butterfly (Parmesan et al., 2001; Hill et al., 2002; Conrad et al., 2004; Rodrigues Filho et al., 2016) and lead to their migration.

4.12 Halyomorpha halys

Halyomorpha halys is a very destructive invasive polyphagous pest for fruit orchards and crops that in the mid-1990s that is first detected in Pennsylvania, USA (Hoebeke & Carter, 2003) is observed in Italy in 2012 (Maistrello et al., 2016). This Pest native to eastern Asia and is recognized as a highly polyphagous pentatomid is spreading rapidly worldwide through human activities (Haye et al., 2015). The adults and nymphs feed stems, leaves, buds, and fruits of host plants such as herbaceous perennials, ornamentals, vegetables, shrubs, and forest trees as well as tree crops (Lee et al., 2013). It is reported that in 2010, *Halyomorpha halys* lead to a > $37 million damage in the USA (United States Apple Association, 2010). This Pest has emerged in Europe and Eurasia and lead to severe damage in hazelnut and pear orchards in northern Italy and western Georgia (Haye & Weber, 2017). In another study, Maistrello et al. (2016), using active techniques, assessed seasonality, abundance, and impact of *Halyomorpha halys* on the pear orchards. They reported that just a few years after first observation in Italy, i.e., 2 years, *Halyomorpha halys* is converted to a season-long pest for pear crops and lead to severe yield losses, e.g., the deformation of more than 50% fruits. It should be noted that available pesticides do not lead to relevant results against this pest (Leskey et al., 2012, 2014). This fact results in grow the pesticide application in order to fight with *Halyomorpha halys*.

It is believed that warmer climatic conditions are significant in bivoltinism and population expansion. Researchers believe that the development of this Pest outside its native causes secondary invasions and accelerates the spread of this Pest in the world (Gariepy et al., 2014, 2015). This Pest also, to overwinter, attacks human-made structures (Inkley, 2012). Kiritani (2006) reported that global warming is responsible for outbreaks of *Halyomorpha halys* due to a rise of 1 °C lead to reduce 15% of winter mortality of adult pests.

In another study, Stoeckli et al. (2020) assessed the response of *Halyomorpha halys* to climate change in Switzerland. They reported that due to the future increase in climate suitability, the pest activity periods expected in spring and late autumn. This expansion leads to the significant growth of the *Halyomorpha halys* population. They cited that warmer temperatures in the late growing season cause more nymphs is wholly developed (Haye et al., 2014). Stoeckli et al. (2020) provided a model that showed after 10 years of low population growth due to the exceptional warm in 2018, the *Halyomorpha halys* population increased. They claimed that convert from one generation to two generations. Consequently, the population growth of *Halyomorpha halys* leads to a rise in crop damages in different regions of Switzerland, e.g., northern and midlands regions.

Kistner (2017) argued that geographic distribution and population growth of *Halyomorpha halys* affected by climate change. They simulated and examined changes in seasonal phenology and voltinism under possible future climate scenarios. They reported that a suitable climate in Europe expands northward, and in north America expands northward into Canada due to enhanced heat stress in the future. They stated that although warmer temperature causes growing seasons of *Halyomorpha halys* to be longer, rising summer temperatures lead to unfavorable conditions for *Halyomorpha halys* growth. On the other hand, they believed that the number of *Halyomorpha halys* generations might annually increase and lead to be multivoltine of this Pest in Europe and north America. They concluded that under possible future climates, southeastern Canada, the northeastern USA, and Europe are suitable for *Halyomorpha halys* growth and development.

4.13 Australiodillo bifrons

In parts of New South Wales, Australia, cereal crops were targeted by *Australiodillo bifrons* in 2006. Researchers believed that this is an emerging pest owing to this Pest usually attacks into decaying organic matter (Paoletti et al., 2008). They reported that *A. bifrons* could reach very high densities and lead to severe damage to wheat. They concluded that increasing *A. bifrons* in parts of New South Wales is caused by climate change and changes in farming practices.

4.14 Tuta absoluta

Tuta absoluta that is currently threatening tomato (Diarra et al., 2014; Guimapi et al., 2016), eggplant (Kanle Satishchandra et al., 2019; Sylla et al., 2019), and potato (El-Naggar & EL-Bassouiny, 2018; A Saad et al., 2020) in Asia (Han et al., 2019), Africa (Abbes et al., 2012), and Europe (Desneux et al., 2011) is recognized an invasive pest from south America (Desneux et al., 2011). This Pest was first seen in 2006 in Spain and quickly expanded its geographical range so that it was observed in African countries in recent years. *Tuta absoluta*'s life cycle is included egg, larva, pupa, adult, which in all four stages attacks different parts of plants (Cuthbertson et al., 2013). It is reported that pest management cost in tomato farms due to *Tuta absoluta* increases about $487 million per year (Desneux et al., 2011).

In an interesting study, Guimapi et al. (2016) simulated the effect of temperature, relative humidity, vegetation cover, and yield of tomato production that directly and indirectly are related to climate change on the spread of *Tuta absoluta* in Africa. They reported *Tuta absoluta* 10 years after observation in Spain would reach South Africa. They also reported that some control applications would increase to 19 years when the insect flies at 75 km (Fig. 2.5).

In another study, Santana et al. (2019) assessed the role of climate change on the global geographic distribution of *Tuta absoluta*. They reported that the climate changes predicted for 2050 and 2100 lead to reduce the high appropriate areas for this Pest in some areas in Asia, sub-Saharan Africa, and large areas in south America. In contrast, they reported that due to climate changes, the midwest and the southeast of the USA and east and north of Europe such as Denmark, Russia, Turkey, and the United Kingdom in 2050 and 2100 would be appropriate living areas for this Pest. They argued that in these regions is predicted an increased temperature and reduction in cold stress. They concluded that near the poles and around the equator are appropriate and inappropriate areas to *Tuta absoluta,* respectively.

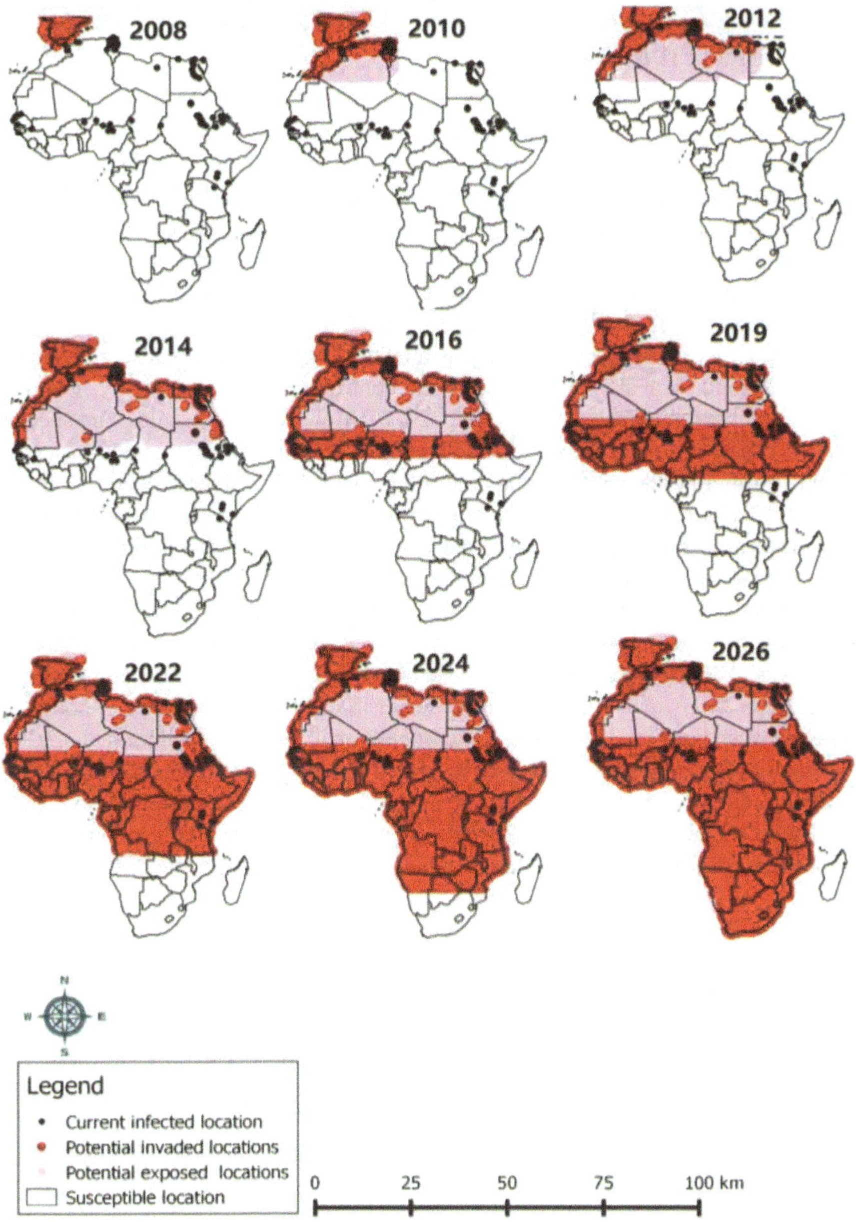

Fig. 2.5 The simulated spread of *Tuta absoluta* in Africa when the insect flies at a 75 km speed due to temperature, humidity, vegetation, and tomato yield (Guimapi et al., 2016). (With permission from Elsevier. Copyright© 2016; License Number: 4924921135309)

5 Strategies to Deal with Climate Change's Adverse Effects on Food Production

In the end, it should be mentioned that agriculture and food production are currently very vulnerable in faced with climate change (Gomez-Zavaglia et al., 2020). In addition to the problems caused by pests, climate change lead to adverse impacts that are reduced food security-safety and are increased food prices (Nelson et al., 2009). It is interesting to know levels of iron, zinc, proteins and amino acids, and other vital nutrients in crops are decreased with elevated levels of carbon dioxide (Dietterich et al., 2014; Myers et al., 2014; Wu et al., 2016).

Ecosystems and the provided services by them to agriculture are also affected by climate change, such as pollination and pest control by the natural enemy, as previously mentioned (Gomez-Zavaglia et al., 2020). Moreover, many wild plant species that are usually used for domestic plant breeding are becoming extinct (Jarvis et al., 2008). Furthermore, warmer temperatures also lead to vulnerable crops due to enhanced water requirements in warm and dry periods (Nelson et al., 2009). Warmer temperatures also result in recede of the world's glaciers that is a critical factor for farmers who for irrigation depend on glacial meltwater (Stocker et al., 2013; Field, 2014). On the other hand, coastal freshwater aquifers that are used for irrigation are salted by salt-water intrusion due to rising sea levels (Backlund et al., 2008). Moreover, changes in rainfall patterns lead to flooding and drought that both threaten agriculture. More specifically, droughts result in dry the soil washing of soil, and flooding washes topsoil that soil fertility depends on it (Gomez-Zavaglia et al., 2020).

Finally, although world leaders have agreed to tackle climate change and global warming, the effects of climate change cannot be eliminated. Therefore, to ensure food security and overcome the adverse consequences of climate change on food production, measures should be taken, which some of them are listed in Table 2.4.

6 Conclusion

Overall, this chapter emphasized that climatic factors have a significant role in the world's insect pest's behavior. It is documented that high temperatures and carbon dioxide could alter the population and distribution, and migration of insect pests. More specifically, due to the high reproduction rate and migration of insect pests caused by increasing temperature and carbon dioxide, different parts of the world will face emerging pests. The emerging pests, caused by climate change, due to the lack of natural enemies, are critical to crop productivity and, subsequently, food security. The emerging pests and their dangerous consequences lead to more application of chemicals to manage pests by farmers. However, the application frequency of pesticides will result in high levels of pesticide resistance. High application of pesticides, in turn, leads to climate change and other environmental impacts and increases the pesticide residue on foods. In many parts of the world, farmers still do

Table 2.4 Strategies to deal with climate change's adverse effects on food production (Gomez-Zavaglia et al., 2020)

Problem	Strategy
Raw crop management	Breeding for drought and temperature tolerance
	Earlier planting
	Increasing the organic content of soils
	Using conservation tillage or no-till to decrease run-off and increase infiltration
	Establishing and maintaining buffers, filter strips, and grassed waterways near water sources
	Shifting to less water-dependent cropping systems
Livestock management	Selecting breeds and types
	Improving nutritional management during periods of high heat load
	Sun-shading, evaporative cooling, mechanical ventilation
	Using rotational grazing systems to minimize damage to range and pasture
	Active management of forage stocks and reduced herd sizes under droughts
Water management	Installing more efficient irrigation systems or making existing systems more efficient
	Water storage in ponds and tanks
	Rational management of water use
	Installing watering facilities to ensure that livestock have access to water
Improving access to information	Using early warning systems for droughts
	Research and development
	Developing programs to help farmers manage risk
	Fostering regional outreach, extension, and education

not know the relationship between climate change and pests. Accordingly, more education of farmers about emerging pests, their development, and their relationship with climate change and proper control should be given special attention. Educating farmers and the government's efforts to deal with climate change can reduce pest behavior change and the spread of emerging pests associated with climate change. It should be noted that before emerging pests arrive in a new area, they should be identified and rapidly eradicated because anticipating emerging pests is more cost-effective. Therefore, further studies should be conducted to predict the occurrence of emerging pests in high-potential areas.

References

A Saad, A. F. S., Tayeb, E. L., Hassan, A. N. T., & Attia, H. G. A. I. (2020). IPM strategy for Tuta absoluta on tomatoes and potatoes. *Alexandria Science Exchange Journal, 41*, 275–283.

Abbes, K., Harbi, A., & Chermiti, B. (2012). The tomato leafminer Tuta absoluta (Meyrick) in Tunisia: Current status and management strategies. *EPPO Bulletin, 42*, 226–233.

Adamo, S. A., Baker, J. L., Lovett, M. M. E., & Wilson, G. (2012). Climate change and temperate zone insects: The tyranny of thermodynamics meets the world of limited resources. *Environmental Entomology, 41*, 1644–1652.

Agovino, M., Casaccia, M., Ciommi, M., Ferrara, M., & Marchesano, K. (2019). Agriculture, climate change and sustainability: The case of EU-28. *Ecological Indicators, 105*, 525–543.

Akbar, S. M. D., Pavani, T., Nagaraja, T., & Sharma, H. C. (2016). Influence of CO_2 and temperature on metabolism and development of Helicoverpa armigera (Noctuidae: Lepidoptera). *Environmental Entomology, 45*, 229–236.

Alagidede, P., Adu, G., & Frimpong, P. B. (2016). The effect of climate change on economic growth: Evidence from Sub-Saharan Africa. *Environmental Economics and Policy Studies, 18*, 417–436.

Aragón, P., & Lobo, J. M. (2012). Predicted effect of climate change on the invasibility and distribution of the Western corn root worm. *Agricultural and Forest Entomology, 14*, 13–18.

Arthur, A. L., Hoffmann, A. A., Umina, P. A., & Weeks, A. R. (2008). Emerging pest mites of grains (Balaustium medicagoense and Bryobia sp.) show high levels of tolerance to currently registered pesticides. *Australian Journal of Experimental Agriculture, 48*, 1126–1132.

Arthur, A. L., Weeks, A. R., Umina, P. A., & Hoffmann, A. A. (2010). Survival and reproduction of the pest mites Balaustium medicagoense and Bryobia spp. on winter grain crops. *Experimental & Applied Acarology, 52*, 141–153.

Arthur, A. L., Weeks, A. R., Hill, M. P., & Hoffmann, A. A. (2011). The distribution, abundance and life cycle of the pest mites Balaustium medicagoense (Prostigmata: Erythraeidae) and Bryobia spp.(Prostigmata: Tetranychidae) in Australia. *Australia Journal of Entomology, 50*, 22–36.

Asplen, M. K., Anfora, G., Biondi, A., Choi, D.-S., Chu, D., Daane, K. M., Gibert, P., Gutierrez, A. P., Hoelmer, K. A., & Hutchison, W. D. (2015). Invasion biology of spotted wing Drosophila (Drosophila suzukii): A global perspective and future priorities. *Journal of Pest Science, 2004*(88), 469–494.

Athanassiou, C. G., & Rumbos, C. I. (2018). Emerging pests in durable stored products. In *Recent advances in stored product protection* (pp. 211–227). Springer.

Audsley, N., Down, R. E., & Isaac, R. E. (2015). Genomic and peptidomic analyses of the neuropeptides from the emerging pest, Drosophila suzukii. *Peptides, 68*, 33–42.

Backlund, P., Janetos, A., & Schimel, D. (2008). *The effects of climate change on agriculture, land resources, water resources, and biodiversity in the United States*. Synthesis and Assessment Products 4.3. U.S. Environmental Protection Agency, Climate Change Science Program. 240 p.

Bailey, S. W. (2004). Climate change and decreasing herbicide persistence. *Pest Management Science: Formerly Pesticide Science, 60*, 158–162.

Bajwa, A. A., Farooq, M., Al-Sadi, A. M., Nawaz, A., Jabran, K., & Siddique, K. H. M. (2020). Impact of climate change on biology and management of wheat pests. *Crop Protection, 137*, 105304.

Bale, J. S., & Hayward, S. A. L. (2010). Insect overwintering in a changing climate. *The Journal of Experimental Biology, 213*, 980–994.

Bale, J. S., Masters, G. J., Hodkinson, I. D., Awmack, C., Bezemer, T. M., Brown, V. K., Butterfield, J., Buse, A., Coulson, J. C., & Farrar, J. (2002). Herbivory in global climate change research: Direct effects of rising temperature on insect herbivores. *Global Change Biology, 8*, 1–16.

Battisti, A., Stastny, M., Netherer, S., Robinet, C., Schopf, A., Roques, A., & Larsson, S. (2005). Expansion of geographic range in the pine processionary moth caused by increased winter temperatures. *Ecological Applications, 15*, 2084–2096.

Bebber, D. P., Ramotowski, M. A. T., & Gurr, S. J. (2013). Crop pests and pathogens move polewards in a warming world. *Nature Climate Change, 3*, 985–988.

Bommarco, R., Vico, G., & Hallin, S. (2018). Exploiting ecosystem services in agriculture for increased food security. *Global Food Security, 17*, 57–63.

Bonato, O. (1999). The effect of temperature on life history parameters of Tetranychus evansi (Acari: Tetranychidae). *Experimental & Applied Acarology, 23*, 11–19.

Boubou, A., Migeon, A., Roderick, G. K., & Navajas, M. (2011). Recent emergence and worldwide spread of the red tomato spider mite, Tetranychus evansi: Genetic variation and multiple cryptic invasions. *Biological Invasions, 13*, 81–92.

Boukhris-Bouhachem, S. (2017). Emerging insects: uncontrolled exchange or climate change? How to prevent? In *Annales de l'INRAT* (pp. 1–21). Institut National de la Recherche Agronomique de Tunisie (INRAT).

Bregaglio, S., Donatelli, M., & Confalonieri, R. (2013). Fungal infections of rice, wheat, and grape in Europe in 2030-2050. *Agronomy for Sustainable Development, 33*, 767–776.

Burrack, H. J., Smith, J. P., Pfeiffer, D. G., Koeher, G., & Laforest, J. (2012). Using volunteer-based networks to track Drosophila suzukii (Diptera: Drosophilidae) an invasive pest of fruit crops. *Journal of Integrated Pest Management, 3*, B1–B5.

Caffarra, A., & Donnelly, A. (2011). The ecological significance of phenology in four different tree species: Effects of light and temperature on bud burst. *International Journal of Biometeorology, 55*, 711–721.

Caffarra, A., Rinaldi, M., Eccel, E., Rossi, V., & Pertot, I. (2012). Modelling the impact of climate change on the interaction between grapevine and its pests and pathogens: European grapevine moth and powdery mildew. *Agriculture, Ecosystems and Environment, 148*, 89–101.

Calabria, G., Máca, J., Bächli, G., Serra, L., & Pascual, M. (2012). First records of the potential pest species Drosophila suzukii (Diptera: Drosophilidae) in Europe. *Journal of Applied Entomology, 136*, 139–147.

Castex, V., Beniston, M., Calanca, P., Fleury, D., & Moreau, J. (2018). Pest management under climate change: The importance of understanding tritrophic relations. *Science of the Total Environment, 616*, 397–407.

Caulfield, F., & Bunce, J. A. (1994). Elevated atmospheric carbon dioxide concentration affects interactions between Spodoptera exigua (Lepidoptera: Noctuidae) larvae and two host plant species outdoors. *Environmental Entomology, 23*, 999–1005.

Chundawat, B. S. (1993). *Advances in horticulture, fruit crops* (p. 763). Malhotra Publishing House.

Cini, A., Ioriatti, C., & Anfora, G. (2012). A review of the invasion of Drosophila suzukii in Europe and a draft research agenda for integrated pest management. *Bulletin of Insectology, 65*, 149–160.

Claver, A., Ormad, P., Rodríguez, L., & Ovelleiro, J. (2006). Study of the presence of pesticides in surface waters in the Ebro river basin (Spain). *Chemosphere, 64*, 1437–1443.

Collier, R. H., Finch, S., Phelps, K., & Thompson, A. R. (1991). Possible impact of global warming on cabbage root fly (Delia radicum) activity in the UK. *The Annals of Applied Biology, 118*, 261–271.

Conrad, K. F., Woiwod, I. P., Parsons, M., Fox, R., & Warren, M. S. (2004). Long-term population trends in widespread British moths. *Journal of Insect Conservation, 8*, 119–136.

Coope, G. R. (1978). Constancy of insect species versus inconstancy of Quaternary environments. In *Diversity of insect faunas* (pp. 176–187). Blackwell Scientific Publications.

Coviella, C. E., & Trumble, J. T. (1999). Effects of elevated atmospheric carbon dioxide on insect plant interactions. *Conservation Biology, 13*, 700–712.

Cuthbertson, A. G. S., Mathers, J. J., Blackburn, L. F., Korycinska, A., Luo, W., Jacobson, R. J., & Northing, P. (2013). Population development of Tuta absoluta (Meyrick)(Lepidoptera: Gelechiidae) under simulated UK glasshouse conditions. *Insects, 4*, 185–197.

Dalton, D. T., Walton, V. M., Shearer, P. W., Walsh, D. B., Caprile, J., & Isaacs, R. (2011). Laboratory survival of Drosophila suzukii under simulated winter conditions of the Pacific

Northwest and seasonal field trapping in five primary regions of small and stone fruit production in the United States. *Pest Management Science, 67*, 1368–1374.
Dar, M. A., Akbar, S. A., & Khan, Z. H. (2019). Acronicta rumicis (Linnaeus, 1758), emerging pest of apple plantation in Kashmir Valley. *National Academy Science Letters, 42*, 287–289.
de Sassi, C., & Tylianakis, J. M. (2012). Climate change disproportionately increases herbivore over plant or parasitoid biomass. *PLoS One, 7*, e40557.
del Pino, M., Bienvenido, C., Boyero, J. R., & Vela, J. M. (2020). Biology, ecology and integrated pest management of the white mango scale, Aulacaspis tubercularis Newstead, a new pest in southern Spain-a review. *Crop Protection, 133*, 105160.
Delcour, I., Spanoghe, P., & Uyttendaele, M. (2015). Literature review: Impact of climate change on pesticide use. *Food Research International, 68*, 7–15.
Demirel, N. (2007). Mortality of false chinch bug, Nysius raphanus (Howard). *Journal of Entomology, 4*, 155–159.
Desneux, N., Luna, M., Guillemaud, T., & Urbaneja, A. (2011). The invasive South American tomato pinworm, Tuta absoluta, continues to spread in Afro-Eurasia and beyond: The new threat to tomato world production. *Journal of Pest Science, 2004*(84), 403–408.
Diarra, K., Sylla, S., Diatte, M., Brevault, T., & Bernadas, G. (2014). Tuta absoluta Meyrick (Lepidoptera: Gelechiidae): A new threat to tomato production in sub-Saharan Africa. *African Entomology, 22*, 441–444.
Dietterich, L. H., Zanobetti, A., Kloog, I., Huybers, P., Leakey, A. D. B., Bloom, A. J., Carlisle, E., Fernando, N., Fitzgerald, G., & Hasegawa, T. (2014). Increasing CO_2 threatens human nutrition. *Nature, 2*, 150036.
Duverney, C., & Ngueye-Ndiaye, A. (2005). Preliminary tests to limit the damage of Tetranychidae on vegetable crops in Sine-Saloum (Senegal). In *Deuxième colloque international sur les acariens des cultures*.
El-Naggar, A. Z., & EL-Bassouiny, H. M. (2018). Monitoring population of tomato leaf miner, Tuta absoluta during winter and summer evergreens of potato filed in Egypt. *Egyptian Academic Journal of Biological Sciences. A, Entomology, 11*, 27–32.
Farrell, J. A., & Stufkens, M. W. (1993). Phenology, diapause, and overwintering of the wheat bug, Nysius huttoni (Hemiptera: Lygaeidae), in Canterbury, New Zealand. *New Zealand Journal of Crop and Horticultural Science, 21*, 123–131.
Field, C. B. (2014). *Climate change 2014-Impacts, adaptation and vulnerability: Regional aspects*. Cambridge University Press.
Fleming, R. A., & Volney, W. J. A. (1995). Effects of climate change on insect defoliator population processes in Canada's boreal forest: Some plausible scenarios. *Water, Air, and Soil Pollution, 82*, 445–454.
Fontaine, J. J., Decker, K. L., Skagen, S. K., & van Riper, C. (2009). Spatial and temporal variation in climate change: A bird's eye view. *Climatic Change, 97*, 305.
Franzén, M., & Johannesson, M. (2007). Predicting extinction risk of butterflies and moths (Macrolepidoptera) from distribution patterns and species characteristics. *Journal of Insect Conservation, 11*, 367–390.
Gani, M. N. A., & Ghosh, M. K. (2018). Impact of climate change on agriculture and sericulture. *Journal of Entomology and Zoology Studies, 6*, 426–429.
García de Cortázar-Atauri, I., Daux, V., Garnier, E., Yiou, P., Viovy, N., Seguin, B., Boursiquot, J. M., Parker, A. K., Van Leeuwen, C., & Chuine, I. (2010). Climate reconstructions from grape harvest dates: Methodology and uncertainties. *The Holocene, 20*, 599–608.
Gariepy, T. D., Haye, T., Fraser, H., & Zhang, J. (2014). Occurrence, genetic diversity, and potential pathways of entry of Halyomorpha halys in newly invaded areas of Canada and Switzerland. *Journal of Pest Science, 2004*(87), 17–28.
Gariepy, T. D., Bruin, A., Haye, T., Milonas, P., & Vétek, G. (2015). Occurrence and genetic diversity of new populations of Halyomorpha halys in Europe. *Journal of Pest Science, 2004*(88), 451–460.

Gautam, H. R., Bhardwaj, M. L., & Kumar, R. (2013). Climate change and its impact on plant diseases. *Current Science, 105*, 1685–1691.

Gibb, A. R., Pinese, B., Tenakanai, D., Kawi, A. P., Bunn, B., Ramankutty, P., & Suckling, D. M. (2007). (Z)-11-Hexadecenal and (3Z, 6Z, 9Z)-tricosatriene: Sex pheromone components of the red banded mango caterpillar Deanolis sublimbalis. *Journal of Chemical Ecology, 33*, 579–589.

Gibert, P., Hill, M., Pascual, M., Plantamp, C., Terblanche, J. S., Yassin, A., & Sgrò, C. M. (2016). Drosophila as models to understand the adaptive process during invasion. *Biological Invasions, 18*, 1089–1103.

Gomez-Zavaglia, A., Mejuto, J. C., & Simal-Gandara, J. (2020). Mitigation of emerging implications of climate change on food production systems. *Food Research International, 134*, 109256.

Gorsky, A. L., Racanelli, G. A., Belvin, A. C., & Chambers, R. M. (2019). Greenhouse gas flux from stormwater ponds in southeastern Virginia (USA). *Anthropocene, 28*, 100218.

Gressitt, J. L. (1942). *Destructive long-horned Beetle Borers at Canton, China*. Lingnan University Canton China.

Grünig, M., Mazzi, D., Calanca, P., Karger, D. N., & Pellissier, L. (2020). Crop and forest pest metawebs shift towards increased linkage and suitability overlap under climate change. *Communications Biology, 3*, 1–10.

Guerenstein, P. G., & Hildebrand, J. G. (2008). Roles and effects of environmental carbon dioxide in insect life. *Annual Review of Entomology, 53*, 161–178.

Guimapi, R. Y. A., Mohamed, S. A., Okeyo, G. O., Ndjomatchoua, F. T., Ekesi, S., & Tonnang, H. E. Z. (2016). Modeling the risk of invasion and spread of Tuta absoluta in Africa. *Ecological Complexity, 28*, 77–93.

Gui-ping, Y. U. (2005). Bionomics of Aromia bungii. *Forest Pests and Diseases, 5*, 5.

Gutbrodt, B., Mody, K., & Dorn, S. (2011). Drought changes plant chemistry and causes contrasting responses in lepidopteran herbivores. *Oikos, 120*, 1732–1740.

Halder, J., Dey, D., & Rai, A. B. (2019). New host record of Rastrococcus iceryoides (Green, 1908) (Hemiptera: Pseudococcidae), an emerging pest of Arecanut from West Bengal, India. *National Academy Science Letters, 42*, 1–3.

Halliday, R. B. (2001). Systematics and biology of the Australian species of Balaustium von Heyden (Acari: Erythraeidae). *Australian Journal of Entomology, 40*, 326–330.

Halliday, R. B., & Paull, C. (2004). Assessment of Chaussieria capensis (Acari: Anystidae) as a predator of Halotydeus destructor (Acari: Penthaleidae). *African Entomology, 12*, 286–290.

Han, P., Bayram, Y., Shaltiel-Harpaz, L., Sohrabi, F., Saji, A., Esenali, U. T., Jalilov, A., Ali, A., Shashank, P. R., & Ismoilov, K. (2019). Tuta absoluta continues to disperse in Asia: damage, ongoing management and future challenges. *Journal of Pest Science, 92*, 1–11.

Harrington, R., Fleming, R. A., & Woiwod, I. P. (2001). Climate change impacts on insect management and conservation in temperate regions: Can they be predicted? *Agricultural and Forest Entomology, 3*, 233–240.

Haye, T., & Weber, D. C. (2017). Special issue on the brown marmorated stink bug, Halyomorpha halys: An emerging pest of global concern. *Journal of Pest Science, 90*, 987–988.

Haye, T., Abdallah, S., Gariepy, T., & Wyniger, D. (2014). Phenology, life table analysis and temperature requirements of the invasive brown marmorated stink bug, Halyomorpha halys, in Europe. *Journal of Pest Science, 2004*(87), 407–418.

Haye, T., Gariepy, T., Hoelmer, K., Rossi, J. P., Streito, J. C., Tassus, X., & Desneux, N. (2015). Range expansion of the invasive brown marmorated stinkbug, Halyomorpha halys: An increasing threat to field, fruit and vegetable crops worldwide. *Journal of Pest Science, 2004*(88), 665–673.

Hill, J. K., Thomas, C. D., Fox, R., Telfer, M. G., Willis, S. G., Asher, J., & Huntley, B. (2002). Responses of butterflies to twentieth century climate warming: Implications for future ranges. *Proceedings of the Royal Society of London. Series B, Biological Sciences, 269*, 2163–2171.

Hill, M. P., Hoffmann, A. A., McColl, S. A., & Umina, P. A. (2012). Distribution of cryptic blue oat mite species in Australia: Current and future climate conditions. *Agricultural and Forest Entomology, 14*, 127–137.

Hill, M. P., Chown, S. L., & Hoffmann, A. A. (2013). A predicted niche shift corresponds with increased thermal resistance in an invasive mite, H alotydeus destructor. *Global Ecology and Biogeography, 22*, 942–951.

Hill, M. P., Bertelsmeier, C., Clusella-Trullas, S., Garnas, J., Robertson, M. P., & Terblanche, J. S. (2016). Predicted decrease in global climate suitability masks regional complexity of invasive fruit fly species response to climate change. *Biological Invasions, 18*, 1105–1119.

Hoebeke, E. R., & Carter, M. E. (2003). Halyomorpha halys (Stål)(Heteroptera: Pentatomidae): A polyphagous plant pest from Asia newly detected in North America. *Proceedings of the Entomological Society of Washington, 105*, 225–237.

Hoffmann, A. A., Weeks, A. R., Nash, M. A., Mangano, G. P., & Umina, P. A. (2008). The changing status of invertebrate pests and the future of pest management in the Australian grains industry. *Australian Journal of Experimental Agriculture, 48*, 1481–1493.

Honek, A. (2013). Geographical variation in thermal requirements for insect development. *EJE, 93*, 303–312.

Inayatullah, C. (1984). Sugar-cane aleurodids, Aleurolobus barodensis (Maskell) and Neomaskellia andropogonis Corbett (Hom.: Aleyrodidae), and their natural enemies in Pakistan. *International Journal of Tropical Insect Science, 5*, 279–282.

Indian Horticulture Database. (2014). Ministry of Agriculture. Government of India.

Inkley, D. B. (2012). Characteristics of home invasion by the brown marmorated stink bug (Hemiptera: Pentatomidae). *Journal of Entomological Science, 47*, 125–130.

Isaacs, R., Hahn, N., Tritten, B., & Garcia, C. (2010). *Spotted wing drosophila-A new invasive pest of Michigan fruit crops, MSU Extension Bulletin E-3140*. Michigan State University.

Jactel, H., Koricheva, J., & Castagneyrol, B. (2019). Responses of forest insect pests to climate change: Not so simple. *Current Opinion in Insect Science, 35*, 103–108.

Jakobs, R., Gariepy, T. D., & Sinclair, B. J. (2015). Adult plasticity of cold tolerance in a continental-temperate population of Drosophila suzukii. *Journal of Insect Physiology, 79*, 1–9.

James, D. G. (1995). Biological control of earth mites in pasture using endemic natural enemies. *Plant Protection Quarterly, 10*, 58–59.

James, D. G., O'Malley, K., & Rayner, M. (1995). Effect of alphacypermethrin and bifenthrin on the survival of five acarine predators of Halotydeus destructor (Acari: Penthaleidae). *Experimental & Applied Acarology, 19*, 647–654.

Jarvis, A., Upadhyaya, H. D., Gowda, C. L. L., Agrawal, P. K., Fujisaka, S., & Anderson, B. (2008). *Climate change and its effect on conservation and use of plant genetic resources for food and agriculture and associated biodiversity for food security*. Food and Agriculture Organization of the United nations.

Javal, M., Terblanche, J. S., Conlong, D. E., & Malan, A. P. (2019). First screening of entomopathogenic nematodes and fungus as biocontrol agents against an emerging pest of sugarcane, Cacosceles newmannii (Coleoptera: Cerambycidae). *Insects, 10*, 117.

Johnson, S. N., & Züst, T. (2018). Climate change and insect pests: Resistance is not futile? *Trends in Plant Science, 23*, 367–369.

Johnson, A. W., Wauchope, R. D., & Burgoa, B. (1995). Effect of simulated rainfall on leaching and efficacy of fenamiphos. *Journal of Nematology, 27*, 555.

Kanle Satishchandra, N., Chakravarthy, A. K., Özgökçe, M. S., & Atlihan, R. (2019). Population growth potential of Tuta absoluta (Meyrick)(Lepidoptera: Gelechiidae) on tomato, potato, and eggplant. *Journal of Applied Entomology, 143*, 518–526.

Kano, M., Nonaka, T., Kiriyama, S., & Iwata, R. (2014). Aromia bungii (Coleoptera: Cerambycidae), an invasive cerambycid, found at Soka, Saitama Pref., Japan, infesting cherry trees, Cerasus× yedoensis "Somei-yoshino". *Forest Pests, 63*, 101–105.

Khaliq, A., Javed, M., Sohail, M., & Sagheer, M. (2014). Environmental effects on insects and their population dynamics. *Journal of Entomology and Zoology Studies, 2*, 1–7.

Kiritani, K. (2006). Predicting impacts of global warming on population dynamics and distribution of arthropods in Japan. *Population Ecology, 48*, 5–12.

Kistner, E. J. (2017). Climate change impacts on the potential distribution and abundance of the brown marmorated stink bug (Hemiptera: Pentatomidae) with special reference to North America and Europe. *Environmental Entomology, 46*, 1212–1224.

Kocmankova, E., Trnka, M., Eitzinger, J., Dubrovský, M., Štěpánek, P., Semeradova, D., Balek, J., Skalak, P., Farda, A., & Juroch, J. (2011). Estimating the impact of climate change on the occurrence of selected pests at a high spatial resolution: A novel approach. *The Journal of Agricultural Science, 149*, 185–195.

Krause, B., & Farina, A. (2016). Using ecoacoustic methods to survey the impacts of climate change on biodiversity. *Biological Conservation, 195*, 245–254.

Kurz, W. A., Dymond, C. C., Stinson, G., Rampley, G. J., Neilson, E. T., Carroll, A. L., Ebata, T., & Safranyik, L. (2008). Mountain pine beetle and forest carbon feedback to climate change. *Nature, 452*, 987–990.

Langille, A. B., Arteca, E. M., & Newman, J. A. (2017). The impacts of climate change on the abundance and distribution of the Spotted Wing Drosophila (Drosophila suzukii) in the United States and Canada. *PeerJ, 5*, e3192.

Lee, D.-H., Short, B. D., Joseph, S. V., Bergh, J. C., & Leskey, T. C. (2013). Review of the biology, ecology, and management of Halyomorpha halys (Hemiptera: Pentatomidae) in China, Japan, and the Republic of Korea. *Environmental Entomology, 42*, 627–641.

Lepetz, V., Massot, M., Schmeller, D. S., & Clobert, J. (2009). Biodiversity monitoring: Some proposals to adequately study species' responses to climate change. *Biodiversity and Conservation, 18*, 3185.

Leskey, T. C., Lee, D.-H., Short, B. D., & Wright, S. E. (2012). Impact of insecticides on the invasive Halyomorpha halys (Hemiptera: Pentatomidae): Analysis of insecticide lethality. *Journal of Economic Entomology, 105*, 1726–1735.

Leskey, T. C., Short, B. D., & Lee, D. H. (2014). Efficacy of insecticide residues on adult Halyomorpha halys (Stål)(Hemiptera: Pentatomidae) mortality and injury in apple and peach orchards. *Pest Management Science, 70*, 1097–1104.

Li, D. H. W., Yang, L., & Lam, J. C. (2012). Impact of climate change on energy use in the built environment in different climate zones-a review. *Energy, 42*, 103–112.

Li, W., Yang, X., Qian, L., An, Y., & Fang, J. (2016). The complete mitochondrial genome of the citrus long-horned beetle, Anoplophora chinensis (Coleoptera: Cerambycidae). *Mitochondrial DNA Part A, 27*, 4665–4667.

Li, L., Wang, M., Pokharel, S. S., Li, C., Parajulee, M. N., Chen, F., & Fang, W. (2019). Effects of elevated CO2 on foliar soluble nutrients and functional components of tea, and population dynamics of tea aphid, Toxoptera aurantii. *Plant Physiology and Biochemistry, 145*, 84–94.

Liang, J., Tang, S., & Cheke, R. A. (2016). Beverton-Holt discrete pest management models with pulsed chemical control and evolution of pesticide resistance. *Communications in Nonlinear Science and Numerical Simulation, 36*, 327–341.

Ma, W. H., Sun, L. Y., Yu, L. G., Wang, J. T., & Chen, J. (2007). Study on the occurrence and life history in Aromia bungii (Faldermann). *Acta Agriculturae Boreali-Sinica, 22*, 247–249.

Macfadyen, S., McDonald, G., & Hill, M. P. (2018). From species distributions to climate change adaptation: Knowledge gaps in managing invertebrate pests in broad-acre grain crops. *Agriculture, Ecosystems and Environment, 253*, 208–219.

Maistrello, L., Dioli, P., Bariselli, M., Mazzoli, G. L., & Giacalone-Forini, I. (2016). Citizen science and early detection of invasive species: Phenology of first occurrences of Halyomorpha halys in southern Europe. *Biological Invasions, 18*, 3109–3116.

Mann, R. S., & Singh, K. (2003). Screening of sugarcane genotype for their reaction against sugarcane whitefly (Aleurolobus barodensis Mask). *Indian Sugar Journal, 23*, 110–111.

Maraseni, T. N., & Qu, J. (2016). An international comparison of agricultural nitrous oxide emissions. *Journal of Cleaner Production, 135*, 1256–1266.

Masse, W. B., Liston, J., Carucci, J., & Athens, J. S. (2006). Evaluating the effects of climate change on environment, resource depletion, and culture in the Palau Islands between AD 1200 and 1600. *Quaternary International, 151*, 106–132.

McDonald, G., Umina, P. A., Macfadyen, S., Mangano, P., & Hoffmann, A. A. (2015). Predicting the timing of first generation egg hatch for the Pest redlegged earth mite Halotydeus destructor (Acari: Penthaleidae). *Experimental & Applied Acarology, 65*, 259–276.

Meisner, M. H., Harmon, J. P., & Ives, A. R. (2014). Temperature effects on long term population dynamics in a parasitoid-host system. *Ecological Monographs, 84*, 457–476.

Micic, S., Hoffmann, A. A., Strickland, G., Weeks, A. R., Bellati, J., Henry, K., Nash, M. A., & Umina, P. A. (2008). Pests of germinating grain crops in southern Australia: an overview of their biology and management options. *Australian Journal of Experimental Agriculture, 48*(12), 1560–1573.

Miraglia, M., Marvin, H. J. P., Kleter, G. A., Battilani, P., Brera, C., Coni, E., Cubadda, F., Croci, L., De Santis, B., & Dekkers, S. (2009). Climate change and food safety: An emerging issue with special focus on Europe. *Food and Chemical Toxicology, 47*, 1009–1021.

Miranda Cabrera, I., Ramos, L., & Fernández, B. M. (2003). Factors influencing the abundance of Steneotarsonemus spinki in rice, in Cuba. *Manejo Integrado de Plagas y Agroecología, 69*, 34–37.

Moradi Vajargah, M., & Parry, H. R. (2017). Environmental and biological drivers of flight initiation in a sporadic pest, Rutherglen bug, Nysius vinitor Bergroth (Hemiptera: Orsillidae). *Austral Entomology, 56*, 225–234.

Moreau, J., Rahme, J., Benrey, B., & Thiéry, D. (2008). Larval host plant origin modifies the adult oviposition preference of the female European grapevine moth Lobesia botrana. *Die Naturwissenschaften, 95*, 317–324.

Mound, L. A., & Halsey, S. H. (1978). *Whitefly of the world. A systematic catalogue of the Aleyrodidae (Homoptera) with host plant and natural enemy data*. Wiley.

Musser, F. R., & Shelton, A. M. (2005). The influence of post exposure temperature on the toxicity of insecticides to Ostrinia nubilalis (Lepidoptera: Crambidae). *Management Science: Formerly Pesticide Science, 61*, 508–510.

Mutthuraju, G. P., Srinivasa, N., & Girish, R. (2014). Rice sheath mite, Steneotarsonemus spinki smiley-An emerging Pest of rice. *Current Biotechnology, 8*, 197–212.

Myers, S. S., Zanobetti, A., Kloog, I., Huybers, P., Leakey, A. D. B., Bloom, A. J., Carlisle, E., Dietterich, L. H., Fitzgerald, G., & Hasegawa, T. (2014). Increasing CO_2 threatens human nutrition. *Nature, 510*, 139–142.

Nagarkatti, S., Tobin, P. C., Saunders, M. C., & Muza, A. J. (2003). Release of native Trichogramma minutum to control grape berry moth. *Canadian Entomologist, 135*, 589.

Nelson, G. C., Rosegrant, M. W., Koo, J., Robertson, R., Sulser, T., Zhu, T., Ringler, C., Msangi, S., Palazzo, A., & Batka, M. (2009). *Climate change: Impact on agriculture and costs of adaptation*. International Food Policy Research Institute.

Newman, J. A. (2004). Climate change and cereal aphids: The relative effects of increasing CO_2 and temperature on aphid population dynamics. *Global Change Biology, 10*, 5–15.

Nikpay, A. (2017). Damage assessment of sugarcane whitefly Neomaskellia andropogonis Corbett and population dynamics on seven commercial varieties in Southwest of Iran. *Sugar Technology, 19*, 198–205.

Nkomwa, E. C., Joshua, M. K., Ngongondo, C., Monjerezi, M., & Chipungu, F. (2014). Assessing indigenous knowledge systems and climate change adaptation strategies in agriculture: A case study of Chagaka Village, Chikhwawa, Southern Malawi. *Physics and Chemistry of the Earth, Parts A/B/C, 67*, 164–172.

Noyes, P. D., McElwee, M. K., Miller, H. D., Clark, B. W., Van Tiem, L. A., Walcott, K. C., Erwin, K. N., & Levin, E. D. (2009). The toxicology of climate change: Environmental contaminants in a warming world. *Environment International, 35*, 971–986.

Oberthür, S., & Ott, H. E. (1999). *The Kyoto Protocol: International climate policy for the 21st century*. Springer Science & Business Media.

Oerke, E., Dehne, H., Schönbeck, F., & Weber, A. (2012). *Crop production and crop protection: Estimated losses in major food and cash crops*. Elsevier.

Ortega-Lopez, V., Amo-Salas, M., Ortiz-Barredo, A., & Diez-Navajas, A. M. (2014). Male flight phenology of the European grapevine moth Lobesia botrana (Lepidoptera: Tortricidae) in different wine-growing regions in Spain. *Bulletin of Entomological Research, 104*, 566–575.

Otieno, P. O., Owuor, P. O., Lalah, J. O., Pfister, G., & Schramm, K.-W. (2013). Impacts of climate-induced changes on the distribution of pesticides residues in water and sediment of Lake Naivasha, Kenya. *Environmental Monitoring and Assessment, 185*, 2723–2733.

Pandya, H. V. (2005). Population status and management of sugarcane whitefly (Aleurolobus barodensis Mask.) in Gujrat. *Cooperative Sugar, 36*, 479–482.

Paoletti, M. G., Tsitsilas, A., Thomson, L. J., Taiti, S., & Umina, P. A. (2008). The flood bug, Australiodillo bifrons (Isopoda: Armadillidae): A potential pest of cereals in Australia? *Applied Soil Ecology, 39*, 76–83.

Parmesan, C., & Yohe, G. (2003). A globally coherent fingerprint of climate change impacts across natural systems. *Nature, 421*, 37–42.

Parmesan, C., Woiwod, I. P., Reynolds, D. R., & Thomas, C. D. (2001). *Insect movement: Mechanisms and consequences*. CABI Publishing.

Pliansinchai, U., Jarnkoon, V., Siengsri, S., Kaenkong, C., Pangma, S., & Weerathaworn, P. (2007). Ecology and destructive behaviour of cane boring grub (Dorysthenes buqueti Guerin) in North Eastern Thailand. *International Society of Sugar Cane Technologists*.

Popov, C., Trotus, E., Vasilescu, S., Barbulescu, A., & Rasnoveanu, L. (2006). Drought effect on pest attack in field crops. *Romanian Agricultural Research, 23*, 43–52.

Porter, J. H., Parry, M. L., & Carter, T. R. (1991). The potential effects of climatic change on agricultural insect pests. *Agricultural and Forest Meteorology, 57*, 221–240.

Rai, A. B., Halder, J., & Kodandaram, M. H. (2014). Emerging insect pest problems in vegetable crops and their management in India: An appraisal. *Pest Management in Horticultural Ecosystems, 20*, 113–122.

Reddy, P. P. (2013). Impact of climate change on insect pests, pathogens and nematodes. *Pest Management in Horticultural Ecosystems, 19*, 225–233.

Reddy, P. V. R., Gundappa, B., & Chakravarthy, A. K. (2018). Pests of mango. In *Pests and their management* (pp. 415–440). Springer.

Reineke, A., & Thiéry, D. (2016). Grapevine insect pests and their natural enemies in the age of global warming. *Journal of Pest Science, 2004*(89), 313–328.

Robinet, C., & Roques, A. (2010). Direct impacts of recent climate warming on insect populations. *Integrative Zoology, 5*, 132–142.

Rodrigues Filho, F. A., Moreira, T. A. A., Valle, R. M., Baêta, J. G. C., Pontoppidan, M., & Teixeira, A. F. (2016). E25 stratified torch ignition engine performance, CO_2 emission and combustion analysis. *Energy Conversion and Management, 115*, 299–307.

Romo, C. M., & Tylianakis, J. M. (2013). Elevated temperature and drought interact to reduce parasitoid effectiveness in suppressing hosts. *PLoS One, 8*, e58136.

Roos, J., Hopkins, R., Kvarnheden, A., & Dixelius, C. (2011). The impact of global warming on plant diseases and insect vectors in Sweden. *European Journal of Plant Pathology, 129*, 9–19.

Rosenzweig, C., Iglesius, A., Yang, X.-B., Epstein, P. R., & Chivian, E. (2001). *Climate change and extreme weather events-Implications for food production, plant diseases, and pests*. Center for Health and the Global Environment, Harward Medical School.

Rosenzweig, C., Tubiello, F. N., Goldberg, R., Mills, E., & Bloomfield, J. (2002). Increased crop damage in the US from excess precipitation under climate change. *Global Environmental Change, 12*, 197–202.

Roth, S. K., & Lindroth, R. L. (1995). Elevated atmospheric CO_2: Effects on phytochemistry, insect performance and insect parasitoid interactions. *Global Change Biology, 1*, 173–182.

Roy, S., Barooah, A. K., Ahmed, K. Z., Baruah, R. D., Prasad, A. K., & Mukhopadhyay, A. (2019). *Impact of climate change on tea pest status in northeast India and effective plans for mitigation* (Vol. 40, pp. 432–442). Acta Ecologica Sinica.

Russo, E., Nugnes, F., Vicinanza, F., Garonna, A. P., & Bernardo, U. (2020). Biological and molecular characterization of Aromia bungii (Faldermann, 1835)(Coleoptera: Cerambycidae), an emerging pest of stone fruits in Europe. *Scientific Reports, 10*, 1–9.

Ryan, G. D., Emiljanowicz, L., Haerri, S. A., & Newman, J. A. (2014). Aphid and host-plant genotype × genotype interactions under elevated CO_2. *Ecological Entomology, 39*, 309–315.

Sabbaghi, M. A., Nazari, M., Araghinejad, S., & Soufizadeh, S. (2020). Economic impacts of climate change on water resources and agriculture in Zayandehroud river basin in Iran. *Agricultural Water Management, 241*, 106323.

Saha, K., Ghosh, S., & Sarkar, P. K. (2016). Rice sheath mite (Steneotarsonemus spinki smiley): An emerging pest and its occurrence in popular rice varieties of West Bengal. *SATSA Mukhaptra Annual Technical Issue, 20*, 154–160.

Santana, P. A., Kumar, L., Da Silva, R. S., & Picanço, M. C. (2019). Global geographic distribution of Tuta absoluta as affected by climate change. *Journal of Pest Science, 2004*(92), 1373–1385.

Scaccini, D., & Furlan, L. (2019a). *Nysius cymoides* (Hemiptera: Lygaeidae), a potential emerging pest: Overview of the information available to implement integrated pest management. *International Journal of Pest Management, 67*, 73–88.

Scaccini, D., & Furlan, L. (2019b). Outbreak of Nysius cymoides on second crop soybean Glycine max and proposal for Integrated Pest Management. *Bulletin of Insectology, 72*, 29–34.

Seppelt, R., Arndt, C., Beckmann, M., Martin, E. A., & Hertel, T. W. (2020). Deciphering the biodiversity-production mutualism in the global food security debate. *Trends in Ecology & Evolution, 35*, 1011–1020.

Sharma, H. C., & Dhillon, M. K. (2020). Climate change effects on arthropod diversity and its implications for pest management and sustainable crop production. *Agroclimatology: Linking Agriculture to Climate, 60*, 595–619.

Sherrard, M. E., Maherali, H., & Latta, R. G. (2009). Water stress alters the genetic architecture of functional traits associated with drought adaptation in Avena barbata. *Evolution: International Journal of Organic Evolution, 63*, 702–715.

Sibanda, T., Dobson, H. M., Cooper, J. F., Manyangarirwa, W., & Chiimba, W. (2000). Pest management challenges for smallholder vegetable farmers in Zimbabwe. *Crop Protection, 19*, 807–815.

Silva, P. (1954). Um novo ácaro nocivo ao tomateiro na Bahia. *Boletim do Instituto Biológico da Bahia, 1*, 1–20.

Singh, H. S., & Kundan, K. (2014). Status and strategies for red banded mango caterpillar (RBMC), Deanolis albizonalis (Hampson)(Lepidoptera: Pyralidae)-an emerging Pest in eastern India. *Journal of Applied Zoological Research, 25*, 11–20.

Singh, S. S., Gupta, N. N., & Rai, M. K. (2010). Pest management in temperate fruits: Opportunities and challenges. *Progressive Horticulture, 42*, 76–83.

Sparks, T. H., Dennis, R. L. H., Croxton, P. J., & Cade, M. (2007). Increased migration of Lepidoptera linked to climate change. *European Journal of Entomology, 104*, 139.

Stejskal, V., Hubert, J., Aulicky, R., & Kucerova, Z. (2015). Overview of present and past and pest-associated risks in stored food and feed products: European perspective. *Journal of Stored Products Research, 64*, 122–132.

Stocker, T. F., Qin, D., Plattner, G.-K., Tignor, M., Allen, S. K., Boschung, J., Nauels, A., Xia, Y., Bex, V., & Midgley, P. M. (2013). *Climate change 2013: The physical science basis. Contribution of Working Group I to the Fifth Assessment Report of the Intergovernmental Panel on Climate Change* (pp. 33–115). Cambridge University Press.

Stoeckli, S., Hirschi, M., Spirig, C., Calanca, P., Rotach, M. W., & Samietz, J. (2012). Impact of climate change on voltinism and prospective diapause induction of a global pest insect-Cydia pomonella (L.). *PLoS One, 7*, e35723.

Stoeckli, S., Felber, R., & Haye, T. (2020). Current distribution and voltinism of the brown marmorated stink bug, Halyomorpha halys, in Switzerland and its response to climate change using a high-resolution CLIMEX model. *International Journal of Biometeorology, 64*, 2019–2032.

Suffert, M., & Ward, M. (2014). Emerging pests of potato in Europe: Early warning, risk analyses and regulation. *Potato Research, 57*, 263–271.

Svobodová, E., Trnka, M., Dubrovský, M., Semerádová, D., Eitzinger, J., Štěpánek, P., & Žalud, Z. K. (2014a). Determination of areas with the most significant shift in persistence of pests in Europe under climate change. *Pest Management Science, 70*, 708–715.

Svobodová, E., Trnka, M., Žalud, Z., Semeradova, D., Dubrovský, M., Eitzinger, J., Štěpánek, P., & Brazdil, R. (2014b). Climate variability and potential distribution of selected pest species in south Moravia and north-east Austria in the past 200 years-lessons for the future. *The Journal of Agricultural Science, 152*, 225–237.

Sylla, S., Brévault, T., Monticelli, L. S., Diarra, K., & Desneux, N. (2019). Geographic variation of host preference by the invasive tomato leaf miner Tuta absoluta: Implications for host range expansion. *Journal of Pest Science, 92*, 1387–1396.

Tubiello, F. N., Salvatore, M., Rossi, S., Ferrara, A., Fitton, N., & Smith, P. (2013). The FAOSTAT database of greenhouse gas emissions from agriculture. *Environmental Research Letters, 8*, 15009.

United States Apple Association. (2010). *Asian Pest inflicting substantial losses, raising alarm in eastern apple orchards*.

Urbanski, J., Mogi, M., O'Donnell, D., DeCotiis, M., Toma, T., & Armbruster, P. (2012). Rapid adaptive evolution of photoperiodic response during invasion and range expansion across a climatic gradient. *The American Naturalist, 179*, 490–500.

Van Dyck, H., Bonte, D., Puls, R., Gotthard, K., & Maes, D. (2015). The lost generation hypothesis: Could climate change drive ectotherms into a developmental trap? *Oikos, 124*, 54–61.

Vetter, S. H., Sapkota, T. B., Hillier, J., Stirling, C. M., Macdiarmid, J. I., Aleksandrowicz, L., Green, R., Joy, E. J. M., Dangour, A. D., & Smith, P. (2017). Greenhouse gas emissions from agricultural food production to supply Indian diets: Implications for climate change mitigation. *Agriculture, Ecosystems and Environment, 237*, 234–241.

Walsh, D. B., Bolda, M. P., Goodhue, R. E., Dreves, A. J., Lee, J., Bruck, D. J., Walton, V. M., O'Neal, S. D., & Zalom, F. G. (2011). Drosophila suzukii (Diptera: Drosophilidae): Invasive Pest of ripening soft fruit expanding its geographic range and damage potential. *Journal of Integrated Pest Management, 2*, G1–G7.

Walther, G.-R., Post, E., Convey, P., Menzel, A., Parmesan, C., Beebee, T. J. C., Fromentin, J.-M., Hoegh-Guldberg, O., & Bairlein, F. (2002). Ecological responses to recent climate change. *Nature, 416*, 389–395.

Wang, X. L., Zwiers, F. W., Swail, V. R., & Feng, Y. (2009). Trends and variability of storminess in the Northeast Atlantic region, 1874-2007. *Climate Dynamics, 33*, 1179.

Watson, G. W., & Mifsud, D. (2017). Invasive mealybugs (Hemiptera: Pseudococcidae) and the threats they present to Mediterranean countries. *Bulletin of the Entomological Society of Malta, 9*, 34–35.

Way, M. J., Conlong, D. E., Rutherford, R. S., Sweby, D. L., Gillespie, D. Y., Stranack, R. A., Lagerwall, G., Grobbelaar, E., & Perissinotto, R. (2017). Cacosceles (Zelogenes) newmannii (Thomson)(Cerambycidae: Prioninae), a new pest in the South African sugarcane industry. In *Proceedings of the Annual Congress-South African Sugar Technologists' Association*, pp. 62–65. South African Sugar Technologists' Association.

Wei, Y. J. (2014). Flight initiation of Nysius huttoni (Hemiptera: Orsillidae) in relation to temperature and wing forms. *Applied Entomology and Zoology, 49*, 119–127.

Woiwod, I. (1997). Detecting the effects of climate change on Lepidoptera. *Journal of Insect Conservation, 1*, 149–158.

Woiwod, I. P., & Harrington, R. (1994). Flying in the face of change: The Rothamsted Insect Survey. In *Long-term experiments in agricultural and ecological sciences* (pp. 321–342). CAB International Wallingford.

Wu, H., Song, Z., Wang, X., Liu, Z., & Tang, S. (2016). Increasing CO_2 differentially affects essential and non-essential amino acid concentration of rice grains grown in cadmium-contaminated soils. *Environmental Pollution, 216*, 86–94.

Xu, T., Yasui, H., Teale, S. A., Fujiwara-Tsujii, N., Wickham, J. D., Fukaya, M., Hansen, L., Kiriyama, S., Hao, D., & Nakano, A. (2017). Identification of a male-produced sex-aggregation pheromone for a highly invasive cerambycid beetle, Aromia bungii. *Scientific Reports, 7*, 1–7.

Yan, Y., Wang, Y. C., Feng, C. C., Wan, P. H. M., & Chang, K. T. T. (2017). Potential distributional changes of invasive crop pest species associated with global climate change. *Applied Geography, 82*, 83–92.

Zhang, Z. Y., Zhang, C. Z., Liu, X. J., & Hong, X. Y. (2006). Dynamics of pesticide residues in the autumn Chinese cabbage (Brassica chinensis L.) grown in open fields. Pest Manag. Sci. Former. *Pesticide Science, 62*, 350–355.

Zhou, X., Harrington, R., Woiwod, I. P., Perry, J. N., Bale, J. S., & Clark, S. J. (1995). Effects of temperature on aphid phenology. *Global Change Biology, 1*, 303–313.

Ziter, C., Robinson, E. A., & Newman, J. A. (2012). Climate change and voltinism in C alifornian insect pest species: Sensitivity to location, scenario and climate model choice. *Global Change Biology, 18*, 2771–2780.

Züst, T., & Agrawal, A. A. (2016). Mechanisms and evolution of plant resistance to aphids. *Nature Plants, 2*, 1–9.

Chapter 3
Future-Proofing Plants Against Climate Change: A Path to Ensure Sustainable Food Systems

Prasanta Kumar Majhi, Basit Raza, Partha Pratim Behera, Shravan Kumar Singh, Aalok Shiv, Suma C. Mogali, Tanmaya Kumar Bhoi, Biswaranjan Patra, and Biswaranjan Behera

Abstract Climate change has altered the pattern of rainfall, temperature, carbon dioxide (CO_2) levels, and emission of greenhouse gases, which result in the frequency and severity of extreme events such as drought, flood, salinity, heavy metal stress, nutrient stress, new diseases, and insect pest. This significantly impacts agriculture production, food security, livelihoods, and nutrition. Worldwide, millions of people are affected due to the consequence of climate change and particularly

P. K. Majhi (✉)
Department of Plant Breeding and Genetics, Regional Research and Technology Transfer Station (RRTTS), Odisha University of Agriculture and Technology (OUAT), Keonjhar, India

B. Raza
Division of Soil Science and Agricultural Chemistry, ICAR-Indian Agricultural Research Institute, New Delhi, India

P. P. Behera
Department of Plant Breeding and Genetics, Assam Agricultural University, Jorhat, India

S. K. Singh
Department of Genetics and Plant Breeding, Institute of Agricultural Sciences, Banaras Hindu University (BHU), Varanasi, India

A. Shiv
Division of Crop Improvement, ICAR-Indian Institute of Sugarcane Research, Lucknow, India

S. C. Mogali
Department of Genetics and Plant Breeding, College of Agriculture, University of Agricultural Sciences, Dharwad, India

T. K. Bhoi
Forest Protection Division, ICFRE-Arid Forest Research Institute (AFRI), Jodhpur, India

B. Patra
Center for Biotechnology, Siksha 'O' Anusandhan Deemed to be University, Bhubaneswar, India

B. Behera
ICAR-Indian Institute of Water Management, Bhubaneswar, India

C. M. Galanakis (ed.), *Biodiversity, Functional Ecosystems and Sustainable Food Production*, https://doi.org/10.1007/978-3-031-07434-9_3

become the most vulnerable, by increasing the frequency and virulence of extreme meteorological events that cause population displacement and reduction in agricultural productivity. A paradigm shift toward more resilient, productive, and sustainable agriculture and food systems is required. The world must act immediately act on it to put an end to hunger and malnutrition. To ensure rapid and advanced agricultural development in a short period, precision farming practices and smart breeding strategies need to follow; such as machine learning, deep learning, big data analysis, remote sensing, artificial intelligence, system biology study, genomic prediction, speed breeding, and haplotype breeding. These techniques can prove the future plants against climate variability with increased yield potential and improved resilience to achieve the goal of resilient climate agriculture.

Keywords Climate change · Food security · Nutrition · Smart breeding · Climate resilient agriculture

1 Introduction

Climate change is now unambiguous, particularly in increasing CO_2 levels, intensifying temperatures, humidity, widespread melting of snow and ice, and rising global average sea level (FAO, 2011). Climate change already has a discernible impact on water resources and freshwater ecosystems in all world regions. It creates significant uncertainty about future water availability (Intergovernmental Panel on Climate Change (IPCC), 2021). It will impact precipitation and runoff, affecting hydrological systems, water quality, and temperature, as well as the groundwater recharge and magnitude and spatiotemporal trends of hydro-climatic parameters such as rainfall (Guo et al., 2020). The salinity of surface and groundwater in coastal areas will be affected by sea-level rise (FAO, 2015a). An increase in interannual variability in rainfall events caused by climate change was the primary cause of disasters such as drought, flood, salinity, heat, and heavy metal stress, as well as a possible increase in the frequency of some biotic stresses (pests and diseases) (Bermúdez et al., 2020; Poulose et al., 2020). However, climate change is not a new phenomenon. The current climate change will already hurt food production, food quality, market demand, income vs. food price, agriculture livelihood, and price (Fig. 3.1). It affects agricultural production systems in both direct and indirect ways. Direct impacts include effects on specific agricultural production systems caused by physical characteristics such as temperature levels and rainfall distribution. Changes in other species, such as pollinators, pests, disease vectors, and invasive species, indirectly affect crop production. These indirect effects can have a significant impact. Because of the large number of interacting parameters and links, many of which are still unknown, they are much more difficult to assess (FAO, 2015a). The poor farmers and developing countries are most vulnerable due to these phenomena. In recent eras, climate change appears to have curtailed the production of major food crops in the countries like;

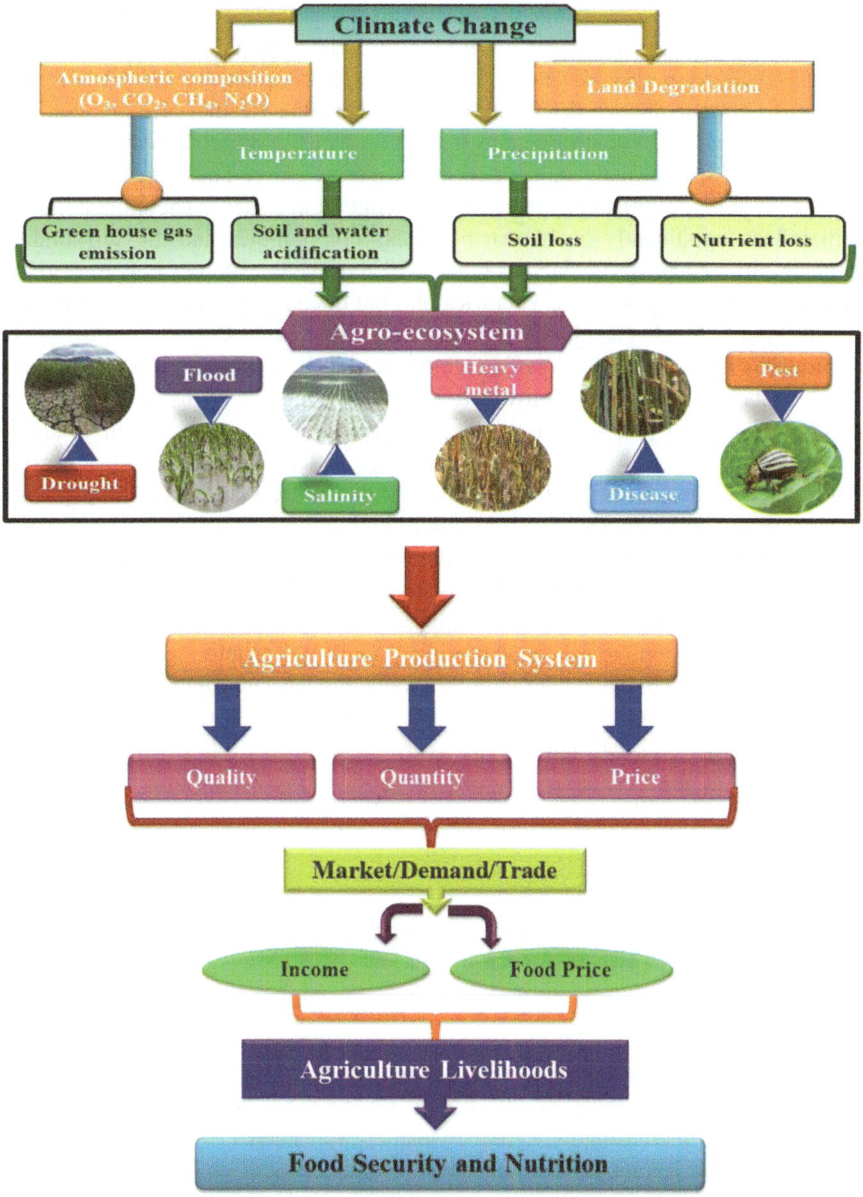

Fig. 3.1 A schematic representation of the impact of climate change on the Agro-ecosystem, agricultural production system, agricultural livelihood, and food security

Sub-Saharan Africa and Europe. This seem like to have had a optimistic influence in Latin America and a variable effect in Asia and North America (Ray et al., 2019).

Furthermore, climate change is expected to cause biodiversity losses, primarily in more marginal environments. Agriculture, dependent on soil characteristics, weather patterns, and biodiversity, is one of the socio-economic sectors most vulnerable to climate change (EEA, 2019, 2020). Thrilling weather phenomenon will become severe and will occur more frequently and the dispersal and richness of different pest species may change (Bocci & Smanis, 2019). These fluctuations will impact on the crop phenology, and yields (Ceglar et al., 2019). The ultimate effect will reflect significantly on human being, its communities and the surrounding existing bio-resources (Yazdandoost et al., 2021). The question of how to feed the world's growing population is centuries old, but due to population growth and the possibility of climate change, it now has an explosive influence. Crop improvement approaches have already addressed significant abiotic and biotic stresses. Climate change adaptation strategies must includes proper matching of the crop phenology with the availability of soil moisture, photoperiod and the response of available temperature. Selection of suitable varieties with different dates of maturity group can escape or avoid the predictable occurrence of stress at critical crop growth cycles. Breeding of improved varieties with higher nutrient use efficiency (NUE), water use efficiency (WUE) and better tolerance ability to heat and cold can provide buffer against such difficulties. Along with the crop improvement strategy, other multi-disciplinary research is also required simultaneously to develop climate-hardy crops and stabilize our food security. Thus, in this chapter, we focused on the impact of climate change on agricultural production systems and food security and how to deal with significant challenges such as abiotic stresses (drought, flood, heat, salinity, heavy metal and nutrient stress). Along with it, we have enlightened the intervention of genetic engineering and genome editing for the nutritional quality improvement of our food crops.

2 Climate Change: Impacts on Agricultural Production System and Our Food Security

Agriculture has faced numerous challenges, including human population growth, climate change, malnutrition, poverty, hunger, and many other stresses. The relentless increase in greenhouse gas emissions is raising the earth's temperature and the consequence of increasing the extreme weather events and a rapid shift in seasons. Climate change has an accelerating pace, combined with global population and income growth, poses a global threat to food security. Higher temperatures eventually reduce desirable crop yields while encouraging weed and insect-pest proliferation and the development of new pathogen races. An alteration in rainfall pattern increases the possibility of long-term production decline and short-term crop failure (Nelson et al., 2009). When temperature increases, evaporation from the soil increases, and the rate of transpiration increases with more moisture loss from the leaves. The collective effect is known as evapotranspiration. Global warming is expected to increase the pattern of rainfall and the net effect of higher temperature on water availability is a race between higher evapotranspiration and more precipitation. Furthermore, the role of higher atmospheric CO_2 concentrations in increasing photosynthetic activity and mitigating the adverse effects of climate change is still being debated (Sultan & Gaetani, 2016). Though there will be gains in some crops in some parts of the world due to climate change, the overall effects of climate change on agriculture are expected to be negative, threatening global food security. Even though, overall agricultural production and productivity is not likely to drop before 2050, but the yearly harvests will become more flexible, appropriate production zones will alter, and the price volatility of agri-commodity will upsurge. This will create a big impact on pattern of crop cultivation, global trade, and provincial markets.

According to FAO, Food security is defined as "all people having physical and economic access to sufficient, safe, and nutritious food that meets their dietary needs and food preferences in order to live an active and healthy life" (FAO, 1996). This explanation of food security includes four dimensions (Fig. 3.2) such as; (1) food availability; (2) accessibility; (3) utilization; and (4) stability of these three dimensions (FAO, 2015a). Variability in climatic factors, and higher intensity of extreme weather events, has great impact on the stability of our food system (FAO, 2015a; Turral et al., 2011). Reassurance of global food security is a major task due

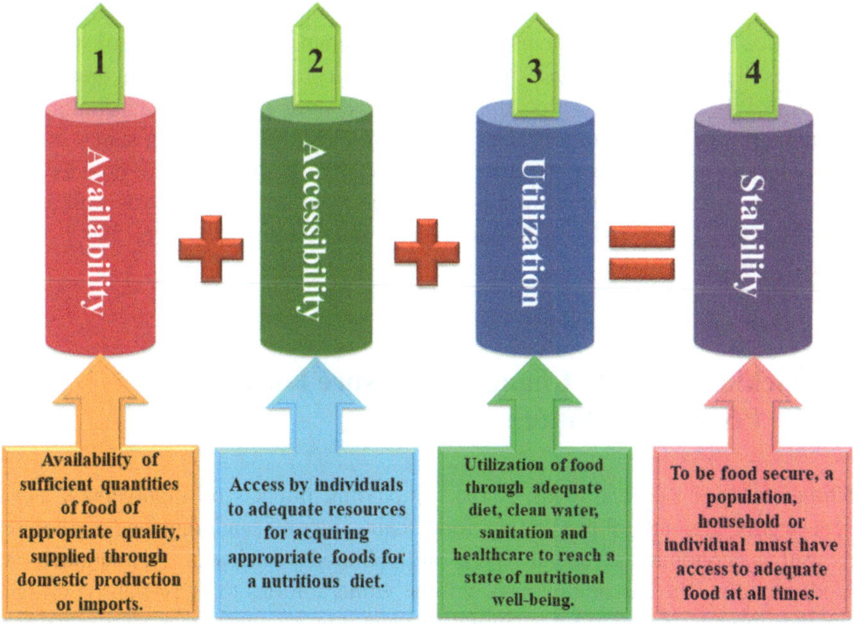

Fig. 3.2 An illustration of four pillars of Food Security. The concept of food stability refers to combining the three dimensions *viz*., food availability, accessibility, and utilization. Climate change has impacted all four dimensions of food security

to the consequence of drastic climate change, diminishing cultivable land area, and rising population growth, which is expected to reach 8.6 billion in 2030 and 10 billion by 2050 (Tomlinson, 2013). Agricultural production indices, livestock ownership indices, and national food balance sheets are examples of some food availability indicators (Renzaho & Mellor, 2010). Food accessibility is defined as "the existence of legal, political, economic, and social resources that an individual requires to access food". The food utilization component of food security on the other hand, refers to consuming food in sufficient quantities through an adequate diet, clean water, sanitation, and healthcare in order to achieve nutritional well-being (FAO, 2008). The requirement that a population, household, or individual always have enough food is referred to as food stability. Therefore, the concept of stability refers to the three dimensions of food security those are available, accessible, and usable. Thus, agricultural production, which provides both food and income to rural households, is inextricably linked to all aspects of food security. Food security vulnerabilities to climate change refer to the food system's proclivity to fail to deliver food security outcomes in the face of climate change, and they include environmental, economic, and social dimensions (FAO, 2016). Climate change has the potential to reverse recent gains in the fight against hunger and malnutrition. Climate change, as highlighted in the most recent assessment report of the Intergovernmental Panel on Climate Change (IPCC), augments and intensifies risks to food security for the most vulnerable countries and populations. The IPCC AR5 has acknowledged four out of the eight critical risks concomitant with climate change which have undeviating allegations for food security: (1) Loss of rural income and livelihoods; (2) Beating of aquatic and coastal ecosystems and their maintenances; (3) Loss of terrestrial and inland water ecosystems; and (4) Food insecurity and food system breakdown (FAO, 2015b). The most vulnerable nations and their populations, including the countries in arid, and semi-arid regions, and small island developing states, are the foremost and most rigorously affected. Climate change will also have broader consequences in terms of trade flows, food markets, and price stability, as well as the introduction of new risks to human health and livelihood (Hope, 2009; Jodie et al., 2009).

3 Tackling Climate Change: To Ensure Food security for Present and Future

Climate change poses a swing of risks, including impacts on biodiversity, agroecosystems, agricultural production, food chains, income, and trade and marketing, as well as economic and social effects on livelihoods, food security, and nutrition. Understanding the cascade of risks and the vulnerabilities to these risks is critical for developing strategies for adapting to changing circumstances (FAO, 2015a). Reducing vulnerabilities is essential for lowering net impacts on food security and nutrition and preventing long-term consequences. Increasing food security resilience in the face of climate change necessitates various interventions, ranging from social protection to agricultural practices and risk management. Agriculture should produce 60–100% more food than it does now by 2050 (Tilman et al., 2011). Nowadays, global population growth rates have outpaced the linear increase in food grain production. To fulfil the requirement, crop productivity should upsurge by 70% in the upcoming decades to sufficiently feed the growing population (FAO, 2009). The fast and exceptional increase in human population demands a 37% increase in annual food production in a linear rates, which is significantly reduced by the effects of climate change and crop production variation (Beddington et al., 2011). Despite climate change, the goal of producing enough food to meet the increasing demand for ensuring food security must be addressed. Therefore, we have to develop climate-resilient crop varieties, which should have improved intake capacity of water and nutrients from resource-poor environments and can have the ability to withstand heat tolerance, salinity, and heavy metal stress tolerance as well as resistance to multiple diseases and insect pests.

4 Climate Resilient Crop Improvement: Future-Proofing Plants Against Abiotic Stresses

Rapid population growth, increasing demand of food, higher demand of agricultural inputs (chemical fertilizers and fertilizers), loss of biodiversity along with climate change affect the crop production and productivity. These production impediments have resulted in significant yield reductions around the world (Begna, 2021). On the other hand, crop improvement strategies played critical roles in overcoming crop plant production constraints by developing high-yielding and climate-resilient crop varieties (biotic and abiotic). Climate adaptation and mitigation are the most important and interdependent strategies for mitigating the effects of climate change and paving the way for sustainable food production (Begna, 2021). Climate adaptation is directly related to reducing the negative impact of climate change through climate-resilient crop plants. In contrast, mitigation is designed to overcome the causes and reduce the potential effects of climate change. Overwhelming these intricate challenges will be tougher in absence of crop genetic improvement to boost the crop productivity by addressing the issue of yield bargain and its links to climate change (Satterthwaite et al., 2010). Looking to the future consequence of climate change, only plant breeding strategy is not enough to develop a resilient crop variety. Still, we have to study the multi-disciplinary approach such as biotechnology, plant physiology, plant pathology, entomology, agronomy, and soil science with advanced crop improvement strategies (Fig. 3.3). Suppose we consider drought as stress because a single stress-resistant mechanism is connected from several angles. The mechanism is linked with genetics, plant breeding, biotechnology, crop physiology, and agronomy. If we understand the problem from multiple directions, then it is

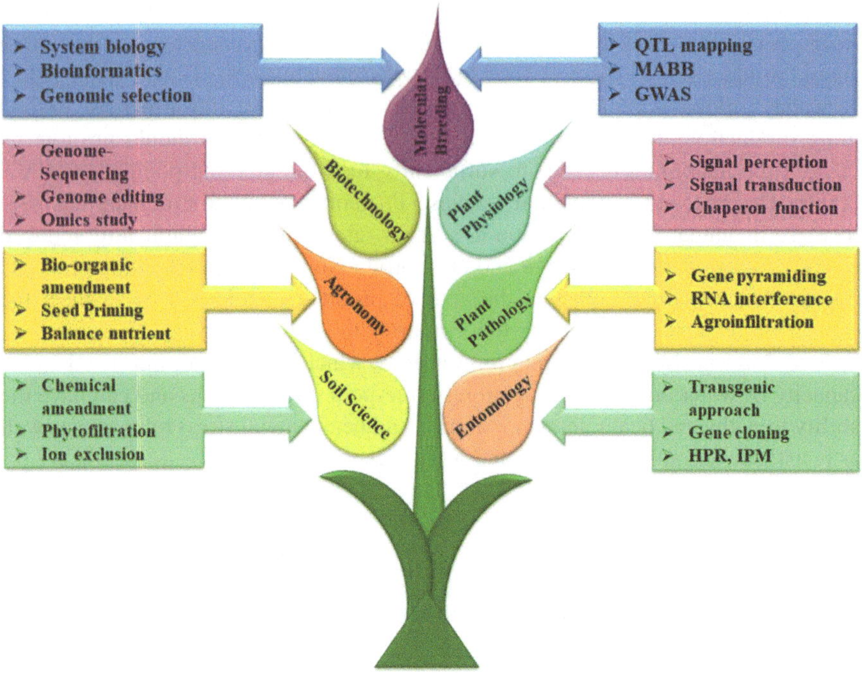

Fig. 3.3 Multidisciplinary research approaches are essential to combat climate change's impact on agricultural production. Different molecular, biochemical, physiological, and agronomical strategies are required to prove our crop plants against several biotic and abiotic stresses developed due to climate change. (Abbreviation: QTL quantitative trait loci, MABB marker-assisted back-cross breeding, GWAS genome-wide association studies, HPR host-plant resistance, IPM Integrated pest management)

possible to develop a robust crop variety for our changing environment to boost agriculture production.

Crop losses due to climate change must be addressed to maintain future food security. Drought, heat waves, cold snaps, and flooding aren't the only effects of climate change affecting agricultural crop production (von der Gathen et al., 2021; Van Houtan et al., 2021). Because of the multi-factor nature of climate change, the yield of important crops and the global food supply may be jeopardized (Zandalinas et al., 2021). The long-term impacts of climate change and the response of a single plant to a particular stress combination are complicated to anticipate (Zandalinas et al., 2020). As a result of drought and heat, the development of drought-prone regions and the growth and production of principal crops are being affected (Zhou et al., 2017). Climate-resilient crop breeding is crucial to minimizing the consequences of climate change on agricultural production. Genetic adaptation of crop cultivars to ongoing and future climate change scenarios is essential to ensure food quality and security for an ever-increasing world population. Breeding or designing plants with high yield potential under changing environmental conditions may be a valuable strategy for climate resilience. Research on climate resistance's molecular and physiological aspects is critical to designing a practical breeding approach (Ali et al., 2020).

4.1 *Future-Proofing Plant with Drought Tolerance*

Drought stress alters plant morphology, physiology, biochemistry, and molecular alterations (Osakabe et al., 2014). Plant growth and development are accelerated by drought, resulting in a reduction in the length of photosynthetic capability and a decrease in seed yield. In order to design drought-tolerance crops, we need an understanding of morphological and physiological responses, molecular pathways connected to drought tolerance, and genetic control at various plant developmental stages for successful crop development. Drought adaptations include drought escape, drought tolerance, and drought resistance. Natural herbaceous populations have complex genetic variation in DE and avoidance, governed by multiple QTL of minor impact and gene × environment interactions, demonstrating genetic limitations prevent concurrent development of both DE and avoidance strategies (Kooyers, 2015). This difficulty has been solved by combining fine-tuned classical breeding methods with marker-assisted breeding with current genomic technologies, including allele mining, genome-wide association studies (GWAS), genomic selection (GS), and genome editing.

4.1.1 Strategies to Develop a Future Proof Plant for Drought Tolerance

Agronomic Approach

A wide range of crop management methods, including soil management and cultural activities, may help mitigate the negative impacts of drought and heat stress. Micronutrients like B, Se, and Mn have been shown to alter the function of stomata and hence increase heat stress tolerance by activating physiological and metabolic mechanisms that preserve high water potential in tissues (Waraich et al., 2012). Due to new ABA formulations, commercial producers can now delay wilting symptoms caused by dryness and increase drought tolerance.

Conventional Breeding Approach

Directly selecting high-yielding cultivars is a common strategy for plant breeding in conventional methods. Through natural selection for grain yield, plant breeders have produced several drought-resistant rice cultivars that are highly heritable (Kumar et al., 2008). A plant's ability to withstand drought may be influenced by various secondary traits such as the thickness, depth, and penetration of its roots, as well as its branching angle and distribution pattern (Comas et al., 2013). Breeding for high yield potential and stability may increase yields in non-stress environments (Tester & Langridge, 2010). Heterosis effects may boost stress performance (Blum, 2013). Hybrids are better able to endure stress and sustain constant photosynthesis due to their higher vigor and yield. Selection under pressure may assist in locating the best alleles for adaptive plasticity interactions between genotype and environment (Des Marais et al., 2014). They will be chosen under advantageous circumstances due to more complicated molecular interactions. The significant limitations of this conventional breeding are low accuracy due to selection based on morphology and time-consuming and labor-intensive methods of screening out large numbers of genotypes. These limitations can be addressed by integrating both conventional and modern breeding approaches. The single-seed descent (SSD) method is frequently used with other advanced molecular breeding approaches to improve rapid generation advanced (RGA).

Crop Wild Relatives and Allele Mining

This hybridization initiative has shown remarkable progress in yield and yield stability by using landraces and crop wild relatives as the significant sources of drought resistance. Future breeding for enhanced root stress response might benefit wild crop species (Calleja-Cabrera et al., 2020). Due to selection or genetic drift, alleles that influenced yield or quality were lost directly or indirectly due to selection (Palmgren et al., 2015). Synthetic hexaploid wheat lines are developed using hybridization of ancestral genomes (SYN-DER) and performed to analyze new wheat drought resistance alleles (Cossani & Reynolds, 2015). Wheat landraces are utilized to improve root architecture and drought resistance, so there is a need to investigate the function of drought-responsive proteins and their expression, abundance, and post-translation changes in drought-tolerant landraces (Hu et al., 2015). Several tropical landraces and inbred lines have been tested for drought to improve WUE and drought tolerance (Xue et al., 2013). Modern genomics research has made it possible to find genes and alleles that are similar to each other in the germplasm of many wild crop species. This has led to the discovery of traits that effect water use efficiency (WUE) and improve drought tolerance. In spite of this, only a handful of these characteristics have really been put into practise in field. (Langridge & Reynolds, 2015). The current challenge is to devise breeding strategies that promote multiple stress adaptation and crop productivity in yield.

Mutation Breeding

Mutation breeding involves exposing plant explants to mutagens, including physical or chemical agents that induce genetic variation. The primary focus of mutation-based plant breeding is improving well-adapted cultivars by modifying important traits like drought tolerance and grain yield, etc. Molecular mutation breeding with high-throughput DNA technologies, including targeting induced local lesions in genomes (TILLING), EcoTILLING, and high-resolution melt analysis (HRM), will considerably boost the effectiveness and efficiency of mutation approaches in crop breeding. The deletion-TILLING, or de-TILLING, a novel knockout approach for obtaining deletion mutants for target genes, was applied in Arabidopsis and rice (Rogers et al., 2009). As drought traits are quantitative, selecting mutants in mutation breeding is usually ineffective (Hallajian, 2016). EMS-induced mutants of N22 having a deep root system with water stress and heat tolerance were characterized (Poli et al., 2013). Placido et al. (2020) found a gene called *LATERAL ROOT DENSITY* (LRD) in the wild wheat relative *Agropyron elongatum* boosted lateral root development under drought.

Designing Future Proof Crop with Drought Tolerance

Root characteristics, notably root length, density, and depth, have long been considered essential for designing a crop with better adaptation to water stress. Desirable specific root architectures for designer crops should be genetically predisposed with mechanisms that allow roots to expand plastically in response to water or nutrient deficits. Domestication and breeding bottlenecks are likely to have reduced modern crop root surface area (RSA) genetic diversity (Voss-Fels et al., 2018). The DRO1 locus in rice shows how to root design may help minimize water and nutrient deficits without affecting the yield (Arai-Sanoh et al., 2014). Newly discovered genes like *QUICK ROOTING 1* and *2* with DRO1genes that enhance root length in rice might be combined to create an arsenal of root ideotypes tailored to particular environments (Kitomi et al., 2018). In simulations for sorghum, a "stay-green" phenotype with decreased tillering or restricted maximum transpiration rate increased yield and delayed leaf senescence in terminal drought stress conditions (Hammer et al., 2016).

RNAi Using MicroRNAs

Small non-coding RNAs (miRNAs) are believed to play a critical role in drought tolerance by regulating gene expression at three levels: translation, transcription, and post-transcription. The target genes must be knocked out or knocked down by overexpression of a miRNA transgene to increase drought tolerance in plants. Using high-throughput next-generation sequencing (NGS) technology, researchers may find drought-induced miRNAs with unprecedented depth and accuracy using high-throughput next-generation sequencing (NGS) technology. The Arabidopsis hydroxypyruvate reductase (AtHPR1) promoter was used to down-regulate FTA expression in canola transgenics (Wang et al., 2009). A long non-coding miRNA was shown to boost yield in rice by decreasing the activity of FT/SQS, an enzyme involved in the production of b-diketone wax. It has been shown that artificial miRNAs (amiRNAs), which are more selective and whose silencing activity is maintained across generations, have been produced to minimize unintentional targeting of other genes by the transgene in potatoes (Pieczynski et al., 2013).

Transgenic Approach

As drought tolerance is quantitative, a successful transgenic technique will likely depend on transforming gene regulators that control vital processes. Transgenic methods comprise transferring target genes encoding TFs, controlling metabolites, and altering protein modification to create drought tolerance in plants. Drought tolerance has been widely used in plants such as Arabidopsis, rice, wheat, potato, soybean, peanut, chrysanthemum, tobacco, and tomato by overexpressing *DREB1/CBF* (Bhatnagar-Mathur et al., 2014). The rate-limiting enzyme in the ABA production pathway, 9-cis-epoxycarotenoid dioxygenase (NCED), was overexpressed constitutively in petunias to improve (Estrada-Melo et al., 2015). It may be possible to break the production plateau associated with adaptation to high temperatures by better understanding the physiological and molecular processes that cause "STAY-GREEN" phenotype or delayed leaf senescence (Abdelrahman et al., 2017). New mechanisms for enhanced production of suitable solutes such as glycine betaine (GB) in plants have increased drought tolerance (Fan et al., 2016). Using systems and synthetic biology discoveries, C3 plants like rice and wheat can be converted to C4 photosynthesis (Wang et al., 2014). However, the widespread use of GM crops, exceptionally drought-tolerant crops, will need proper safety testing and public acceptability.

Marker-Assisted Selection Approaches with Particular Emphasis on QTLs Mapping

Early generation selection using molecular markers has enhanced selection accuracy and genetic gain for desirable characteristics. It enables the rapid selection of new genetic combinations in top cultivars. Because drought tolerance is complex and relies on interactions with the environment and other genes and characters, conventional breeding is less successful than expected. Drought tolerance genes are usually introgressed from wild or unadapted germplasm using marker-assisted backcross breeding (MABC) or gene pyramiding. Quantitative trait locus (QTL) mapping can identify gene combinations or genomic areas linked to complex characteristics like drought tolerance (Jiang et al., 2012). Drought-related high grain yield QTLs in rice have been found for application in rice cultivation (Venuprasad et al., 2012). The QTLs $qDTY_{3.2}$ and $qDTY_{1.1}$ (Vikram et al., 2011) were strongly associated with grain yield during drought. In rice, a QTL (*qtl12.1*) boosted HI, biomass production, and plant height while decreasing the number of days to blooming. The use of SSR markers to study rice genetic diversity and drought tolerance was described by Verma et al. (2019). The QTLs from wild emmer wheat were used to increase drought tolerance in top durum and bread wheat cultivars using marker-assisted selection (Merchuk-Ovnat et al., 2016).

Epigenetic Perspective

Postgenomics has been driven by the introduction of next-generation sequencing (NGS) which facilitates the visualization of various genes, alternative splice variants, cis or trans-regulatory regions, and chromosomal changes (Lister et al., 2009). DNA methylation and histone alterations are examples of epigenetic or chromatin control that is heritable. The changes in histone modifications involve the methylation, acetylation, phosphorylation, sumoylation, and ubiquitination process. This can rapidly and reversibly alter gene expression of regulatory genes governing hormones, small RNAs, stress protectants, and their combined complex interactions response to drought stress to develop drought tolerance cultivar (Kim et al., 2012). Using CRISPR/Cas9 (Li et al., 2015) and tissue-specific chromatin profiling under drought stress improve the single base-pair resolution of N-terminal histone modifier mutants. The stability and permanence of the epigenetic alterations may be necessary for adaptation to changing climates, but their usage in breeding programs is essential.

Omics Approaches with Particular Emphasis on Genomics

Omics study large-scale multidimensional data sets with extensive research on their complete genetic, structural, or functional components. Many plant molecular drought tolerance processes involve changes in omic profiles, such as transcriptomics, proteomics, and metabolomics (Jing et al., 2014). Exploring these ideas will help us better understand how plants withstand drought. These studies rely on advanced sequencing tools, specifically next-generation sequencing. Drought-tolerant species like desiccation-tolerant (DT) or resurrection plants may serve as models for developing drought-tolerant crops (Giarola et al., 2017). Genomics-assisted breeding (GAB) integrates modern genomics technologies with an upgraded green super rice breeding strategy (GSR-BT), is one of the most dependable methods for developing CSRVs. A designed QTL pyramiding (DQP) strategy was also developed for stacking traits/genes resulting from derived trait-specific introgression lines (ILs) (Ali et al., 2020). Genome-wide association studies and other gene-mapping methods with high throughput phenotyping may now be evaluated with nucleotide-level accuracy. Using high-throughput field-based phenotyping data gathering technologies may considerably enhance resource usage efficiency and lead to more significant genetic gain via better selection efficiency (Araus et al., 2018). Climate-smart crops are being developed using genomic approaches in cereal crops (Serba et al., 2020), pulse crops (De Ron et al., 2019), oilseeds (Kole, 2019), fruit crops (Kole, 2020), and vegetable crops (Chittaranjan, 2020). These polymorphisms can be restored through targeted gene editing while reducing off-target effects and wild species DNA integration. CRISPR-Cas-based targeted genome editing is improving spatial and temporal control over the gene and network alterations (Mega et al., 2019) or the drought-induced suppression of negative regulators of ABA signal transduction. The loci for root features associated with drought resistance have proved challenging to uncover due to their quantitative character and poor heredity (Bray & Topp, 2018).

Transcriptomics, which examines an organism's whole transcriptome, could help to provide insights into the regulation of plant stress responses. For entire transcriptome profiling, high-throughput methods like microarrays and RNA sequencing (RNA-seq) are used, together with expressed sequence tags and suppression subtractive hybridization. Many transcription factors (TFs) under drought stress may be used in transgenic crops using these sophisticated technologies and previous "functional gene" investigations (Barbosa et al., 2013). In addition, transcriptome sequencing has offered vital information on complex drought-responsive genes due to the joint action of various genes in line with environmental factors. Proteomics is the large-scale analysis of protein sets expressed in specific cell types or organisms. A single gene's transcription, translation, and posttranslational modification may result in a varied proteome (Arc et al., 2011). The Rice Proteome Database allows researchers to compare proteome maps under water stress and look for elevated or down-regulated proteins. For example, wheat proteins associated with carbon metabolism, photosynthesis, and amino acid metabolism change during water stress. Metabolomics is the study related to quantifying and identifying all primary and secondary metabolites in a biological process. It is a straightforward technique to investigate an organism's metabolic pathways and functional genes. The development of metabolomics tools such as gas chromatography-mass spectrometry (GC-MS), liquid chromatography-mass spectrometry (LC-MS), capillary electrophoresis mass spectrometry (CE-MS), and Fourier transform ion cyclotron resonance mass spectrometry (FT-ICR-MS) has been used for metabolic profiling under drought conditions. Semel et al. (2007) employed GC-MS to detect amino acids, including proline, which postulated to function in drought stress. Epigenomics focuses on epigenetic mechanisms such as heritable DNA methylation, histone modification, and the impacts of short RNAs, which are thought to influence a broad range of activities in response to a variety of stresses, both biotic and abiotic. For example, in a simulated drought, (González et al., 2013) found that the tomato *Asr2* protein decreased and the upstream regulatory region of its gene lacked cytosine methylation, coupling epigenetics and drought adaptation in plants.

4.2 Future-Proofing Plant with Flood Tolerance

Although water is required for plant development, flooding hinders the exchange of gases such as oxygen and carbon dioxide and diminishes light intensity (Voesenek & Sasidharan, 2013). Due to climate change-induced floods, there was a need to improve agricultural yields by 70%. However, there remained significant challenges in feeding a population predicted to exceed 9 billion by 2050 (Godfray et al., 2010). Flooding-tolerant crops are urgently required as global climate change increases the threat of agricultural yield loss (Bailey-Serres et al., 2019). The idea of plant cross-breeding is to create new plants with distinctive traits by choosing desirable accessions. Through traditional breeding, Ramiah (1940) identified two duplicated genes named ef1 and ef2 (elongation factor) that influence internode elongation. The SSD approach with rapid generation advance (RGA) is the most suitable and is used routinely to increase breeding efficiency (Collard et al., 2017). As marker-assisted selection/backcross breeding is not sufficient to increase breeding efficiency and complement traditional breeding efforts, new germplasm must be exploited, new efficient and precise molecular markers developed, and effective phenotyping technologies developed. The recent developments in marker technologies and molecular biology have paved the way for fine mapping, QTL analysis, and subsequent cloning of the submergence tolerance gene. To build a more successful MAS breeding plan for the introgression of flood tolerance characteristics from landraces to commercial cultivars, new DNA diagnostic markers firmly connected to the traits of interest must be found. The poor predictive value of QTL markers for performance in plant breeding is attributable to several reasons, including high and low gene expression, interplay between genotype and environmental factors, and gene recombination (Oladosu et al., 2017). Marker-assisted backcrossing has the major advantage of retaining all of the good characteristics, including yield and quality, of the recurrent parent in SUB1 cultivars. The AGP QTL qAG-9-2 was introduced into the *SUB1* background cultivar Ciherang-Sub1 to boost GSOD tolerance without compromising submergence tolerance (Toledo et al., 2015). Kim and Reinke (2018) found that combining three QTLs (*qAG1b* + *qAG1a* + *qAG8*) increased rice underwater establishment compared to connecting two QTLs. IRRI has introgressed Sub1 QTL into the high-yielding cultivar Swarna, which is currently adopted in several Indian states (Dar et al., 2017). A deep-water rice variety, C9285, has two ethylene response factor genes, *SNORKEL1* (*SK1*) and *SNORKEL2* (*SK2*), responsible for shoots elongation in this floating rice (Tamang & Fukao, 2015). The QTL *qAG9-2* (AG1) from Khao Hlan On was used to initiate anaerobic germination (AG) tolerance breeding in rice (Kato et al., 2019).

Root anatomical characteristics help plants adapt to drought and floods by balancing water and oxygen transport rates. The aerenchyma and radial oxygen loss (ROL) barrier improves oxygen transport in primary and lateral roots (Pedersen et al., 2021b). According to Pedersen et al. (2021a), thick roots lose less oxygen from the root surface than thin roots; that is why large roots are more adaptable to soil flooding. A higher cortex to stele ratio (CSR) provides more excellent room for cortical cell death (*i.e.*, aerenchyma development). It avoids anoxia in stelar cells during soil flooding (Yamauchi et al., 2019). Sorghum, maize, and barley may also be relevant for investigations into root anatomical features since they have previously undergone QTL mapping and GWAS (Kajiya-Kanegae et al., 2020). Many experts have suggested root system architecture (RSA) ideotypes for flood tolerance (Lynch, 2019). In this instance of flooding stress, surface rooting and adventitious roots may be practical RSAs to escape hypoxia (Pedersen et al., 2021a, b). RSA has been studied using QTL and genome-wide association analyses (Deja-Muylle et al., 2020). Many researchers have concentrated on these characteristics and conducted QTL studies to develop novel flood-tolerant rice (Singh et al., 2017). One of these Bulu rice types (Kitomi et al., 2020) has been identified as having a QTL associated with soil-surface roots (SOR), namely SOIL *SURFACE ROOTING 1* (*qSOR1*), which has been identified from one of these Bulu rice types (Kitomi et al., 2020). SOR phenotypes may help crops resist hypoxic stress regardless of crop species. Descriptive studies on the molecular activities of RSA-associated genes in maize and rice. A survey of DRL2 mutant lines revealed that DRL2 is important in root growth angle (Kitomi et al., 2020). These novel methods enable functional gene alteration by creating site-directed point mutations, deletions, or insertions (Mishra & Zhao, 2018). *SUB1A-1*, *OsTPP7*, *SK1/2*, and *SD1-DW* may be used in genome editing for better submergence tolerance. The omics techniques reveal that seeds germinating under submergence have higher transcription levels and protein activities of ethanol-producing enzymes such as pyruvate decarboxylase (PDC) and alcohol dehydrogenase (ADH) (Hsu & Tung, 2017).

4.3 Future-Proofing Plant with Heat Tolerance

The yield-risk trade-off is a severe issue for farmers in developed and developing countries (Kruseman et al., 2020). The projected warming by 2050 increases the frequency of adverse temperatures around flowering, causing yield loss (Lobell et al., 2015). Heat stress is causing enormous production losses in numerous crops, putting global food and nutritional security at risk. Furthermore, due to the increase in temperature, changes in plant phenology raise risks of direct high-temperature shock, such as pollen sterility (Nguyen et al., 2013), and accelerate crop growth (Zheng et al., 2016). Heat stress is a primary concern for plant physiologists and plant breeders since the pattern and extent of yield loss change considerably from year to year due to its complex inheritance. As a result, improvements in heat tolerance breeding are still needed to improve food security for hundreds of millions of rural poor people who rely on rain-fed agriculture.

The limited access to accurate phenotyping tools and, most importantly, significant G x E effects hamper advances in breeding for improved heat tolerance (Cossani & Reynolds, 2012). Nevertheless, it is possible to build a future-proof plant with heat tolerance using breeding methods that maximize germplasm resources and existing genetic diversity integrated with advanced forward and reverse genetic tools and phenomics platforms. Using these modern genomic approaches, this substantial phenotypic data might be used to breed new "climate-resilient" cultivars. Contrary to expectations, genetic diversity in transpiration efficiency (TE) correlated with variability in stomatal conductance under high-temperature conditions has enhanced the photosynthetic capacity (Geetika et al., 2019). Consistent selection and intermating in crop breeding have resulted in a lack of genetic variation, particularly for economically essential features that have been domesticated or selected. There appears to be a significant possibility of improving genetic gains for stress adaptable attributes among elite lines using a more strategic approach (Crespo-Herrera et al., 2017). Accelerating crop improvement needs considerable genetic research on both cultivated and wild species. In this context, heat-tolerant genes/QTLs and component characteristics must be comprehensively investigated throughout the whole gene pool, focusing on non-adapted and neglected crop wild relatives (CWRs) and landraces (Fernie et al., 2006). Also, the advanced Gayabyeo/N22 cross line gave HS tolerance and high yield (Manigbas et al., 2014). In 2050 climatic scenarios, it is predicted that crop cycle extension (delayed maturity) increased yields for maize (Parent et al., 2018) and wheat (Zheng et al., 2016).

Physiological trait-based breeding outperforms traditional breeding, so genetic manipulation of physiological attributes may be a possible way to incorporate genes/QTLs that influence complex abiotic stress tolerance in crop plants (Reynolds & Trethowan, 2007). Physiological traits relating to canopy structure, delayed senescence, photosynthetic efficiency, lower respiration rates, reproductive qualities, and harvest index should be used to develop heat stress cultivars (Gupta et al., 2012). Vivitha et al. (2017) identified QTLs were shown to be responsible for preserving membrane integrity and yield at high temperatures. In three-line hybrid rice breeding, heat-susceptible restorer lines with high yield are often used (Guan-fu et al., 2015). A newly detected primary dominant locus, OsHTAS (*Oryza sativa* heat tolerance at the seedling stage), provided high-temperature tolerance at 48 °C during seedling and grain filling phases (Wei et al., 2013). The stay-green trait has been shown to increase grain yield in hot weather, and it has been reported in several crops. Genomic selection is an emerging method for predicting the phenotype of lines based on the genomic estimated breeding value (GEBV) of genotypes using high-density markers and advanced phenotyping platforms. Using genomic selection to improve heat and drought tolerance in wheat has shown promising results, especially when paired with modern high-throughput phenotyping approaches (Rutkoski et al., 2016). The prediction accuracy for drought and heat stress yields was 0.56 and 0.62, respectively (Juliana et al., 2019).

4.4 Techniques to Address Salinity Stress

4.4.1 Effect of Salinity Stress

Salinity is one of the most difficult challenges that harm plant growth and development. Salt-affected soils are found all over the world, with Australia, Asia, and the Pacific accounting for more than half of the global salt-affected land (Hoang et al., 2016). These areas generally are low lying with shallow water table and are characterized by a low precipitation to evaporation ratio (≤ 0.75) (Brady & Weil, 2015). Due to osmotic, ionic, and oxidative stressors, salinity can drastically reduce crop development and output (Isayenkov & Maathuis, 2019). Salt stress can also disrupt cellular homeostasis, denature proteins and nucleic acids, increase reactive oxygen species (ROS), and cause lipid peroxidation. In higher plants, ROS accumulates owing to disruption, overflow, or even interruption of the electron transport chain in mitochondria and chloroplasts. Lipid peroxidation, reduced photosynthesis, increased photorespiration, and electrolyte leakage might be consequences of ROS production (El-Mashad & Mohamed, 2012; Alkharabsheh et al., 2021). Because of the defective ionic selectivity of the cell membranes, the accumulation of Na^+ at greater concentrations in plant tissues produces alterations in the Na^+ to K^+ ratio. As a result, it may limit the absorption of essential food components. High salinity substantially impacts plant development, resulting in lower germination rates, dry matter, photosynthetic pigments, nutrient intake, and, most importantly, lower crop yields. High salt concentrations induce irreversible damage to the photosynthetic machinery, altering the structure of the chloroplasts, degrading the chloroplast membrane, and triggering chloroplast protrusions at any stage of plant growth (Ha-Tran et al., 2021). Plants should use adaptive strategies such as lowering Na^+ ion absorption and maintaining internal osmotic balance, increasing osmolyte storage, apoplastic acidification, hormonal control, and scavenging ROS antioxidant defense system to prevent these negative impacts (Negrão et al., 2017).

4.4.2 Modern Strategies for Enhancing Salinity Stress Tolerance

Traditional Agronomic Practices

The only approach to deal with the salinity problem in the past was to focus on replacing the shortfalls by adding nutrients and water rather than eliminating the accumulated salts. In earlier days, the only approach to deal with the salinity problem was to replace the shortfalls by adding nutrients and water rather than eliminating the accumulated salts (Flowers & Yeo, 1989). However, given that higher levels of salt or alkalinity in the soil can limit or completely preclude plant development, these soils must be recovered to make them productive. Long-established practices for achieving this goal include the leaching of soluble salts below the root zone (to control osmotic effects on plant growth), the provision of adequate drainage facilities, the replacement of Na^+ ions from the exchange site of the soil-exchange complex, and the availability of good-quality water for successful reclamation. Apart from these, there are some successful interventions such as physical (deep plowing, sub-soiling, sanding, profile inversion, and scrapping), hydro-technical (leaching salts from root zone), and chemical amelioration methods (reduce high pH, exchangeable Na^+).

Modern Techniques

Modification of conventional procedures and more sophisticated and long-term viable ways are all part of modern management strategies. These solutions employ an integrated approach to address the salinity problem, which was not adequately discussed in the past. Some of the promising methods are as follows:

(i) ***Bio-organic amendments***

The term "bio-organic amendment" refers to the combined use of beneficial microorganisms and organic sources of nutrients in crop cultivation for higher yield. Through increased soil organic matter, vital nutrients, water availability, stable soil structure, and enhanced microbial activity, bio-organic amendments can significantly boost both soil and crop production. The beneficial microbes may include plant growth-promoting rhizobacteria (PGPR), arbuscular mycorrhizal fungi (AMF), and other nutrient solubilizing or fixing microbes. Generally, plant root-colonizing microorganisms (e.g., fungi and bacteria) form symbiotic associations with plants to confer tolerance under different stress conditions, such as salinity. This is achieved by improving the root system development for an increased uptake of water, essential nutrients, and other organic compounds to counteract the negative impact of Na^+ in the rhizosphere. The dual inoculation of AMF and PGPR has been demonstrated to promote salt tolerance in maize by increasing nutrient accumulation, AMF colonization, and leaf proline production (Krishnamoorthy et al., 2016). Mycorrhizal association significantly improves nutrient and water uptake, helps to maintain ionic homeostasis, prevents oxidative damage of cell constituents by superoxide radicals and regulates plant hormones thus minimizing salt stress in plants (Kapoor et al., 2019). Salinity stimulates the buildup of glomalin in mycorrhizal plants which gives them an advantage over non-mycorrhizal plants. Glomalin, a heat shock protein 60 (HSP60) homolog is thought to reduce cytosolic damage caused by Na^+ mediated protein misfolding. (Gadkar & Rillig, 2006).

Under salt stress, plants associated with mycorrhiza have been shown to possess higher osmotic potential than their non-mycorrhizal counterparts due to the accumulation of more osmolytes, thus imparting tolerance to higher salt concentration (Navarro et al., 2014). The application of salt-tolerant PGPR has shown remarkable success in enhancing the agricultural productivity of saline soils through various mechanisms depicted in Fig. 3.4 (Numan et al., 2018; Kumar et al., 2021). Under salt stress, PGPR can control the expression of aquaporins, which improves the plant-water uptake. An increased salt tolerance was seen in barley in association with *Azospirillum brasilense* caused by expression of HvPIP2;1 transcript in barley roots (Zawoznik et al., 2011). Inoculation of maize with *Pantoea agglomera* increased PIP2;1 gene expression and promoted plant development in saline conditions (Gond et al., 2015). Besides osmotic stress, excessive intracellular Na^+ and Cl^- concentrations can cause ion toxicity and disrupt several cellular activities. Plants employ salt overly sensitive (SOS) signaling pathways to maintain ion homeostasis and minimize Na^+ influx. To reduce ion toxicity, PGPR can also upregulate the genes of the SOS pathway (Mishra et al., 2021). Ethylene plays a significant role in plant growth and development. PGPR controls ethylene levels in plants via ACC deaminase, which converts the ethylene precursor ACC to ammonia and α-ketobutyrate, facilitating plant development and conferring stress tolerance (Kumar et al., 2020). The ability of halophilic microorganisms to preserve protein structure and enzyme function for diverse metabolic pathways even at high salt concentrations is a significant adaptation (Ruppel et al., 2013).

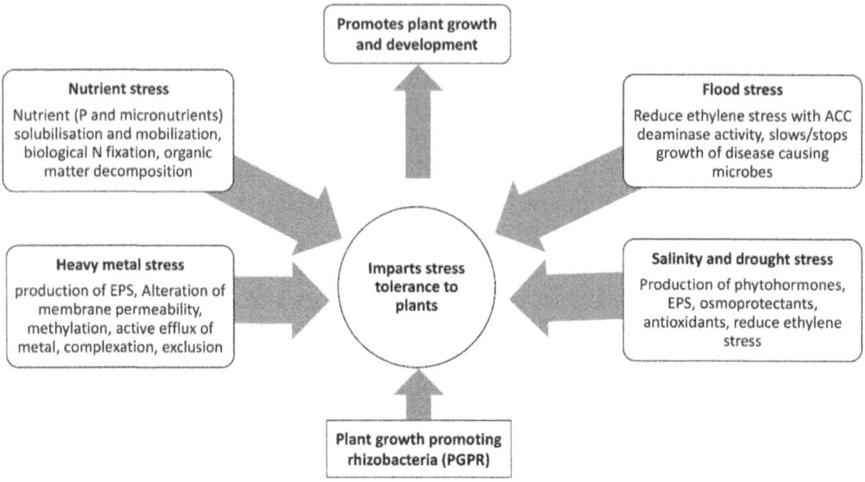

Fig. 3.4 Role of plant growth-promoting rhizobacteria in providing stress tolerance in plants

(ii) ***Exogenous phytohormone application and seed priming***

Phytohormones are the most significant endogenous molecules involved in influencing physiological responses, which eventually lead to adaptation to an adverse environment. Several phytohormones, including abscisic acid (ABA), gibberellins (GA), ethylene, auxins (IAA), cytokinins (CKs), and brassinosteroids, regulate various physiological and biochemical processes, are active and may aid in predicting plant tolerance or susceptibility processes. Exogenous administration of phytohormones has been recommended as a rational way of dealing with salt stress and has been implicated in several studies with varying degrees of effectiveness in mitigating the adverse effects of salinity (Iqbal et al., 2012; Sharma et al., 2013; Iqbal & Ashraf, 2013; Amjad et al., 2014). Exogenous application of ABA at 100 M to Indica rice seedlings increased survival by 20% and caused proline buildup via expression of the OsP5CS1 gene in rice (Sripinyowanich et al., 2013). Seed priming with several auxin sources (IAA, IBA, and tryptophane) substantially reduced the deleterious effects of salt stress on endogenous ABA levels in a salt-intolerant wheat cultivar (Iqbal & Ashraf, 2013). Hormo-priming is the process of treating seeds with plant growth hormones before planting (Sahab et al., 2021). Cytokinin has also shown to reduce the negative effects of salt on plant development. Plant resistance to salt stress was improved by seed priming with CKs (Iqbal et al., 2006). Reduced ABA content in plants emerging from kinetin primed seeds was probably responsible for wheat salt stress relief. Seed priming is a new approach that effectively addresses salt stress, enhances germination and seedling development. Several types of priming *viz.* hydro-priming, halo-priming, osmo-priming, thermo-priming, bio-priming, and solid matrix priming can be adopted to enhance seed germination and seedling growth under adverse environmental stress conditions, in this case, salinity (Banerjee & Roychoudhury, 2018; Farooq et al., 2019; Sahab et al., 2021).

(iii) ***Nanoparticles***

Nanoparticles (NPs) have a wide range of significant implications on plants' morphological, physiological, and biochemical features. The function of NP is determined by their physicochemical qualities, concentration, and application technique. Several studies have found that NPs help plants cope with salt stress (Karimi et al., 2020; Abdoli et al., 2020; Baz et al., 2020; Nejatzadeh, 2021). NPs are applied at various doses by different modes such as foliar spray, soil application, priming, and solution culture. Sen et al. (2020) found that priming *Vigna radiata* with chitosan NPs reduced H_2O_2 and MDA levels compared to controls, resulting in improved growth, chlorophyll content, and metabolism. In addition, plants with NPs have increased K^+ absorption, K^+/Na^+ ratio, antioxidant enzyme activity, trapped energy flow and electron transport flux, sucrose biosynthesis, enhanced morphological characteristics, proline, and phenolic content (Abdoli et al., 2020; Faizan et al., 2021; Shah et al., 2021).

4.5 *Techniques to Address Heavy Metal Stress*

4.5.1 Sources of Heavy Metal Pollution

Heavy metals are any metallic chemical element characterized by high atomic mass and density. In a broad sense, they must have a density greater than 5 g cc^{-1} and an atomic number (Z) > 20 or atomic mass (A) > 23 (Koller & Saleh, 2018). This criteria of delineation of elements in a heavy metal category are prone to error, as a result, is not worth reliance. Poring over the periodic table indicates the inclusion of many non-metals if the above criteria are considered. Heavy metals also include metalloid elements (e.g., arsenic), often adjudged to cause ill effects on human health even under low concentrations. The menace of heavy metal pollution is increasing due to rampant industrial activities and mining (Jarup, 2003; Duruibe et al., 2007). Because of their non-biodegradability, they pose a more significant long-term threat to human health and environmental quality. Negligence shown by the industries in the disposal of heavy metal(loid) loaded waste is an important cause of pollution of the agriculture farms and water bodies situated nearby. The most common heavy metal contaminants are arsenic (As), cadmium (Cd), chromium (Cr), copper (Cu), mercury (Hg), lead (Pb), and nickel (Ni). These have severe implications for human health if present above certain critical limits (Table 3.1). Sources of heavy metal(loid) contamination can be classified into two broad categories involving natural and anthropogenic sources (Sodango et al., 2018). In the natural sources, ultramafic igneous rock is the primary source of heavy metal in soil, although other commonly found igneous and sedimentary rocks also contribute substantially (Garrett, 2000). Sedimentary rocks covering a significant portion of the top 5 km of the earth crust play a substantial role in deciding the inherent supplying capacity of heavy metals to soil (Alloway, 2013). Heavy metals can be found in various forms, viz. hydroxides, oxides, sulfides, sulfates,

Table 3.1 Sources and harmful effects of some significant heavy metals on human health

Heavy metal	Sources	Harmful effects	Permissible value (mg L^{-1})[a]	References
Arsenic	Mining, smelting, and coal-fire power plants, arsenical agrochemicals	Skin, kidney, liver, and lung cancer, fetal loss	0.02	Kapaj et al. (2006), Garelick et al. (2009)
Cadmium	By-product of zinc production, welding, batteries	Renal damage, bone fracture, endocrine disruptor, nephrosis, cancer	0.06	Bernard (2008), Godt et al. (2006)
Chromium	Metal plating tanneries, fertilizers, sewage, metallurgy	Mutagenic, hair loss, carcinogenic, reproductive, genotoxic	0.05	Ghani and Ghani (2011), Wilbur (2000)
Copper	Textile processing, paper milling, steel mining, and coal conversion	Brain, liver, and kidney damage, Wilson's Disease, stomach and demyelination	0.1	Wuana and Okieimen (2011), Zahra and Kalim (2017)
Mercury	Forest fires, discharge from hydroelectric, mining, pulp, and paper industries	Anemia, autoimmune diseases, mononucleosis, Hodgkin's disease, memory loss, cardiomyopathy	0.01	Rice et al. (2014)
Nickel	Electroplating,	Nickel itch, inhalation cause lungs cancer, throat cancers, fibrosis, pneumonia, hyperplasia of pulmonary cells	–	Zahra and Kalim (2017)
Lead	Mining, nervous dysfunction, gastrointestinal problems	Reduced intelligence, short term memory loss, learning disabilities	0.1	Singh et al. (2011)

[a]Singh et al. (2011)

phosphates, and silicates. These can be added to the soil by different processes, including volcanic eruptions, sea-salt sprays, forest fires, rock weathering, biogenic sources, and wind-borne soil particles (Alloway, 2012; Bu et al., 2016; Gautam et al., 2016).

4.5.2 Phytoremediation of Heavy Metals

Phytoremediation is based on the use of natural or genetically modified (transgenic) plants ability to extract harmful substances from the surrounding environment, which including heavy metals, pesticides, radionuclides, polychlorinated biphenyls, and polynuclear aromatic hydrocarbons, and transforming them into relatively

Fig. 3.5 Mechanisms of phytoremediation of inorganic pollutant elements (heavy metals) in soil-plant-atmosphere continuum

secure substances. The techniques of Phytoremediation includes phytoextraction (or phyto-accumulation), phyto-volatilization, phyto-stabilization, phyto-filtration, and phyto-degradation (Fig. 3.5; Alkorta et al., 2004). Phytoextraction is the process by which detrimental byproducts from the soil and water are overtaken by roots system and translocated; and finally accrued in the aboveground plant parts, like shoots. Phytoextraction is also otherwise known as phytoaccumulation or phytoabsorption, or phytosequestration (Rafati et al., 2011). This entails growing plants chosen for their ability to concentrate one or more heavy metals on contaminated soil and the plants are then harvested and incinerated, and the ash is either restricted or heavy metals are removed from it. As the harvesting of root biomass is difficult, metal translocation to shoots is an significant biochemical process that is desirable in effective phytoextraction (Zacchini et al., 2009). When a plant takes up heavy metals against the concentration gradient from the soil solution to the cell cytoplasm, it is regarded as a hyperaccumulator. This has the capacity to accumulate very high metal concentrations in the plant tissues without interfering with general growth and metabolic processes of the plant body. The phenomenon is considered as an evolutionary selection process which safeguards the plants from the outbreak of herbivores and pathogens. The efficiency of phytoextraction can be quantified by calculating bioconcentration factor and translocation factor. Figure 3.6 shows the characters that are prerequisite of being a hyperaccumulator plant.

Several hyperaccumulator plants play role in soil remediation but they have low biomass production. Trees with strong rate of transpiration, deeper root organizations, and rapid growth and more productivity, have been suggested as potential phytoextractors (Pulford & Watson, 2003). The heavy metals like Cd and Zn are extracted extensively by the plant species; *Salix* spp. and *Populus* spp. from the soberly contaminated soil (Pulford & Watson, 2003). The cultivation of fast growing plants (woody trees) near polluted industrial areas are advantageous for phytoextraction as well as provide renewable energy (Pulford & Watson, 2003; Tlustoš

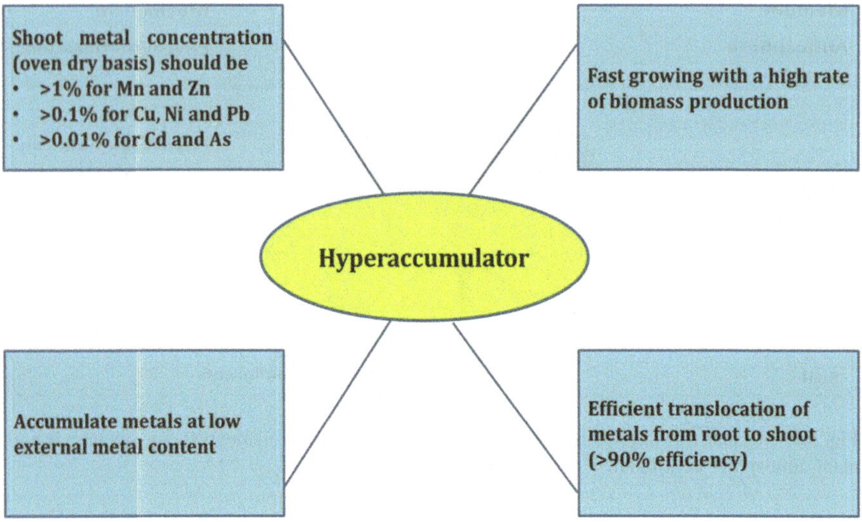

Fig. 3.6 Parameters for being a hyperaccumulating plant

et al., 2006). Lists of major hyper accumulators of some significant heavy metals are presented in Table 3.2. The process of Phytofiltration is considered as the exclusion of water contaminants from contaminated surface waters or waste waters by the use of plants (Sangeeta & Maiti, 2010). Phytofiltration may be characterized as; blasto-filtration (use of seedlings), or rhizofiltration (use of plant roots) or caulofiltration (use of excised plant shoots; Latin caulis = shoot). In phytofiltration, the movement of pollutants to underground water is reduced by the process of adsoption or absoption by plant roots. Artificially produced marshes planted with plant species are capable for absorbing or adsorbing the heavy metals. The usage of certain plants to stabilize pollutants in contaminated soils is called Phyto-immobilization or Phytostabilization (Singh, 2012). Plants have the capacity to immobilize the heavy metals in soils through sorption by roots, or precipitation and complexation. Consequently the toxicity levels of different heavy metals vary according to their valences and plant can immobilize the metals by reducing the valences in the rhizosphere zone of the plant. For an example, the reduction of Cr(VI) to Cr(III) is extensively studied where the latter is less mobile and less toxic (Wu et al., 2010). Plants can secrete special redox enzymes to transfigure the dangerous heavy metals to relatively less toxic to overcome the stress in a skillful manner. Though, phytostabilization is not a long lasting solution because it only limits the movement of the heavy metals but not totally eradicate from the soil. This only is a managing approach for steadying (inactivating) the potential toxic contaminants from soil and water (Vangronsveld et al., 2009). The processes involved in phytostabilization have been illustrated in the Fig. 3.7. These processes can be broadly categorized into (i) prevention of soil movement mediated contaminant transport (ii) immobilization of

Table 3.2 List of some efficient hyperaccumulators for some significant heavy metals

Metal	Hyperaccumulator	Reported concentration (mg kg^{-1})	References
Zinc	*Thlaspi caerulescens*	52,000	Brown et al. (1994)
	Streptanthus polygaloides	6000	Boyd and Davis (2001)
	Eleocharis acicularis	11200	Sakakibara et al. (2011)
Copper	*Ipomoea alpine Eleocharis acicularis*	12,300 20,200	Baker and Walker (1990), Sakakibara et al. (2011)
Manganese	*Schima superba*	62,412	Yang et al. (2008)
Cadmium	*Thlaspi caerulescens*	1800	Baker and Walker (1990)
	Azolla pinnata	740	Rai (2008)
	Eleocharis acicularis	239	Sakakibara et al. (2011)
Lead	*Thlaspi rotundifolium*	8200	Baker and Walker (1990)
	Sonchus arvensis	3664	Surat et al. (2008)
Nickel	*Alyssum lesbiacum*	47,500	Küpper et al. (2001)
	Alyssum bracteatum	2300	Ghaderian et al. (2007)
	Alyssum corsicum	18,100	Li et al. (2003)
	Alyssum markgrafii	19,100	Bani et al. (2010)
	Alyssum murale	4730–20,100	Bani et al. (2010)
	Alyssum serpyllifolium	10,000	Prasad (2005)

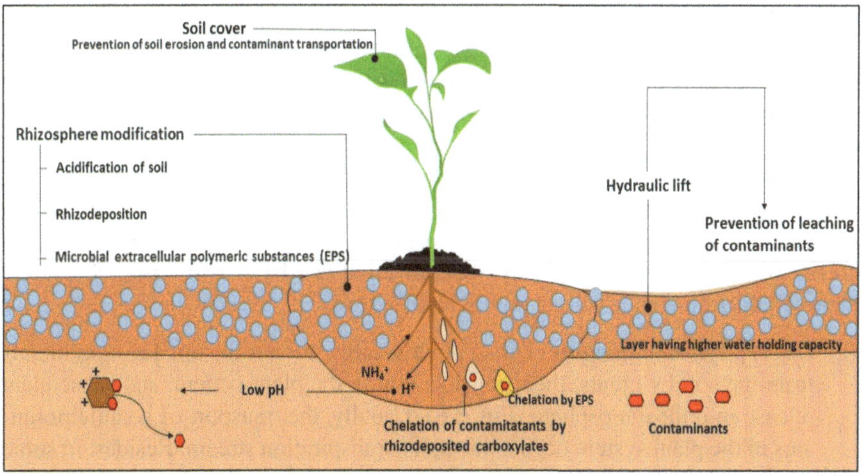

Fig. 3.7 Mechanistic illustration of phytostabilization of inorganic contaminants in plants

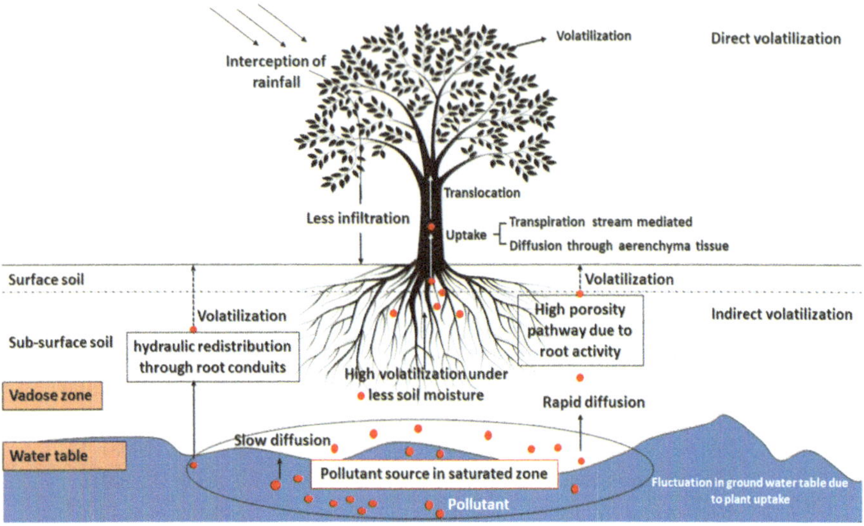

Fig. 3.8 Depiction of phytovolatilization process in remediating heavy metals from soil and water. (Note: solid red dot indicate inorganic pollutant elements)

contaminants through rhizospheric modification (iii) prevention of contaminant leaching to the groundwater.

Phytovolatilization is the process by which plants absorb pollutants from the soil, convert them to a volatile state, and then release them into the atmosphere. This method may be used to remove organic pollutants as well as heavy metals such as Hg and Se. However, its use is restricted since it does not remove the pollutant; rather, it transfers it from one matrix (soil) to another (atmosphere), where it might be redeposited, thus making it one of the most contentious phytoremediation processes (Padmavathiamma & Li, 2007). Phytodegradation, on the other hand, is the breakdown of organic pollutants by plants using enzymes such as dehalogenase and oxygenase and is not dependent on rhizospheric microbes (Vishnoi & Srivastava, 2007). Based on the pathway, phytovolatilization can be of two types, *viz.*, direct phytovolatilization and indirect phytovolatilization.

(a) ***Direct phytovolatilization***: Direct phytovolatilization is the uptake of pollutants from the soil by plants, their translocation in the plant system, and subsequent release into the atmosphere (Fig. 3.8). Usually, the transport of volatile pollutants in the plant system occurs through a transpiration stream. Besides, in some wetland plants, volatile contaminants may diffuse through the aerenchyma. The porous aerenchyma channel connecting roots to shoot is developed after the death of cortex parenchyma cells due to lack of oxygen under wetland. Root pressure also plays a vital role in transporting volatile pollutants through convective gas flow in the plant system. The uptaken pollutants are usually released through stomata and lenticels, sometimes through the epidermis (hydrophobic contaminants).

(b) ***Indirect phytovolatilization***: In indirect phytovolatilization, plant root-modified rhizosphere provides a conducive environment for the volatilization of pollutants (Fig. 3.5). The depletion of soil water due to plant use lowers the groundwater level and increases soil aeration and thickness of the vadose zone (Wang et al., 2017). Furthermore, it has been found that the diffusion of volatile compounds is faster in the air than water (LaPack et al., 1990). Therefore, the flexible boundary (modified by rhizosphere) between the vadose zone and groundwater table decides the volatilization of soil pollutants.

Plants are regarded as 'Green Liver' for the biosphere due to their ability to metabolically detoxify the organic xenobiotics after acquiring from contaminated environments. However, phytodegradation is limited to removal of organic pollutants only because heavy metals are non-biodegradable. Scientists have lately expressed an interest in the study of phytodegradation of numerous organic contaminants, such as synthetic herbicides and insecticides. Some studies have even suggested the use of genetically engineered plants (e.g., transgenic poplars) for this purpose (Doty et al., 2000).

4.5.3 Microbial Bioremediation of Heavy Metals

Microorganisms adopt different mechanisms to interact and survive in the presence of a xenobiotic substance, which has the potential of causing harm to environmental health and functioning. As far as degradation of organic pollutants is concerned, these act as a carbon source for microbes consumed either in the presence or absence of oxygen-producing less or non-toxic substrate products. Heavy metals being the primary group of inorganic contaminants, are remediated by microbes harboring a series of mechanisms, including oxidation, methylation, enzymatically, metal-organic complexation, metal-ligand degradation, metal efflux pumps, intracellular and extracellular metal sequestration, and metal exclusion (Ramasamy & Banu, 2007). Tolerant microbes show potential to survive at high heavy metal(loid) contents by rendering survival mechanisms such as biotransformation, extrusion, use of enzymes, production of exopolysaccharide (EPS), and synthesis of metallothioneins (Wu et al., 2010). Microbe transformation of metal(loid)s could be broadly categorized into two types, redox conversions (oxidation and reduction) to less toxic state and conversions from inorganic to organic form and vice versa. In addition, oxidation of metal(loid)s could help gain energy. On the contrary, the reduction can occur through dissimilatory reduction where microorganisms utilize metal(loid)s as a terminal electron acceptor for anaerobic respiration. Therefore, the reduction process mainly could be methylation and demethylation. Microbial methylation is crucial because volatile products formed during the process lead to complete phase change of inorganic contaminants from solid (soil) to gas (atmosphere), resulting in complete amelioration.

The use of PGPR and biochar, along with a hyperaccumulator plant, have shown a significant increase in the Cd content and bioaccumulation factor of an accumulator, about 412% and 403% higher than that of control, respectively (Wu et al., 2019). This system also increased the fresh and dry biomass of the metal accumulator plant by 227.2% and 178.3%, respectively. Khan and Bano (2016) identified catalase and oxidase enzymes secreted from the PGPR isolated from municipal wastewater that solubilizes insoluble bound phosphate and exhibits antifungal and antibacterial activities. Furthermore, endophytes reside in the living plants sharing a mutualistic relationship without causing apparent negative symptoms of infection and showing quantitative gene expression of xenobiotic catabolic genes and genetic engineering of the catabolic pathway, further enhancing the efficacy of bioremediation by plants (Huo et al., 2012). Some mercury (Hg) resistant endophytic bacteria have also shown to express the MerB gene which encodes organomercuriallyase enzyme that breaks organomercurials into mercuric ion (Hg^{2+}) (Brown et al., 2003). Luo et al. (2011) found that *Serratia sp.* LRE07 could remove over 65% of Cd and 35% of Zn from metal solutions, thus alleviating the phytotoxic effects of the metals.

Genetically engineered microbes (GEMs) can improve the overall efficiency and efficacy of bioremediation techniques. Microbes can be genetically modified to have desirable traits such as the capacity to endure metal stress, the overexpression of metal-binding proteins, and the ability to accumulate metal. There are many reports of using many transgenic bacteria to treat inorganic contaminated wastes (Chen & Wilson, 1997; Ruiz et al., 2011; Dash et al., 2014). A genetically engineered organism permits the transfer of certain functional genes into the genome of a specific organism (Tozzini, 2000). Thus genes having desirable traits *viz.* volatilization, absorption, degradation can be incorporated in the engineered microbe to make it suitable for use in contaminated sites for remediation purposes. Different researches have noted successful exploitation of micro-organisms potential in different environments. For example, Frederick et al. (2013) engineered microorganisms to produce trehalose and established a successful reduction of Cr (VI) to Cr (III). A similar study was also conducted by Srivastava et al. (2010) where they used *Alcaligenes eutrophus* AE104 for remval of Cr from wastewater. Genetically modified *Chlamydomonas reinhardtii* showed significant tolerance to Cd toxicity (Ibuot et al., 2017). Engineered *Corynebacterium glutamicum* showed overexpression of ars operons (ars 1 and ars 2) resulting in successful remediation of As contaminated soil (Mateos et al., 2017). *Rhodopseudomonas palustris* was developed by recombinant technology for Hg^{2+} removal from Hg contaminated water by overexpression of metallothionein producing gene (Deng & Jia, 2011).

4.5.4 Nanoparticles

Nanoparticles (NPs) can be synthesized by applying chemical methods, such as the hydrothermal process, sol-gel method, gas phase method, thermolysis, and hydrolysis. In addition, nanoparticles are also produced through UV irradiation, aerosol techniques, laser ablation, and photochemical reduction. These processes are costly and result in harmful by-products. It is also challenging to control the nanoparticles' surface chemistry, size, and structure through these methods. As a result, green synthesis, a bottom-up approach, has recently become a popular procedure for NP synthesis, in which extracts from natural plant parts such as leaves, fruit juice, or microbial products, are used as reducing agents or stabilisers, rather than costly chemical reagents, to synthesise metal or metal oxide NPs. The biogenic reduction of metal precursors to the corresponding NPs is eco-friendly, sustainable, chemical-free, less expensive, and compliant with large-scale processing (Chandra et al., 2020). Nano zero-valent iron (nZVI) particles are synthesized using plant-derived anti-oxidant extracts, which are used to reduce ferric or ferrous iron to Fe(0) state, making it more reactive towards pollutants. Green synthesized Fe nanoparticles have been used to remove many contaminants. For example, nanoscale zero-valent iron/Cu was prepared by the green synthesis method and was used to remediate Cr(VI) contaminated groundwater quite effectively at pH 5 with an efficiency of 94.7% (Zhu et al., 2018).

4.6 *Techniques to Address Nutrient Stress*

Due to improper management and intensive cultivations, soil degradation has posed severe challenges towards sustainable crop yield worldwide. The rate of yield increase per unit addition of fertilizer input has been decreasing faster after the advent of the Green Revolution. Therefore, with efficient fertilizers input, superior nutrient efficient genetic materials are vital to achieving higher yields and high-quality food materials at a notably lower expense of fertilizers input. Nutrient use efficiency (NUE) distinguishes plant species based on their capacity to absorb nutrients for maximum dry matter production. The efficiency of nutrient uptake and utilization for producing yield depends on various factors, broadly classified into the plant, soil, fertilizer, agronomic management, and biotic stress factor (Baligar & Fageria, 2015). Nutrient interactions are crucial because a lack of one inhibits the absorption and utilization. Therefore, application of nutrients with proper rate, place and time is considered as a best management practice (BMP) and this is essential for attaining optimal nutrient use efficiency. Some of the advanced conservation practices such as minimum or zero tillage, laser land leveling, residue cover (permanent or semi-permanent) to reduce moisture and nutrient loss, sowing of nutrient use efficient varieties, FIRB (furrow irrigated raised bed) method, system of rice intensification (SRI), direct-seeded rice (DSR), dry-DSR, precision farming by using drone, leaf color chart (LCC) based fertilizer application, GreenSeeker, and chlorophyll meter, etc. have been shown to increase NUE as well as the yield of the crop (Dass et al., 2015). By following laser land leveling, it was observed that the nitrogen-use efficiency was increased upto 6–7%, reduced the cost of production, and productivity enhanced (Jat et al., 2009). Islam et al. (2007), recorded that the in wheat, application of 30 kg N ha^{-1} each time with an LCC score of 4 with a total 120 kg N application per hectare has resulted higher partial factor productivity of N

(PFP_N), N uptake, and NUE than consuming the same extent of N in three fixed time splits.

4.6.1 Strategies for Increasing Fertilizer Nutrient Use Efficiency

Nutrient use efficiency (NUE) can be enhanced by reducing fertilizer nutrient losses while prolonging the duration of nutrient availability from fertilizers. For this, fertilizer source, additives to retard the release of nutrients to maintain synchrony with the plant nutrient demand, and fertilizer application method are crucial. Single- or multi-nutrient slow-release fertilisers (SRF) and controlled release fertilisers (CRF) have additional benefits in improving plant nutrient recovery efficiency by lowering the nutrient release rate, thereby limiting volatilization (N_2O, NH_3) and leaching losses (NO_3^-−N, K), reducing P fixation, and providing constant nutrient availability throughout the plant growing season (Baligar & Fageria, 2015). The most effective strategy for correcting a particular nutritional disorder and improving plant growth, and NUE is the timely delivery of nutrients. The quantity of phosphorus (P) fertilization is critical for increasing PUE. Proper P fertilizer application management is critical for optimizing agricultural productivity and protecting water quality. Banding (band placement method), increasing organic matter through manure applications, conservation tillage measures, and delivering fertilizers as near to the peak crop nutrient demand are all examples of management strategies that improve P availability (Datta et al., 2015). Soil factors such as the nature of soil clays, organic matter content, soil pH, moisture, temperature, and interaction with other plant nutrients in soil influence the availability, transformation, and fixation (sorption) of cationic and anionic micronutrients (Srivastava & Gupta, 1996). Inorganic nutrient formulations (organo-complexes, nutrient-enriched organic wastes from industries), chelated (EDTA, HEDTA, EDDHA), or polymer-coated form with slower nutrient release rates over the conventional more soluble fertilizer sources, making it less prone to being fixation or leaching (Srivastava et al., 2016). Microorganisms thrive in soil due to organic matter, which serves as a source of carbon and energy. They are closely associated with the nutrient cycling process and play a critical role in releasing plant nutrients hooked with organic matter. In order to sustain agriculture production and ensure food security operation of microbes in close association with chemical and organic nutrient sources is one of the economically feasible and environmentally sound methods. Microorganisms that are beneficial for plants' growth and accountable for high-yield production, bio-fertilizers as they are termed, are harmless and could be substituted to minimize the use of chemical fertilizers. They enhance NUE and reduce environmental pollution. Conjoint use of organic manure, chemical fertilizer, and bio-fertilizer has beneficial effect on crop plants. According to the findings of Nguyen et al. (2021), N-fertilizer was lowered with bio-fertilizer, and a substantial improvement in grain production was seen, which may result in significant advantages to producers in the form of lower input costs and greater grain yield. Biofertilizers may increase the number of beneficial bacteria that can make nutrients available to plants (Vessey, 2003). Organic and bio-fertilizers impact crop development and output and may reduce N inorganic fertilizer input (Son et al., 2001). Bio-fertilizer application in various crops could be minimized to about 25–40% N and 15–30% P_2O_5, respectively (Naher et al., 2016; Panhwar et al., 2014).

5 Conclusion and Way Forward

Climate change is a global phenomenon characterized by changes in climate composition or elements (temperature, precipitation, and wind), and it is caused by changes in each of the climate components, which include the atmosphere, hydrosphere, biosphere, cryosphere, and lithosphere, or by the complex interaction of all of these components. There are two significant cause for climate change like; (1) natural causes (changes in solar activity, volcanic eruptions, seawater temperature); and (2) artificial causes (deforestation, carbon dioxide emissions from industrial and agricultural field activity, acid rain, and depletion of ozone layer). Agriculture and climate change strongly correlate, implying that climate change harms global agricultural production. Agricultural production faces a variety of challenges, particularly in the world's semi-arid regions, including moisture stress, flood and submergence, salinity, heat stress, heavy metal stress, and increased disease and insect-pest infestation. All of these production constraints directly impact yield stability, leading to global food insecurity. Genetic diversity plays a critical role in the survival of a species by providing adaptation mechanisms to biotic and abiotic environmental stresses and enabling changes in genetic composition to cope with environmental changes. The ability of a plant or crop to survive and recover from climate change is referred to as climate resilience. Climate-smart breeding considers not only climatic requirements but also producer and end-user preferences. Climate-smart crop varieties are able to produce an excellent return even under unfavorable environment marked by altered unfavorable environment. Thus, food production must be improved to feed the increasing population of the world despite the anticipated decline in arable land and unpredictable rainfall and temperature. As the arable land is shrinking day by day, this inviting more and more modern agricultural practices to increase the level of productivity which is required to feed approximately 10 billion people by 2050. In human history, plant breeding always played a vital role starting from crop evolution to eradicate the hunger by green revolution with the development of genetically superior crop varieties. But looking to the future demands, recent advances in genomics, combined with high-throughput genotyping and precision phenotyping are need to join hands with plant breeding strategies to identify the genes controlling critical agronomic traits. The advancement of precise plant phenotyping and genotyping provides tremendous opportunities to develop crop varieties that are better suited to changing environment. This will aid in boosting plant breeding activities for developing climate-resilient high-yielding and hybrid varieties. The novel gene discovery strategy can now be combined with genome editing techniques to rapidly build climate-resilient crops, such as plants with improved biotic and abiotic stress tolerance and high nutritional

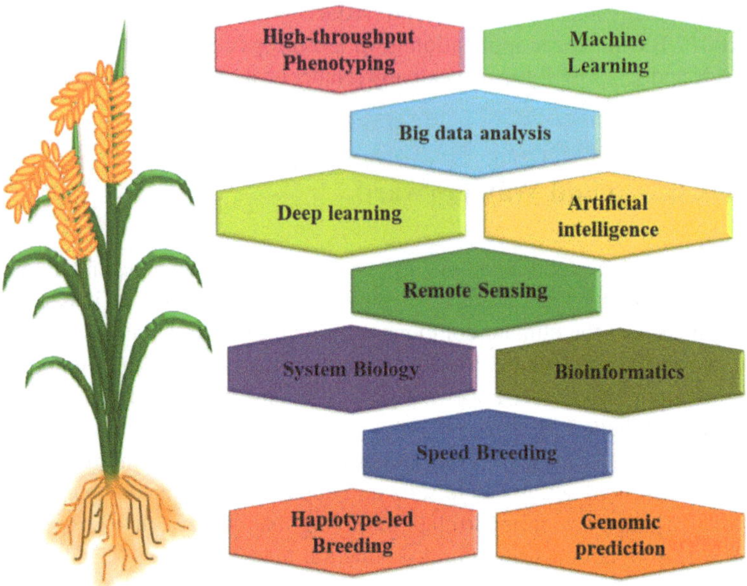

Fig. 3.9 The future crop improvement strategies need precision molecular breeding and biotechnological tools

value. To prove our future crops against climate change, we must have to follow the precision molecular approaches and novel agricultural practices such as machine learning, big data analysis, deep learning, artificial intelligence, remote sensing, system biology study, genomic prediction, speed breeding, and haplotype breeding (Fig. 3.9). To overcome food security issues, plant breeding approaches should be aided and integrated with advanced biotechnological tools, plant physiology, plant pathology, entomology, soil science, and agronomical developments to accelerate crop genetic improvements to avoid the consequence of ongoing and projected climate change.

References

Abdelrahman, M., El-Sayed, M., Jogaiah, S., Burritt, D. J., & Tran, L. S. P. (2017). The "STAY-GREEN" trait and phytohormone signaling networks in plants under heat stress. *Plant Cell Reports, 36*(7), 1009–1025.

Abdoli, S., Ghassemi-Golezani, K., & Alizadeh-Salteh, S. (2020). Responses of ajowan (Trachyspermum ammi L.) to exogenous salicylic acid and iron oxide nanoparticles under salt stress. *Environmental Science and Pollution Research, 27*(29), 36939–36953.

Ali, J., Mahender, A., Prahalada, G. D., Sevilla, M. A. L., Galang, A., De Asis, E. J., et al. (2020). Genomics-assisted breeding of climate-smart inbred and hybrid rice varieties. In *Genomic designing of climate-smart cereal crops* (pp. 1–43). Springer.

Alkharabsheh, H. M., Seleiman, M. F., Hewedy, O. A., Battaglia, M. L., Jalal, R. S., Alhammad, B. A., et al. (2021). Field crop responses and management strategies to mitigate soil salinity in modern agriculture: A review. *Agronomy, 11*(11), 2299.

Alkorta, I., Hernández-Allica, J., Becerril, J. M., Amezaga, I., Albizu, I., Onaindia, M., & Garbisu, C. (2004). Chelate-enhanced phytoremediation of soils polluted with heavy metals. *Reviews in Environmental Science and Biotechnology, 3*(1), 55–70.

Alloway, B. J. (2012). *Heavy metals in soils: trace metals and metalloids in soils and their bioavailability* (Vol. 22). Springer Science & Business Media.

Alloway, B. J. (2013). Sources of heavy metals and metalloids in soils. In *Heavy metals in soils* (pp. 11–50). Springer.

Amjad, M., Akhtar, J., Haq, M. A. U., Imran, S., & Jacobsen, S. E. (2014). Soil and foliar application of potassium enhances fruit yield and quality of tomato under salinity. *Turkish Journal of Biology, 38*(2), 208–218.

Arai-Sanoh, Y., Takai, T., Yoshinaga, S., Nakano, H., Kojima, M., Sakakibara, H., et al. (2014). Deep rooting conferred by *DEEPER ROOTING 1* enhances rice yield in paddy fields. *Scientific Reports, 4*(1), 1–6.

Araus, J. L., et al. (2018). Translating high-throughput phenotyping into genetic gain. *Trends in Plant Science, 23*, 451–466.

Arc, E., Galland, M., Cueff, G., Godin, B., Lounifi, I., Job, D., & Rajjou, L. (2011). Reboot the system thanks to protein post-translational modifications and proteome diversity: How quiescent seeds restart their metabolism to prepare seedling establishment. *Proteomics, 11*(9), 1606–1618.

Bailey-Serres, J., Parker, J. E., Ainsworth, E. A., Oldroyd, G. E., & Schroeder, J. I. (2019). Genetic strategies for improving crop yields. *Nature, 575*(7781), 109–118.

Baker, A. J., & Walker, P. L. (1990). Ecophysiology of metal uptake by tolerant plants. In *Heavy metal tolerance in plants: Evolutionary aspects* (Vol. 2, pp. 155–165). CRC Press.

Baligar, V. C., & Fageria, N. K. (2015). Nutrient use efficiency in plants: An overview. In A. Rakshit, H. B. Singh, & A. Sen (Eds.), *Nutrient use efficiency: From basics to advances* (Vol. 417). Springer.

Banerjee, A., & Roychoudhury, A. (2018). Seed priming technology in the amelioration of salinity stress in plants. In *Advances in seed priming* (pp. 81–93). Springer.

Bani, A., Pavlova, D., Echevarria, G., Mullaj, A., Reeves, R. D., Morel, J. L., & Sulçe, S. (2010). Nickel hyperaccumulation by the species of Alyssum and Th laspi (Brassicaceae) from the ultramafic soils of the Balkans. *Botanica Serbica, 34*(1), 3–14.

Barbosa, E. G. G., Leite, J. P., Marin, S. R. R., Marinho, J. P., Carvalho, J. D. F. C., Fuganti-Pagliarini, R., et al. (2013). Overexpression of the ABA-dependent AREB1 transcription factor from Arabidopsis thaliana improves soybean tolerance to water deficit. *Plant Molecular Biology Reporter, 31*(3), 719–730.

Baz, H., Creech, M., Chen, J., Gong, H., Bradford, K., & Huo, H. (2020). Water-soluble carbon nanoparticles improve seed germination and post-germination growth of lettuce under salinity stress. *Agronomy, 10*(8), 1192.

Beddington, J. R., Asaduzzaman, M., Bremauntz, F. A., Clark, M. E., Guillou, M., et al. (2011). *Achieving food security in the face of climate change: Summary for policy makers from the Commission on Sustainable Agriculture and Climate Change*.

Begna, T. (2021). Global role of plant breeding in tackling climate change. *Journal of Agricultural Science and Food Technology, 7*(2), 223–229.

Bermúdez, M., Cea, L., Van Uytven, E., Willems, P., Farfán, J. F., & Puertas, J. (2020). A robust method to update local river inundation maps using global climate model output and weather typing based statistical downscaling. *Water Resources Management, 34*(14), 4345–4362.

Bernard, A. (2008). Cadmium & its adverse effects on human health. *Indian Journal of Medical Research, 128*(4), 557.

Bhatnagar-Mathur, P., Rao, J. S., Vadez, V., Dumbala, S. R., Rathore, A., Yamaguchi-Shinozaki, K., & Sharma, K. K. (2014). Transgenic peanut overexpressing the DREB1A transcription factor has higher yields under drought stress. *Molecular Breeding, 33*(2), 327–340.

Blum, A. (2013). Heterosis, stress, and the environment: A possible road map towards the general improvement of crop yield. *Journal of Experimental Botany, 64*(16), 4829–4837.

Bocci, M., & Smanis, T. (2019). *Assessment of the impacts of climate change on the agriculture sector in the southern Mediterranean: Foreseen developments and policy measures*. Union for the Mediterranean. Accessed 18 Nov 2020.

Boyd, R. S., & Davis, M. A. (2001). Metal tolerance and accumulation ability of the Ni hyperaccumulator Streptanthus polygaloides Gray (Brassicaceae). *International Journal of Phytoremediation, 3*(4), 353–367.

Brady, N. C., & Weil, R. R. (2015). Soil acidity. In *The nature and properties of soils* (14th ed., pp. 423–468). Pearson Education, Inc.

Bray, A. L., & Topp, C. N. (2018). The quantitative genetic control of root architecture in maize. *Plant and Cell Physiology, 59*(10), 1919–1930.

Brown, S. L., Chaney, R. L., Angle, J. S., & Baker, A. J. M. (1994). Phytoremediation potential of *Thlaspi caerulescens* and bladder campion for zinc-and cadmium-contaminated soil. *Journal of Environmental Quality, 23*(6), 1151–1157.

Brown, N. L., Stoyanov, J. V., Kidd, S. P., & Hobman, J. L. (2003). The MerR family of transcriptional regulators. *FEMS Microbiology Reviews, 27*(2–3), 145–163.

Bu, J., Sun, Z., Zhou, A., Xu, Y., Ma, R., Wei, W., & Liu, M. (2016). Heavy metals in surface soils in the upper reaches of the Heihe River, northeastern Tibetan Plateau, China. *International Journal of Environmental Research and Public Health, 13*(3), 247.

Calleja-Cabrera, J., Boter, M., Oñate-Sánchez, L., & Pernas, M. (2020). Root growth adaptation to climate change in crops. *Frontiers in Plant Science, 11*, 544.

Ceglar, A., et al. (2019). Observed northward migration of agro-climate zones in Europe will further accelerate under climate change. *Earth's Future, 7*(9), 1088–1101.

Chandra, H., Kumari, P., Bontempi, E., & Yadav, S. (2020). Medicinal plants: Treasure trove for green synthesis of metallic nanoparticles and their biomedical applications. *Biocatalysis and Agricultural Biotechnology, 24*, 101518.

Chen, S., & Wilson, D. B. (1997). Genetic engineering of bacteria and their potential for Hg_2þ bioremediation. *Biodegradation, 8*, 97–103.

Chittaranjan, K. (2020). *Genomic designing of climate-smart vegetable crops*. Springer International Publishing.

Collard, B. C., Beredo, J. C., Lenaerts, B., Mendoza, R., Santelices, R., Lopena, V., et al. (2017). Revisiting rice breeding methods–evaluating the use of rapid generation advance (RGA) for routine rice breeding. *Plant Production Science, 20*(4), 337–352.

Comas, L. H., Becker, S. R., Cruz, V. M. V., Byrne, P. F., & Dierig, D. (2013). Root traits contributing to plant productivity under drought. *Frontiers in Plant Science, 4*, 442.

Cossani, C. M., & Reynolds, M. P. (2012). Physiological traits for improving heat tolerance in wheat. *Plant Physiology, 160*(4), 1710–1718.

Cossani, C. M., & Reynolds, M. P. (2015). Heat stress adaptation in elite lines derived from synthetic hexaploid wheat. *Crop Science, 55*(6), 2719–2735.

Crespo-Herrera, L. A., Crossa, J., Huerta-Espino, J., Autrique, E., Mondal, S., Velu, G., et al. (2017). Genetic yield gains in CIMMYT's international elite spring wheat yield trials by modeling the genotype × environment interaction. *Crop Science, 57*, 789.

Dar, M. H., Chakravorty, R., Waza, S. A., Sharma, M., Zaidi, N. W., Singh, A. N., et al. (2017). Transforming rice cultivation in flood prone coastal Odisha to ensure food and economic security. *Food Security, 9*(4), 711–722.

Dash, H. R., Mangwani, N., & Das, S. (2014). Characterization and potential application in mercury bioremediation of highly mercury-resistant marine bacterium Bacillus thuringiensis PW-05. *Environmental Science and Pollution Research, 21*, 2642–2653.

Dass, A., Jat, S. L., & Rana, K. S. (2015). Resource conserving techniques for improving nitrogen-use efficiency. In A. Rakshit, H. B. Singh, & A. Sen (Eds.), *Nutrient use efficiency: From basics to advances* (Vol. 417). Springer.

Datta, A., Shrestha, S., Ferdous, Z., & Win, C. C. (2015). Strategies for enhancing phosphorus efficiency in crop production systems. In A. Rakshit, H. B. Singh, & A. Sen (Eds.), *Nutrient use efficiency: From basics to advances* (Vol. 417). Springer.

De Ron, A. M., Kalavacharla, V., Álvarez-García, S., Casquero, P. A., Carro-Huerga, G., Gutiérrez, S., et al. (2019). *Genomic designing of climate-smart pulse crops*. Springer.

Deja-Muylle, A., Parizot, B., Motte, H., & Beeckman, T. (2020). Exploiting natural variation in root system architecture via genome-wide association studies. *Journal of Experimental Botany, 71*(8), 2379–2389.

Deng, X., & Jia, P. (2011). Construction and characterization of a photosynthetic bacterium genetically engineered for Hg^{2+} uptake. *Bioresource Technology, 102*(3), 3083–3088.

Des Marais, D. L., Auchincloss, L. C., Sukamtoh, E., McKay, J. K., Logan, T., Richards, J. H., & Juenger, T. E. (2014). Variation in MPK12 affects water use efficiency in Arabidopsis and reveals a pleiotropic link between guard cell size and ABA response. *Proceedings of the National Academy of Sciences, 111*(7), 2836–2841.

Doty, S. L., Shang, T. Q., Wilson, A. M., Tangen, J., Westergreen, A. D., Newman, L. A., et al. (2000). Enhanced metabolism of halogenated hydrocarbons in transgenic plants containing mammalian cytochrome P450 2E1. *Proceedings of the National Academy of Sciences, 97*(12), 6287–6291.

Duruibe, J. O., Ogwuegbu, M. O. C., & Egwurucywu, Y. N. (2007). Heavy metal pollution and human biotoxic effects. *International Journal of Physical Sciences, 2*, 112–118.

EEA. (2019). *Climate change adaptation in the agriculture sector in Europe* (EEA report No 4/2019). European Environment Agency. Accessed 19 Nov 2020.

EEA. (2020). *The European environment-state and outlook 2020*. European Environment Agency. Accessed 19 Nov 2020.

El-Mashad, A. A. A., & Mohamed, H. I. (2012). Brassinolide alleviates salt stress and increases antioxidant activity of cowpea plants (*Vigna sinensis*). *Protoplasma, 249*(3), 625–635.

Estrada-Melo, A. C., Reid, M. S., & Jiang, C. Z. (2015). Overexpression of an ABA biosynthesis gene using a stress-inducible promoter enhances drought resistance in petunia. *Horticulture Research, 2*(1), 1–9.

Faizan, M., Bhat, J. A., Chen, C., Alyemeni, M. N., Wijaya, L., Ahmad, P., & Yu, F. (2021). Zinc oxide nanoparticles (ZnO-NPs) induce salt tolerance by improving the antioxidant system and photosynthetic machinery in tomato. *Plant Physiology and Biochemistry, 161*, 122–130.

Fan, W., Wang, H., & Zhang, P. (2016). Engineering glycinebetaine metabolism for enhanced drought stress tolerance in plants. In *Drought stress tolerance in plants* (Vol. 2, pp. 513–530). Springer.

FAO. (1996). *Rome declaration on world food security and World Food Summit Plan of Action*. World Food Summit 13–17 November 1996. https://bit.ly/3rb70fK

FAO. (2008). Challenges for sustainable land management (SLM) for food security in Africa. In *25th Regional Conference for Africa, Nairobi Kenya, Information Paper*.

FAO. (2009). How to feed the world 2050: High-level expert forum. Global agriculture towards.

FAO. (2011). *Climate change, water and food security* (pp. 1–200).

FAO. (2015a). *Climate change and food security: Risks and responses* (pp. 1–122).

FAO. (2015b). *Coping with climate change - The roles of genetic resources for food and agriculture*. http://www.fao.org/3/a-i3866e.pdf

FAO. (2016). *Climate change and food security risks and responses* (pp. 1–46).

Farooq, M. A., Shakeel, A., Atif, R. M., & Saleem, M. F. (2019). Genotypic variations in salinity tolerance among BT cotton. *Pakistan Journal of Botany, 51*(6), 1945–1953.

Fernie, A. R., Tadmor, Y., & Zamir, D. (2006). Natural genetic variation for improving crop quality. *Current Opinion in Plant Biology, 9*(2), 196–202.

Flowers, T. S., & Yeo, A. R. (1989). Effects of salinity on plant growth and crop yields. In *Environmental stress in plants* (pp. 101–119). Springer.

Frederick, T. M., Taylor, E. A., Willis, J. L., Shultz, M. S., & Woodruff, P. J. (2013). Chromate reduction is expedited by bacteria engineered to produce the compatible solute trehalose. *Biotechnology Letters, 35*(8), 1291–1296. https://doi.org/10.1007/s10529-013-1200-z

Gadkar, V., & Rillig, M. C. (2006). The arbuscular mycorrhizal fungal protein glomalin is a putative homolog of heat shock protein 60. *FEMS Microbiology Letters, 263*(1), 93–101.

Garelick, H., Jones, H., Dybowska, A., & Valsami-Jones, E. (2009). Arsenic pollution sources. In *Reviews of environmental contamination* (Vol. 197, pp. 17–60). Springer.

Garrett, R. G. (2000). Natural sources of metals to the environment. *Human and Ecological Risk Assessment, 6*(6), 945–963.

Gautam, P. K., Gautam, R. K., Banerjee, S., Chattopadhyaya, M. C., & Pandey, J. D. (2016). Heavy metals in the environment: Fate, transport, toxicity and remediation technologies. *Nova Science Publishers, 60*, 101–130.

Geetika, G., van Oosterom, E. J., George-Jaeggli, B., Mortlock, M. Y., Deifel, K. S., McLean, G., & Hammer, G. L. (2019). Genotypic variation in whole-plant transpiration efficiency in sorghum only partly aligns with variation in stomatal conductance. *Functional Plant Biology, 46*(12), 1072–1089.

Ghaderian, S. M., Mohtadi, A., Rahiminejad, M. R., & Baker, A. J. M. (2007). Nickel and other metal uptake and accumulation by species of Alyssum (Brassicaceae) from the ultramafics of Iran. *Environmental Pollution, 145*(1), 293–298.

Ghani, A., & Ghani, A. (2011). Effect of chromium toxicity on growth, chlorophyll and some mineral nutrients of brassica juncea L. *Egyptian Academic Journal of Biological Sciences, H. Botany, 2*(1), 9–15.

Giarola, V., Hou, Q., & Bartels, D. (2017). Angiosperm plant desiccation tolerance: Hints from transcriptomics and genome sequencing. *Trends in Plant Science, 22*(8), 705–717.

Godfray, H. C. J., Beddington, J. R., Crute, I. R., Haddad, L., Lawrence, D., Muir, J. F., et al. (2010). Food security: The challenge of feeding 9 billion people. *Science, 327*(5967), 812–818.

Godt, J., Scheidig, F., Grosse-Siestrup, C., Esche, V., Brandenburg, P., Reich, A., & Groneberg, D. A. (2006). The toxicity of cadmium and resulting hazards for human health. *Journal of Occupational Medicine and Toxicology, 1*(1), 22.

Gond, S. K., Torres, M. S., Bergen, M. S., Helsel, Z., & White, J. F., Jr. (2015). Induction of salt tolerance and up-regulation of aquaporin genes in tropical corn by rhizobacterium P antoea agglomerans. *Letters in Applied Microbiology, 60*(4), 392–399.

González, R. M., Ricardi, M. M., & Iusem, N. D. (2013). Epigenetic marks in an adaptive water stress-responsive gene in tomato roots under normal and drought conditions. *Epigenetics, 8*(8), 864–872.

Guan-fu, F., Cai-xia, Z., Yong-jie, Y., Jie, X., Xue-qin, Y., Xiu-fu, Z., et al. (2015). Male parent plays more important role in heat tolerance in three-line hybrid rice. *Rice Science, 22*(3), 116–122.

Guo, J., Yan, Y., Chen, D., Lv, Y., Han, Y., Guo, X., Liu, L., Miao, Y., Chen, T., Nie, J., & Zhai, P. (2020). The response of warm-season precipitation extremes in China to global warming: An observational perspective from radiosonde measurements. *Climate Dynamics, 54*(9–10), 3977–3989.

Gupta, P. K., Balyan, H. S., Gahlaut, V., & Kulwal, P. L. (2012). Phenotyping, genetic dissection, and breeding for drought and heat tolerance in common wheat: Status and prospects. *Plant Breeding Reviews, 36*, 85–168.

Hallajian, M. T. (2016). Mutation breeding and drought stress tolerance in plants. In *Drought stress tolerance in plants* (Vol. 2, pp. 359–383). Springer.

Hammer, G., McLean, G., Doherty, A., van Oosterom, E., & Chapman, S. (2016). Sorghum crop modeling and its utility in agronomy and breeding. *Sorghum: State of the art and future perspectives* (agronmonogr58).

Ha-Tran, D. M., Nguyen, T. T. M., Hung, S. H., Huang, E., & Huang, C. C. (2021). Roles of plant growth-promoting rhizobacteria (PGPR) in stimulating salinity stress defense in plants: A review. *International Journal of Molecular Sciences, 22*(6), 3154.

Hoang, T. M. L., Tran, T. N., Nguyen, T. K. T., Williams, B., Wurm, P., Bellairs, S., & Mundree, S. (2016). Improvement of salinity stress tolerance in rice: Challenges and opportunities. *Agronomy, 6*(4), 54.

Hope, K. R. (2009). Climate change and poverty in Africa. *International Journal of Sustainable Development & World Ecology, 16*, 451–461. https://bit.ly/2Ua7gzG

Hsu, S. K., & Tung, C. W. (2017). RNA-Seq analysis of diverse rice genotypes to identify the genes controlling coleoptile growth during submerged germination. *Frontiers in Plant Science, 8*, 762.

Hu, J., Rampitsch, C., & Bykova, N. V. (2015). Advances in plant proteomics toward improvement of crop productivity and stress resistancex. *Frontiers in Plant Science, 6*, 209.

Huo, W., Zhuang, C. H., Cao, Y., Pu, M., Yao, H., Lou, L. Q., & Cai, Q. S. (2012). Paclobutrazol and plant-growth promoting bacterial endophyte Pantoea sp. enhance copper tolerance of guinea grass (*Panicum maximum*) in hydroponic culture. *Acta Physiologiae Plantarum, 34*(1), 139–150.

Ibuot, A., Dean, A. P., McIntosh, O. A., & Pittman, J. K. (2017). Metal bioremediation by CrMTP4 over-expressing *Chlamydomonas reinhardtii* in comparison to natural waste watertolerant microalgae strains. *Algal Research, 24*, 89–96.

IPCC. (2021). Climate change 2021: The physical science basis. Contribution of Working Group I to the Sixth Assessment Report of the Intergovernmental Panel on Climate Change. In V. Masson-Delmotte, P. Zhai, A. Pirani, S. L. Connors, C. P'ean, S. Berger, N. Caud, Y. Chen, L. Goldfarb, M. I. Gomis, M. Huang, K. Leitzell, E. Lonnoy, J. B. R. Matthews, T. K. Maycock, T. Waterfield, O. Yelekçi, R. Yu, & B. Zhou (Eds.). Cambridge University Press. In Press.

Iqbal, M., & Ashraf, M. (2013). Alleviation of salinity-induced perturbations in ionic and hormonal concentrations in spring wheat through seed preconditioning in synthetic auxins. *Acta Physiologiae Plantarum, 35*(4), 1093–1112.

Iqbal, N., Ashraf, M. Y., Javed, F., Martinez, V., & Ahmad, K. (2006). Nitrate reduction and nutrient accumulation in wheat grown in soil salinized with four different salts. *Journal of Plant Nutrition, 29*(3), 409–421.

Iqbal, K. J., Qureshi, N. A., Ashraf, M., Rehman, M. H. U., Khan, N., Javid, A., et al. (2012). Effect of different salinity levels on growth and survival of Nile tilapia (*Oreochromis niloticus*). *Journal of Animal and Plant Sciences, 22*(4), 919–932.

Isayenkov, S. V., & Maathuis, F. J. (2019). Plant salinity stress: Many unanswered questions remain. *Frontiers in Plant Science, 10*, 80.

Islam, Z., Bagchi, B., & Hossain, M. (2007). Adoption of leaf color chart for nitrogen use efficiency in rice: Impact assessment of a farmer-participatory experiment in West Bengal, India. *Field Crops Research, 103*(1), 70–75.

Jarup, L. (2003). Hazards of heavy metal contamination. *British Medical Bulletin, 68*(1), 167–182.

Jat, M. L., Gathala, M. K., Ladha, J. K., Saharawat, Y. S., Jat, A. S., Kumar, V., et al. (2009). Evaluation of precision land leveling and double zero-till systems in the rice–wheat rotation: Water use, productivity, profitability and soil physical properties. *Soil and Tillage Research, 105*(1), 112–121.

Jiang, Y., Cai, Z., Xie, W., Long, T., Yu, H., & Zhang, Q. (2012). Rice functional genomics research: Progress and implications for crop genetic improvement. *Biotechnology Advances, 30*(5), 1059–1070.

Jing, Z., Jian, Z., Xi-Lin, H., Feng, W., & Ai-Sheng, X. (2014). Transcriptomic, proteomic, metabolomic and functional genomic approaches for the study of abiotic stress in vegetable crops. *Critical Reviews in Plant Sciences, 33*(2–3), 225–237. https://doi.org/10.1080/07352689.2014.870420

Jodie, K., Page, S., Kergna, A., & Kennan, J. (2009). *Climate change and developing country agriculture: An overview of expected impacts, adaptation and mitigation challenges, and funding requirements*. https://bit.ly/3z0MCkb

Juliana, P., Montesinos-López, O. A., Crossa, J., Mondal, S., Pérez, L. G., Poland, J., et al. (2019). Integrating genomic-enabled prediction and high-throughput phenotyping in breeding for climate-resilient bread wheat. *Theoretical and Applied Genetics, 132*(1), 177–194.

Kajiya-Kanegae, H., Takanashi, H., Fujimoto, M., Ishimori, M., Ohnishi, N., Wacera, W. F., et al. (2020). RAD-seq-based high-density linkage map construction and QTL mapping of biomass-related traits in sorghum using the Japanese landrace Takakibi NOG. *Plant and Cell Physiology, 61*(7), 1262–1272.

Kapaj, S., Peterson, H., Liber, K., & Bhattacharya, P. (2006). Human health effects from chronic arsenic poisoning–a review. *Journal of Environmental Science and Health, Part A, 41*(10), 2399–2428.

Kapoor, D., Singh, S., Kumar, V., Romero, R., Prasad, R., & Singh, J. (2019). Antioxidant enzymes regulation in plants in reference to reactive oxygen species (ROS) and reactive nitrogen species (RNS). *Plant Gene, 19*, 100182.

Karimi, S., Tavallali, V., Ferguson, L., & Mirzaei, S. (2020). Developing a nano-Fe complex to supply iron and improve salinity tolerance of pistachio under calcium bicarbonate stress. *Communications in Soil Science and Plant Analysis, 51*(14), 1835–1851.

Kato, Y., Collard, B. C., Septiningsih, E. M., & Ismail, A. M. (2019). Increasing flooding tolerance in rice: Combining tolerance of submergence and of stagnant flooding. *Annals of Botany, 124*(7), 1199–1209.

Khan, N., & Bano, A. (2016). Role of plant growth promoting rhizobacteria and Ag-nano particle in the bioremediation of heavy metals and maize growth under municipal wastewater irrigation. *International Journal of Phytoremediation, 18*(3), 211–221.

Kim, S. M., & Reinke, R. F. (2018). Identification of QTLs for tolerance to hypoxia during germination in rice. *Euphytica, 214*(9), 1–10.

Kim, J. M., To, T. K., Ishida, J., Matsui, A., Kimura, H., & Seki, M. (2012). Transition of chromatin status during the process of recovery from drought stress in Arabidopsis thaliana. *Plant and Cell Physiology, 53*(5), 847–856.

Kitomi, Y., Nakao, E., Kawai, S., Kanno, N., Ando, T., Fukuoka, S., et al. (2018). Fine mapping of QUICK ROOTING 1 and 2, quantitative trait loci increasing root length in rice. *G3: Genes, Genomes, Genetics, 8*(2), 727–735.

Kitomi, Y., Hanzawa, E., Kuya, N., Inoue, H., Hara, N., Kawai, S., et al. (2020). Root angle modifications by the DRO1 homolog improve rice yields in saline paddy fields. *Proceedings of the National Academy of Sciences, 117*(35), 21242–21250.

Kole, C. (Ed.). (2019). *Genomic designing of climate-smart oilseed crops*. Springer.

Kole, C. (Ed.). (2020). *Genomic designing of climate-smart fruit crops*. Springer Nature.

Koller, M., & Saleh, H. M. (2018). Introductory chapter: Introducing heavy metals. *Heavy Metals, 1*, 3–11.

Kooyers, N. J. (2015). The evolution of drought escape and avoidance in natural herbaceous populations. *Plant Science, 234*, 155–162.

Krishnamoorthy, R., Kim, K., Subramanian, P., Senthilkumar, M., Anandham, R., & Sa, T. (2016). Arbuscular mycorrhizal fungi and associated bacteria isolated from salt-affected soil enhances the tolerance of maize to salinity in coastal reclamation soil. *Agriculture, Ecosystems & Environment, 231*, 233–239.

Kruseman, G., Bairagi, S., Komarek, A. M., Molero Milan, A., Nedumaran, S., Petsakos, A., et al. (2020). CGIAR modeling approaches for resource-constrained scenarios: II. Models for analyzing socioeconomic factors to improve policy recommendations. *Crop Science, 60*(2), 568–581.

Kumar, A., Bernier, J., Verulkar, S., Lafitte, H. R., & Atlin, G. N. (2008). Breeding for drought tolerance: Direct selection for yield, response to selection and use of drought-tolerant donors in upland and lowland-adapted populations. *Field Crops Research, 107*(3), 221–231.

Kumar, A., Singh, S., Gaurav, A. K., Srivastava, S., & Verma, J. P. (2020). Plant growth-promoting bacteria: Biological tools for the mitigation of salinity stress in plants. *Frontiers in Microbiology, 11*, 1216.

Kumar, A., Singh, S., Mukherjee, A., Rastogi, R. P., & Verma, J. P. (2021). Salt-tolerant plant growth-promoting Bacillus pumilus strain JPVS11 to enhance plant growth attributes of rice and improve soil health under salinity stress. *Microbiological Research, 242*, 126616.

Küpper, H., Lombi, E., Zhao, F. J., Wieshammer, G., & McGrath, S. P. (2001). Cellular compartmentation of nickel in the hyperaccumulators Alyssum lesbiacum, Alyssum bertolonii and Thlaspi goesingense. *Journal of Experimental Botany, 52*(365), 2291–2300.

Langridge, P., & Reynolds, M. P. (2015). Genomic tools to assist breeding for drought tolerance. *Current Opinion in Biotechnology, 32*, 130–135.

LaPack, M. A., Tou, J. C., & Enke, C. G. (1990). Membrane mass spectrometry for the direct trace analysis of volatile organic compounds in air and water. *Analytical Chemistry, 62*(13), 1265–1271.

Li, Y. M., Chaney, R., Brewer, E., Roseberg, R., Angle, J. S., Baker, A., et al. (2003). Development of a technology for commercial phytoextraction of nickel: Economic and technical considerations. *Plant and Soil, 249*(1), 107–115.

Li, J. F., Zhang, D., & Sheen, J. (2015). Targeted plant genome editing via the CRISPR/Cas9 technology. In *Plant functional genomics* (pp. 239–255). Humana Press.

Lister, R., Gregory, B. D., & Ecker, J. R. (2009). Next is now: New technologies for sequencing of genomes, transcriptomes, and beyond. *Current Opinion in Plant Biology, 12*(2), 107–118.

Lobell, D. B., Hammer, G. L., Chenu, K., Zheng, B., McLean, G., & Chapman, S. C. (2015). The shifting influence of drought and heat stress for crops in northeast Australia. *Global Change Biology, 21*(11), 4115–4127.

Luo, S. L., Chen, L., Chen, J. L., Xiao, X., Xu, T. Y., Wan, Y., & Zeng, G. M. (2011). Analysis and characterization of cultivable heavy metal-resistant bacterial endophytes isolated from Cd-hyperaccumulator *Solanum nigrum* L. and their potential use for phytoremediation. *Chemosphere, 85*(7), 1130–1138.

Lynch, J. P. (2019). Root phenotypes for improved nutrient capture: An underexploited opportunity for global agriculture. *New Phytologist, 223*(2), 548–564.

Manigbas, N. L., Lambio, L. A. F., Luvina, B., & Cardenas, C. C. (2014). Germplasm innovation of heat tolerance in rice for irrigated lowland conditions in the Philippines. *Rice Science, 21*(3), 162–169.

Mateos, L. M., Villadangos, A. F., de la Rubia, A. G., Mourenza, A., Marcos-Pascual, L., Letek, M., & Gil, J. A. (2017). The arsenic detoxification system in corynebacteria. *Advanced Applied Microbiology, 99*, 103–137. https://doi.org/10.1016/bs.aambs.2017.01.001

Mega, R., Abe, F., Kim, J. S., Tsuboi, Y., Tanaka, K., Kobayashi, H., et al. (2019). Tuning water-use efficiency and drought tolerance in wheat using abscisic acid receptors. *Nature Plants, 5*(2), 153–159.

Merchuk-Ovnat, L., Barak, V., Fahima, T., Ordon, F., Lidzbarsky, G. A., Krugman, T., & Saranga, Y. (2016). Ancestral QTL alleles from wild emmer wheat improve drought resistance and productivity in modern wheat cultivars. *Frontiers in Plant Science, 7*, 452.

Mishra, R., & Zhao, K. (2018). Genome editing technologies and their applications in crop improvement. *Plant Biotechnology Reports, 12*(2), 57–68.

Mishra, P., Mishra, J., & Arora, N. K. (2021). Plant growth promoting bacteria for combating salinity stress in plants–Recent developments and prospects: A review. *Microbiological Research, 252*, 126861.

Naher, U. A., Panhwar, Q. A., Othman, R., Ismail, M. R., & Berahim, Z. (2016). Biofertilizer as a supplement of chemical fertilizer for yield maximization of rice. *Journal of Agriculture Food and Development, 2*, 16–22.

Navarro, J. M., Pérez-Tornero, O., & Morte, A. (2014). Alleviation of salt stress in citrus seedlings inoculated with arbuscular mycorrhizal fungi depends on the rootstock salt tolerance. *Journal of Plant Physiology, 171*(1), 76–85.

Negrão, S., Schmöckel, S. M., & Tester, M. (2017). Evaluating physiological responses of plants to salinity stress. *Annals of Botany, 119*(1), 1–11.

Nejatzadeh, F. (2021). Effect of silver nanoparticles on salt tolerance of Satureja hortensis l. during in vitro and in vivo germination tests. *Heliyon, 7*(2), e05981.

Nelson, G. C., Mark, W. R., Jawoo K., Richard, R., Timothy, S., Tingju, Z., Claudia, R., Siwa, M., Amanda, P., Miroslav, B., Marilia, M., Rowena, V., Mandy, E., & David, L. (2009). *Climate change: Impact on agriculture and costs of adaptation* (pp. 1–8). International Food Policy Research Institute. https://doi.org/10.2499/0896295354

Nguyen, C. T., Singh, V., van Oosterom, E. J., Chapman, S. C., Jordan, D. R., & Hammer, G. L. (2013). Genetic variability in high temperature effects on seed-set in sorghum. *Functional Plant Biology, 40*(5), 439–448.

Nguyen, D. N., Wang, S. L., Nguyen, A. D., Doan, M. D., Tran, D. M., Nguyen, T. H., et al. (2021). Potential application of rhizobacteria isolated from the central highland of Vietnam as an effective biocontrol agent of Robusta coffee nematodes and as a bio-fertilizer. *Agronomy, 11*(9), 1887.

Numan, M., Bashir, S., Khan, Y., Mumtaz, R., Shinwari, Z. K., Khan, A. L., et al. (2018). Plant growth promoting bacteria as an alternative strategy for salt tolerance in plants: A review. *Microbiological Research, 209*, 21–32.

Oladosu, Y., Rafii, M. Y., Abdullah, N., Magaji, U., Miah, G., Hussin, G., & Ramli, A. (2017). Genotype× Environment interaction and stability analyses of yield and yield components of established and mutant rice genotypes tested in multiple locations in Malaysia. *Acta Agriculturae Scandinavica, Section B-Soil & Plant Science, 67*(7), 590–606.

Osakabe, Y., Osakabe, K., Shinozaki, K., & Tran, L. S. P. (2014). Response of plants to water stress. *Frontiers in Plant Science, 5*, 86.

Padmavathiamma, P. K., & Li, L. Y. (2007). Phytoremediation technology: Hyper-accumulation metals in plants. *Water, Air, and Soil Pollution, 184*(1–4), 105–126.

Palmgren, M. G., Edenbrandt, A. K., Vedel, S. E., Andersen, M. M., Landes, X., Østerberg, J. T., et al. (2015). Are we ready for back-to-nature crop breeding? *Trends in Plant Science, 20*(3), 155–164.

Panhwar, Q. A., Naher, U. A., Radziah, O., Shamshuddin, J., & Razi, I. M. (2014). Bio-fertilizer, ground magnesium limestone and basalt applications may improve chemical properties of Malaysian acid sulfate soils and rice growth. *Pedosphere, 24*(6), 827–835.

Parent, B., Leclere, M., Lacube, S., Semenov, M. A., Welcker, C., Martre, P., & Tardieu, F. (2018). Maize yields over Europe may increase in spite of climate change, with an appropriate use of the genetic variability of flowering time. *Proceedings of the National Academy of Sciences, 115*(42), 10642–10647.

Pedersen, O., Nakayama, Y., Yasue, H., Kurokawa, Y., Takahashi, H., Heidi Floytrup, A., et al. (2021a). Lateral roots, in addition to adventitious roots, form a barrier to radial oxygen loss in Zea nicaraguensis and a chromosome segment introgression line in maize. *New Phytologist, 229*(1), 94–105.

Pedersen, O., Sauter, M., Colmer, T. D., & Nakazono, M. (2021b). Regulation of root adaptive anatomical and morphological traits during low soil oxygen. *New Phytologist, 229*(1), 42–49.

Pieczynski, M., Marczewski, W., Hennig, J., Dolata, J., Bielewicz, D., Piontek, P., et al. (2013). Down-regulation of CBP 80 gene expression as a strategy to engineer a drought-tolerant potato. *Plant Biotechnology Journal, 11*(4), 459–469.

Placido, D. F., Sandhu, J., Sato, S. J., Nersesian, N., Quach, T., Clemente, T. E., et al. (2020). The LATERAL ROOT DENSITY gene regulates root growth during water stress in wheat. *Plant Biotechnology Journal, 18*(9), 1955–1968.

Poli, Y., Basava, R. K., Panigrahy, M., Vinukonda, V. P., Dokula, N. R., Voleti, S. R., et al. (2013). Characterization of a Nagina22 rice mutant for heat tolerance and mapping of yield traits. *Rice, 6*(1), 1–9.

Poulose, J., Rao, A. D., & Dube, S. K. (2020). Mapping of cyclone induced extreme water levels along Gujarat and Maharashtra coasts: A climate change perspective. *Climate Dynamics, 55*(11–12), 3565–3581.

Prasad, M. N. V. (2005). Nickelophilous plants and their significance in phytotechnologies. *Brazilian Journal of Plant Physiology, 17*(1), 113–128.

Pulford, I. D., & Watson, C. (2003). Phytoremediation of heavy metal-contaminated land by trees-A review. *Environment International, 29*(4), 529–540.

Rafati, M., Khorasani, N., Moattar, F., Shirvany, A., Moraghebi, F., & Hosseinzadeh, S. (2011). Phytoremediation potential of Populus alba and Morus alba for cadmium, chromuim and nickel absorption from polluted soil. *International Journal of Environmental Research, 5*(4), 961–970.

Rai, P. K. (2008). Phytoremediation of Hg and Cd from industrial effluents using an aquatic free floating macrophyte *Azolla pinnata*. *International Journal of Phytoremediation, 10*(5), 430–439.

Ramasamy, K., & Banu, S. P. (2007). Bioremediation of metals: Microbial processes and techniques. In *Environmental bioremediation technologies* (pp. 173–187). Springer.

Ramiah, K. (1940). Floating habit in rice. *Indian Journal of Agricultural Sciences, 11*, 1–8.

Ray, D. K., et al. (2019). Climate change has likely already affected global food production. *PLoS One, 14*(5), e0217148.

Renzaho, A. M., & Mellor, D. (2010). Food security measurement in cultural pluralism: Missing the point or conceptual misunderstanding. *Nutrition, 26*, 1–9. https://bit.ly/3BbUh11

Reynolds, M. P., & Trethowan, R. M. (2007). Physiological interventions in breeding for adaptation to abiotic stress. *Frontis, 21*, 127–144.

Rice, K. M., Walker, E. M., Jr., Wu, M., Gillette, C., & Blough, E. R. (2014). Environmental mercury and its toxic effects. *Journal of Preventive Medicine and Public Health, 47*(2), 74.

Rogers, C., Wen, J., Chen, R., & Oldroyd, G. (2009). Deletion-based reverse genetics in Medicago truncatula. *Plant Physiology, 151*(3), 1077–1086.

Ruiz, O. N., Alvarez, D., Gonzalez-Ruiz, G., & Torres, C. (2011). Characterization of mercury bioremediation by transgenic bacteria expressing metallothionein and polyphosphate kinase. *BMC Biotechnology, 11*, 82. https://doi.org/10.1186/1472-6750-11-82

Ruppel, S., Franken, P., & Witzel, K. (2013). Properties of the halophyte microbiome and their implications for plant salt tolerance. *Functional Plant Biology, 40*(9), 940–951.

Rutkoski, J., Poland, J., Mondal, S., Autrique, E., Pérez, L. G., Crossa, J., et al. (2016). Canopy temperature and vegetation indices from high-throughput phenotyping improve accuracy of pedigree and genomic selection for grain yield in wheat. *G3: Genes, Genomes, Genetics, 6*(9), 2799–2808.

Sahab, S., Suhani, I., Srivastava, V., Chauhan, P. S., Singh, R. P., & Prasad, V. (2021). Potential risk assessment of soil salinity to agroecosystem sustainability: Current status and management strategies. *Science of the Total Environment, 764*, 144164.

Sakakibara, M., Ohmori, Y., Ha, N. T. H., Sano, S., & Sera, K. (2011). Phytoremediation of heavy metal-contaminated water and sediment by *Eleocharis acicularis*. *CLEAN–Soil, Air, Water, 39*(8), 735–741.

Sangeeta, M., & Maiti, S. K. (2010). Phytoremediation of metal enriched mine waste: A review. *American-Eurasian Journal of Agricultural & Environmental Sciences, 9*(5), 560–575.

Satterthwaite, D., McGranahan, G., & Tacoli, C. (2010). Urbanization and its implications for food and farming. *Philosophical Transactions of the Royal Society B: Biological Sciences, 365*, 2809–2820. https://bit.ly/3wDHtNm

Semel, Y., Schauer, N., Roessner, U., Zamir, D., & Fernie, A. R. (2007). Metabolite analysis for the comparison of irrigated and non-irrigated field grown tomato of varying genotype. *Metabolomics, 3*(3), 289–295.

Sen, S. K., Chouhan, D., Das, D., Ghosh, R., & Mandal, P. (2020). Improvisation of salinity stress response in mung bean through solid matrix priming with normal and nano-sized chitosan. *International Journal of Biological Macromolecules, 145*, 108–123.

Serba, D. D., Yadav, R. S., Varshney, R. K., Gupta, S., Mahalingam, G., Srivastava, R. K., et al. (2020). *Genomic designing of climate-smart cereal crops*. Springer.

Shah, T., Latif, S., Saeed, F., Ali, I., Ullah, S., Alsahli, A. A., et al. (2021). Seed priming with titanium dioxide nanoparticles enhances seed vigor, leaf water status, and antioxidant enzyme

activities in maize (*Zea mays* L.) under salinity stress. *Journal of King Saud University-Science, 33*(1), 101207.
Sharma, I., Ching, E., Saini, S., Bhardwaj, R., & Pati, P. K. (2013). Exogenous application of brassinosteroid offers tolerance to salinity by altering stress responses in rice variety Pusa Basmati-1. *Plant Physiology and Biochemistry, 69*, 17–26.
Singh, S. (2012). Phytoremediation: A sustainable alternative for environmental challenges. *International Journal of Green and Herbal Chemistry, 1*, 133–139.
Singh, R., Gautam, N., Mishra, A., & Gupta, R. (2011). Heavy metals and living systems: An overview. *Indian Journal of Pharmacology, 43*(3), 246.
Singh, A., Septiningsih, E. M., Balyan, H. S., Singh, N. K., & Rai, V. (2017). Genetics, physiological mechanisms and breeding of flood-tolerant rice (Oryza sativa L.). *Plant and Cell Physiology, 58*(2), 185–197.
Sodango, T. H., Li, X., Sha, J., & Bao, Z. (2018). Review of the spatial distribution, source and extent of heavy metal pollution of soil in China: Impacts and mitigation approaches. *Journal of Health and Pollution, 8*(17), 53–70.
Son, T. T. N., Van Thu, V., Man, L. H., & Kobayashi, H. (2001). Effect of organic and bio-fertilizer on soybean and rice under rice based cropping system. In *Proceedings of the 2001 annual workshop of JIRCAS Mekong Delta Project* (pp. 43–53).
Sripinyowanich, S., Klomsakul, P., Boonburapong, B., Bangyeekhun, T., Asami, T., Gu, H., et al. (2013). Exogenous ABA induces salt tolerance in indica rice (Oryza sativa L.): The role of OsP5CS1 and OsP5CR gene expression during salt stress. *Environmental and Experimental Botany, 86*, 94–105.
Srivastava, P. C., & Gupta, U. C. (1996). *Trace elements in crop production*. Science Publishers Inc.
Srivastava, N. K., Jha, M. K., Mall, I. D., & Singh, D. (2010). Application of genetic engineering for chromium removal from industrial wastewater. *International Journal of Chemical and Biological Engineering, 3*, 3.
Srivastava, V., De Araujo, A. S. F., Vaish, B., Bartelt-Hunt, S., Singh, P., & Singh, R. P. (2016). Biological response of using municipal solid waste compost in agriculture as fertilizer supplement. *Reviews in Environmental Science and Biotechnology, 15*(4), 677–696.
Sultan, B., & Gaetani, M. (2016). Agriculture in West Africa in the twenty-first century: Climate change and impacts scenarios, and potential for adaptation. *Frontiers in Plant Science, 7*, 1262.
Surat, W., Kruatrachue, M., Pokethitiyook, P., Tanhan, P., & Samranwanich, T. (2008). Potential of *Sonchus arvensis* for the phytoremediation of lead-contaminated soil. *International Journal of Phytoremediation, 10*(4), 325–342.
Tamang, B. G., & Fukao, T. (2015). Plant adaptation to multiple stresses during submergence and following desubmergence. *International Journal of Molecular Sciences, 16*(12), 30164–30180.
Tester, M., & Langridge, P. (2010). Breeding technologies to increase crop production in a changing world. *Science, 327*(5967), 818–822.
Tilman, D., Balzer, C., & Hill, J., Befort, B. L. (2011). Global food demand and the sustainable intensifi cation of agriculture. *Proceedings of the National Academy of Sciences of the United States of America, 108*, 20260–20264. https://bit.ly/3xNTMIq
Tlustoš, P., Pavlíková, D., Száková, J., Fischeroá, Z., & Balík, J. (2006). Exploitation of fast growing trees in metal remediation. In *Phytoremediation rhizoremediation* (pp. 83–102). Springer.
Toledo, A. M. U., Ignacio, J. C. I., Casal, C., Gonzaga, Z. J., Mendioro, M. S., & Septiningsih, E. M. (2015). Development of improved Ciherang-Sub1 having tolerance to anaerobic germination conditions. *Plant Breeding and Biotechnology, 3*, 77–87.
Tomlinson, I. (2013). Doubling food production to feed the 9 billion: A critical perspective on a key discourse of food security in the UK. *Journal of Rural Studies, 29*, 81–90.
Tozzini, A. C. (2000). Semi-quantitative detection of genetically modified grains based on CaMv 35S promoter amplification. *Electronic Journal of Biotechnology, 32*(2), 0717–3458.
Turral, H., Burke, J., & Faurès, J. M. (2011). *Climate change, water and food security*. Food and Agriculture Organization of the United Nations. https://bit.ly/3wHV3zd

Van Houtan, K. S., Tanaka, K. R., Gagné, T. O., & Becker, S. L. (2021). The geographic disparity of historical greenhouse emissions and projected climate change. *Science Advances, 7*(29), eabe4342.

Vangronsveld, J., Herzig, R., Weyens, N., Boulet, J., Adriaensen, K., Ruttens, A., et al. (2009). Phytoremediation of contaminated soils and groundwater: Lessons from the field. *Environmental Science and Pollution Research, 16*(7), 765–794.

Venuprasad, R., Bool, M. E., Quiatchon, L., Cruz, M. S., Amante, M., & Atlin, G. N. (2012). A large-effect QTL for rice grain yield under upland drought stress on chromosome 1. *Molecular Breeding, 30*(1), 535–547.

Verma, H., Borah, J. L., & Sarma, R. N. (2019). Variability assessment for root and drought tolerance traits and genetic diversity analysis of rice germplasm using SSR markers. *Scientific Reports, 9*(1), 1–19.

Vessey, J. K. (2003). Plant growth promoting rhizobacteria as biofertilizers. *Plant and Soil, 255*(2), 571–586.

Vikram, P., Swamy, B. M., Dixit, S., Ahmed, H. U., Cruz, M. T. S., Singh, A. K., & Kumar, A. (2011). qDTY$_{1.1}$, a major QTL for rice grain yield under reproductive-stage drought stress with a consistent effect in multiple elite genetic backgrounds. *BMC Genetics, 12*(1), 1–15.

Vishnoi, S. R., & Srivastava, P. N. (2007). Phytoremediation–green for environmental clean. In *Proceedings of Taal2007: The 12th World lake conference* (Vol. 1016, p. 1021).

Vivitha, P., Raveendran, M., & Vijayalakshmi, D. (2017). Introgression of QTLs controlling spikelet fertility maintains membrane integrity and grain yield in improved white Ponni derived progenies exposed to heat stress. *Rice Science, 24*(1), 32–40.

Voesenek, L. A. C. J., & Sasidharan, R. (2013). Ethylene–and oxygen signalling–drive plant survival during flooding. *Plant Biology, 15*(3), 426–435.

von der Gathen, P., Kivi, R., Wohltmann, I., Salawitch, R. J., & Rex, M. (2021). Climate change favours large seasonal loss of Arctic ozone. *Nature Communications, 12*(1), 1–17.

Voss-Fels, K. P., Snowdon, R. J., & Hickey, L. T. (2018). Designer roots for future crops. *Trends in Plant Science, 23*(11), 957–960.

Wang, Y., Beaith, M., Chalifoux, M., Ying, J., Uchacz, T., Sarvas, C., et al. (2009). Shoot-specific down-regulation of protein farnesyltransferase (α-subunit) for yield protection against drought in canola. *Molecular Plant, 2*(1), 191–200.

Wang, L., Czedik-Eysenberg, A., Mertz, R. A., Si, Y., Tohge, T., Nunes-Nesi, A., et al. (2014). Comparative analyses of C4 and C3 photosynthesis in developing leaves of maize and rice. *Nature Biotechnology, 32*(11), 1158–1165.

Wang, W., Zhang, Z., Yeh, T. C. J., Qiao, G., Wang, W., Duan, L., et al. (2017). Flow dynamics in vadose zones with and without vegetation in an arid region. *Advances in Water Resources, 106*, 68–79.

Waraich, E.A., Ahmad, R., Halim, A., & Aziz, T. (2012). Alleviation of temperature stress by nutrient management in crop plants: A review. *Journal of Soil Science and Plant Nutrition, 12*(2): 221–244.358. Springer.

Wei, H., Liu, J., Wang, Y., Huang, N., Zhang, X., Wang, L., et al. (2013). A dominant major locus in chromosome 9 of rice (*Oryza sativa* L.) confers tolerance to 48 C high temperature at seedling stage. *Journal of Heredity, 104*(2), 287–294.

Wilbur, S. B. (2000). *Toxicological profile for chromium*. US Department of Health and Human Services, Public Health Service, Agency for Toxic Substances and Disease Registry.

Wu, G., Kang, H., Zhang, X., Shao, H., Chu, L., & Ruan, C. (2010). A critical review on the bio-removal of hazardous heavy metals from contaminated soils: Issues, progress, eco-environmental concerns and opportunities. *Journal of Hazardous Materials, 174*(1–3), 1–8.

Wu, B., Wang, Z., Zhao, Y., Gu, Y., Wang, Y., Yu, J., & Xu, H. (2019). The performance of biochar-microbe multiple biochemical material on bioremediation and soil micro-ecology in the cadmium aged soil. *Science of the Total Environment, 686*, 719–728.

Wuana, R. A., & Okieimen, F. E. (2011). Heavy metals in contaminated soils: A review of sources, chemistry, risks and best available strategies for remediation. *ISRN Ecology, 2011*, 1–20.

Xue, Y., Warburton, M. L., Sawkins, M., Zhang, X., Setter, T., Xu, Y., et al. (2013). Genome-wide association analysis for nine agronomic traits in maize under well-watered and water-stressed conditions. *Theoretical and Applied Genetics, 126*(10), 2587–2596.

Yamauchi, T., Abe, F., Tsutsumi, N., & Nakazono, M. (2019). Root cortex provides a venue for gas-space formation and is essential for plant adaptation to waterlogging. *Frontiers in Plant Science, 10*, 259.

Yang, S. X., Deng, H., & Li, M. S. (2008). Manganese uptake and accumulation in a woody hyperaccumulator, Schima superba. *Plant, Soil and Environment, 54*(10), 441–446.

Yazdandoost, F., Moradian, S., Izadi, A., & Aghakouchak, A. (2021). Evaluation of CMIP6 precipitation simulations across different climatic zones: Uncertainty and model intercomparison. *Atmospheric Research, 250*, 105369. https://doi.org/10.1016/j.atmosres.2020.105369

Zacchini, M., Pietrini, F., Mugnozza, G. S., Iori, V., Pietrosanti, L., & Massacci, A. (2009). Metal tolerance, accumulation and translocation in poplar and willow clones treated with cadmium in hydroponics. *Water, Air, and Soil Pollution, 197*(1–4), 23–34.

Zahra, N., & Kalim, I. (2017). Perilous effects of heavy metals contamination on human health. *Pakistan Journal of Analytical & Environmental Chemistry, 18*(1), 1–17.

Zandalinas, S. I., Fichman, Y., Devireddy, A. R., Sengupta, S., Azad, R. K., & Mittler, R. (2020). Systemic signaling during abiotic stress combination in plants. *Proceedings of the National Academy of Sciences, 117*(24), 13810–13820.

Zandalinas, S. I., Sengupta, S., Fritschi, F. B., Azad, R. K., Nechushtai, R., & Mittler, R. (2021). The impact of multifactorial stress combination on plant growth and survival. *New Phytologist, 230*(3), 1034–1048.

Zawoznik, M. S., Ameneiros, M., Benavides, M. P., Vázquez, S., & Groppa, M. D. (2011). Response to saline stress and aquaporin expression in Azospirillum-inoculated barley seedlings. *Applied Microbiology and Biotechnology, 90*(4), 1389–1397.

Zheng, B., Chenu, K., & Chapman, S. C. (2016). Velocity of temperature and flowering time in wheat–assisting breeders to keep pace with climate change. *Global Change Biology, 22*(2), 921–933.

Zhou, R., Yu, X., Ottosen, C. O., Rosenqvist, E., Zhao, L., Wang, Y., et al. (2017). Drought stress had a predominant effect over heat stress on three tomato cultivars subjected to combined stress. *BMC Plant Biology, 17*(1), 1–13.

Zhu, F., Ma, S., Liu, T., & Deng, X. (2018). Green synthesis of nano zero-valent iron/Cu by green tea to remove hexavalent chromium from groundwater. *Journal of Cleaner Production, 174*, 184–190.

Chapter 4
The Role of Integrated Pest Management for Sustainable Food Production: The Soybean Example

Rodrigo Mendes Antunes Maciel and Adeney de Freitas Bueno

Abstract Agriculture is the most important human activity to produce food for an increasing global population that is expected to reach 10 billion by 2050. However, this food production must be performed with environmental preservation. One of the essential strategies to accomplish this sustainable agriculture, fostering equitable, secure, sufficient, and stable flows of both food and ecosystem services is by the adoption of integrated pest management (IPM). IPM is based upon the principle that some degree of plant injury is tolerable without requiring pest control. In addition, the most environment-friendly pest control tools should be combined in order to have a longer lasting pest solution. Those tools include biological control, transgenic plants, among others that are discussed in more details in this book chapter and are commercially available, combining environmental, economic, and social benefits.

Keywords Sustainability · Biological control · Pesticide mitigation · Agriculture · Crop protection

1 Introduction

Agriculture is essential for humanity's survival. There is an enormous need for food that must be produced inexpensively to feed a growing number of consumers. The global population is expected to reach 10 billion by 2050 (Ezeh et al., 2012; ONU, 2019). Despite this increasing demand, the agroecosystem is homogenized when eatable plants are cultivated over large areas in conventional agricultural systems. Consequently, we end up favoring arthropods that also feed on this same plant species, creating what is known as "agricultural pests" (Smith, 2021; Yactayo-Chang

R. M. A. Maciel (✉)
Universidade Federal do Paraná, Curitiba, Paraná, Brazil

A. de Freitas Bueno
Embrapa Soybean, Londrina, Paraná, Brazil
e-mail: adeney.bueno@embrapa.br

C. M. Galanakis (ed.), *Biodiversity, Functional Ecosystems and Sustainable Food Production*, https://doi.org/10.1007/978-3-031-07434-9_4

et al., 2020). In order to avoid those pests from consuming our crops, which could trigger food shortages, pest outbreaks must be managed (Cook et al., 2007; Deutsch et al., 2018; Rohde et al., 2006; Thomson & Hoffmann, 2007). One of the simplest and cheapest pest control methods is the spray of chemical insecticides. Despite pesticide importance to protect our crops from pests, whenever chemicals are overused, especially the most toxic ones, it harms the agroecosystem sustainability due to a misleading use of this vital tool (Pedlowski et al., 2012; Tooker & Pearsons, 2021). Abusive use of synthetic chemical insecticides can trigger significant negative side-effects upon different species of beneficial organisms which live in agroecosystems. Among those beneficial organisms, the natural biological control agents (predators, parasitoids, and entomopathogens) (Alexandre, 2010; Carmo et al., 2010; Fernandes et al., 2010) as well as pollinators (Kuldna et al., 2009), especially the honeybee *Apis mellifera* (Linnaeus, 1758) (Hymenoptera: Apidae) (Desneux et al., 2007; Greig-Smith et al., 1994; Shires et al., 2006) are essential for sustainable food production and are extremely fragile to the threat posed by an increasing load of chemical insecticides used in more conventional agricultural systems. Global pesticide production is expected to reach around ten million metric tons by 2050 (Tilman et al., 2001; Wagner et al., 2010).

Biocontrol agents are essential to keep pests under natural control (Parra et al., 2021; Torres, & Bueno, 2018), while pollinators are crucial for promoting high yields of cropped plants. In the world, 87.7% of the flowering plant species are dependable from pollination performed by animals. Among many pollinator species, *A. mellifera* must be highlighted as a cosmopolitan, super-generalist pollinator that provides a vital ecosystem service (Keng-Lou et al., 2018). Therefore, a sustainable food production system should aim at agricultural practices that combine environmental safety with efficient pest management, that neither imposes harm to the ecological balance of the agroecosystem nor reduces profits of the yield (Bueno et al., de Bueno et al., 2011). It is undoubtedly challenging since arthropods englobe both beneficial and pest species. Nevertheless, eliminating pests with insecticides while preserving beneficial organisms is possible, at least to some extent. It is discussed in more details ahead in this chapter during the subitem "Moderate use of selective insecticides."

The search for more selective pest control tools has been constant and currently emphasized when regarded as bees. Since the Collapse of Disorder of Colonies phenomenon (van Engelsdorp et al., 2009, 2017, 2007) global decrease of bees is an eminent concern regarding its possible impact on the reduction of world food production due to a potential lack of pollination. The most plausible cause of this disorder is the indiscriminate use of synthetic chemical insecticides. Some sublethal and lethal pesticide side-effects of common pesticides are known upon bees (Abati et al., 2021; Desneux et al., 2007; Tosi & Nieh, 2017). Therefore, the use of the most selective insecticides, restricted to when necessary, combined with other more environmental-friendly pest management tools, will form the fundamental foundation of the most accepted and sustainable way of controlling pests in agriculture named integrated pest management (IPM) (Bueno et al., 2021).

Therefore, integrated pest management is a set of tools used on pest management of different crop systems to maintain the agroecosystem as close as possible to a biological balance, mitigating any potential harmful effect upon agriculture. Those tools might need to be adapted to different necessities of each crop in a way that fosters long-term sustainability of that specific crop system, in contrast to an old concept that advocates a vision of immediate profitability at any cost (Bueno & Bueno, 2012; Prokopy & Kogan, 2009).

The first principle of IPM is that plants tolerate some levels of injuries triggered by pests without reducing yield (Fig. 4.1) (Higley & Peterson, 1996; Peshin et al., 2009; Prokopy & Kogan, 2009). Eventually, even overcompensation might occur under low injury, slightly increasing yield.

When output is reduced, it is essential to compare yield reduction costs versus control costs before deciding to use insecticides. In order to facilitate performing such comparison, Stern et al. (1959) termed the smallest population of pests that can cause economic damage to cropped plants as economic injury level (EIL). However, since the progress of the injury curve (Fig. 4.1) will not be stopped immediately after the control is taken, to avoid pest population from reaching EIL, management action should be taken earlier to EIL, at the economic threshold (ET), thus preventing pest number from reaching EIL (Bueno et al., 2013; Panizzi, 2013; Pedigo, 1986), that is when an economic injury occurs (Fig. 4.2).

In order to have a better knowledge of the importance of IPM to sustainable food production, it is crucial to go over the concepts of this technology as well as the essential pest management strategies that might be associated with the lowest impact possible upon the agroecosystem. Those points are discussed in more detail in the following subitems of this chapter.

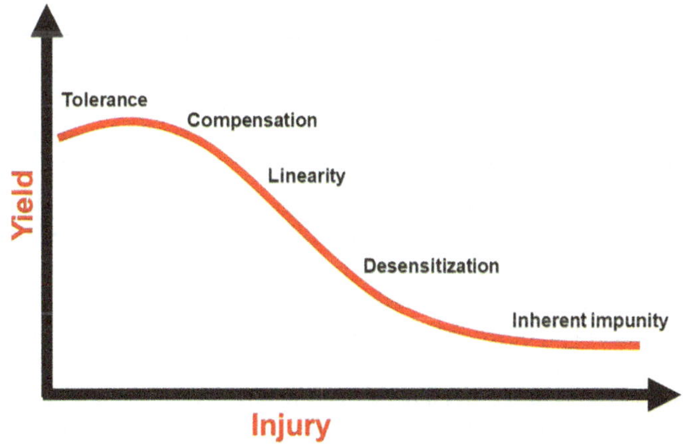

Fig. 4.1 Yield curve response of plants to pest injury

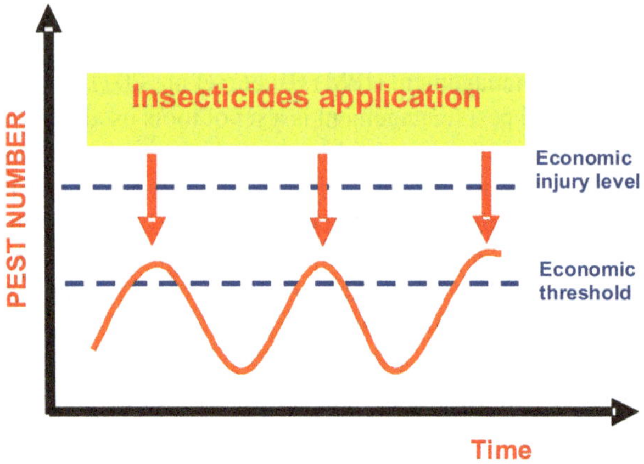

Fig. 4.2 The rational use of insecticides inside integrated pest management with chemical control only occurs when economic thresholds are reached or surpassed by the insect numbers

2 The History and Concepts of Integrated Pest Management (IPM) as a Sustainable Crop Management Strategy

The beginning of IPM was strengthened when organophosphate insecticides started being banned for agricultural use in different countries (Ehler, 2006; Peshin & Dhawan, 2009). Without this frequently-used tool to control pests, there was an increasing need to go back to researching for new pest management strategies that were not only efficient but also environmentally safe and sound (Ha, 2014). Such demand triggered the development of IPM in a way that in the late 1950s, the IPM terminology was first established and used as well as other similar terminologies such as pest management or integrated control (Kogan, 1998; Prokopy & Kogan, 2009). This scenario fostered a scientific discussion that helped spread the IPM concepts around the world. The first scientific publication which proposed the replacement of the so far used "identify and spray strategy" to a "sustainable pest management strategy" for agricultural use, to support a better environmental quality of the agroecosystem, was by Geier and Clark (1960). The proposed strategy was initially named Pest Management and later termed IPM (Ehler, 2006; Ha, 2014; Kogan, 1998; Pedigo, 1995).

This history of IPM development was also later marked for the classic "Silent Spring" written by Rachel Carson (Carson, 1962). The author reported over her science book the environmental contamination due to the overuse of pesticides in North America. Carson was an activist author at her time, and in her book, she accused the pesticide industry of spreading disinformation while public authorities accepted those industry marketing claims unquestioningly. "Silent Spring" illustrated the importance of a complete change in pest control, which was happening slowly at that time with the beginning of IPM all over the world. A change that urged for rational and sustainable use of synthetic pesticides in agriculture in association with the help of other more environmental-compatible pest control strategies (Kogan, 1998) such as biological control, use of transgenic plants, among other tools, which is discussed in detail in this chapter.

IPM concept supports the harmonious use of different control strategies based on the EIL and ET criteria as previously mentioned and on the correct sampling and pest identification. It is also crucial to IPM's success the measuring of pest natural mortality causes (biotic and abiotic) (Bueno et al., 2021; Panizzi, 2013). Only with the association of ETs, pest numbers, and insect mortality causes is possible a precise diagnosis of field scenario. Then, just after that, the best decision of which pest management strategy or strategies should be adopted can then be taken. Among the different pest control strategies that might be combined inside IPM, the ones nowadays available for farmers that should be highlighted in this chapter are: (1) Biological control, (2) Chemical insecticides, (3) Insect resistant varieties or cultivars (which also includes the genetically modified plants such as the Bt plants), (4) Pheromones (which is more restricted to sampling, at least for crops cultivated over large areas), (5) Genetic manipulation of pests (which is a new technology that will probably increase in the future), and (6) Manipulation of the environment and cultural methods (which is more restricted to stored products than cropped fields) (Fig. 4.3).

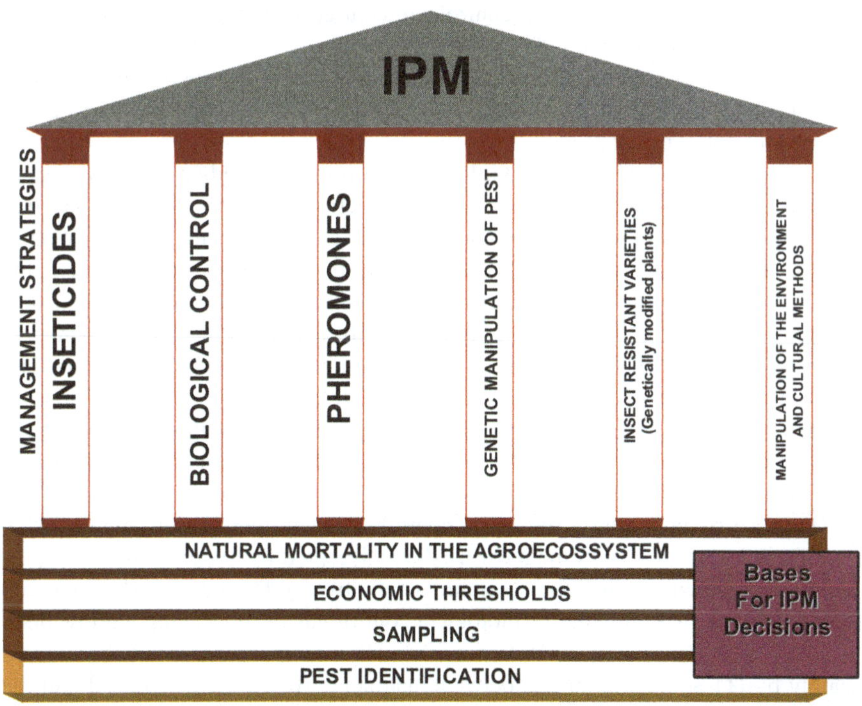

Fig. 4.3 Different pest management strategies combined into the concept of Integrated Pest Management

2.1 Biological Control

In the simplest way possible, biological control is the action of a living organism (biocontrol agent) controlling the population of another species of a living organism (pest) (Altieri, 1999). It is considered one of the most sustainable and compatible pest management strategies with environmental preservation. Despite only more recently being overspread around the world and commercially explored, biological control is, in fact, a very ancient pest control strategy. It has been globally used for more than 100 years (Gurr et al., 2000) over more than 30 million hectares (van Lenteren, 2012).

Biological control can be didactically divided into natural, conservative, classical, and augmentative biological control (van Lenteren et al., 2018). Conventional biological control is based on preserving natural characteristics of the environment to support a higher number of biocontrol agents of pests in the fields, which can naturally prevent pest outbreaks (Baker et al., 2020). In addition, it seeks to preserve the environment to integrate beneficial insects back into crop systems for natural pest control. A natural biological control refers to the naturally existent biocontrol agents (predators, parasitoids, and entomopathogens) that occur without human interference (Bale et al., 2008). Despite its importance to agroecosystem sustainability, when natural biological control is insufficient to keep pest numbers below EIL, the augmentative biological control can be used with massive releases of biocontrol agents (Barratt et al., 2017). On the other hand, classical biological control refers to introducing an exotic natural biocontrol agent in the environment, which, if established, is expected to keep the pest population under control (Parra et al., 2021; van Lenteren, 2012).

As you might see, all types of biological control are related and might co-occur in the field, with the division only related to didactic matters. In practice, conservative biological control, classic biological control, and augmentative biological control are ways to improve the efficacy of natural biological control (Barratt et al., 2017). While the conservative biological control is the group of strategies to enhance the agroecosystem to support a higher number of biocontrol agents, augmentative biological control will speed the rise of the natural biological control population by massive releasing biocontrol agents that could already be present in the environment in low numbers. Similarly, classical biological control will import beneficial exotic species to improve natural biological control when existing organisms are not efficient against a specific pest.

Despite different types of biological control and their interrelation, augmentative biological control has been the most used and accepted biological control strategy among IPM adopters since the beginning of the twenty-first century (Lacey et al., 2015). The use of augmentative biological control has been growing worldwide at around 10–20% per year (van Lenteren et al., 2018). This growth has been fostered mainly by (1) Some similarities between augmentative biological control and the traditional use of insecticides where farmers experience a fast decrease of pest population after the "application" of a commercial product. These similarities make farmers more comfortable and confident with the use of augmentative biological control; (2) Demands from the international market for more sustainable products, with less use of synthetic chemical insecticides. Those demands have been used as barriers for closing deals, making farmers more open to using other pest management strategies different from chemical insecticides; (3) Increasingly strict environmental legislation that restricts the approval of new chemical products, imposing criteria more stringent in terms of toxicity and persistence in the environment and, (4) Cases of pest resistance to chemical insecticides increasingly common in the field and, consequently, failure of control, creating a demand for new pest management tools which biological control products have fulfilled.

In addition, the production of biological control products that were previously produced in small factories, often in a more artisanal way, has been professionalizing, becoming a big business and, consequently, attracting the attention of large industries in the agricultural sector, together with the quality of formulation and the possibility of joint use with other pesticides (Maciel et al., 2021, 2022), thus encouraging the adoption of biological products. As a result, the augmentative biological control market reached USD 5.2 billion in 2020 worldwide. Furthermore, the increase of biological control sellers in the field, working as consultants, has also helped the market rise and, therefore, the biological control adoption.

A great example of the successful use of biological control inside the IPM concept occurs in sugarcane. In this crop in Brazil, biological control is essential and extensively used in IPM (Parra, 2014). Different biological control agents are used, and the association of varying biocontrol agents seems to be efficient in managing some important pest species. Among the biocontrol agents used, releases of the larval parasitoid *Cotesia flavipes* Cameron, 1891 (Hymenoptera: Braconidae) associated with releases of the egg parasitoid *Trichogramma galloi* Zucchi, 1988 (Hymenoptera: Trichogrammatidae) are used in 2.2 million hectares. In addition, another 1.3 million hectares receive the release of only *C. flavipes* to control the sugarcane borer *Diatraea saccharalis* Fabricius, 1794 (Lepidoptera: Crambidae) (Bezerra et al., 2021; Parra, 2014). Also, in sugarcane, the entomopathogenic fungus *Metarhizium anisopliae* is used to control the leafhopper *Mahanarva fimbriolata* Stal, 1854 (Hemiptera: Cercopidae) in more than one million hectares (Bale et al., 2008; Mascarin et al., 2019). Hubner, 1818

Indeed, this high use of biological control in sugarcane is somehow a consequence of difficulties of spraying insecticides over the crop that requires most of the time the use of airplanes. It makes insecticide use more expensive, favoring biological control in comparing both tools usually performed by farmers before deciding on each management tool they will adopt. Of course, biological control use in sugarcane just increased because it combined both efficiency and economic benefits. In addition, the consequent low use of insecticide in the crop creates a better environment for the success of the used biocontrol agents favoring, even more, the biological control success.

Another excellent example of a successful augmentative biological control program in the world happened in soybean. It was the use of baculovirus to control *Anticarsia gemmatalis* Hubner, 1818 (Lepidoptera: Erebidae) in the 90s in soybean fields in the State of Paraná, Brazil, totaling an area of two million hectares treated with this entomopathogen in a very effective way (Moscardi, 1999). It was considered the highest augmentative biological control program by its time. However, this area has decreased more recently due to unfair competition of the baculovirus with cheaper chemical insecticides and the lack of IPM adoption and pest sampling in the field. For the augmentative biological control to succeed in its use, biocontrol agents must be released or applied in areas where IPM is also adopted. IPM provides a more stable and ecologically balanced agroecosystem, which favors the success of the biocontrol agents (Prokopy & Kogan, 2009).

Not only is IPM essential to augmentative biological control success, but also biological control is crucial for IPM to succeed. Therefore, it is necessary to avoid excessive pesticides (insecticides, fungicides, and herbicides) to preserve biocontrol. For example, in the soybean field in Brazil, before 2001, fungicides used to be sprayed only once over the crop season. However, after the first occurrence of *Phakopsora pachyrhizi* in the country, a fungal disease that affected soybean plants, fungicide sprays increased to 3 or more sprays over the crop season. This increase in fungicide use harmed the natural incidence of *Metarhizium rileyi*, an entomopathogenic fungus that causes epizootics in populations of lepidopteran pests. As an adverse side-effect, it triggered the occurrence of the caterpillar *Chrysodeixis includens* Walker, 1858 (Lepidoptera: Noctuidae), which used to be a secondary pest controlled by the fungus. That is why it is essential to emphasize that insecticides and fungicides, herbicides, and any other type of chemical applied over the crop should be sprayed only when necessary to preserve biocontrol agent action in the field. This rational use of chemicals increases natural control of pest outbreaks (Sosa-Gómez, 2017) what will be discussed in more detail in the following subitems.

2.2 Moderate Use of Selective Insecticides

At the same speed that the demand for food grows globally, so makes the loss of crop yield caused by pests. By 2050, it is estimated that crop losses in European wheat will increase from 50% to 100%, reaching 16 million tons. Similarly, the same study indicates a 30–40% increase in yield losses in North American corn. A loss of 27 million tons of rice in China is also expected due to pest outbreaks (Deutsch et al., 2018). In Brazil, yield losses caused by Asian soybean rust alone increased from USD 177 million in the 2001/2002 crop season to USD 2.38 billion in the 2007/2008 crop season. It is estimated that pests cause annual losses of approximately USD 12 billion to Brazilian farmers only in Brazil (Rangel, 2015). Thus, it is clear that despite the growing availability of sustainable food production tools, synthetic chemical insecticides can still be needed, at least for a short-term scenario. Therefore, to improve the sustainability of this conventional agricultural tool, the use of more selective pesticides is essential for healthier and more sustainable food production around the world.

Chemical insecticides are, in general, the first line of defense used by farmers to control agricultural pests. This farmer's preference for chemical control is usually a consequence of (1) The ease of chemical insecticide application; (2) low prices of most insecticides; (3) Effective control, at least in a short-term scenario (Bueno, 2017). However, it is essential to consider that those chemicals trigger significant negative side-effects on beneficial organisms. This side-effects can cause resurgence of key pests, emergence of secondary pests (Bueno et al., 2021), or insect resistance to the insecticides used (Carvalho et al., 2013; Diez-Rodriguez & Omoto, 2001; Sosa-Gómez et al., 2001; Sosa-Gómez & Silva, 2010) in addition to the already mentioned negative consequences of misusing chemicals in agriculture. Those negative impacts of chemical pesticides certainly have an economic impact usually not accounted for by farmers who usually only compare market prices of chemicals versus biological insecticides.

Therefore, the best insecticide to be sprayed should have the highest efficiency against the target pest associated with the lowest impact on non-target beneficial organisms (Bueno et al., 2017; Torres, & Bueno, 2018). This is possible with the use of selective insecticides herein discussed.

Selectivity of a chemical product is its ability to act upon the arthropods inhabiting a specific agroecosystem, killing only the target species (pest) while causing the most negligible impact possible on beneficials (biocontrol agents) (Collier et al., 2016). Among the most selective insecticides used, the biological control products are the most important group, followed by the Insecticide Growth Regulators (IGRs) and then diamides and spinosyns (Torres, & Bueno, A. de F., 2018). Those products should be prioritized over the most harmful insecticides, usually in the groups of pyrethroids carbamates, among others. Unfortunately, when those harmful insecticides are the cheapest ones available, farmers still choose them to be used in their fields. Whenever selective insecticides are not available, non-selective products can still be used in a particular way, called ecological selectivity.

An important example of ecological selectivity in soybean is the mixture of insecticides with 0.5% NaCl (sodium chloride) sprayed on the field's border to control stink bugs. This recommendation is effective at the beginning of insect infestation because stink bugs always initiate outbreaks concentrated on the edges of the crop. In addition, the use of sodium chloride has an arresting effect on the stink bugs, making them stay longer on the applied border. Thus, they will be more contaminated by the insecticides (Corso & Gazzoni, 1998). This strategy allows an ecological separation of the pest species (insecticide target) and part of biocontrol agents living in the agroecosystem, at least beneficial arthropods that can still inhabit the center of the crop, which do not need to face the adverse effects of the pesticide.

It is essential to point out that the negative impact of pesticides above preservation of biological control can also happen, despite the minor intensity, with all types of products, even biological-based products. That is why all the products used in agriculture, more selective or harmful, must be used only, when necessary, with the adoption of IPM and ETs (Bueno et al., 2021).

2.3 Transgenic Plants

Integrated pest management has been improved in commercial fields by introducing biotechnology of engineered plants, expressing insecticide protein from *Bacillus thuringiensis* (Bt) at high levels all season long (Miklos et al., 2007). Those Bt plants have enormous potential to significantly contribute to sustainable agriculture food production (Godfray et al., 2010). Brazil was one of the first countries to adopt Bt cotton and Bt maize and the first to adopt Bt soybean. During the 2020/21 crop season, more than 30 million hectares have been cultivated with this technology only with soybean in Brazil (Sparks, 2021).

The adoption of Bt plants has benefited the integrated management of pests and the whole sustainability of food production in different agricultural systems (especially soybean, maize, and cotton). In addition to protecting plants from pest attack, the use of Bt plants reduces insecticide application, preserves natural enemies, increases profitability for farmers, and can cause regional pest suppression (Dively et al., 2018; Hutchison et al., 2010), which reduces, even more, the use of insecticides. Thus, Bt crops are essential for preserving natural biological control action (Romeis et al., 2019).

Only in Parana State, Brazil, the adoption of Bt soybean reduced an average of one insecticide spray per crop season during 2018/19 (Table 4.1). It might be considered just a slight reduction. However, taking more than 40 million hectares cropped in the country (Brazil) into consideration, it is possible to realize that the amount of insecticide saved each season is enormous. Furthermore, Bt adoption reduced the insecticide use and increased in more than 12 days, the first insecticide spray in the field. Twelve days more than the action of natural biological control is allowed in the area without the negative impacts of chemical insecticides. Therefore, it brings substantial help to sustainable food production. Moreover, Bt adoption reduced the insecticide control costs (Bt royalties were not considered here), and increased yield. Thus it is a promising result because it combines environmental and financial benefits (Table 4.1).

Table 4.1 Soybean Bt adoption results (Mean), Paraná State, Brazil

	Crop season 2018/19		Crop season 2019/20	
Variable	non-Bt	Bt	non-Bt	Bt
Number of insecticide sprayings over the crop season	3.1	2.1	2.3	2.2
Days until first insecticide spraying	52.5 days	64.8 days	51.2 days	64.9 days
Pest control costs (kg/ha)	228.0	150.0	180.0	132
Yield (kg/ha)	2976.0	3018.0	3774.0	3876.0

Adapted from CONTE et al. (2019, 2020)

3 Soybean-IPM as a Successful Set of Tools for Sustainable Food Production

Soybean is considered one of today's most important crops, supplying most of the global demand for vegetable oil and protein (FAO, 2020). Unfortunately, conventional soybean production where IPM is not adopted is still planted on millions of hectares worldwide, usually requiring large amounts of chemicals (Zalucki et al., 2009). However, it is essential toward sustainable food production, efficiently managing soybean pests and preserving the agroecosystem, adopting ecologically and economically sound integrated pest management (IPM) practices (Bueno et al., 2021). The use of IPM recommendations can create a more profitable business since it reduces the costs of insect control (due to the reduction of insecticide use) and provides more sustainable agriculture. In addition, IPM fosters equitable, secure, sufficient, and stable flows of both food and ecosystem services (Bueno et al., 2021; Castle & Naranjo, 2009; Ellsworth et al., 2017; Peterson et al., 2018).

One of the most significant examples of sustainable soybean production throughout IPM adoption is in the Parana Stage, South Brazil. It is a joint project carried out by State extensionists from Paraná Rural Development Institute (IDR-Paraná) and researchers from Embrapa Soybean. The State extensionists offer IPM guidance to different soybean farmers inside the project. The project has been running since the 2013/2014 crop season. As a result, the farmers assisted by State extensionists have conquered a reduced use of insecticides compared with the average use of insecticides in the State. It was a consequence of the adoption of ET to trigger insecticide sprays. Insecticides were only sprayed over the crops when ET was reached or surpassed.

Moreover, farmers assisted by the State extensionist have adopted the recommendation of prioritizing the use of the most selective pesticides available to natural biological control agents whenever chemical control is necessary. It helped preserve natural biological control and reduce the demand for new sprays even more. With the adoption of those practices, farmers have reached promising results in reducing insecticide use without any yield reduction.

The results of 7 years of the IPM project show a reduction in insecticide use between 44.7% (2016/2017 season) and 55.9% (2017/2018 season) in areas with IPM. An average decrease over the 7 years of 51.3% less insecticide used without yield reduction. This saving in insecticides is equivalent to a value of 124.1 kg of soybean/ha. It is essential to mention that despite these savings in insecticide use, the average yield over those 7 years was still slightly higher in IPM fields (74.2 kg). Thus, adopting IPM over those years meant a higher profit of at least 198.3 kg/ha per year (124.1 kg of savings in insecticide use +74.2 kg of higher yield) (Table 4.2), highlighting its economic and environmental importance toward more profitable and sustainable food production. This was only possible due to the adoption of ETs, which are one of the essential keystones to a successful IPM.

The adoption of soybean-IPM reduced the overall use of insecticides and delayed the first insecticide spray. Although on average, the first insecticide spray was

Table 4.2 Soybean-IPM results (Mean). Program[1] was carried out in 2013 in Parana State, South Brazil

Variable	Comparison	Crop season							
		2013/14	2014/15	2015/16	2016/17	2017/18	2018/19	2019/20	Average from 2013 to 2020
Number of insecticide sprayings over the crop season	IPM	2.3 (46 growers)	2.1 (106 growers)	2.1 (123 growers)	2.0 (141 growers)	1.5 (196 growers)	1.7 (241 growers)	1.7 (255 growers)	**1.9**
	Non-IPM	5.0 (333 growers)	4.7 (330 growers)	3.8 (314 growers)	3.7 (390 growers)	3.4 (615 growers)	3.4 (773 growers)	3.0 (553 growers)	**3.9**
Days until first insecticide spraying	IPM	60 days	66 days	66.8 days	70.8 days	78.7 days	74.0 days	62.5 days	**68.4 days**
	Non-IPM	33 days	34 days	36 days	40.5 days	43.6 days	40.3 days	45.1 days	**38.9 days**
Pest control costs (kg/ha)	IPM	144.6	120.0	120.0	138.0	84.6	126.0	108.0	**120.2**
	Non-IPM	301.8	300.0	240.0	246.0	196.2	246.0	180.0	**244.3**
Yield (kg/ha)	IPM	2953.8	3612.0	3426.0	3870.0	3702.0	3006.0	3864.0	**3490.5**
	Non-IPM	2920.2	3516.0	3282.0	3852.0	3.624.0	2916.0	3804.0	**3416,3**

Adapted from Conte et al. (2016, 2018, 2020, 2019, 2017, 2015)

[a]Program where public consultants (from IDR – Paraná) sampled pests over the seasons and took all the decisions about pest management in IPM areas in selected farmers. At the end of the season, the results of IPM areas were compared with other non-IPM areas of Parana, Brazil

performed 38.9 days after soybean sowing in the State of Parana, IPM adopters first sprayed insecticides only 68.4 days after sowing (Table 4.2). This more extended period without insecticides in the crop favors more remarkable preservation of biological control agents. This preservation of the natural biological control helps keep pests naturally under control, sometimes even dismissing the need for chemical sprays. Furthermore, it is essential to understand a pest management strategy should not cause 100% pest control. Without pests, their natural biological control agents will also die or migrate from the area due to the lack of food. Therefore, a more sustainable and durable pest management solution will only reduce pest populations to a level below EIL (Dara, 2019). Consequently, it is crucial to preserve food and biocontrol agents in the cropped area. Therefore, natural biological control preservation is vital to prevent pest outbreaks

To give better figures of the biological control importance, *Geocoris* sp.; a small predator of 3–4 mm in length by 1–2 mm in width (Tamaki & Weeks, 1972), can kill and consume around 9 eggs Lepidoptera per day (Corrêa-Ferreira & Moscardi, 1985). *Nabis* sp., another small predator between 6.8 mm and 10 mm in length (Cislaghi, 1986), can daily feed 21.16 eggs or 3.29 caterpillars (Corrêa-Ferreira & Moscardi, 1985). Larvae of *Callida* sp. can vanish around 65.6 caterpillars to reach the adult stage (Corrêa-Ferreira & Pollato, 1989). Similar to predators, parasitoids and entomopathogens are also important. For example, the egg parasitoid of the genus *Trichogramma* can parasitize and kill around 50 eggs of the of different pests of order Lepidoptera in a few days (Bueno et al., 2012), illustrating the importance and economic value of preserving those beneficial species into the goal of enhancing the natural biological control. This is only possible with the adoption of IPM. Those figures presented here illustrate the crucial value of IPM adoption to preserve biological control and, therefore, to accomplish a sustainable food production system.

3.1 Stink Bug Management

Stink bugs are the most critical pest species that attack soybean in South America, seriously reducing yield, especially at latitudes from 0° to 23° (de Bueno et al., 2015; Panizzi & Correa-Ferreira, 1997). They are pests that injure soybean during the reproductive stage, feeding directly on soybean pods, seriously reducing crop yields besides seed quality because of physiological and sanitary injuries (Corrêa-Ferreira & De Azevedo, 2002).

Where stink bugs occur, they are the major drivers to insecticide sprays, harming crop sustainability the most. The repeated use of the same mode of action group associated with the abusive use of these insecticides, which in some regions can reach up to 3 or more applications per season, led to the selection of resistant stink bug populations (Sosa-Gómez et al., 2020). The occurrence of resistant stink bugs to most of the available insecticides is one of the most significant problems for farmers in Brazil. However, from 2020/2021 crop season, farmers started to count on new tools to manage this pest in soybean fields. Among these novelties, the following stand out:

1. A new insecticide came into the Brazilian market recently, bringing a different mode of action. This facilitates the management of stink bug resistance to insecticides. Furthermore, with the availability of this new insecticide (ethiprole) with a different mode of action (GABA-mediated chloride channel blocker) from the other used insecticides, farmers can now more easily rotate the mode of action in the use of insecticides. Therefore, this rotation reduces the likelihood of selection of resistant stink bug population in the area, favoring the sustainability of the pest management in the region.
2. Development of soybean cultivars tolerant to stinking bugs (Block® technology);
3. Registration for the use of the first biological control product based on an egg parasitoid species [*Telenomus podisi* Ashmead, 1893 (Hymenoptera: Scelionidae)] for the management of stink bug eggs.
4. When used inside IPM concepts, these new tools will help foster sustainable soybean production over the country.

3.2 Soybean Cultivar Tolerance to Stink Bugs (Block® Cultivars)

Especially for areas with a recurrent history of problems with stink bugs, it is essential that, at the time of sowing, soybean farmers can choose to use cultivars that are more tolerant to the attack of this pest. Then, these cultivars are beneficial in reducing damage caused by the insects and managing resistant populations since tolerant plants do not impose selection pressure because the plant allows the insect to feed. In addition, however, it presents compensation mechanisms for the injury, maintaining the yield even under injury (Peterson et al., 2017). To make this possible, Embrapa, through conventional genetic improvement, selected soybean cultivars with greater tolerance to stink bugs. Thus, in 2019, the Block® technology was launched in the Brazilian market. This technology identifies soybean cultivars with less yield loss when subjected to intense stink bug outbreaks than cultivars without Block® technology (Lucini et al., 2021). Therefore, these cultivars have high productive potential compared to the best market standards. In addition to increased productivity, when subjected to higher populations of stink bugs, their yield will not be compromised with damaged grains in the same intensity as cultivars without the technology.

These Block® cultivars are availabilities associated or not with other technologies. Therefore, it is now possible to find Block® cultivars such as Bt soybean, RR soybean, or even conventional soybean (for more information about Block® Technology, see https://www.embrapa.br/en/soja/block).

3.3 *The Use of the Egg Parasitoid* Telenomus podisi

Telenomus podisi is a parasitoid insect of stink bug eggs, naturally found in the environment. The adult of this insect is a small wasp (around 1.0 mm) that develops (from egg to adult) inside the eggs of the host, killing it and preventing the emergence of nymphs of the pest (Bueno et al., 2012; Pacheco & Corrêa-Ferreira, 2000). Thus, in the parasitized egg, instead of the emergence of a stink bug nymph, an adult of the parasitoid emerges, which will continue to parasitize and control new stink bug eggs during its adult life. This parasitoid is a very effective biological control agent in controlling stink bugs, and an adult female of the parasitoid can parasitize up to 100 eggs of this pest (Silva et al., 2018).

Considering that the amount of *T. podisi* in nature may be insufficient to keep the stink bug population at low levels, farmers can now buy this parasitoid and release it on his cropped field. After years of research, the first bioproduct with this biological control agent in Brazil was registered in December 2019. Therefore, from the 2020/2021 crop onwards, the availability and use of this new management tool have gradually increased in different regions, strengthening the sustainability of soybean production in the country.

The soybean farmer who uses this wasp in the field must be aware of some of its peculiarities. As this parasitoid does not control the stink bug but the eggs that will give rise to the stink bugs, the right moment of its use in the field is crucial for the technology to succeed. The research recommends starting the release of this parasitoid when the presence of the first adults of the stink bugs is detected in the field, carrying out two to three releases within a 7-day interval of 6500 parasitoids per hectare. This will increase the chances of matching the parasitoid's presence in the field with its hosts, the stink bug eggs.

The release of the parasitoid is usually done at the pupal stage. These pupae can be distributed in the field separately or protected inside capsules containing eggs parasitized by the wasp (Braz et al., 2021) that must be strategically distributed in the area at central points. This release must be carried out as close as possible to the emergence of adults, avoiding sweltering days (late afternoon is the most indicated period) to reduce parasitoid mortality. In Brazil, Bueno et al. (2020) recorded 70% parasitism of *E. heros* eggs after 3 releases of *T. podisi* at weekly intervals, while the natural parasitism rate was only close to 10% (Fig. 4.4). However, since *T. podisi* adult lifespan is around 14 days (Silva et al., 2018), field release intervals could be increased for 2 weeks, what will probably improve parasitoid efficacy.

Also, it is essential to consider that these wasps are living organisms, which is why insecticide applications should be avoided in the area where the parasitoids were released, at least 10 days before and 2 weeks after the release of these parasitoids.

Soon after their emergence in the field, females of the parasitoid will search for hosts and lay their eggs in them, interrupting the development of the pest and starting the development of a new parasitoid. The parasitized eggs take about 2 weeks to give rise to a new population of "wasps," whose females are copulated and go out in

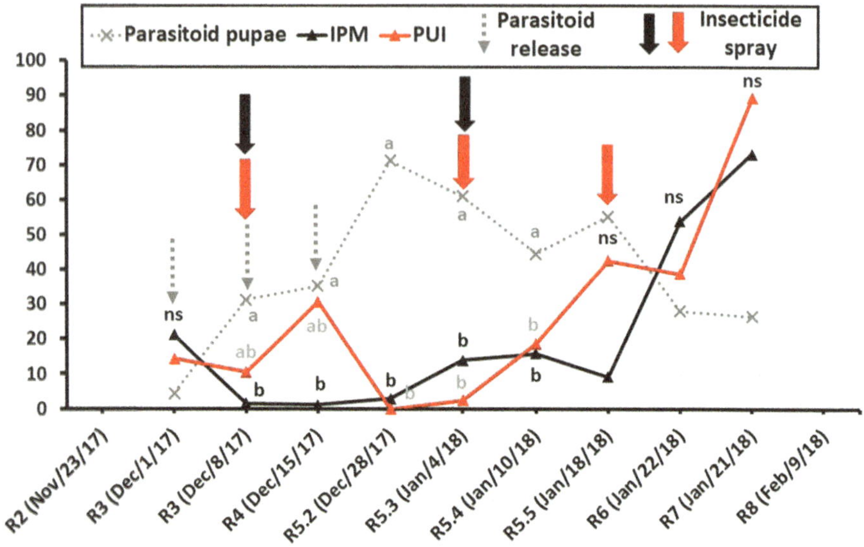

Fig. 4.4 Parasitism (%) of *Euschistus heros* eggs during the 2017/18 crop season. Arrows indicate the moment of application of *T. podisi* pupae or insecticides in each treatment. Means followed by the same letter in each soybean phenological stage did not differ statistically (Tukey test at 5% probability). Insecticides were applied only in Integrated Pest Management (**IPM**) when reaching the pest economic threshold or surpassing it and in Prophylactic Use of Insecticides (**PUI**) on a calendar basis following usual farmer's management

search of fresh eggs of stink bugs to parasitize them, preventing the increase of pest population in the field.

3.4 *Management of Resistance of Stink Bugs to Insecticides*

It is natural and expected that resistant stink bugs exist in nature. However, when these insects are managed correctly, those considered resistant will be a minority in the population, and their presence will not compromise the efficiency of control. However, when an insecticide with the same mode of action is used repeatedly and abusively on the crop, the resistant insects will be the only ones to survive. Crossing with each other, these resistant insects will multiply and become the majority in the population, at which point the insecticide will no longer work. Therefore, to avoid this selection of resistant insects, it is necessary to adopt some resistance management measures.

The primary measure is the rotation of products of different modes of action. This, with the availability of ethiprole on the market, from 2021/22 crop season, we have another distinct group of the method of action that can be used in rotation with other products on the market and thus helping the management of resistance of stink bugs to insecticides. Ideally, for each required application in the field, the soybean farmer would adopt an insecticide with a different mode of action.

4 Final Remarks

IPM adoption emphasizing augmentative biological control, transgenic plants, and other environment-friendly pest control tools is urgently needed toward sustainable food production. These pest management tools and recommendations are currently possible and combine environmental, economic, and social benefits.

The adoption of biological control is critical in sustainable food production, especially more critical recently since it has been an increasing requirement of different international markets, often even used as non-tariff barriers in commercialization. That is why for the success of agriculture, more augmentative biological control inside the IPM concept must be adopted, which will offer sustainable crop management and profitable results.

Most of the time, the international trademark requires certifications of sustainable production of trades being bought, biological control adoption, and reduced chemical pesticide use have been parameters often measured by certifiers for this matter. The international food market is increasingly going toward this tendency. Therefore, farmers must adapt themselves to these stricter requirements, knowing and quickly learning the use of biological control in practice. This is a one-way path to the future of modern agriculture.

Policymakers could help the adoption of IPM and more sustainable agriculture by passing laws to support the research, development, and field use of biological control and other environmentally friendly tools to manage the pest in the area. It essencial to a a more sustainable and profitable food production system.

Acknowledgments The authors thank Embrapa Soybean, Universidade Federal do Parana, and the National Council for Scientific and Technological Development (CNPq grant 302645/2018-7) for all the support that made this work possible.

References

Abati, R., Sampaio, A. R., Maciel, R. M. A., Colombo, F. C., Libardoni, G., Battisti, L., et al. (2021). Bees and pesticides: The research impact and scientometrics relations. *Environmental Science and Pollution Research, 28*(25), 32282–32298. https://doi.org/10.1007/s11356-021-14224-7

Alexandre, T. M. (2010). Estratégias para o manejo integrado de *Pseudoplusia includens* (Walker) (Lepidoptera: Noctuidae) em soja. 104f. Tese (Doutorado em Entomologia) – Universidade Federal do Paraná, Curitiba, PR.

Altieri, M. A. (1999). The ecological role of biodiversity in agroecosystems. *Agriculture Ecosystems & Environment, 74*, 19–31. https://doi.org/10.1016/S0167-8809(99)00028-6

Baker, B. P., Green, T. A., & Loker, A. J. (2020). Biological control and integrated pest management in organic and conventional systems. *Biological Control, 140*(February 2019), 104095. https://doi.org/10.1016/j.biocontrol.2019.104095

Bale, J. S., van Lenteren, J. C., & Bigler, F. (2008). Biological control and sustainable food production. *Philosophical Transactions of the Royal Society B: Biological Sciences, 363*(1492), 761–776. https://doi.org/10.1098/rstb.2007.2182

Barratt, B. I. P., Moran, V. C., Bigler, F., & van Lenteren, J. C. (2017). The status of biological control and recommendations for improving uptake for the future. *BioControl, 63*(1), 155–167. https://doi.org/10.1007/s10526-017-9831-y

Bezerra, J. I. M., Molter, A., Rafikov, M., & Frighetto, D. F. (2021). Biological control of the chaotic sugarcane borer-parasitoid agroecosystem. *Ecological Modelling, 450*(March), 109564. https://doi.org/10.1016/j.ecolmodel.2021.109564

Braz, É. C., de Bueno, A. F., Colombo, F. C., & de Queiroz, A. P. (2021). Temperature impact on Telenomus podisi emergence in field releases of unprotected and encapsulated parasitoid pupae. *Neotropical Entomology, 50*(3), 462–469. https://doi.org/10.1007/s13744-021-00857-3

Bueno, A. F., Paula-Moraes, S. V., Gazzoni, D. L., & Pomari, A. F. (2013). Economic thresholds in soybean-integrated pest management: Old concepts, current adoption, and adequacy. *Neotropical Entomology, 42*(5), 439–447. https://doi.org/10.1007/s13744-013-0167-8

Bueno, A. de F., Carvalho, G. A., Santos, A. C. d., Sosa-Gómez, D. R., da Silva D. M. (2017). Pesticide selectivity to natural enemies: challenges and constraints for research and field recommendation. *Ciência Rural, 47*(6). https://doi.org/10.1590/0103-8478cr20160829

Bueno, A. F., Braz, E. C., Favetti, B., França-Neto, M., & J. B. Silva. G. V. (2020). Release of the egg parasitoid *Telenomus podisi* to manage the Neotropical Brown stink bug, *Euschistus heros*, in soybean production. *Crop Protection, 137*, 105310.

Bueno, A. F., Panizzi, A. R., Hunt, T. E., Dourado, P. M., Pitta, R. M., & Gonçalves, J. (2021). Challenges for adoption of Integrated Pest Management (IPM): The soybean example. *Neotropical Entomology, 50*, 5–20. https://doi.org/10.1007/s13744-020-00792-9

Carmo, E. L., Bueno, A., de Bueno, F., & R. C. O. de F. (2010). Pesticide selectivity for the insect egg parasitoid *Telenomus remus*. *BioControl, 55*(4), 455–464. https://doi.org/10.1007/s10526-010-9269-y

Carson, R. L. (1962). Silent spring.

Carvalho, R. A., Omoto, C., Field, L. M., Williamson, M. S., & Bass, C. (2013). Investigating the molecular mechanisms of organophosphate and pyrethroid resistance in the fall armyworm *Spodoptera frugiperda*. *PLoS One, 8*, e62268. https://doi.org/10.1371/journal.pone.0062268

Castle, S., & Naranjo, S. E. (2009). Sampling plans, selective insecticides and sustainability: The case for IPM as "informed pest management.". *Pest Management Science, 65*(12), 1321–1328. https://doi.org/10.1002/ps.1857

Cislaghi, R. (1986). Representantes da Familia Nabidae (Hemiptera: Gymnocerata) no Rio Grande do Sul, *Brasil*. Pontifícia Universidade Católica do Rio Grande do Sul.

Collier, R., Jukes, A., Daniel, C., & Hommes, M. (2016). Ecological selectivity of pesticides and pesticide application methods. *Integrated Protection in Field Vegetables IOBC-WPRS Bulletin, 118*, 94–98.

Conte, O., Oliveira, F. T., Harger, N., Correa-Ferreira, B. S., & Roggia, S. (2015). *Resultados do manejo integrado de pragas da Soja na safra 2014/15 no Paraná* (361st ed.). Embrapa-CNPSo.

Conte, O., Oliveira, F. T., Harger, N., Correa-Ferreira, B. S., Roggia, S., Prando, A. M., & Seratto, C. D. (2016). *Resultados do manejo integrado de pragas da Soja na safra 2015/16 no Paraná* (375th ed.). Embrapa-CNPSo.

Conte, O., Oliveira, F., Harger, N., Corrêa-Ferreira, B., Roggia, S., Prando, A., & Serrato, C. (2017). *Resultados do manejo integrado de pragas da Soja na safra 2016/17 no Paraná* (402nd ed.). Embrapa Soja.

Conte, O., Oliveira, F. T., Harger, N., Correa-Ferreira, B. S., Roggia, S., Prando, A. M., & Seratto, C. D. (2018). *Resultados do manejo integrado de pragas da Soja na safra 2017/18 no Paraná* (402nd ed.). Embrapa-CNPSo.

Conte, O., Harger, N., Corrêa-Ferreira, B., Roggia, S., Prando, A., & Serrato, C. (2019). Resultados do manejo integrado de pragas da Soja na safra 2018/19 no Paraná. Retrieved from https://www.embrapa.br/busca-de-publicacoes/-/publicacao/1111771/resultados-do-manejo-integrado-de-pragas-da-soja-na-safra-201819-no-parana

Conte, O., Oliveira, F. T., Harger, N., Silva, G. C., Reis, E. A., Gomes, E. C., et al. (2020). *Resultados do manejo integrado de pragas da Soja na safra 2019/20 no Paraná* (431st ed.). Embrapa-CNPSo.

Cook, S. M., Khan, Z. R., & Pickett, J. A. (2007). The use of push-pull strategies in integrated pest management. *Annual Review of Entomology, 52*, 375–400. https://doi.org/10.1146/annurev.ento.52.110405.091407

Corrêa-Ferreira, B. S., & De Azevedo, J. (2002). Soybean seed damage by different species of stink bugs. *Agricultural and Forest Entomology, 4*(2), 145–150. https://doi.org/10.1046/j.1461-9563.2002.00136.x

Corrêa-Ferreira, B.S, & Moscardi, F. (1985). Potencial de consumo dos principais insetos predadores ocorrentes na cultura da soja . *Resultados de Pesquisa de Soja 1984/85*, 79.

Corrêa-Ferreira, B. S., & Pollato, S. L. B. (1989). Biologia e consumo do predador Callida sp. (Coleoptera: Carabidae) criado em *Anticaria gemmatalis* Hübner, 1818. *Pesquisa Agropecuaria Brasileira, 24*(8), 923–927.

Corso, I. C., & Gazzoni, D. L. (1998). Sodium chloride: An insecticide enhancer for controlling pentatomids on soybeans. *Pesquisa Agropecuaria Brasileira, 33*(10), 1563–1571.

Dara, S. K. (2019). The new integrated pest management paradigm for the modern age. *Journal of Integrated Pest Management, 10*(1). https://doi.org/10.1093/jipm/pmz010

de Bueno, A. F., Batistela, M. J., de Bueno, R. C. O. F., de França-Neto, J. B., Nishikawa, M. A. N., & Filho, A. L. (2011). Effects of integrated pest management, biological control and prophylactic use of insecticides on the management and sustainability of soybean. *Crop Protection, 30*(7), 937–945. https://doi.org/10.1016/j.cropro.2011.02.021

de Bueno, R. C. O. F., Parra, J. R. P., & de Bueno, A. F. (2012b). *Trichogramma pretiosum* parasitism of *Pseudoplusia includens* and *Anticarsia gemmatalis* eggs at different temperatures. *Biological Control, 60*(2), 154–162. https://doi.org/10.1016/j.biocontrol.2011.11.005

de Bueno, A. F., Bortolotto, O. C., Pomari-Fernandes, A., & de França-Neto, J. B. (2015). Assessment of a more conservative stink bug economic threshold for managing stink bugs in Brazilian soybean production. *Crop Protection, 71*, 132–137. https://doi.org/10.1016/j.cropro.2015.02.012

de Bueno, A. F., Carvalho, G. A., dos Santos, A. C., Sosa-Gómez, D. R., & da Silva, D. M. (2017). Pesticide selectivity to natural enemies: Challenges and constraints for research and field recommendation. *Ciência Rural, 47*(6). https://doi.org/10.1590/0103-8478cr20160829

Desneux, N., Decourtye, A., & Delpuech, J. M. (2007). The sublethal effects of pesticides on beneficial arthropods. *Annual Review of Entomology, 52*, 81–106. https://doi.org/10.1146/annurev.ento.52.110405.091440

Deutsch, C. A., Tewksbury, J. J., Tigchelaar, M., Battisti, D. S., Merrill, S. C., Huey, R. B., & Naylor, R. L. (2018). CLIMATE CHANGE increase in crop losses to insect pests in a warming climate downloaded from. *Science, 361*(August), 31. Retrieved from http://science.sciencemag.org/

Diez-Rodriguez, G. I., & Omoto, C. (2001). Herança da resistência de *Spodoptera frugiperda* (J. E. Smith) (Lepidoptera: Noctuidae) à lambda-cialotrina. *Neotropical Entomology, 30*, 311–316. https://doi.org/10.1590/S1519-566X2001000200016

Dively, G. P., Dilip Venugopal, P., Bean, D., Whalen, J., Holmstrom, K., Kuhar, T. P., et al. (2018). Regional pest suppression associated with widespread Bt maize adoption benefits vegetable growers. *Proceedings of the National Academy of Sciences of the United States of America, 115*(13), 3320–3325. https://doi.org/10.1073/pnas.1720692115

Ehler, L. E. (2006). Integrated pest management (IPM): Definition, historical development and implementation, and the other IPM. *Pest Management Science, 63*, 787–789. https://doi.org/10.1002/ps.1247

Ellsworth, P. C., Fournier, A., Frisvold, G., & Naranjo, S. E. (2017). Chronicling the socio-economic impact of integrating biological control, technology, and knowledge over 25 years of IPM in Arizona. In P. G. Mason, D. R. Gillespie, & C. Vincent (Eds.), *Proceedings of the 5th international symposium on biological control of arthropods* (pp. 11–15). CABI.

Ezeh, A. C., Bongaarts, J., & Mberu, B. (2012). Global population trends and policy options. *The Lancet, 380*(9837), 142–148. https://doi.org/10.1016/S0140-6736(12)60696-5

FAO. (2020). Soybean crop description. available: https://www.fao.org/home/en

Fernandes, F. L., Bacci, L., & Fernandes, M. S. (2010). Impact and selectivity of insecticides to predators and parasitoids. *EntomoBrasilis, 3*, 1–10. https://doi.org/10.12741/ebrasilis.v3i1.52

Geier, P. W., & Clark, L. R. (1960). An ecological approach to pest control. *Proceedings of the Eighth Technical Meeting. International Union of Conservation of Nature and Natural Resource*, 10–18.

Godfray, C. H. J., Beddington, J. R., Crute, I. R., Lawrence, H., Lawrence, D., Muir, J. F., et al. (2010). Food security: The challenge of Green revolution. Increases in production will have an important part to play, but they will be constrained as never before by the finite resources feeding 9 billion people. *Science, 327*(February), 812–818. https://doi.org/10.1109/CIS.2016.52

Greig-Smith, P. W., Thompson, H. M., Hardy, A. R., Bew, M. H., Findlay, E., & Stevenson, J. H. (1994). Incidents of poisoning of honey bees (Apis mellifera) by agricultural pesticides in Great Britain 1981–1991. *Crop Protection, 13*, 567–581. https://doi.org/10.1016/0261-2194(94)90002-7

Gurr, G. M., Barlow, N. D., Memmott, J., Wratten, S. D., & Greathead, D. J. (2000). A history of methodological, theoretical and empirical approaches to biological control. In G. M. Gurr & S. D. Wratten (Eds.), *Biological control: Measures of success* (pp. 3–37). Springer. https://doi.org/10.1007/978-94-011-4014-0_1

Ha, T. M. (2014). A review on the development of integrated Pest management and its integration in modern agriculture. *Asian Journal of Agriculture and Food Science, 2*(4), 336–340.

Higley, L. G., & Peterson, R. K. D. (1996). The biological basis of the EIL. In L. G. Higley & L. P. Pedigo (Eds.), *Economic thresholds for integrated pest management* (pp. 22–40). Nebraska Press.

Hung, K-L, J., Kingston, J, M., Albrecht, M., Holway, D, A., Kohn, J, R. (2018). The worldwide importance of honey bees as pollinators in natural habitats. Proceedings of the Royal Society B: Biological Sciences 285(1870) 20172140-10.1098/rspb.2017.2140.

Hutchison, W. D., Burkness, E. C., Mitchell, P. D., Moon, R. D., Leslie, T. W., Fleischer, S. J., et al. (2010). Areawide suppression of European corn borer with Bt maize reaps savings to non-Bt maize growers. *Science, 330*(6001), 222–225. https://doi.org/10.1126/science.1190242

Kogan, M. (1998). Integrated pest management: Historical perspectives and contemporary developments. *Annual Review of Entomology, 43*(1), 243–270. https://doi.org/10.1146/annurev.ento.43.1.243

Kuldna, P., Peterson, K., Poltimäe, H., & Luig, J. (2009). An application of DPSIR framework to identify issues of pollinator loss. *Ecological Economics, 69*(1), 32–42. https://doi.org/10.1016/j.ecolecon.2009.01.005

Lacey, L. A., Grzywacz, D., Shapiro-Ilan, D. I., Frutos, R., Brownbridge, M., & Goettel, M. S. (2015). Insect pathogens as biological control agents: Back to the future. *Journal of Invertebrate Pathology, 132*, 1–41. https://doi.org/10.1016/j.jip.2015.07.009

Lucini, T., Panizzi, A. R., & de Bueno, A. F. (2021). Evaluating resistance of the soybean block technology cultivars to the Neotropical brown stink bug, Euschistus heros (F.). *Journal of Insect Physiology, 131*(November 2020). https://doi.org/10.1016/j.jinsphys.2021.104228.

Maciel, R. M. A., Amaro, J. T., Colombo, F. C., Neves, P. M. O. J., & Bueno, A. F. (2021). Mixture compatibility of ChinNPV baculovirus with herbicides and fungicides used in soybean. *Semina: Ciencias Agrarias, 42*(5), 2629–2637. https://doi.org/10.5433/1679-0359.2021v42n5p2629

Maciel, R. M. A., Amaro, J. T., Colombo, F. C., Neves, P. M. O. J., & Bueno, A. F. (2022). Mixture compatibility of *Anticarsia gemmatalis* nucleopolyhedrovirus (Agmnpv) with pesticides used in soybean. *Ciencia Rural, 52*(2). https://doi.org/10.1590/0103-8478CR20210027

Mascarin, G. M., Lopes, R. B., Delalibera, Í., Fernandes, É. K. K., Luz, C., & Faria, M. (2019). Current status and perspectives of fungal entomopathogens used for microbial control of arthropod pests in Brazil. *Journal of Invertebrate Pathology, 165*(August 2017), 46–53. https://doi.org/10.1016/j.jip.2018.01.001

Miklos, J. A., Alibhai, M. F., Bledig, S. A., Connor-Ward, D. C., Gao, A. G., Holmes, B. A., et al. (2007). Characterization of soybean exhibiting high expression of a synthetic Bacillus thuringiensis cry1A transgene that confers a high degree of resistance to lepidopteran pests. *Crop Science, 47*(1), 148–157. https://doi.org/10.2135/cropsci2006.07.0463

Moscardi, F. (1999). Assessment of the application of Baculoviruses for control of Lepidoptera. *Annual Review of Entomology, 44*(1), 257–289. https://doi.org/10.1146/annurev.ento.44.1.257

ONU, U. N. D. of economic and S. A. P. division. (2019). *World population prospects 2019 – Volume II: Demographic profiles. World population prospects 2019 – Volume II: Demographic profiles* (Vol. II). https://doi.org/10.18356/7707d011-en

Pacheco, D. J. P., & Corrêa-Ferreira, B. S. (2000). Parasitismo de *Telenomus podisi* Ashmead (Hymenoptera: Scelionidae) em Populações de Percevejos Pragas da Soja. *Anais Da Sociedade Entomológica Do Brasil, 29*(2), 295–302. https://doi.org/10.1590/S0301-80592000000200011

Panizzi, A. R. (2013). History and contemporary perspectives of the integrated pest management of soybean in Brazil. *Neotropical Entomology, 42*(2), 119–127. https://doi.org/10.1007/s13744-013-0111-y

Panizzi, A. R., & Correa-Ferreira, B. S. (1997). Dynamics in the insect fauna adaptation to soybean in the tropics. *Trends in Entomology, 1*(1), 71–88. Retrieved from https://www.scienceopen.com/document?vid=06fbcd4c-5ad1-406b-a273-ec3334bfa113

Parra, J. R. P. (2014). Biological control in Brazil. *Scientia Agricola, 71*(October), 345–355. https://doi.org/10.1590/0103-9016-2014-0167

Parra, J. R. P., Pinto, A. S., Nava, D. E., de Oliveira, R. C., & Diniz, A. J. F. (2021). *Controle biológico com parasitoides e predadores na agricultura brasileira. (J. R. P. Parra, A. S. Pinto, D. E. Nava, R. C. de Oliveira, & A. J. F. Diniz, Eds.)* (1st ed.). FEALQ.

Pedigo, L. (1986). Economic injury levels in theory and practice. *Annual Review of Entomology, 31*(1), 341–368. https://doi.org/10.1146/annurev.ento.31.1.341

Pedigo, L. P. (1995). Closing the gap between IPM theory and practice. *Journal of Agricultural Entomology, 12*, 171–181.

Pedlowski, M. A., Canela, M. C., da Costa Terra, M. A., & Ramos de Faria, R. M. (2012). Modes of pesticides utilization by Brazilian smallholders and their implications for human health and the environment. *Crop Protection, 31*(1), 113–118. https://doi.org/10.1016/j.cropro.2011.10.002

Peshin, R., & Dhawan, A. K. (2009). Integrated pest management: Concept, opportunities and challenges. In R. Peshin & A. K. Dhawan (Eds.), *Integrated pest management: Innovation-development process* (Vol. 1, 1st ed., pp. 51–81). Springer Science. https://doi.org/10.1007/978-1-4020-8992-3

Peshin, R., Bandral, R. S., Zhang, W., Wilson, L., & Dhawan, A. K. (2009). Integrated Pest management: A global overview of history, programs and adoption. In R. Peshin & A. K. Dhawan (Eds.), *Integrated pest management: Innovation-development process* (Vol. 1, 1st ed., pp. 1–49). Springer Science. https://doi.org/10.1007/978-1-4020-8992-3

Peterson, R. K. D., Varella, A. C., & Higley, L. G. (2017). Tolerance: The forgotten child of plant resistance. *PeerJ, 2017*(10), 1–16. https://doi.org/10.7717/peerj.3934

Peterson, R. K. D., Higley, L. G., & Pedigo, L. P. (2018). Whatever happened to IPM? *American Entomologist, 64*, 146–150.

Prokopy, R., & Kogan, M. (2009). Integrated pest management. In V. H. Resh & R. T. Cardé (Eds.), *Encyclopedia of insects* (2nd ed., pp. 523–528). Academic. https://doi.org/10.1016/B978-0-12-374144-8.00148-X

Rangel, L. E. (2015). Perdas e danos para o agronegócio. Defesa Agropecuária. Política Agrícola. DSV/MAPA.

Rohde, C., Alves, L. F. A., Neves, P. M. O. J., Alves, S. B., da Silva, E. R. L., de Almeida J. E. M. (2006). Seleção de isolados de Beauveria bassiana (Bals.) Vuill. e Metarhizium anisopliae (Metsch.)

Sorok. contra o cascudinho Alphitobius diaperinus (Panzer) (Coleoptera: Tenebrionidae). *Neotropical Entomology 35*(2) 231–240. https://doi.org/10.1590/S1519-566X2006000200012
Romeis, J., Naranjo, S. E., Meissle, M., & Shelton, A. M. (2019). Genetically engineered crops help support conservation biological control. *Biological Control, 130*, 136–154. https://doi.org/10.1016/j.biocontrol.2018.10.001
Silva, G. V., Bueno, A. D. F., Neves, P. M. O. J., & Favetti, B. M. (2018). Biological characteristics and parasitism capacity of *Telenomus podisi* (Hymenoptera: Platygastridae) on eggs of *Euschistus heros* (Hemiptera: Pentatomidae). *Journal of Agricultural Science, 10*(8), 210. https://doi.org/10.5539/jas.v10n8p210
Smith, C. M. (2021). Conventional breeding of insect-resistant crop plants: Still the best way to feed the world population. *Current Opinion in Insect Science, 45*, 7–13. https://doi.org/10.1016/j.cois.2020.11.008
Sosa-Gómez, D. R. (2017). *Microbial control of soybean pest insects and mites* (pp. 199–208). From Theory to Practice. https://doi.org/10.1016/B978-0-12-803527-6.00013-5
Sosa-Gómez, D. R., & Silva, J. J. D. (2010). Neotropical brown stink bug (*Euschistus heros*) resistance to methamidophos in Paraná, Brazil. *Pesquisa Agropecuária Brasileira, 45*, 767–769. https://doi.org/10.1590/S0100-204X2010000700019
Sosa-Gómez, D. R., Corso, I. C., & Morales, L. (2001). Insecticide resistance to endosulfan, monocrotophos and methamidophos in the neotropical brown stink bug, *Euschistus heros* (F.). *Neotropical Entomology, 30*, 317–320. https://doi.org/10.1590/S1519-566X2001000200017
Sosa-Gómez, D. R., Corrêa-Ferreira, B. S., Kraemer, B., Pasini, A., Husch, P. E., Delfino Vieira, C. E., et al. (2020). Prevalence, damage, management and insecticide resistance of stink bug populations (Hemiptera: Pentatomidae) in commodity crops. *Agricultural and Forest Entomology, 22*(2), 99–118. https://doi.org/10.1111/afe.12366
Sparks. (2021). BIP soja 2021.
Stern, V. M., Smith, R. F., van den Bosch, R., & Hagen, K. S. (1959). The integrated control concept. *Journal of Agricultural Science, 29*, 81–101.
Tamaki, G., & Weeks, R. E. (1972). Efficiency of three predators, *Geocoris bullatus*, *Nabis americoferus* and *Cocinella transversogutata*, used alone or in combination against three insect prey species, Myzus persicae, Ceramica picta, and Manastra configurata, in a greenhouse study. *Environmental Entomology, 1*, 258–263.
Thomson, L. J., & Hoffmann, A. A. (2007). Ecologically sustainable chemical recommendations for agricultural pest control? *Journal of Economic Entomology, 100*(6), 1471–1750. https://doi.org/10.1093/jee/100.6.1741
Tilman, D., Fargione, J., Wolff, B., D'Antonio, C., Dobson, A., Howarth, R., Schindler, D., Schlesinger, W. H., Simberloff, D., & Swackhamer, D. (2001). Forecasting agriculturally driven global environmental change. *Science, 292*, 281–284. https://doi.org/10.1126/science.1057544
Tooker, J. F., & Pearsons, K. A. (2021). Newer characters, same story: Neonicotinoid insecticides disrupt food webs through direct and indirect effects. *Current Opinion in Insect Science, 46*, 50–56. https://doi.org/10.1016/j.cois.2021.02.013
Torres, J. B., & Bueno, A. de F. (2018). Conservation biological control using selective insecticides – A valuable tool for IPM. *Biological Control, 126*, 53–64. https://doi.org/10.1016/j.biocontrol.2018.07.012
Tosi, S., & Nieh, J. C. (2017). A common neonicotinoid pesticide, thiamethoxam, alters honey bee activity, motor functions, and movement to light. *Scientific Reports, 7*(1), 1–13. https://doi.org/10.1038/s41598-017-15308-6
van Lenteren, J. C. (2012). The state of commercial augmentative biological control: Plenty of natural enemies, but a frustrating lack of uptake. *BioControl, 57*(1), 1–20. https://doi.org/10.1007/s10526-011-9395-1
van Lenteren, J. C., Bolckmans, K., Köhl, J., Ravensberg, W. J., & Urbaneja, A. (2018). Biological control using invertebrates and microorganisms: Plenty of new opportunities. *BioControl, 63*(1), 39–59. https://doi.org/10.1007/s10526-017-9801-4

Vanengelsdorp, D., Underwood, R., Caron, D., & Hayes, J. (2007). An estimate of managed colony losses in the winter of 2006–2007: A report commissioned by the apiary inspectors of America. American Bee Journal, 147(7), 599–603.

VanEngelsdorp, D., Evans, J. D., Saegerman, C., Mullin, C., Haubruge, E., Nguyen, B. K., et al. (2009). Colony collapse disorder: A descriptive study. *PLoS One, 4*(8). https://doi.org/10.1371/journal.pone.0006481

VanEngelsdorp, D., Traynor, K. S., Andree, M., Lichtenberg, E. M., Chen, Y., Saegerman, C., & Cox-Foster, D. L. (2017). Colony Collapse Disorder (CCD) and bee age impact honey bee pathophysiology. *PLoS One, 12*(7), 1–24. https://doi.org/10.1371/journal.pone.0179535

Wagner, T. C., Rauf, A., & Schwarze, S. (2010). Pesticides and tropical biodiversity. *Frontiers in Ecology and the Environment, 4*, 178–179. https://doi.org/10.1890/10.WB.015

Yactayo-Chang, J. P., Tang, H. V., Mendoza, J., Christensen, S. A., & Block, A. K. (2020). Plant defense chemicals against insect pests. *Agronomy, 10*(8). https://doi.org/10.3390/agronomy10081156

Zalucki, M. P., Adamson, D., & Furlong, M. J. (2009). The future of IPM: Whither or wither? *Australian Journal of Entomology, 48*, 85–96.

Chapter 5
(Alternative Approaches to Pesticide Use): Plant-Derived Pesticides

Marie-Noëlle Sylvestre, Ahissan Innocent Adou, Allan Brudey, Muriel Sylvestre, Ludovic Pruneau, Sarra Gaspard, and Gerardo Cebrian-Torrejon

Abstract The indiscriminate use of pesticides over the years has caused a number of problems, including pest resistance and contamination of important global sources such as water, air and soil. Plant-based pesticides can therefore be an ecological alternative to synthetic pesticides to improve the efficiency of agricultural production and sustainably reduce the food crisis, while protecting the health of consumers. They are cheap, biodegradable, environmentally friendly and act more specifically through multiple mechanisms of action, suggesting that they are less dangerous to humans and the environment. In general, these compounds have important ecological activities in nature, such as: antifoedant, attractant, nematicide, fungicide, repellent, insecticide, growth regulator.

Keywords Biopesticides · Phytocompounds · Chemical ecology · Pest control · Natural products · Phytochemistry · Pest control · Insecticides · Green chemistry

M.-N. Sylvestre (✉) · L. Pruneau
Institut de Systématique, Evolution, Biodiversité (ISYEB), Muséum National d'Histoire Naturelle, CNRS, Sorbonne Université, EPHE, UFR des Sciences Exactes et Naturelles, Université des Antilles, Pointe-à-Pitre, Guadeloupe
e-mail: Marie-Noelle.Sylvestre@univ-antilles.fr

A. I. Adou · M. Sylvestre · S. Gaspard · G. Cebrian-Torrejon (✉)
COVACHIM-M2E Laboratory EA 3592, University of the French West Indies, Fouillole Campus, UFR SEN, Department of Chemistry, Pointe-à-Pitre Cedex, France
e-mail: gerardo.cebrian-torrejon@univ-antilles.fr

A. Brudey
Institut de Systématique, Evolution, Biodiversité (ISYEB), Muséum National d'Histoire Naturelle, CNRS, Sorbonne Université, EPHE, UFR des Sciences Exactes et Naturelles, Université des Antilles, Pointe-à-Pitre, Guadeloupe

COVACHIM-M2E Laboratory EA 3592, University of the French West Indies, Fouillole Campus, UFR SEN, Department of Chemistry, Pointe-à-Pitre Cedex, France

C. M. Galanakis (ed.), *Biodiversity, Functional Ecosystems and Sustainable Food Production*, https://doi.org/10.1007/978-3-031-07434-9_5

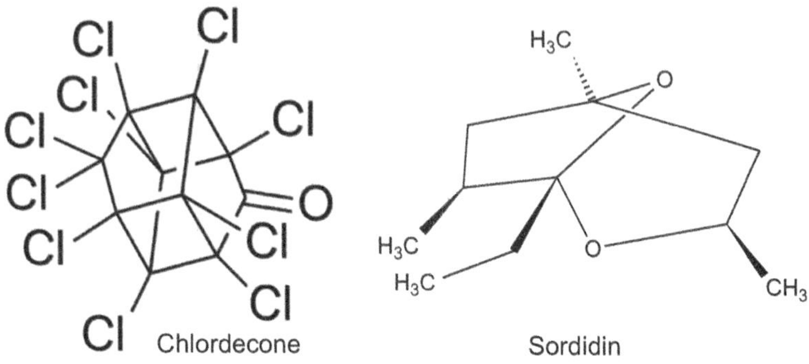

Fig. 5.1 Structure of chlordecone and sordidin

The synthetic organochlorated pesticide chlordecone (CLD, Fig. 5.1) was used from 70's to 90's in French lesser Antilles (Guadeloupe and Martinique) to control *Cosmopolitans sordidus* a banana's very aggressive pest. Due to its high stability and persistence, soils, groundwater, surface waters and coastal marine waters, crops, the entire food chain, and consequently reared animals in the banana-producing areas are polluted by this molecule (Cabidoche et al., 2009; Devault et al., 2016). After CLD use had been banned, it was replaced by using a weevil trap containing a pheromone, sordidin (Fig. 5.1) (Beauhaire et al., 1995).

Nevertheless, as a consequence of foodstuff contamination, the population of Guadeloupe and Martinique is exposed to chlordecone contamination through contaminated food and drinking waters consumption (Dubuisson et al., 2007). It has been shown (Dereumeaux et al., 2020) that 92.5% of the people from Martinique and 94.9% of people from Guadeloupe have detectable CLD concentrations in their blood. Several epidemiologic studies have been done to determine the impact of this exposure on health. CLD exposure is associated with an elevated risk of prostate cancer (Multigner et al., 2010) and of type 2 diabetes (Han et al., 2020) in the population. In addition, CLD is known to have endocrine-disrupting properties (Multigner et al., 2016). The TIMOUN study shows a correlation between pre- and postnatal exposure to chlordecone and short-term memory and fine motor abilities of young infants (Boucher et al., 2013; Dallaire et al., 2012). Recent studies show that CLD exposure may be associated with altering epigenetic marks (Legoff et al., 2020) and autoimmune disease (Khan et al., 2020).

The problem of chlordecone in the French West Indies is proof of the importance of awareness regarding pest environmental problems. Therefore, it is essential to invest in research of green pesticides to avoid these disasters (Hermawan et al., 1997; Miresmailli and Isman, 2014). In this review, we will focus on biopesticides from plant origin acting by several original mechanisms, including nervous, respiratory, endocrine, or hydric systems, and insect behavior (for example, attractive, repellent, or antifeedant biopesticides) (Souto et al., 2021).

1 Nervous System

1.1 Plant Derived Insecticides that Affect the Nervous System

The insect nervous system ensures the rapid functioning and coordination of the effector organs based on the information received by the sensory receptors. It is made up of neurons formed during the embryonic development of the ectoderm, whose axons ensure the propagation of nervous impulses. All areas of the insect brain work in synergy.

The insect Central Nervous System (CNS) consists of the brain (Fig. 5.2a) and the ventral nerve cord that stretches over the thoracic segments and several abdominal body segments (Fig. 5.2b). Among the insects, the CNS of *Drosophila* has been most thoroughly studied in terms of architecture and development. Although less is known in other insect models like the honey bee or the moth *Manduca sexta*, basic brain architecture is thought to be quite similar (Ribi et al., 2008; Rybak, 2012; Ito et al., 2014; Rittschof & Schirmeier, 2017).

1.2 Voltage-Gated Sodium Channels

The voltage-gated sodium channel of the insect nervous system was structurally and functionally homologous with the α-subunit of mammalian. This α-subunit is a single polypeptide chain with four internal homologous domains (I – IV) linked by intracellular linkers. Each part carries six transmembrane helices (S1 – S6) connected by intracellular or extracellular loops (P-loops). Inside the axonal membrane, the domains associate and form an aqueous pore for ion-conducting (PD), bounded

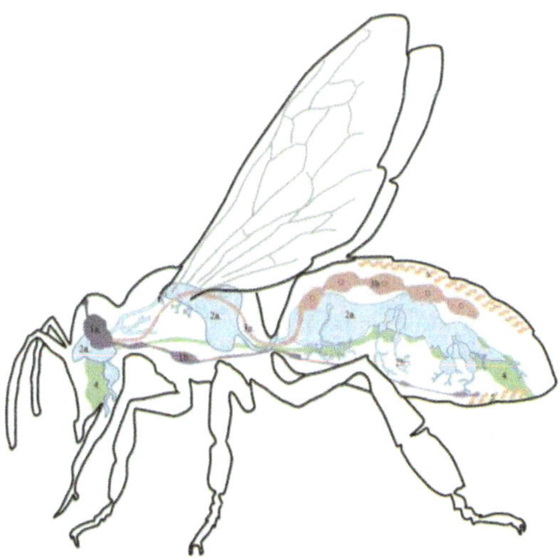

Fig. 5.2 Metabolism systems in insects (as modeled in the honey bee). Brain (**1a**); Ventral nerve cord with segmented ganglia (**1b**); Respiratory system (**2a, b**); Dorsal heart (**3a, b**); Digestive system (**4**); Fat body (**5**) (Rittschof & Schirmeier, 2017)

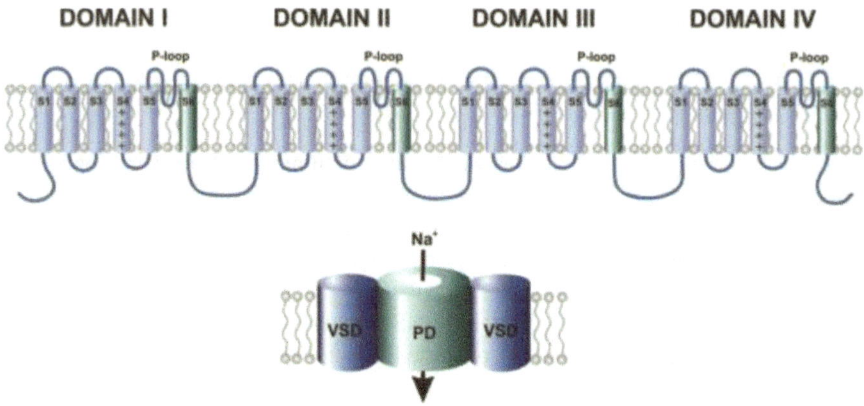

Fig. 5.3 Structure of transmembrane voltage-gated sodium channel (Davies et al., 2007)

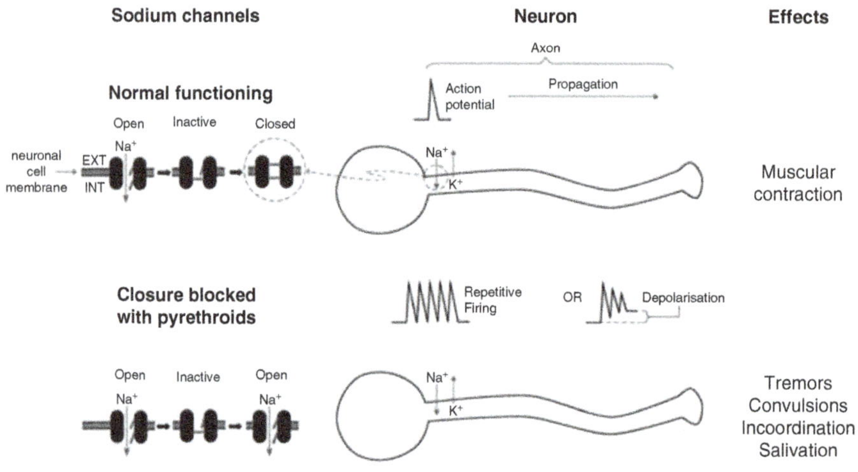

Fig. 5.4 Mechanism of action of pyrethrins on neurons (Hénault-Ethier, 2016)

by the different helix and P-loops (S5 and S6). The voltage-sensing part of the channel (VSD) was formed by the helix S1 and four cathionique segments of S4 (Catterall, 2000; Davies et al., 2007) (Fig. 5.3).

In physiological function, the flow of sodium ions into and out of the insect synapse occurs through the sodium channel present on the neuron's cell membrane. This flow is controlled by the regular movements of the insect's muscles. When the sodium channel is open, the muscle is activated, whereas when it is closed, the muscle can relax. This opening/closing allows the insect to move normally (Fig. 5.4, Hénault-Ethier, 2016). The insect axonal membrane contains a high concentration of sodium (Na+) and a low concentration of potassium (K^+), the nerve stimulation opens de sodium channels and increases the membrane permeability to this ion.

This gradient of ions is the motor force of the membrane action potential (Davies et al., 2007).

Insecticides that modify cell action potential *via* voltage-gated channels (sodium and potassium) are axonic poisons because they interfere with the axonal conduction of the nerve impulse.

Phytocompounds Acting via Voltage-Gated Sodium Channels

-Terpenoids: The pyrethrin and pyrethroid insecticides are terpenoids isolated in *Tanacetum cinerariifolium* (Trevir.) Sch. Bip. These natural products affect both insects' peripheral and central nervous systems. Pyrethrin terpenoids disrupt the sodium and potassium membrane potential and interrupts the correct transmission of nervous impulses.

A pyrethrin binds the sodium channel blocking it (Fig. 5.4). This ineffective closing produces an overstimulation of the muscles of the insect and different symptoms like tremors, convulsion, involuntary movements, salivation, and sublethal effect, known as 'knockdown.' The insect will eventually die if the exposure is high enough (Casida, 1980; Soderlund et al., 2002; Ware & Withacre, 2004; Davies et al., 2007; Hénault-Ethier, 2016; Oguh et al., 2019).

Pyrethrins I and II (Fig. 5.5) are both contact and stomach poison. Pyrethrins insecticides are fast-acting and cause an immediate "knockdown" paralysis, particularly in flying insects (Casida, 1980; Isman, 2006; Soderlund et al., 2002; Soderlund, 2012).

However, many insects can quickly metabolize pyrethrins and recover rather than die. In addition, when populations of house flies were chronically exposed to pyrethrins we observed phenotype of 'knockdown-resistant' flies, presenting modification in the amino acid sequence of the loop between helix S4 and S5 of the second subdomain the chanel (Vais et al., 2000; Böttger et al., 2018). In order to block this feedback effect, two synergistic adjuvants (piperonyl butoxide (PBO[1]) and MGK-264) are added in the preparations containing pyrethrins. Each of them has intrinsic toxicity and particular physicochemical characteristics that can increase

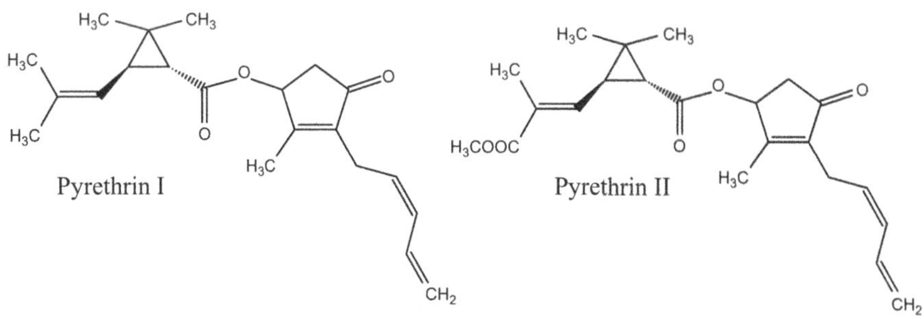

Fig. 5.5 Structure of pyrethrin I and II

[1] The piperonyl butoxide (PBO) is a semisynthetic derivative of safrol a phenylpropanoid isolated from *Sassafras albidum* (Nutt.) Nees.

the toxicity of pyrethrins by inhibiting the enzymes ordinarily responsible for their degradation in the body (Hénault-Ethier, 2016; Oguh et al., 2019; EPA Pyrethroids and Pyrethrins. http://www.epa.gov/oppsrrd1/reevaluation/pyrethroids-pyrethrins.html). Various pyrethroids pesticides (against houseflies) are formulated with piperonyl butoxide due to its synergetic effect related to its oxidase-inhibiting action, as well as the increase of the biodisponibility for the cuticule penetration (Joffe et al., 2012). In parallel, this combination increases toxicity by inhibitionnof detoxification by cytochrome P450-mediated d (Demeneix et al., 2020). Pyrethrins don't preent a acute mammalian toxicity,mainly because their absortoon is not effective and are easily degradated by digestive enzymes (Shivanandappa & Rajashekar, 2014). Additionally mamals'sodium-channels are less sensitive to this insecticides (Vais et al., 2000). Authors' work in rats shows that a unique mutation due to substituting a single amino acid (isoleucine for methionine at position 874) in the α-subunit of S4-S5 linker in domain II increases the sensitivity of the neuronal sodium channel to pyrethroid insecticides by 100 fold. Another terpene acting by this pathway is the β-himachalene isolated from the essential oil of *Cedrus deodara* (Roxb. ex D.Don) G.Don.

-Alkaloids: Sabadilla, named as well Veratrine, is an alkaloid extract from *Schoenocaulon officinale* A. Gray. This extract is composed mainly of two different alkaloids, the cevadine and the veratridine (being cevadine the most abundant (Fig. 5.6).

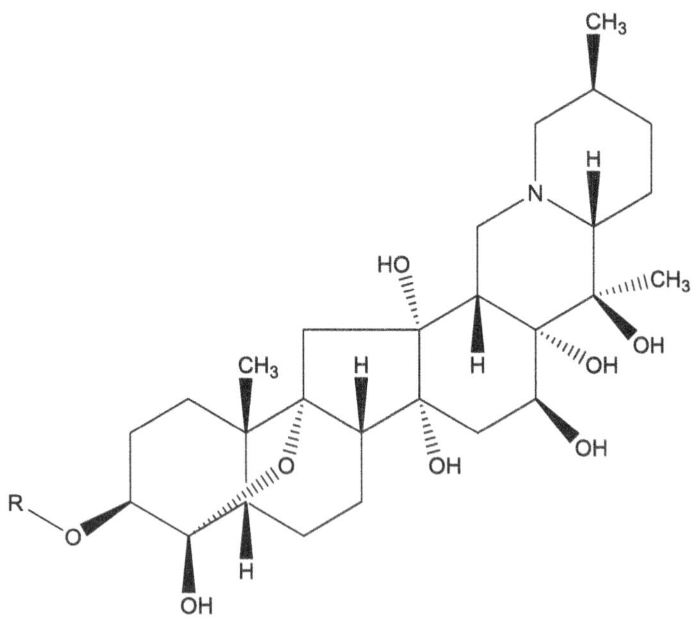

Cevadine R = (z)-CH3CH=C(CH3)CO
Veratridine R = 3,4-(CH3O)2PhCO

Fig. 5.6 Structure of cevadine et veratridine

Sabadilla acts as a contact and a stomach poison, and the sabadilla alkaloids (veratridine and cevadine, Fig. 5.6) action mode seems similar to that of the pyrethrins on voltage-sensitive sodium channels of nerve, heart, and skeletal muscle cell membranes. The action of veratridine is closely associated with the prior opening of single normal sodium channels, which is different from the case of pyrethrins (Barnes & Hille, 1988; Ujváry, 2010).

-***Phenylpropanoids***: The piperonyl butoxide (PBO, Fig. 5.7) is a synergic agent working in combination with many different BPs, for example, the pyrethrins or with ryanodine, enhancing the effect and the time of action.

-***Sugars***: Decaleside I and II (Fig. 5.8) are isolated from the roots of *Decalepis hamiltonii* Wight & Arn. These trisacharides are toxic to insects by contact *via* the

Fig. 5.7 Structure of piperonyl butoxide (PBO)

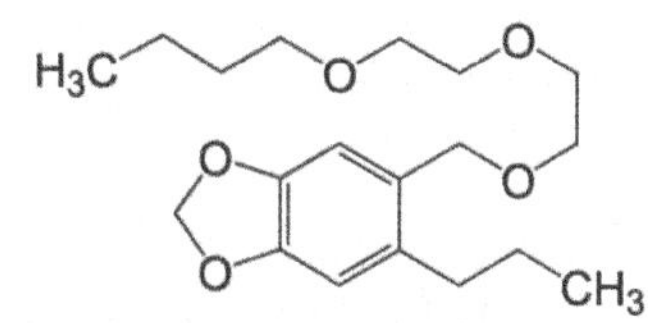

Piperonyl butoxide

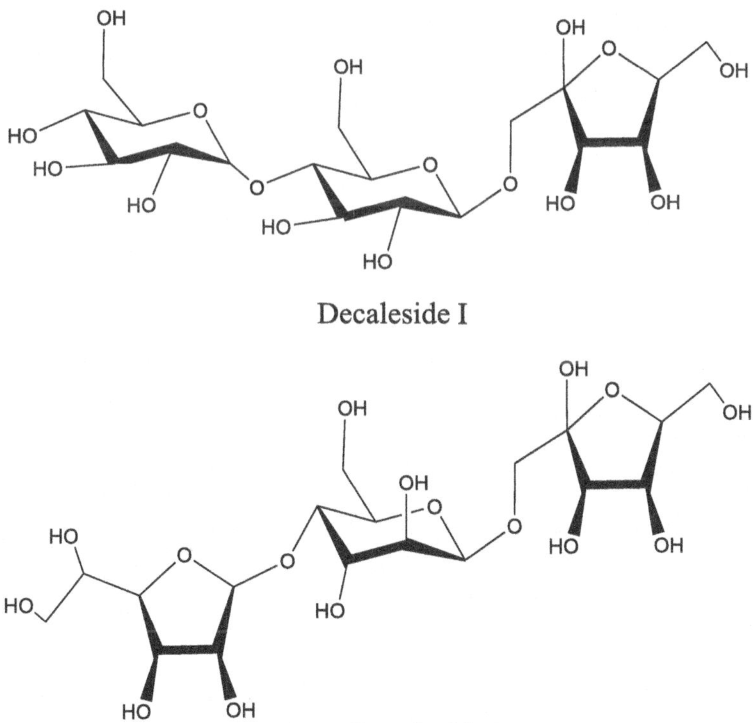

Decaleside I

Decaleside II

Fig. 5.8 Structure of decaleside I and II

tarsal gustatory chemosensillar but not harmful by oral or topical application (Rajashekar et al., 2010, 2012; Rajashekar & Shivanandappa, 2017).

In Drosophila a direct reporting is observed between the axons of the gustatory receptor neurons from the chemosensilla and the thoracic-abdominal and subesophageal ganglion (Inoshita & Tanimura, 2006). Works demonstrate that Na^+, K^+-ATPase is the target of decaleside, and toxicity of decaleside involves pump sodium inhibition by interacting with the ATP binding site of the enzyme (house fly: both head and thorax; cockroaches: nervous (ganglion) and muscle tissue). Furthermore, the abolition of toxicity of Decaleside II by hydrolysis by the salivary enzymes indicates that only the intact trisaccharide molecules exhibit insecticidal activity (Rajashekar & Shivanandappa, 2017). Like the knockdown effect of pyrethrins, decalesides act on neuronal sodium channels; however, pyrethrins act on the opening of these channels and are toxic by contact at any point of the insect‘s body decalesides inhibit the sodium pump. Their action is only mediated by contact with the tarsus.

1.3 Voltage-Gated Calcium Channels

Calcium (Ca^{2+}) plays a key role in cellular signalization (Gu et al., 1998; Bootman et al., 2001; Quintavalle, 2013; De Mandal et al., 2019). Indeed, Ca^{2+} signals initiate more significant global signals that propagate throughout the cells. Therefore, the control of calcium homeostasis is essential for normal cell function (Bootman et al., 2001). Control of Ca^{2+} homeostasis is modulated by the sarcoplasmic reticulum (SR) in striated muscles and the endoplasmic reticulum (ER) of neurons in other cells, which are the main intracellular calcium reservoirs. Passive Ca^{2+} transport across the plasma membrane, when the intracellular concentration is low, regulates calcium homeostasis. However, ER and SR also rapidly trigger Ca^{2+} signals when needed (Fill & Copello, 2002). Membrane depolarization, neurotransmitters, and hormones lead to a release of Ca^{2+} from the intracellular calcium-storing organelles (ER/SR) (Toprak et al., 2021), made possible in insects by the presence of two-channel proteins expressed in ER or SR and associated with the ER/SR membrane: ① inositol 1,4,5-trisphosphate receptor (IP3R), activated by the secondary messenger inositol 1,4,5-trisphosphate (IP3); ② and ryanodine (Fig. 5.9) receptor (RyR), mainly triggered by Ca^{2+}(Sattelle et al., 2008; Toprak et al., 2021).

The ryanodine receptor forms a tetrameric chanel for intracellular Ca^{2+} release (Fig. 5.10, Meissner, 1994; Sattelle et al., 2008; Lahm et al., 2009; Isaacs et al., 2012). This receptor is encoded by a single gene in insects. The RyRs are mainly expressed in embryo, hypodermal, somatic and visceral muscle (Takeshima et al., 1994; Wang et al., 2015).

Insect RyRs were studied in different insects (Sattelle et al., 2008; Toprak et al., 2021 review). These receoptors are composed of an amino-terminal region including a MIR domain (Mannosyltransferase, IP3R, and RyR), two RIH domains (RyR and IP3R Homology), three SPRY domains (splA kinase and ryanodine receptor), four RyR repeat domains, one RIH-associated domain, and a carboxy-terminal

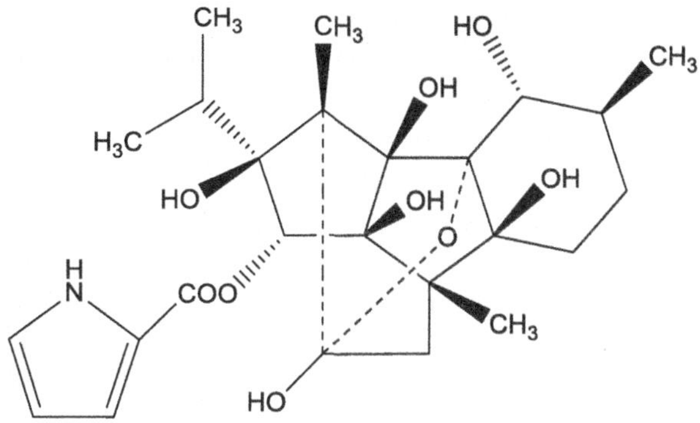

Fig. 5.9 Structure of ryanodine

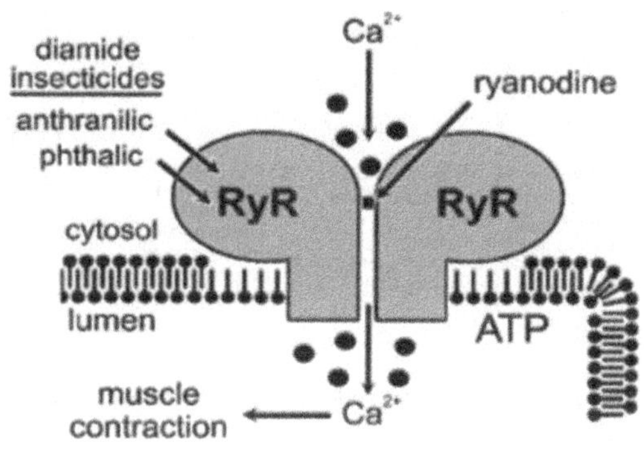

Fig. 5.10 Schema of the ryanodine receptor (associated proteins that interact directly with the RyR are not shown (Isaacs et al., 2012))

region including six transmembrane domains and two calcium-binding EF-hand domains (Catterall, 2000; D'Cruz et al., 2013; Quintavalle, 2013; Lin et al., 2020; Toprak et al., 2021 review). These RyRs are the largest ion channels (Fill & Copello, 2002; Sun et al., 2016).

The flight muscle of insects is a striated muscle with the regulation of its contraction and relaxation identical to that of the skeletal muscle of vertebrates. In muscle, SR stores a high concentration of calcium. The t-tubules invagination of the sarcolemma (plasma membrane of skeletal muscle) presents receptors that allow action potentials to propagate toward the cell's interior, leading to the opening of the RyRs, which are embedded in the SR membrane and coupled to these receptors. Once opened, following excitation of the plasma membrane, calcium from the sarcoplasm flows into the cell's cytosol, diffuses into the cytoplasm, and binds to troponin, a regulatory protein located on the actin thin filaments (G actin monomers). Troponin activates another regulatory protein, tropomyosin, by binding to it. Tropomyosin, then physically moves from its inhibitory position to binding sites on actin, and myosin cross-bridges, moving actin filaments toward the center of the sarcomere. This transport is an active mechanism against a concentration gradient and therefore consumes energy (Iwamoto, 2011). Thereby, muscle contraction is initiated (Fig. 5.11).

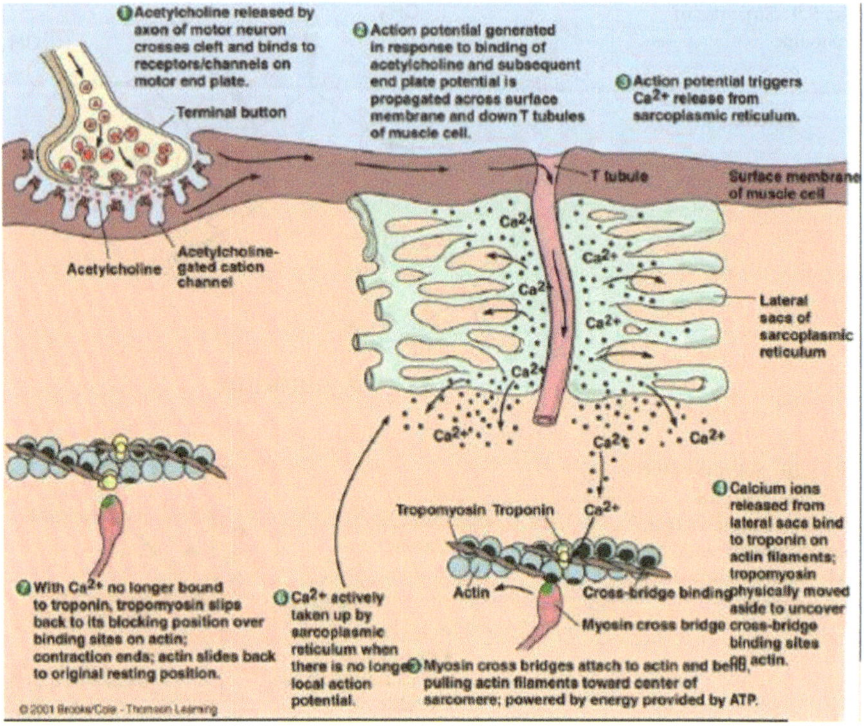

Fig. 5.11 Contraction/relaxation of the muscle (http://www.as.wvu.edu/~rbrundage/chapter8a / sld021.htm)

In contrast, relaxation occurs when the Ca^{2+} pump returns Ca^{2+} ions to the SR lumen (https://youtu.be/3Wc7I-H5stQ; Ellington, 1985; Lanner et al., 2010; Iwamoto, 2011; Feher, 2017). RyRs play a central role in Ca^{2+} pathways, especially in excitation/relaxation coupling (Coronado et al., 1994; Takeshima et al., 1994; Sullivan et al., 2000; Xu et al., 2000). Ca^{2+} opens RyR channel. At cytosolic Ca^{2+} concentrations (millimolar) the RyR channel is inhibited (Meissner, 1986, 1994). Regarding IP3Rs few research work can be found.

The RyRs calcium channel protein was named after its target insecticide, ryanodine (Nauen, 2006). The insecticidally active alkaloids contained in the plant species Ryania, ryanodine, and 9,21-dehydroryanodine (Jefferies et al., 1992) are the most active. These molecules interfere with the release of calcium into muscle tissue, thus blocking neuromuscular junctions, and are effective by either contact or ingestion against insects (Jefferies et al., 1992; Regnault-Roger, 2012). Ryanodine is a slow-acting stomach poison, and insects stop feeding soon after ingestion (Dimetry et al., 1993; Grdiša & Grsic, 2013). The antifeedant and insecticidal effects of two groups of ryanoid diterpenes: ryanodol/isoryanodol-type (nonalkaloidal type) and ryanodine-type (alkaloidal type) were studied.. Most of the nonalkaloidal ryanoids were antifeedants and toxic compared to alkaloidal ryanoids. Additionally the antifeedant effects did not parallel their toxic action. Ryanodol, the hydrolysis product of ryanodine, is also a botanical insecticide (González-Coloma et al., 1999; Ujváry, 2010). Ryania is effective against the fruit moth, coddling moth and corn earworm, European corn borer, and citrus thrips. Still, it is ineffective against the cabbage maggot cauliflower worms or the boll weevil (Isman, 2006; Sattelle et al., 2008). Ryania is effectively synergized PBO (Dimetry et al., 1993).

Ebbinghaus-Kintscher et al.'s (2006) study shows that phthalic acid diamides selectively activate ryanodine-sensitive intracellular calcium release channels (RyR) in larvae of the tobacco budworm *Heliothis virescens*. This causes Ca^{2+} channels to remain partially open, which leads to uncontrolled calcium release in muscle, resulting in feeding cessation, uncoordinated muscle contraction, paralysis, and death (Lahm et al., 2009; Jeanguenat, 2013; Lümmen, 2013; Yuan et al., 2017; Ma et al., 2020). Samurkas et al. (2020) discovered a potential species-specific green insecticide targeting the lepidopteran RyR N-terminal domain. This pesticides can be a solution for the problems related with the resistance to ecofriendly diamide insecticides.

Phytocompounds Acting via Voltage-Gated Calcium Channels
-Alkaloids: The ryanodine (Fig. 5.9) is an alkaloid isolated from *Ryania speciosa* Vahl; PBO synergizes Ryania products.

1.4 Acetylcholinesterase Enzyme (AChE)

Acetylcholine allows the trasmision of nerve impulses. Insect AcetylCholinesterase (AChE) is widely distributed and plays a crucial role in neuromuscular and neuronal communication in insects (Gnagey et al., 1987; Marcel et al., 1998). This protein, associated with cholinergic synapses, is a crucial enzyme of the cholinergic system in which it regulates the level of acetylcholine. AChE has two sites at the molecular level: the esterase site containing the catalytic triad and the anionic site, which is the choline-binding site (Mohamed, 2014; Marrs & Maynard, 2013). AChE catalyzes the hydrolysis of the neurotransmitter acetylcholine, releasing only choline, which is to be brought back into the presynaptic neuron, briefly depolarising the postsynaptic cell membrane and thus ending nerve impulses (Praveena & Sanjayan, 2011) (Fig. 5.12).

Anticholinesterases are substances that bind to and thereby inhibit the enzyme acetylcholinesterase, causing toxic neurotransmitter effects on insect pests by the membrane disruption of the postsynaptic junction. Furthermore, the inability of AChE to hydrolyze acetylcholine leads to an increase in the concentration of acetylcholine in the synapse, inducing a repetitive activity of the neurons by the prolonged binding of ACh to its postsynaptic receptor and finally, the death of the insect (Fig. 5.13).

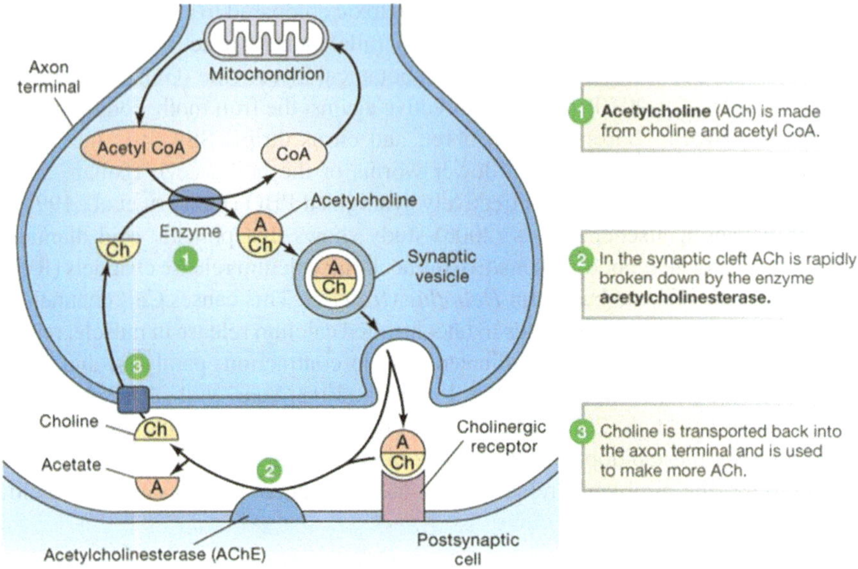

Fig. 5.12 Action mechanism of Acetylcholinesterase Enzyme (google image)

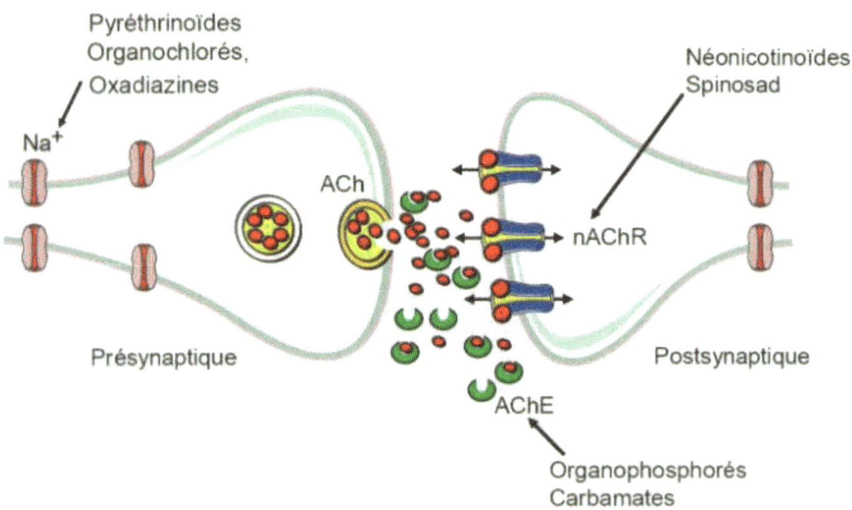

Fig. 5.13 Principal targets of insecticides on a cholinergic synapse (ACh acétylcholine, AChE acétylcholinestérase) (from the thesis of Mohamed (2014))

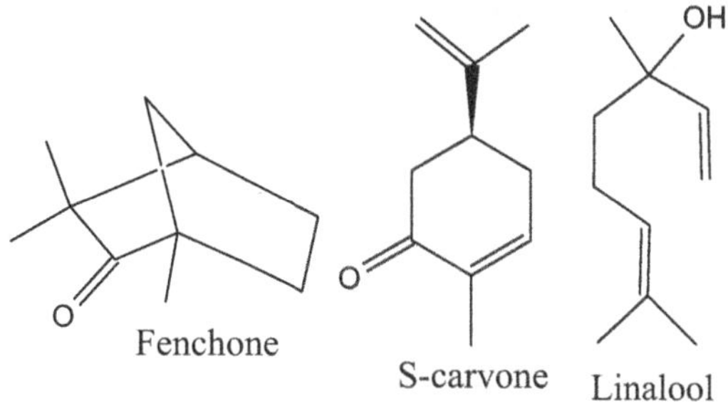

Fig. 5.14 Structure of fenchone, S-carbone et linalool

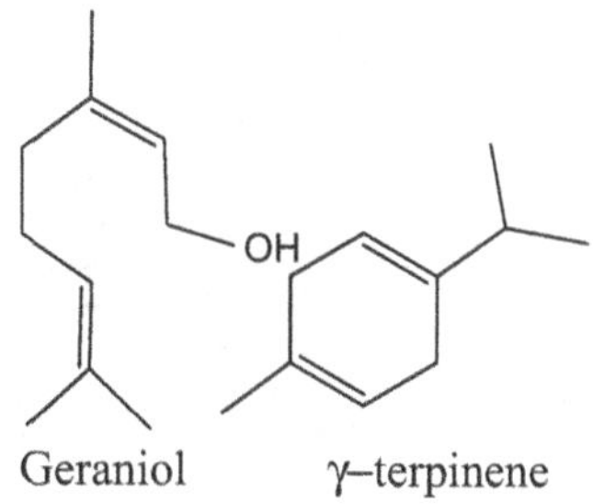

Fig. 5.15 Structure of γ-terpinene and geraniol

Phytocompounds Acting via Acetylcholinesterase Enzyme

-*Terpenoids*: The Essential oils components inhibit the AChE activity (Dassanayake et al., 2021; Wang et al., 2019). Monoterpenoids were the first inhibitors from plants considered to have anticholinesterase properties. Works of Lόpez and Pascual-Villalobos (2010) on three stored-product insect pests, *Sitophilus oryzae* L. (Coleoptera: Curculionidae), *Rhyzoper thadominica* Fabricius (Coleoptera: Bostrichidae), and *Cryptolestes pusillus* Schönherr (Coleoptera: Cucujidae), show that the majority of monoterpenoids tested inhibited the enzyme AChE with fenchone, S-carvone, and linalool producing the highest inhibition and highest insect mortality (Fig. 5.14).

It should be noted that fenchone, γ-terpinene, geraniol, and linalool (Figs. 5.14 and 5.15) showed reversible competitive inhibition, at least at the hydrophobic active site of the enzyme. At the same time, an inhibition for this enzyme is produced by estragole, camphor and S-carvone (Lopez-Hernandez et al., 2009; Lόpez & Pascual-Villalobos, 2010).

-*2,3-dihydro benzofurans*: Coumaran (Fig. 5.16, 2,3-dihydro benzofuran, 2,3-DHB), an active ingredient found in *Lantana camara* L., acts by fumigation on respiratory pigments (spiracles) (Rajashekar et al., 2014). As a result, the insects become hyperactive, indicating that the affected targets are neural/neuromuscular sites (neuronal excitation), and then a 'knockdown' effect is observed, leading to mortality.

Coumaran does not appear to affect Na^+, K^+ATPase, whereas AChE is inhibited in insects (houseflies, stored grain insects) exposed to Coumaran. Inhibition is found in the head and thorax, and nervous tissue.

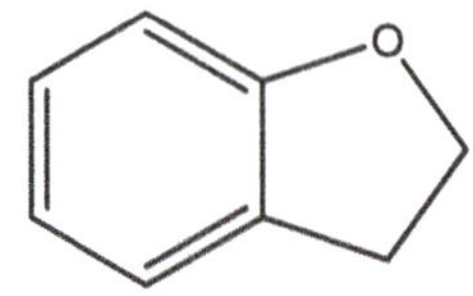

(Coumaran)2,3-dihydrobenzofuran

Fig. 5.16 Structure of coumaran

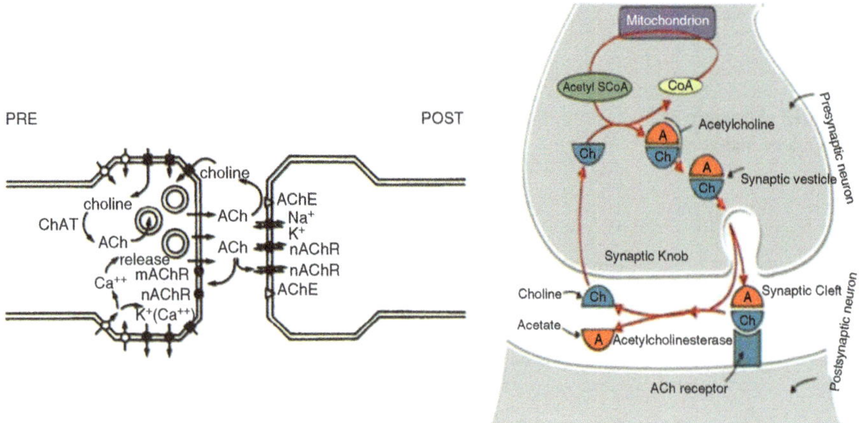

Fig. 5.17 Insect synaptic communication (Breer and Sattelle, 1987; Oguh et al., 2019). AChR acetylcholine receptor (n nicotinic, m muscarinic)

1.5 Nicotinic Acethylcholine Receptors

The nervous system of invertebrates is different from vertebrates by the inversion of neurotransmitters. In the insect brain, acetylcholine ensures communication, while in humans, it is glutamate (Millar and Denholm, 2007). Acetylcholine transferase (AChT), by releasing acetylcholine, generates an action potential that induces Ca^{2+} entry. The acetylcholine (Ach) released into the synaptic cleft will bind to cholinergic receptors on the postsynaptic neuron (POST) and produce an influx of sodium and potassium ions (Na^+/K^+) which depolarizes the neuron. The signal is stopped by acetylcholine esterase (AChE) (Breer and Sattelle, 1987) (Fig. 5.17).

Nicotinic acetylcholine receptors (nAChR) are composed of five subunits, which assemble either in a homomeric or in a heteromeric combination of subunits α and β (Cabirol and Haase, 2019) (Fig. 5.17). These are the most numerous ionotropic receptors, which explains why many neurotoxic compounds act on the nervous system of insects (Narahashi et al., 2000).

Phytocompounds Acting via Nicotinic Acethylcholine Receptors
-Alkaloids: These receptors can be activated by several molecules, in particular muscarine and nicotine (alkaloid isolated from *Nicotiana tabacum* L.), which act on two different surfaces of the receptor (Millar and Denholm, 2007) (Fig. 5.18).

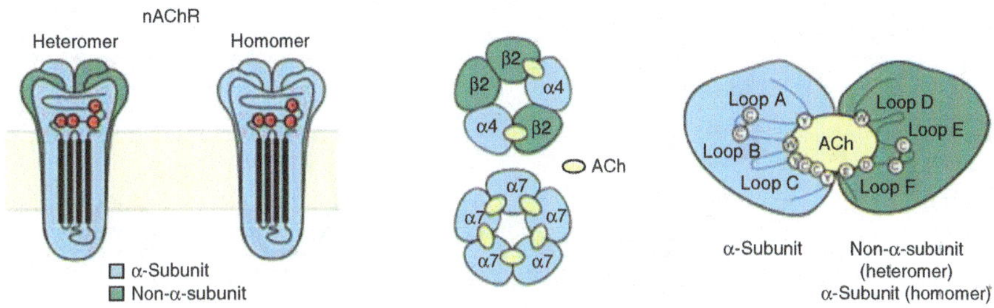

Fig. 5.18 Nicotinic acetylcholine receptor structure (Matsuda et al., 2001)

Fig. 5.19 Structure of nicotine

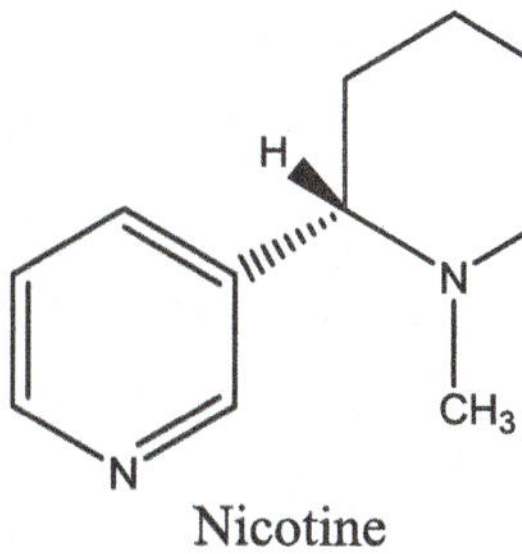

Nicotine (Fig. 5.19, *Nicotiana tabacum*), present an agonist effect on acetylcholine receptor mimicking acetylcholine, leading to an influx of sodium ion and generation of action potentials. Under normal conditions, the synaptic action of acetylcholine is terminated by AChE.

Neonicotinoids include several molecules, such as imidacloprid, acetamiprid, or thiamethoxam. These molecules have a strong affinity for insects and have various activities (Matsuda et al., 2001). Nicotine and neonicotinoid insecticides bind to nicotinic acetylcholine receptors mimicking the effect of acetylcholine opening the ion channels (EFSA Journal, 2013)

Neonicotinoids induce different responses in the brain, and the enzyme that degrades acetylcholine and completes the reaction is ineffective against these compounds. Imicloprid by binding to the receptor (Fig. 5.20) will have an impact on the cerebral function of the bee, in particular the recognition of odors which is associated with the process of memorization (Cabirol and Haase, 2019). Moreover, this molecule blocks the transmission of acetylcholine in the synapse (Matsuda et al., 2001). Acetamiprid, one of the acetylcholine agonists toxic to insects and mammals, has the same binding site as imidacloprid. Insects are resistant to these compounds, which is linked to their high levels of detoxifying enzymes (Abdel-Haleem et al., 2020). Bees, on the contrary, are much less endowed, and the link to these pesticides causes locomotor disorders in them (El Hassani et al., 2007).

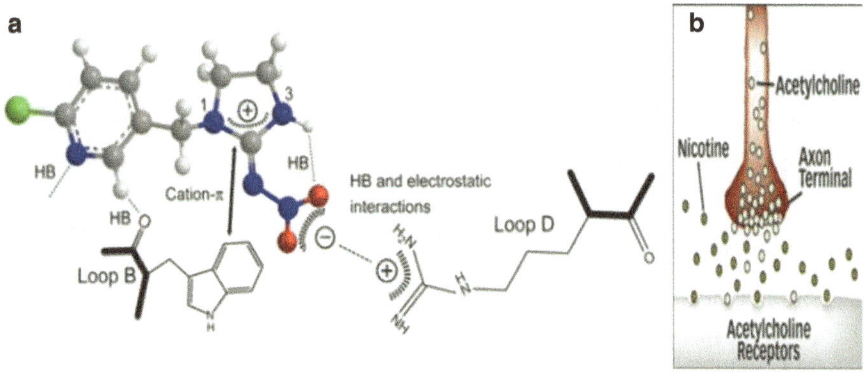

Fig. 5.20 (**a**) Imidacloprid possible binding with D and B loop (Ihara et al., 2007) (**b**) Schematic representation of action mode of nicotine in the nerve synapse (Oguh et al., 2019)

1.6 GABA-Gated Chloride Channels

GABA (gamma-aminobutyric acid) is the primary inhibitory neurotransmitter of the insect's central and peripheral nervous system. GABA allows the flow of chloride in the neurons by opening chloride channels by binding to its specific transmembrane receptors (GABA-gated chloride channels). Resulting in a negative charge on the transmembrane potential causing a reduction in membrane entry resistance by hyperpolarization (Wafford et al., 1989; Sattelle et al., 1991; Bloomquist, 1994, 2003; Johnston, 2005; Tong, 2010; Tong & Coats, 2010; Soderlund, 2012; Dassanayake et al., 2021). GABA-gated chloride channels are expressed in the central nervous system and peripheral nerves of insects, where they regulate neuronal activity and muscle relaxation (Nauen & Bretschneider, 2002). The GABA receptors have a pentameric structure in which the different types of subunits are organized around a chloride ion channel (Nauen & Bretschneider, 2002; Bloomquist, 2003). Pre- or postsynaptic inhibition by the two primary receptors $GABA_AR$ (G protein-coupled receptor, relatively slow-acting) and $GABA_BR$ (ligand-gated ion channel, fast-acting receptor of the pentameric superfamily) depends on its subunits composition and the invertebrate species (Bowery et al., 1980; Casida, 1993). They are, however, the preferred targets for many insects. The $GABA_AR$ has different ligand binding sites: ① GABA site (an allosteric channel site responsible of GABA recognition); ② Benzodiazepine site (positive allosteric modulation site); ③ Barbiturate site (regulates the GABA site); ④ Picrotoxin site (negative allosteric modulation) and others sites (Sattelle et al., 1991; Casida, 1993; Olsen, 2018) (Fig. 5.21).

Phytocompounds Acting via GABA-Gated Chloride Channels

-***Terpenoids***: These insecticides bind to insect GABA receptors, decrease or increase chlorure influx into neurons, and kill insects by causing excessive excitation or inhibition of the nervous system (hyperactivity, hyperexcitability, convulsions,

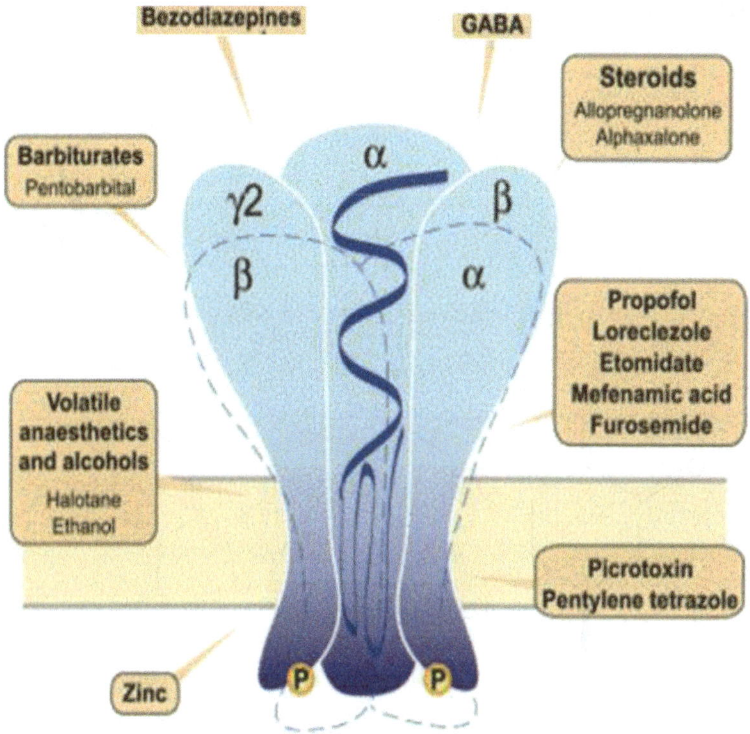

Fig. 5.21 Schematic representation of subunits of GABA and its binding sites (google image from Bowery et al., 2002)

Fig. 5.22 Structure of thujone

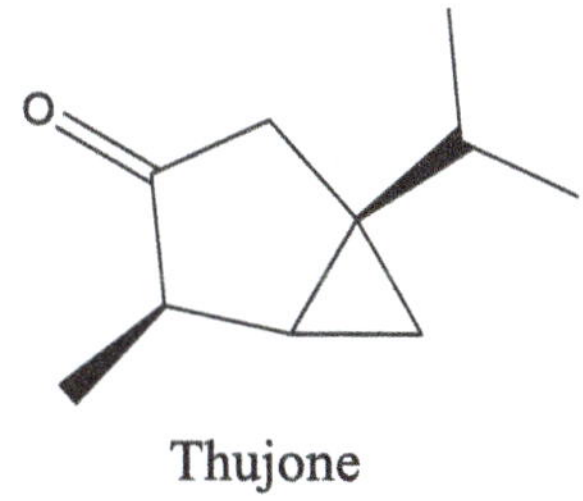

production of prolonged high-frequency discharges, …) (Casida, 1993; Bloomquist, 2003). It was found that the toxin alpha-thujone (Fig. 5.22) blocks brain GABA receptor (Patočka & Plucar, 2003; Jankowska et al., 2017; Böttger et al., 2018).

Without access to GABA, a natural inhibitor of nerve impulses, neurons fire too quickly, and their signaling goes out of control. The non-competitive antagonist picrotoxinin counteracts the inhibitory effect of GABA by blocking the flow of chloride activated by GABA. It suppresses both the inhibitory effects of GABA and the actions of inhibitory neurons and thus acts as a channel blocker (Olsen, 2018; Smelt et al., 2021).

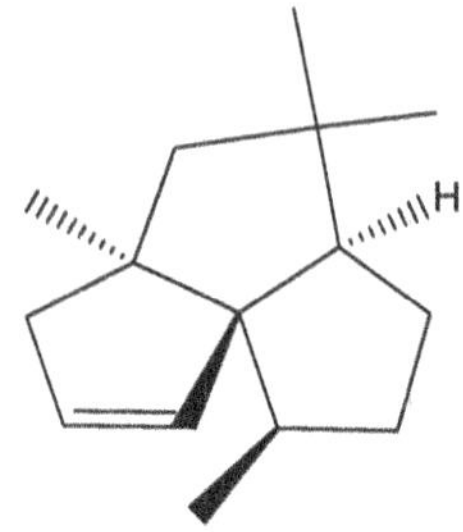

Fig. 5.23 Structure of silphinene

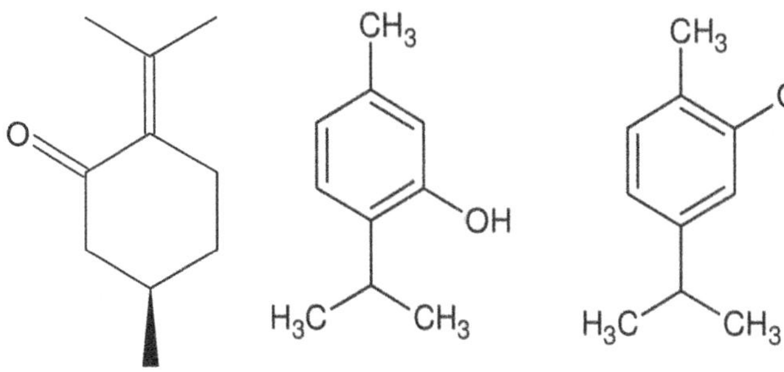

Fig. 5.24. Structure of carvacrol, pulegone, and thymol

Another natural products that antagoniwe the action og GABA are the silphinene-type sesquiterpenes (Fig. 5.23). These terpenoids stabilize non-conducting conformations of the chloride channel. Other ligands are positive allosteric modulators at GABA receptors: carvacrol, pulegone, thymol (Fig. 5.24) on bed bugs, house fly and cockroaches (Tong & Coats, 2010; Gaire et al., 2020).

1.7 Octopamine Receptor

Octopamine, a biogenic monoamine, is an invertebrate structural analog of vertebrate norepinephrine, which acts as a neurohormone (release in the hemolymph for a lipid mobilizing during flight and long-lasting motor behaviors), a neuromodulator, and a neurotransmitter in invertebrates (Roeder, 1999; Farooqui, 2012). Present in high concentrations in most invertebrates' central and peripheral nervous tissues, octopamine is a multipotent substance that regulates many behaviors in insects as reported in the review of (Evans, 1978; Armstrong & Robertson, 2006; Farooqui, 2012) and presented in Fig. 5.25.

Octopamine is a non-peptide neurotransmitter exclusive of invertebrates. Depolarization of octopamergic neurons by increased potassium concentration or

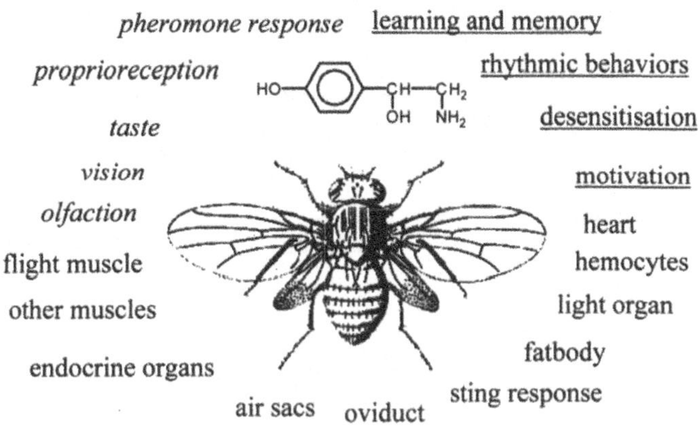

Fig. 5.25 Physiological and behavioral systems modulated by octopamine in an insect. Italic (sense organs) underline (central systems), not underlined (others peripheral tissues) (Roeder, 1999)

electrical stimulation releases octopamine into the extracellular space. However, to maintain octopamine homeostasis and octopamergic neurotransmission, the excess of octopamine removed into the extracellular space during exocytosis is taken up by endocytosis vesicles (transporters). This recycling system is a means of protection for the insect and, therefore, the target of insecticides. The primary enzymatic pathway of octopamine inhibition in insects is the conversion of octopamine to N-acetyl-octopamine by N-acetylation (N-acetyltransferase). Octopamine binds to its octopaminergic receptors, belonging to the G protein-coupled receptor superfamily, to trigger a physiological response (Bischof & Enan, 2004; Evans & Maqueira, 2005; Rattan, 2010; Farooqui, 2012). Binding of octopamine to the adrenergic receptor stimulates activation of phospholipase C (PLC) *via* the Gq protein (G), causing hydrolysis of phosphatidylinositol-4,5-bis-phosphate (PIP2) to Inositol-1,4,5-tri-phosphate plus diacylglycerol (DGA). Inositol-1,4,5-tri-phosphate elicits the closure of calcium channels, releasing CA^{2+} into the cytoplasm. DGA and Ca^{2+}, activate protein kinase C (PKC) by binding to its receptor in the endoplasmic reticulum (ER) and PKC phosphorylates various signaling proteins and ion channels, ans as consecuence regulates the physiological response. Adenylate cyclase (AC) is also activated by octopamine-Octα–R binding *via* Gs proteins, which induces intracellular cAMP release that stimulates protein kinase A (PKA), responsible for the cellular response phosphorylation of signaling proteins. Octopamine can also bind to the β-adrenergic octopamine receptor (Octβ-R) associated with elevated intracellular cAMP levels. Inositol-1,4,5-tri-phosphate, DGA, cAMP, and Ca^{2+} are intracellular second messengers (Fig. 5.26) (Farooqui, 2012; Dassanayake et al., 2021).

Phytocompounds Acting via Octopamine Receptor

-Terpenoids: Octopamine inhibitors disrupt the octopaminergic nervous system by blocking the octopamine receptors, altering physiological regulation at the muscle junction and body fluid homeostasis of insects. Work of Kostyukovsky et al. (2002)

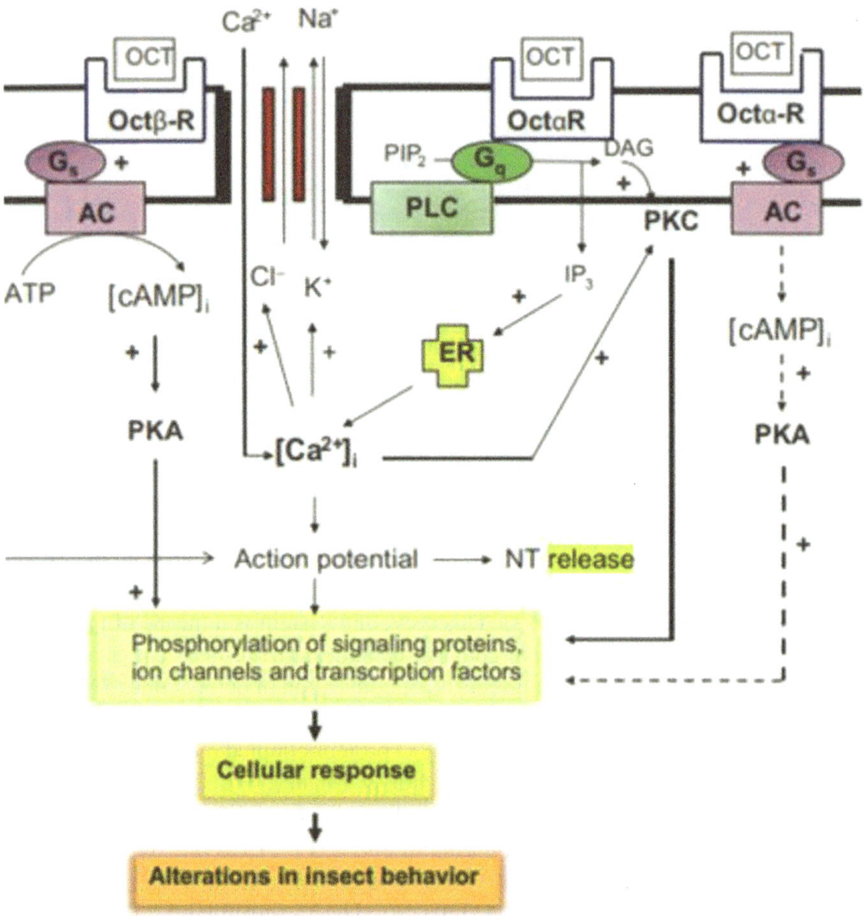

Fig. 5.26 Octopamine regulatory pathways (Farooqui, 2012)

in *Helicoverpa armigera* Hübn shows a significant increase in intracellular levels of cAMP in abdominal epidermal tissue following exposure to essential oil constituents, similar to that observed during treatment with octopamine. The same toxicity effect was observed when using honokiol (Fig. 5.26) extracted from *Magnolia denudata* against four species of mosquitoes (Wang et al., 2019). The use of an octopaminergic antagonist, phentolamine, effectively inhibits this increase. Reynoso et al. (2020) observed the same inhibitory effect when pre-treated with phentolamine hydrochloride on the hyperactivating effect of octopamine and eugenol (Fig. 5.27) on nymphs of Blood-Sucking Bug *Triatoma infestans* (Hemiptera: Reduviidae), confirming that the octopamine receptor is a target site for eugenol action (Enan, 2001).

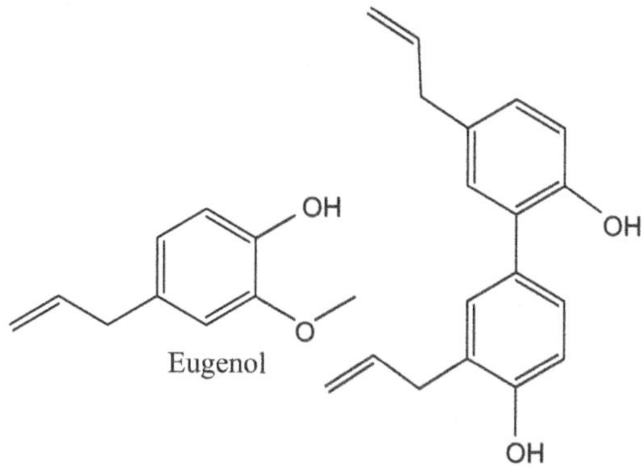

Fig. 5.27 Structure of eugenol and honokiol

2 Respiratory System

Insects have a particular respiratory system focusing on a trachea system that opens to the outside through stigmata allowing gas exchange between the interior and exterior environments. To supply the tissues with oxygen, the trachea will branch out into smaller and smaller ducts to form tracheoles at the heart of the tracheolar cells, which provide a direct supply of oxygen to the tissues. The end of the tracheoles is filled with tracheolar fluid. Then, the respiratory gases dissolve (which is necessary for their transmembrane diffusion), which varies according to the metabolic activity of the cells (Fig. 5.28 Lafont & Toullec, 2016).

Oxidative phosphorylation is a complex process in the mitochondria, which converts nutrients into energy (Adenosine triphosphate (ATP)). This conversion is done by the mitochondrial respiratory chain composed of four multi enzymatic complexes. By oxidizing the cofactors, Nicotinamide Adenine Dinucleotide (NADH, H+) Flavin Adenine Dinucleotide ($FADH_2$) generates a proton gradient and a proton gradient from ATP. Three different membrane protein complexes of the respiratory chain are implicated in the proton gradient across the membrane. This complexes are: Complex I (NADH/ubiquinone oxidoreductase)/Complex III (cytochrome c reductase)/Complex IV (cytochrome c oxidase). The complex II (Succinate dehydrogenase) does not generate protons (Fukami, 1985; Voet et al., 2016).

Figure 5.29 shows a schematic representation of the respiratory chain (Kühlbrandt, 2015). Complex I (in blue) catalyzes the transfer of two electrons from NADH, H^+ *via* flavin mononucleotide and a set of [Fe–S] centers to intramembrane ubiquinol (UQ). This exothermic reaction pumps four protons from the matrix. The electrons from ubiquinol were taken by the complex III and transferred to the cytochrome c. The complex IV transfers electrons from cytochrome c to O_2 and contributes to the creation of the proton gradient using reducing the molecule of O_2 into water. Complex II transfers electrons from succinate directly to ubiquinol and does not contribute to the proton gradient. Then the ATP synthetase will procudes ATP from ADP and inorganic phosphate (Pi) (Kühlbrandt, 2015).

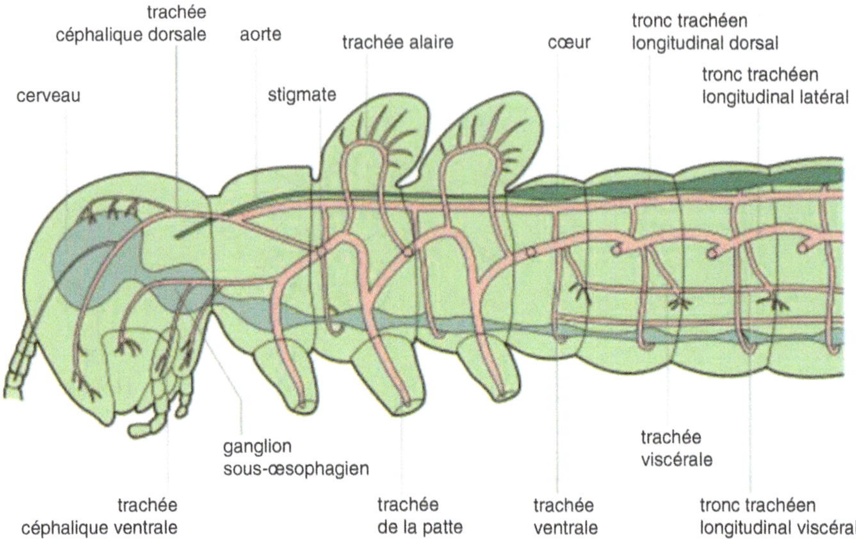

Fig. 5.28 Insect respiratory system (Lafont & Toullec, 2016)

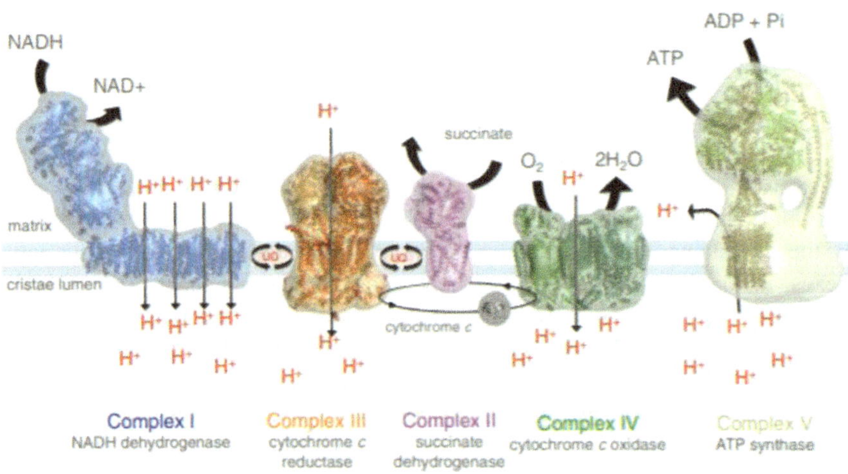

Fig. 5.29 Respiratory chain (Kühlbrandt, 2015)

The mitochondria are also the place of production of free oxygen radicals (RSO) (Fig. 5.30-Scialò et al., 2017), superoxide (O_2^-), hydrogen peroxide (H_2O_2), and hydroxyl radical (OH), produced mainly within Complexes I and III of the mitochondria (Romana Fato et al., 2009). These Radicals at physiological concentrations are harmless and beneficial to the cell by functioning as redox messengers in intracellular signaling and regulation (McLennan & Esposti, 2000). On the other hand, Rotenone (ROT), by disturbing the balance NAD^+/NADH, H^+, accentuates

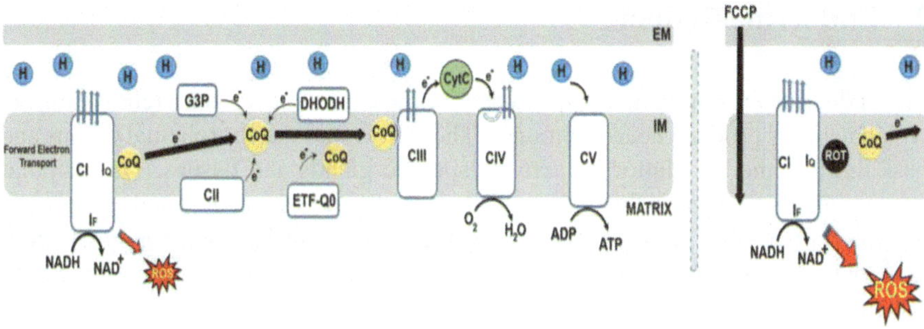

Fig. 5.30 Production of ROS by electron transport chain (Scialò et al., 2017)

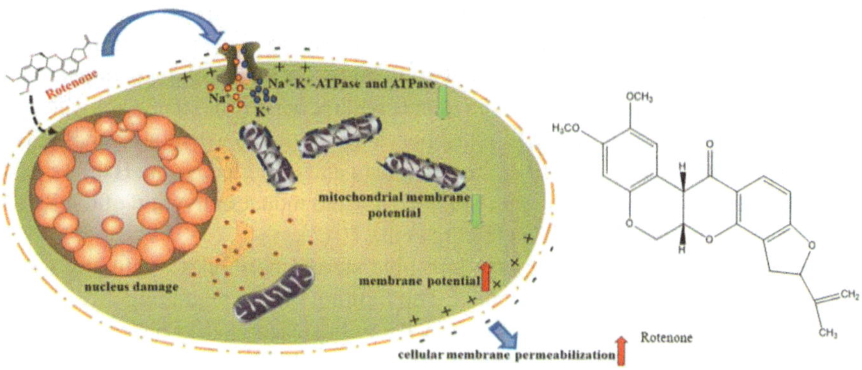

Fig. 5.31 Rotenone action on cellular membrane permeabilization (Sun et al., 2021)

ROS production, particularly hydrogen peroxide and superoxide anion by complex I, which will be released in the mitochondrial matrix or the cytosol. These very reactive radicals can alter the structure of macromolecules such as DNA proteins and can also modify the plasma membranes or even cause tissue apoptosis following radical reactions (Thany et al., 2013; Firoze and Hui, 2020).

Phytocompounds Targeting the Respiratory System

-Isoflavonoids. Rotenone (Fig. 5.30) is the most common natural product among rotenoids, a type of isoflavonoid, and usually found in species from *Lonchocarpus, rhododendron*, and *Derris* genera, spread throughout East Indies, Malaya, and South America.

Rotenone (Fig. 5.31) inhibites cellular respiration, and blocks the converstion of nutrient compounds into energy. Inhibition of complex I by rotenone destabilizes the [NAD^+/NADH, H^+] ratio and leads to a halt in the oxidation of the NADH, H^+ molecule to NAD^+. As a result, the other metabolic pathways requiring an NAD^+ supply for their functioning are disrupted. Blockage at this site reduces energy (ATP) production, leads to the depolymerization of microtubules (Choi et al., 2011), permeabilization of the membrane, and induces whole animal toxicity: inactivity, paralysis, and death (Fig. 5.5) (Sun et al., 2021).

3 Endocrine System

The endocrine system is made up entirely of the central nervous system (brain), in which neurons produce neurohormones. These flow into the neurohemal organs and pass through the circulation to activate specific glands and tissues. The unipolar neurons are distributed in the brain and all along the sub-esophageal and prothoracic ganglia (Fig. 5.32a). They are extended by their axons and discharge secretory products (neurohormones or neuropeptides) into the various target organs (Fig. 5.32b) (Chapman et al., 2012).

The different metamorphoses undergone by insects to pass from one stage to another (larval stage to metamorphosis) are called molts (ecdysis). Molting is controlled by two hormones: ① steroid hormone (ecdysone), phytosterol responsible for molting (Truman, 2019); ② and the juvenile hormone (JH), lipids, responsible for inhibiting the steroid hormone, avoiding the premature ecdysis, maintaining the larval state of the insect (Riddiford, 2012). Juvenile hormone binds to an unidentified transmembrane receptor, leading *via* phospholipase C (PLC) to the activation of cam kinase-2 (CaMKII). CaMKII, phosphorylates, in the nucleus, the complex JH/Metreceptor (previously formed in the cytoplasm and imported by the Hsp83 transporter in the case of Drosophila into the nucleus) and induce activation of the Kr-h1 molting inhibitor gene (Fig. 5.33) (Jindra et al., 2015).

The metamorphosis is caused by insoluble steroid hormones, including ecdysone, secreted by the prothoracic gland. The limiting step in regulating this mechanism requires that: ① the invertebrates reach a critical size; ② they have a sufficient supply of nutrients regulated by insulin (Brogiolo et al., 2001). Under the action of the prothoracic hormone (PPTH), produced by the *corpora cardiaca*, and the insect insulin, the secretion of ecdysone (HM α-ecdysone: E) is activated then released into the hemolymph and transported, in the case of the tarantula, by a protein hemocyanin (Spindler et al., 2009). On reaching the target tissues, it is transformed into 20-hydroxyecdysone (β-ecdysone: 20E), the active form of the hormone in peripheral tissues, inducing molting (Fig. 5.34). At the same time, HJ decreases to give way to molt (Niwa & Niwa, 2014; Truman, 2019).

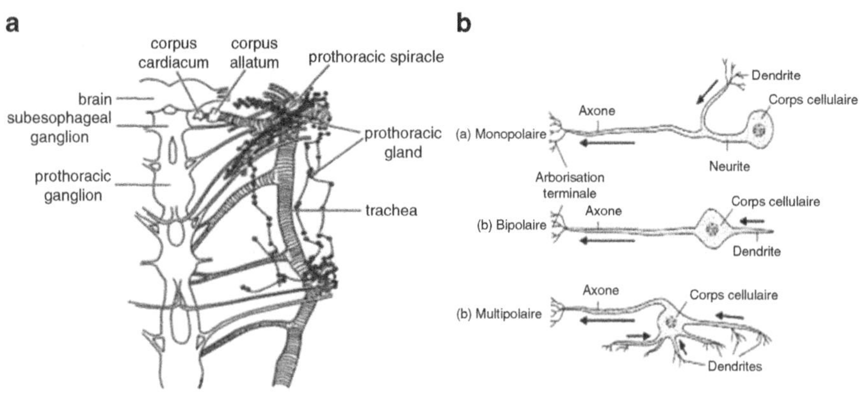

Fig. 5.32 Insect endocrine system with secretor neuron (Chapman et al., 2012)

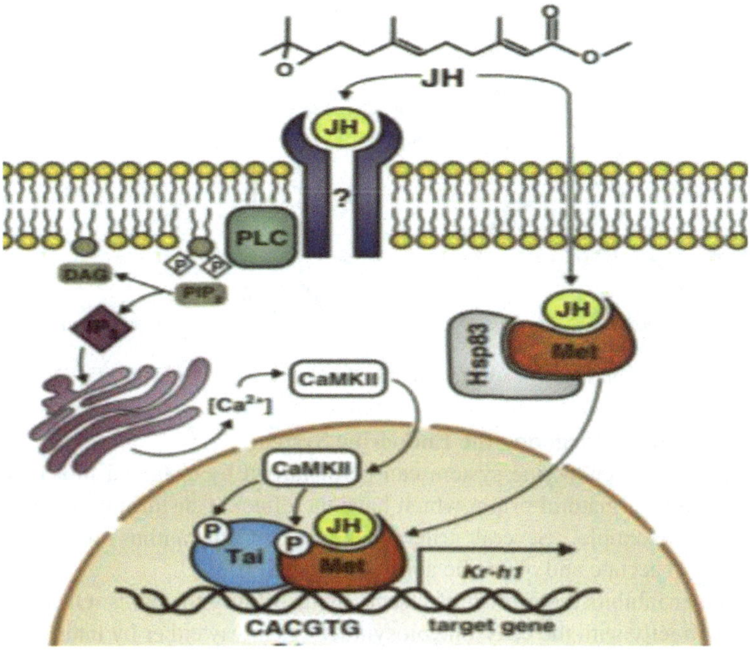

Fig. 5.33 Action of ecdysone on insect molting (Jindra et al., 2015)

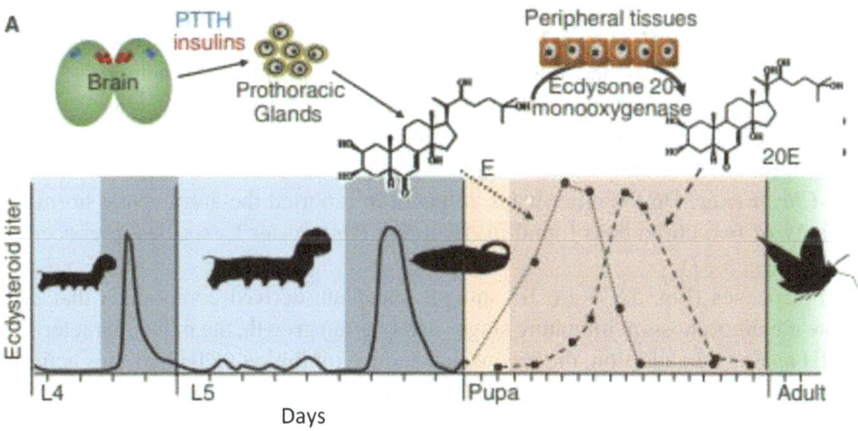

Fig. 5.34 Mode action of ecdysone system in *Manduca sexta* (Truman, 2019)

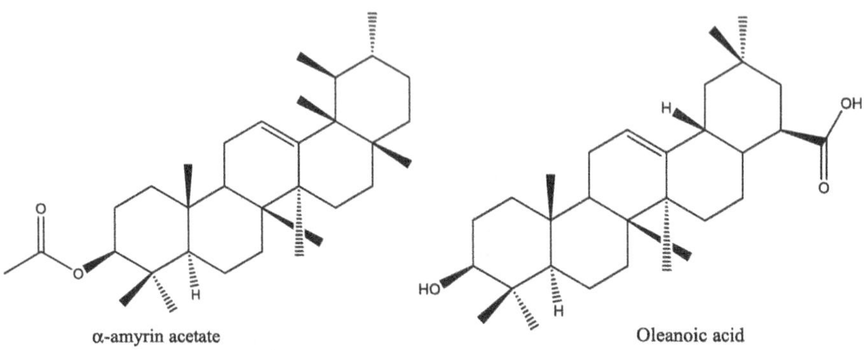

Fig. 5.35 Structure of α-amyrin acetate and oleanoid acid

Phytocompounds Targeting the Endocrine System

-Terpenoids. The endocrine system can be disturbed by chemical molecules, but also molecules of natural origin which have the effect of disturbing the growth of the insect (example: Several triterpenes from Catharanthus roseus (Linn.) G. (α-amyrin acetate and oleanolic acid, see Fig. 5.35)).

The latter inhibits the growth of insects by its affinity for lipids. Oleanolic acid interacts directly with the ecdysone biosynthetic pathway either by interfering with phytosterols taken up by the insect or by interacting directly with the ecdysone receptor (Cai et al., 2017; Tian et al., 2021).

Azadirachtin (Fig. 5.36), a complex tetranortriterpenoidlimonoid, majorly found in the seeds of *Azadirachta indica* A. Juss., is involved in the secretion of the prothoracic hormone PPTH, which will induce the release of ecdysone, causing a slowdown in the synthesis and release of PPTH. As a result, the functioning of the nucleus of secretory neurons and endocrine glands is affected, and the insect can no longer molt. In addition, azadirachtin also modifies the production and stop of the growth functions (Luntz et al., 2005).

-Chromenes. On the other hand, it has been reported the antijuvenile hormone activity of two chromenes found in *Ageratum conyzoydes* L. species, precocene I and II (Fig. 5.37).

Precocenes (Fig. 5.37) are JH-antagonistic plant-derived compounds that promote metamorphosis at immature stages and ovarian growth, the main characteristic of JH analogs. In addition, the precocenes induce inhibition of JH and thus activate the maturation and the precocity of the passages between the different stages (Bowers et al., 1976; Graf, 1993).

4 Hydric System: Water Balance

The cuticle of insects (arthropods) is an acellular exoskeleton composed of several layers secreted by epithelial cells. It consists of chitin, a polymer of N-Acetylglucosamine, which forms glycoprotein complexes with Resilin,

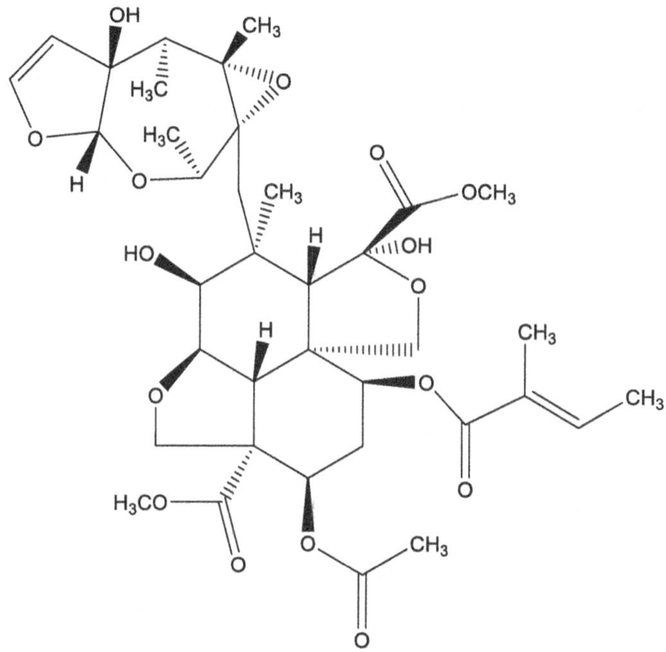

Azadirachtin

Fig. 5.36 Structure of azadirachtin

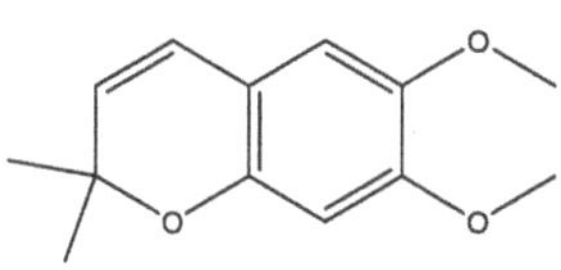

Precocene II (6,7-dimethoxy-2,2-dimethylchromene)

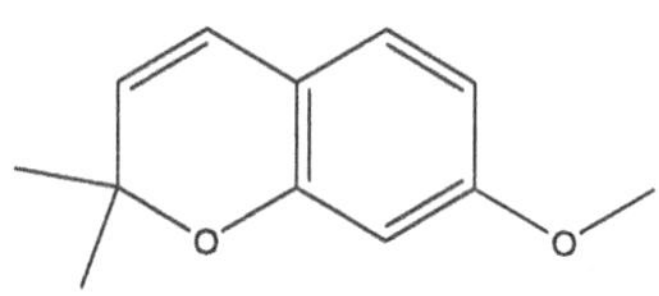

Precocene I (7-methoxy-2,2-dimethylchromene)

Fig. 5.37 Structure of precocenes I and II

Arthropodin, and Sclerotin (Fig. 5.38). Chitin is located at the level of the procuticle (Jeuniaux, 1975). The cuticle comprises several layers, but there are 3 fundamental layers: the epicuticle and the exocuticle, and the endocuticle, which forms the procuticle (Moussian, 2010).

The cuticle is formed by the action of the hormones Ecdyson and Bursicon (neuropeptide inducing polarization, that is to say, the hardness of the cuticle and the tanning, which makes it hydrophobic). Sclerization and tanning use the hydroxylation pathway of L-tyrosine to dopamine (DOPA) by tyrosine hydroxylase. This pathway is activated by Buriscon, which binds to a receptor linked to a G protein and then follows a cascade of reactions (Warren et al., 2002; Flaven-Pouchon et al., 2020). The principal protection of the insect against water loss is the epicuticle, more precisely, the waxy layer that regulates water loss. This cuticle has been

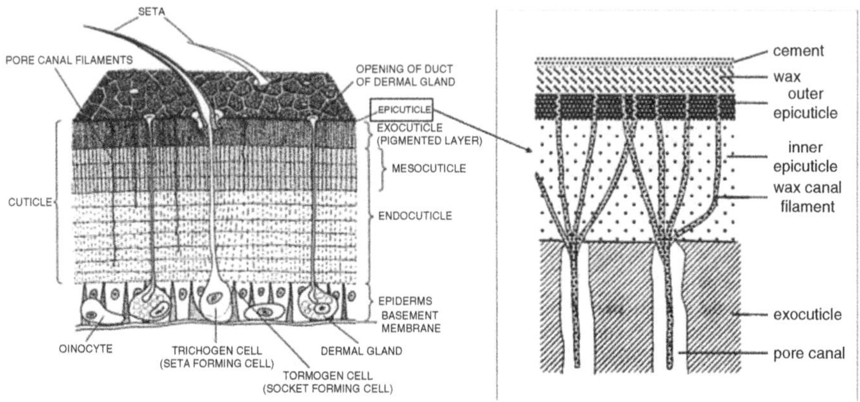

Fig. 5.38 Insect cuticle focuses on epicuticle (Chapman et al., 2012)

described as a conduit measuring between 60 and 130 Å. It is made of phospholipids with reversed polarity allowing water passage. The transpiration rate of insects varies according to conditions. Lipids having a melting temperature increase in temperature removes the wax and leaves a free quote of water in the ducts (Wigglesworth, 1945; Locke, 1965). The loss of this impermeable layer leads to transpiration which is lethal for the insect (Wigglesworth, 1945).

Phytocompounds Targeting the Hydric System

-Saponins. Molecules such as saponin can cause wax abrasion and, through its amphiphilic nature, cross water transport channels (Chaieb, 2010; Cui et al., 2019). Saponins are amphiphilic glycosides of a hydrophilic carbohydrate chain and another aglycone chain, generally triterpene or fat-soluble steroid. Classified in 2 groups: ① Saponosides with a steroid aglycone group; ② Saponosides with a triterpene aglycone group. The plant toxicity is determined by the synthesized groups (R1/R2) of the aglycone part; thus, actions against their target organisms will depend on the plant selected. Saponins affect different aspects of the insect at the physiological, biochemical, and anatomical levels (De Geyter et al., 2007; Chaieb, 2010; Mugford & Osbourn, 2013).

The alterations observed in the skin and the intestine are due to the saponin affinity for lipids, especially for cholesterol. However, the insect has a sterol supply *via* its diet. The saponin molecule interferes with the ecdysone biosynthesis pathway by complexing with cholesterol, so there is no longer any lipid for the insect's molt, which affects its growth (Qasim et al., 2020). Membranes are made of lipids, and saponin has a hydrophobic end that can easily slip into the membrane, destabilize it, and cause the permeabilization of the membrane and the leakage of cell contents (Fig. 5.39-De Geyter et al., 2007).

-Phenylpropanoids. The 2-benzoyloxycinnamaldehyde (2-BCA, Fig. 5.40), a natural product isolated from the roots of *Pleuropterus ciliinervis* Nakai (Polygonaceae), produces inhibition of chitin synthase, a β-(1,4)-linked homopolymer of N-acetyl-D-glucosamine, one of the essential structural components of

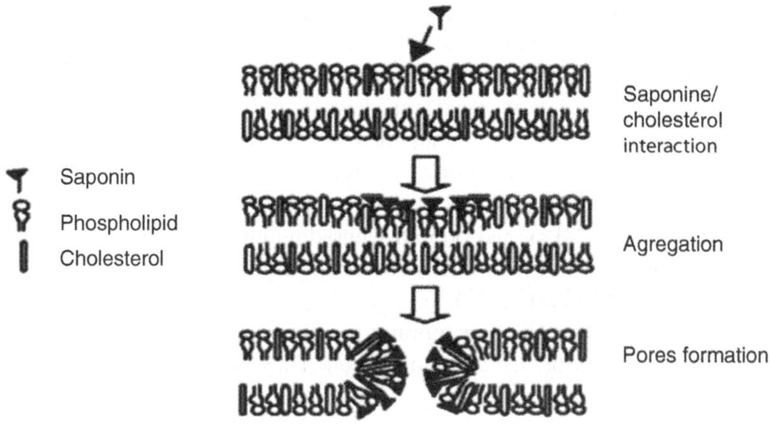

Fig. 5.39 Theoretical interaction saponin/membrane (De Geyter et al., 2007)

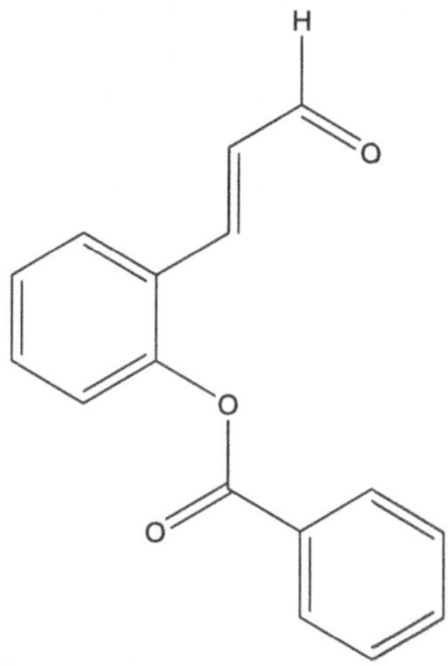

2`-Benzoyloxycinnamaldehyde (2`-BCA)

Fig. 5.40 Structure of 2-benzoyloxycinnamaldehyde (2-BCA)

nearly all fungi cell walls, and also a protective component of the insect exoskeleton, an a important pest target

5 Repellents

Nowadays, several molecules extracted from plants can act as repellents against insects, but little is known about their mechanisms of action. It is well established that these repellent molecules interact with the insect olfactory system. This system is formed by tiny hairs called sensilla located in the maxillary palps and antennae of insects. Inside of the sensilla is olfactory receptor neurons. The behavior of insects towards the repellents will depend on which olfactory receptor neurons are explicitly activated. For initiation of signal transduction, the repellents go inside the sensilla by the pore tubules; they stimulate the odorants Receptors proteins (ORs) and other proteins related with the odorate (as for example, the Odorant-binding proteins (OBPs), ionotropic receptors (IRs), sensory neuron membrane proteins, chemosensory proteins (CSPs), and odorant-degrading enzymes (ODEs)) (Fan et al., 2011; Breer, 2003; Leal, 2012). OBPs ensure the connection between the external environment and ORs (Leal, 2005).

These ORs are a family of membrane protein receptors responsible for most insect olfactory perception and communication. They belong to a diverse family of proteins across insect orders (Hansson & Stensmyr, 2011) and present specificity for diverse repellents.

These ORs are associated with a co-receptor that is a conserved domain to form odorant-gated ion channels (Wicher et al., 2008; Sato et al., 2008)

According to Clark and Ray (2016), there are four general repellence mechanisms. The first mechanism concerns the activation of ORs dedicated to aversion. The second mechanism is based on the activation of pheromone receptors that cause repulsion. The pheromone receptors regroup a specific family of proteins that bind and transport pheromones. The third concerns the inhibition of odorant receptors dedicated to attraction, and the fourth is a prolonged activation of odorant receptors participating in attraction.

Phytocompounds Acting as Repellents

-***Terpenoids***. Monoterpenes are acting as repellents, for example, the nepetalactone (Fig. 5.41) primary compound of essential oil of *Nepeta cataria* L., and is considered a highly effective repellent of insects.

Recent studies on some plant-derived repellents indicated that these could activate or inhibit the ORs (Grison et al., 2020). It seems that repellents can be specific to different ORs from different insects. Various repellents molecules can be activated on the same ORs but with more or less strong repelling power. For example, the volatile organic compound methyl jasmonate (Fig. 5.42), which is synthetized

Fig. 5.41 Structure of nepelactone

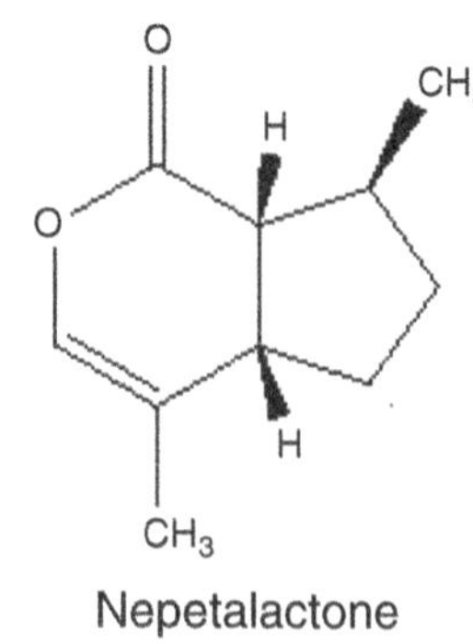

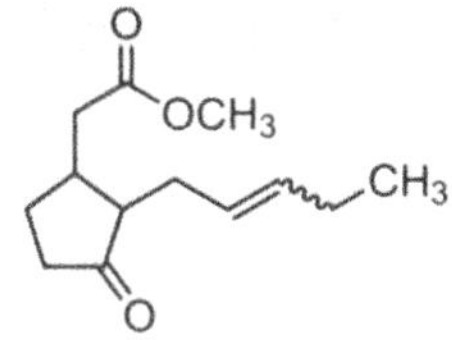

Fig. 5.42 Structure of methyl jasmonate

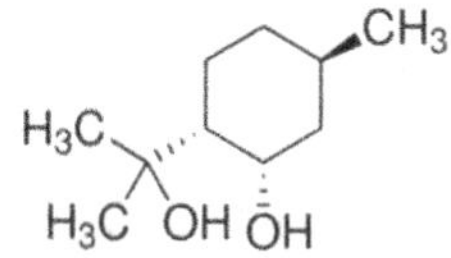

Fig. 5.43 Structure of *para*-menthane-3,8-diol (PMD)

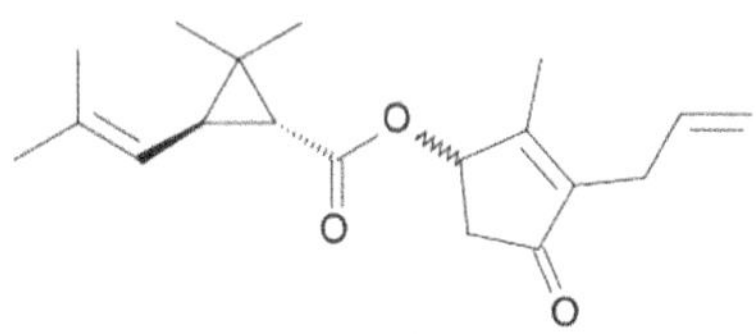

Fig. 5.44 Structure of bioallethrin

by plants like *Rosmarinus officinalis* and *Jasminum grandiflorum*, has been demonstrated to act on the same ORs as synthetic molecules N,N-diethyl-meta-toluamide (DEET), and Ethyl Butylacetylaminopropionate (IR3535) in the mosquito *Culex quinquefasciatus* (Zeng et al., 2018; Corbel et al., 2009).

In a similar study, authors showed that the terpene *para*-menthane-3,8-diol (PMD, Fig. 5.43), a natural repellent, inhibited activation of two ORs in *Aedes aegypti* (Bohbot et al., 2011), whereas, in another study on the same molecule, authors showed activation of two ORs in *C. quinquefasciatus*.

However, this activation was observed only without antagonists (Xu et al., 2019). These studies tend to show that repellents can act as activators or inhibitors of Odorants Receptors and induce different behavioral responses in insects.

Furthermore, repellents could also act directly on Odorant-binding proteins (OBPs) by fixation with more or less affinity. This is the case for PMD, which can link two OBPs in *C. quinquefasciatus* with an affinity difference of 10-fold (Yin et al., 2015).

Finally, some repellents could be activated in different signalization pathways. For example, it is well documented that synthetic repellent DEET act on ORs, but also it is an inhibitor of acetylcholinesterase, an enzyme involved in the nervous system. More recently, Valbon and co-authors showed the double action of Bioallethrin (Fig. 5.44) in repellency (Valbon et al., 2022). Bioallethrin is a pyrethroid derived from natural insecticides pyrethrins extracted from *Chrysanthemum* species. Action of Bioallethrin on voltage-gated sodium channels was known. Still, their results showed that Bioallethrin activated specific ORs in *A. aegypti*, and this activation took part in the repellency effect of Bioallethrin.

6 Attractants

Plant-derived molecules identified as attractants interact with the insect olfactory system as repellents. These molecules, considered like plant volatiles, are synthetized under particular conditions and are specific to insect species (Proffit et al., 2020). Their mechanisms of action are not yet elucidated, but several recent studies indicate a potential role of Odorant-binding proteins (OBPs) in the interaction with attractant molecules.

Phytocompounds Acting as Attractants

Wang and co-authors identified a specific OBP and its volatile organic compound associated between plant *Ficus pumila var. pumila* and the pollinator *Wiebesia pumilae* (Wang et al., 2021). Interestingly, they also identified a set of genes in *F. pumila var. pumila* involved in the biosynthesis regulation of both attractants and repellents delivered by this plant, thus suggesting a synchronous synthesis (Wang et al., 2021).

7 Antifeedants

Insect survival requires recognition of edible substances and compounds that may be toxic. Therefore, plants are faculty to synthesize a class of secondary metabolites that can have antifeedant or deterrent effects for insects. These metabolites usually have a chemical impact called antifeedants (Bernays et al., 2000). Depending on the chemosensory system, insects detect feeding stimulants and feeding deterrents in plant tissues with different levels. This detection seems to be based on chemosensory proteins (CSPs) located in the sensillar lymph (Yoshizawa et al., 2011).

Phytocompounds Acting as Antifeedants

Until now, only a few CSPs have been identified and studied. As a result, the action mechanisms for antifeedants are not elucidated but are of interest to the scientific community. For example, in a recent study on *Locusta migratoria*, authors were allowed to identify the specific CSP involved in detecting antifeedant alpha-amyl cinnamaldehyde (Fig. 5.45) (Jiang et al., 2021).

They showed the high affinity of CSP to this antifeedant. At least, antifeedant molecules seem to act on the activity of some key enzymes in insects, such as acetylcholinesterase, carboxylesterase, mixed-function oxidase, and glutathione S-transferases (Pan et al., 2016; Yang et al., 2018). Plant-derived antifeedants could act simultaneously on these enzymes by more or less strongly inhibiting enzymes

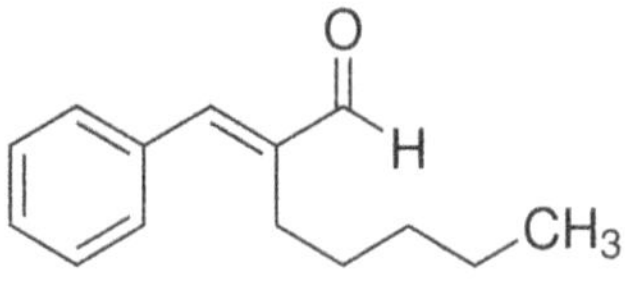

Fig. 5.45 Structure of *alpha*-amyl cinnamaldehyde

activities (Liu et al., 2020). These inhibitions are based on direct interactions between enzymes and antifeedants.

8 Conclusion

Plant biopesticides are a real alternative to pest management. The insecticidal activities of these natural products and crude extracts act via different pathways. This versatility is fascinating to avoid the development of pest resistances (similarly to antibiotics). Additionally, they are cheaper, less hazardous, and environmentally friendly (used to be biodegradable); therefore, they may improve crop production, increase sustainability, and enhance the benefits through a green approach.

Acknowledgments This research was funded by "Plan Chlordecone III en Guadeloupe et Martinique" in the framework of the DICHOAL-MAREC (DIREC) project.

References

Abdel-Haleem, D. R., Gad, A. A., Farag, S. M. (2020). 'Larvicidal, biochemical and physiological effects of acetamiprid and thiamethoxam against Culex pipiens L. (Diptera: Culicidae)'. *Egyptian Journal of Aquatic Biology and Fisheries, 24*(3), 271–283. https://doi.org/10.21608/ejabf.2020.91119

Armstrong, G. A. B., & Robertson, R. M. (2006). A role for octopamine in coordinating thermoprotection of an insect nervous system. *Journal of Thermal Biology, 31*(1–2), 149–158.

Barnes, S., & Hille, B. (1988). Veratridine modifies open sodium channels. *The Journal of General Physiology, 91*, 421–443.

Beauhaire, J., Ducrot, P. H., Malosse, C., Rochat, D., Ndiege, I. O., & Otieno, D. O. (1995). Identification and synthesis of sordidin, a male pheromone emitted by Cosmopolites sordidus. *Tetrahedron Letters, 36*, 1043–1046.

Bernays, E. A., Chapman, R. F., & Singer, M. S. (2000). Sensitivity to chemically diverse phagostimulants in a single gustatory neuron of a polyphagous caterpillar. *Journal of Comparative Physiology. A, 186*, 13–19.

Bischof, L. J., & Enan, E. E. (2004). Cloning, expression and functional analysis of an octopamine receptor from *Periplaneta Americana*. *Insect Biochemistry and Molecular Biology, 34*, 511–521.

Bohbot, J. D., Fu, L., Le, T. C. K., Chauhan, R., Cantrell, C. L., Dickens, J. C. (2011). Multiple activities of insect repellents on odorant receptors in mosquitoes, 25(4), 436–444. https://doi.org/10.1111/j.1365-2915.2011.00949.x

Bowery, N. G., Bettler, B., Froestl, W., Gallagher, J. P., Marshall, F., Raiteri, M., Bonner, T. I., Enna S. J. (2002). International union of pharmacology. XXXIII. Mammalian GABAB receptors: structure and function. *Pharmacol. Rev. 54*(2), 247e264. https://doi.org/10.1124/pr.54.2.247

Bloomquist, J. R. (1994). Cyclodiene resistance at the insect GABA receptor/chloride channel complex confers broad cross resistance to convulsants and experimental phenylpyrazole insecticides. *Archives of Insect Biochemistry and Physiology, 26*(1), 69–79.

Bloomquist, J. R. (2003). Chloride channels as tools for developing selective insecticides. *Archives of Insect Biochemistry and Physiology: Published in Collaboration with the Entomological Society of America, 54*(4), 145–156.

Bootman, M. D., Lipp, P., & Berridge, M. J. (2001). The organisation and functions of local Ca2+ signals. *Journal of Cell Science, 114*(12), 2213–2222.

Böttger, A., Vothknecht, U., Bolle, C., & Wolf, A. (2018). *Lessons on caffeine, cannabis & co: Plant-derived drugs and their interaction with human receptors (learning materials in biosciences) (English Edition)* (1st ed.). Springer.. https://doi.org/10.1007/978-3-319-99546-5_6

Boucher, O., Simard, M. N., Muckle, G., Rouget, F., Kadhel, P., & Bataille, H. (2013). Exposure to an organochlorine pesticide (chlordecone) and development of 18- month-old infants. *Neurotoxicology, 35*, 162–168.

Bowers, W., Ohta, T., Cleere, J., & Marsella, P. (1976). Discovery of insect anti-juvenile hormones in plants. Plants yield a potential fourth-generation insecticide; 2U. *Science, 193*(4253), 542–547.

Bowery, N. G., Hill, D. R., Hudson, A. L., Doble, A., Middlemiss, D. N., Shaw, J., & Turnbull, M. (1980). (−)Baclofen decreases neurotransmitter release in the mammalian CNS by an action at a novel GABA receptor. *Nature, 283*(5742), 92–94.

Breer, H. (2003). Olfactory receptors: Molecular basis for recognition and discrimination of odors. *Analytical and Bioanalytical Chemistry, 377*, 427–433.

Breer, H., & Sattelle, D. B. (1987). Molecular properties and functions of insect acetylcholine receptors. *J. Insect Physiol. 33*(11), 771–790.

Brogiolo, W., Stocker, H., Ikeya, T., Rintelen, F., Fernandez, R., & Hafen, E. (2001). An evolutionarily conserved function of the Drosophila insulin receptor and insulin-like peptides in growth control. *Current Biology, 11*(4), 213–221.

Cabidoche, Y. M., Achard, R., Cattan, P., Clermont-Dauphin, C., Massat, F., & Sansoulet, J. (2009). Long-term pollution by chlordecone of tropical volcanic soils in the French West Indies: A simple leaching model accounts for current residue. *Environmental Pollution, 157*(5), 1697–1705.

Cabirol, A., & Haase, A. (2019). The Neurophysiological Bases of the Impact of Neonicotinoid Pesticides on the Behaviour of Honeybees. *Insects, 10*(10), 344–. https://doi.org/10.3390/insects10100344

Casida, J. E. (1980). Pyrethrum flowers and pyrethroid insecticides. *Environmental Health Perspectives, 34*, 189–202.

Casida, J. E. (1993). Insecticide action at the GABA-gated chloride channel: Recognition, progress, and prospects. *Archives of Insect Biochemistry and Physiology, 22*(1–2), 13–23.

Catterall, W. A. (2000). Structure and regulation of voltage-gated calcium channels. *Annual Review of Cell and Developmental Biology, 16*, 521–555.

Chaieb, I. (2010). Saponins as insecticides: A review. *Tunisian Journal of Plant Protection, 5*(1), 39–50.

Chapman, R. F., Simpson, S. J., & Douglas, A. E. (2012). *The insects: Structure and function (Insects)* (959p). Cambridge University Press.

Choi, W. S., Palmiter, R. D., & Xia, Z. (2011). Loss of mitochondrial complex I activity potentiates dopamine neuron death induced by microtubule dysfunction in a Parkinson's disease model. *The Journal of Cell Biology, 192*(5), 873–882.

Clark, J. T., & Ray, A. (2016). Olfactory mechanisms for discovery of odorants to reduce insect-host contact. *Journal of Chemical Ecology, 42*, 919–930.

Corbel, V., Stankiewicz, M., Pennetier, C., Fournier, D., Stojan, J., Girard, E., Dimitrov, M., Molgó, J., Hougard, J. M., & Lapied, B. (2009). Evidence for inhibition of cholinesterases in insect and mammalian nervous systems by the insect repellent deet. *BMC Biology, 7*(1), 47.

Coronado, R., Morrissette, J., Sukhareva, M., & Vaughan, D. M. (1994). Structure and function of ryanodine receptors. *American Journal of Physiology-Cell Physiology, 266*(6), C1485–C1504.

Cui, C., Yang, Y., Zhao, T., Zou, K., Peng, C., Cai, H., Wan, X., & Hou, R. (2019). Insecticidal activity and insecticidal mechanism of total saponins from *Camellia oleifera*. *Molecules, 24*(24), 4518.

D'Cruz, A. A., Babon, J. J., Norton, R. S., Nicola, N. A., & Nicholson, S. E. (2013). Structure and function of the SPRY/B30.2 domain proteins involved in innate immunity. *Protein Science, 22*(1), 1–10.

Dallaire, R., Muckle, G., Rouget, F., Kadhel, P., Bataille, H., & Guldner, L. (2012). Cognitive, visual, and motor development of 7-month-old Guadeloupean infants exposed to chlordecone. *Environmental Research, 118*, 79–85.

Dassanayake, M. K., Chong, C. H., Khoo, T.-J., Figiel, A., Szumny, A., & Choo, C. M. (2021). Synergistic field crop pest management properties of plant-derived essential oils in combination with synthetic pesticides and bioactive molecules: A review. *Food, 10*(9), 19.

Davies, T. G. E., Field, L. M., Usherwood, P. N. R., & Williamson, M. S. (2007). DDT, pyrethrins, pyrethroids and insect sodium channels. *Life, 59*(3), 151–162.

De Geyter, E., Lambert, E., Geelen, D., & Smagghe, G. (2007). Novel advances with plant saponins as natural insecticides to control pest insects. *Pest Technology, 1*(2), 96–105.

De Mandal, S., Shakeel, M., Prabhakaran, V. S., Karthi, S., Xu, X., & Jin, F. (2019). Alternative splicing and insect ryanodine receptor. *Archives of Insect Biochemistry and Physiology, 102*(3), e21590.

Demeneix, B., Leemans, M., & Couderq, S. (2020). Pyrethroid exposure: Not so harmless after all. *The Lancet Diabetes & Endocrinology, 8*(4), 266–268.

Dereumeaux, C., Saoudi, A., Guldner, L., Pecheux, M., Chesneau, J., Thomé, J., Ledrans, M., Le Tertre, A., Denys, S., & Fillol, C. (2020). Chlordecone and organochlorine compound levels in the French West Indies population in 2013–2014. *Environmental Science and Pollution Research, 27*(33), 41033–41045.

Devault, D. A., Laplanche, C., Pascaline, H., Bristeau, S., Mouvet, C., & Macarie, H. (2016). Natural transformation of chlordecone into 5b-hydrochlordecone in French West Indies soils: Statistical evidence for investigating long-term persistence of organic pollutants. *Environmental Science and Pollution Research, 23*(1), 81–97.

Dimetry, N. Z., Amer, S. A. A., & Reda, A. S. (1993). Biological activity of two neem seed kernel extracts against the two-spotted spider mite *Tetranychus urticae* Koch. *Journal of Applied Entomology, 116*(1–5), 308–312.

Dubuisson, C., Heraud, F., Leblanc, J. C., Gallotti, S., Flamand, C., & Blateau, A. (2007). Impact of subsistence production on the management options to reduce the food exposure of the Martinican population to chlordecone. *Regulatory Toxicology and Pharmacology, 49*(1), 5–16.

Ebbinghaus-Kintscher, U., Luemmen, P., Lobitz, N., Schulte, T., FunkeC, F. R., Masaki, T., Yasokawa, N., & Tohnishi, M. (2006). Phthalic acid diamides activate ryanodine-sensitive Ca^{2+} release channels in insects. *Cell Calcium, 39*(1), 21–33.

El Hassani, A. K., Dacher, M., Gary, V., Lambin, M., Gauthier, M., Armengaud, C. (2007). Effects of Sublethal Doses of Acetamiprid and Thiamethoxam on the Behavior of the Honeybee (Apis mellifera), *Arch Environ Contam Toxicol 54*, 653–661. https://doi.org/10.1007/s00244-007-9071-8

Ellington, C. P. (1985). Power and efficiency of insect flight muscle. *The Journal of Experimental Biology, 115*(1), 293–304.

Enan, E. (2001). Insecticidal activity of essential oils: Octopaminergic sites of action. *Comparative Biochemistry and Physiology Part C, 130*, 325–337.

EPA Pyrethroids and Pyrethrins. *Revised ecological risk mitigation and response to comments on the ecological risk mitigation proposal for 23 chemicals*. Docket Number EPA-HQ-OPP-2008-0331 http://www.epa.gov/oppsrrd1/reevaluation/pyrethroids-pyrethrins. html

EFSA. (2013). Scientific Opinion on the developmental neurotoxicity potential of acetamiprid and imidacloprid, *EFSA Journal, 11*(12), 3471.

Evans, P. D. (1978). Octopamine distribution in the insect nervous system. *Journal of Neurochemistry, 30*, 1009–1030.

Evans, P. D., & Maqueira, B. (2005). Insect octopamine receptors: A new classification scheme based on studies of cloned *Drosophila* G-protein coupled receptors. *Invertebrate Neuroscience, 5*(3–4), 111–118.

Fan, J., Francis, F., Liu, Y., Chen, J. L., & Cheng, D. F. (2011). An overview of odorant- binding protein functions in insect peripheral olfactory reception. *Genetics and Molecular Research, 10*, 3056–3069.

Farooqui, T. (2012). Review of octopamine in insect nervous systems. *Open Access Insect Physiology, 4*, 1–17.

Fato, R., Bergamini, C., Bortolus, M., Maniero, A. L., Leoni, S., Ohnishi, T., & Lenaz, G. (2009). Differential effects of mitochondrial complex I inhibitors on production of reactive oxygen species. *Biochim Biophys Acta Bioenergetics, 1787*(5), 384–392.

Feher, J. (2017). The neuromuscular junction and excitation–contraction coupling. In *Quantitative human physiology* (pp. 318–333). https://doi.org/10.1016/B978-0-12-800883-6.00029-X

Fill, M., & Copello, J. A. (2002). Ryanodine receptor calcium release channels. *Physiological Reviews, 82*(4), 893–922.

Firoze, M. K., & Hui, W. (2020). Environmental exposures and autoimmune diseases: Contribution of gut microbiome. *Frontiers in Immunology, 10*, 3094.

Flaven-Pouchon, J., Alvarez, J. V., Rojas, C., & Ewer, J. (2020). The tanning hormone, bursicon, does not act directly on the epidermis to tan the Drosophila exoskeleton. *BMC Biology, 18*(1), 17.

Fukami, J. I. (1985). Insecticides as inhibitors of respiration. In *Approaches to new leads for insecticides* (Vol. 1985, pp. 47–69). Springer.

Gaire, S., Scharf, M. E., & Gondhalekar, A. D. (2020). Synergistic Toxicity Interactions between Plant Essential Oil Components against the Common Bed Bug (*Cimex lectularius* L.). *Insects, 11*, 133.

Gnagey, A. L., Forte, M., & Rosenberry, T. L. (1987). Isolation and characterization of acetylcholinesterase from *Drosophila. The Journal of Biological Chemistry, 262*(27), 13290–13298.

González-Coloma, A., Gutiérrez, C., Hübner, H., Achenbach, H., Terrero, D., & Fraga, B. M. (1999). Selective insect antifeedant and toxic action of ryanoid diterpenes. *Journal of Agricultural and Food Chemistry, 47*(10), 4419–4424.

Graf, J. F. (1993). The role of insect growth regulators in arthropod control. *Parasitology Today, 9*(12), 471–474.

Grdiša, M., & Grsic, K. (2013). Botanical insecticides in plant protection. *Agriculturae Conspectus Scientificus, 78*(2), 85–93.

Grison, C., Carrasco, D., Pelissier, F., & Moderc, A. (2020). Reflexion on bio-sourced mosquito repellents: Nature, activity, and preparation. *Frontiers in Ecology and Evolution, 8*, 8.

Gu, S. H., Chow, Y. S., & O'Reilly, D. R. (1998). Role of calcium in the stimulation of ecdysteroidogenesis by recombinant prothoracicotropic hormone in the prothoracic glands of the silkworm, *Bombyx mori. Insect Biochemistry and Molecular Biology, 28*, 861–867.

Han, X., Zhang, F., Meng, L. L., Xu, Y., Li, Y., Turyk, A. E., Yang, R., Wang, P., Zhang, J., Zhang, Q., & Jiang, L. (2020). Exposure to organochlorine pesticides and the risk of type 2 diabetes in the population of East China. *Ecotoxicology and Environmental Safety, 190*, 110125.

Hansson, B. S., & Stensmyr, M. C. (2011). Evolution of insect olfaction. *Neuron, 72*, 698–711.

Hénault-Ethier, L. (2016). *Document d'information: Les pyréthrinoïdes, utilisés à la maison, mais non sans dangers*. Technical Report. https://doi.org/10.13140/RG.2.1.2725.0803

Hermawan, W., Nakajima, S., Tsukuda, R., Fujisaki, K., Nakasuji F. (1997). Isolation of an antifeedant compound from Andrographis paniculata (Acanthaceae) against the diamondback moth, Plutella xylostella (Lepidoptera: Yponomeutidae). *Appl. Entomol. Zool. 32*, 551–5.

Ihara, M., Shimomura, M., Ishida, C., Nishiwaki, H., Akamatsu, M., Sattelle, D. B., Matsuda, K. (2007). A hypothesis to account for the selective and diverse actions of neonicotinoid insecticides at their molecular targets, nicotinic acetylcholine receptors: catch and release in hydrogen bond networks. *Invert Neurosci, 7*, 47–51, https://doi.org/10.1007/s10158-006-0043-x

Inoshita, T., & Tanimura, T. (2006). Cellular identification of water gustatory receptor neurons and their central projection pattern in *Drosophila. Proceedings of the National Academy of Sciences, 103*(4), 1094–1099.

Isaacs, A. K., Qi, S., Sarpong, R., & Casida, J. E. (2012). Insect ryanodine receptor: Distinct but coupled insecticide binding sites for [*N*-C^3H_3]Chlorantraniliprole, Flubendiamide, and [^{3}H]Ryanodine. *Chemical Research in Toxicology, 25*(8), 1571–1573.

Isman, M. B. (2006). Botanical insecticides, deterrents, and repellents in modern agriculture and an increasingly regulated world. *Annual Review of Entomology, 51*(1), 45–66.

Ito, K., Shinomiya, K., Ito, M., Armstrong, J. D., Boyan, G., Hartenstein, V., Harzsch, S., Heisenberg, M., Homberg, U., Jenett, A., Keshishian, H., Restifo, L. L., Rössler, W., Simpson, J. H., Strausfeld, N. J., Strauss, R., & Vosshall, L. B. (2014). A systematic nomenclature for the insect brain. *Neuron, 81*(4), 755–765.

Iwamoto, H. (2011). Structure, function and evolution of insect flight muscle. *Biophysics, 7*, 21–28.

Jankowska, M., Rogalska, J., Wyszkowska, J., & Stankiewicz, M. (2017). Molecular targets for components of essential oils in the insect nervous system—A review. *Molecules, 23*(1), 34.

Jeanguenat, A. (2013). The story of a new insecticidal chemistry class: The diamides. *Pest management science, 69*(1), 7–14.

Jefferies, P. R., Toia, R. F., Brannigan, B., Pessah, I., & Casida, J. E. (1992). Ryania insecticide: Analysis and biological activity of 10 natural ryanoids. *Journal of Agricultural and Food Chemistry, 40*(1), 142–146.

Jeuniaux, C. (1975). Insectes: Téguments, système nerveux, organes sensoriels. *Traité de Zoologie, 8*(3), 910.

Jiang, X., Xu, H., Zheng, N., Yin, X., & Zhang, L. A. (2021). Chemosensory protein detects antifeedant in locust (*Locusta migratoria*). *Insects, 12*, 1.

Jindra, M., Bellés, X., & Shinoda, T. (2015). Molecular basis of juvenile hormone signaling. *Current Opinion in Insect Science, 11*, 39–46.

Joffe, T., Gunning, R. V., Allen, G. R., Kristensen, M., Alptekin, S., Field, L. M., & Moores, G. D. (2012). Investigating the potential of selected natural compounds to increase the potency of pyrethrum against houseflies *Musca domestica* (Diptera: Muscidae). *Pest Management Science, 68*(2), 178–184.

Johnston, G. A. R. (2005). $GABA_A$ receptor channel pharmacology. *Current Pharmaceutical Design, 11*, 1867–1885.

Khan, M. F., & Wang, H. (2020). Environmental Exposures and Autoimmune Diseases: Contribution of Gut Microbiome. *Frontiers in Immunology, 10*. https://doi.org/10.3389/fimmu.2019.03094

Kostyukovsky, M., Rafaeli, A., Gileadi, C., Demchenko, N., & Shaaya, E. (2002). Activation of octopaminergic receptors by essential oil constituents isolated from aromatic plants: Possible mode of action against insect pests. *Pest Management Science, 58*, 1101–1106.

Kühlbrandt, W. (2015). Structure and function of mitochondrial membrane protein complexes. *Kühlbrandt BMC Biology, 13*, 89.

Lafont, R., & Toullec, J. Y. (2016). *Insects*. Encyclopædia Universalis.

Lahm, G. P., Cordova, D., & Barry, J. D. (2009). New and selective ryanodine receptor activators for insect control. *Bioorganic & Medicinal Chemistry, 17*(12), 4127–4133.

Lanner, J. T., Georgiou, D. K., Joshi, A. D., & Hamilton, S. L. (2010). Ryanodine receptors: Structure, expression, molecular details, and function in calcium release. *Cold Spring Harbor Perspectives in Biology, 2*(11), a003996.

Leal, W. S. (2005). Pheromone reception. In S. Schulz (Ed.), *The chemistry of pheromones and other semiochemicals II* (pp. 1–36). Springer.

Leal, W. S. (2012). Odorant reception in insects: Roles of receptors, binding proteins, and degrading enzymes. *Annual Review of Entomology, 58*, 373–391.

Legoff, L., D'Cruz, S. C., Bouchekhchoukha, K., Monfort, C., Jaulin, C., Multigner, L., & Smagulova, F. (2020). In utero exposure to chlordecone affects histone modifications and activates LINE-1 in cord blood. *Life Science Alliance, 4*(6), e202000944.

Lin, L., Hao, Z., Cao, P., & Yuchi, Z. (2020). Homology modeling and docking study of diamondback moth ryanodine receptor reveals the mechanisms for channel activation, insecticide binding and resistance. *Pest Management Science, 76*(4), 1291–1303.

Liu, C., Tian, J., An, T., Lyu, F., Jia, P., Zhou, M., Liu, Z., & Feng, Y. (2020). Secondary metabolites from *Solanum rostratum* and their antifeedant defense mechanisms against *Helicoverpa armigera*. *Journal of Agricultural and Food Chemistry, 68*(1), 88–96.

Locke, M. (1965). Permeability of insect cuticle to water and lipids. *Science, 147*(3655), 295–298.

López, M. D., & Pascual-Villalobos, M. J. (2010). Mode of inhibition of acetylcholinesterase by monoterpenoids and implications for pest control. *Industrial Crops and Products, 31*, 284–288.
Lopez-Hernandez, G. Y., Thinschmidt, J. S., Zheng, G., Zhang, Z., Crooks, P. A., Dwoskin, L. P., & Papke, R. L. (2009). Selective inhibition of acetylcholine-evoked responses of 7 neuronal nicotinic acetylcholine receptors by novel tris- and tetrakis-Azaaromatic quaternary ammonium antagonists. *Molecular Pharmacology, 76*(3), 652–666.
Lümmen, P. (2013). Calcium channels as molecular target sites of novel ınsecticides. In *Advances in insect physiology* (Vol. 44, pp. 287–347). Academic Press..
Luntz, A. J. M., Morgan, E. D., & Nisbet, A. J. (2005). *Azadirachtin, a natural product in insect control* (pp. 117–135). Elsevier BV.
Ma, R., Haji-Ghassemi, O., Ma, D., Jiang, H., Lin, L., Yao, L., Samurkas, A., Li, Y., Wang, Y., Cao, P., Wu, S., Zhang, Y., Murayama, T., Moussian, B., Van Petegem, F., & Yuchi, Z. (2020). Structural basis for diamide modulation of ryanodine receptor. *Nature Chemical Biology, 16*, 1246–1254.
Marcel, V., Palacios, L. G., Pertuy, C., Masson, P., & Fournier, D. (1998). Two invertebrate acetylcholinesterases show activation followed by inhibition with substrate concentration. *The Biochemical Journal, 329*(2), 329–334.
Marrs, T. C., & Maynard, R. L. (2013). Neurotranmission systems as targets for toxicants: A review. *Cell Biology and Toxicology, 29*(6), 381–396.
Matsuda, K., Buckingham, S. D., Kleier, D., Rauh, J. J., Grauso, M., Sattelle D. B. (2001). Neonicotinoids: insecticides acting on insect nicotinic acetylcholine receptors. *TRENDS in Pharmacological Science. 22*(11), 0–580. https://doi.org/10.1016/s0165-6147(00)01820-4
McLennan, H. R., & Esposti, M. D. (2000). The contribution of mitochondrial respiratory complexes to the production of reactive oxygen species. *Journal of Bioenergetics and Biomembranes, 32*(2), 153–162.
Meissner, G. (1986). Ryanodine activation and inhibition of the Ca^{2+} release channel of sarcoplasmic reticulum. *The Journal of Biological Chemistry, 261*(14), 6300–6306.
Meissner, G. (1994). Ryanodine receptor/Ca2+ release channels and their regulation by endogenous effectors. *Annual Review of Physiology, 56*(1), 485–508.
Miresmailli, S., Isman, M. B. (2014). Botanical insecticides inspired by plant-herbivore chemical interactions. *Trends in Plant Science, 19*(1), 29–35. https://doi.org/10.1016/j.tplants.2013.10.002
Mohamed, A. A. E. (2014). *Etude du mode d'action neurotoxique d'un répulsif, le DEET utilisé seule et en association avec un insecticide sur l'Acétylcholinestérase des DUM neurones d'un insecte la blatte Periplaneta americana* (Thesis), 178p.
Moussian, B. (2010). Recent advances in understanding mechanisms of insect cuticle differentiation. *Insect Biochemistry and Molecular Biology, 40*(5), 363–375.
Mugford, S. T., & Osbourn, A. (2013). Saponin synthesis and function. In T. J. Bach & M. Rohmer (Eds.), *Isoprenoid synthesis in plants and microorganisms. New concepts and experimental approaches* (pp. 405–424). Springer Science+Business Media.
Multigner, L., Ndong, J. R., Giusti, A., Romana, M., Delacroix-Maillard, H., Cordier, S., Jegou, B., Thome, J. P., & Blanchet, P. (2010). Chlordecone exposure and risk of prostate cancer. *Journal of Clinical Oncology, 28*(21), 3457–3462.
Multigner, L., Kadhel, P., Rouget, F., Blanchet, P., & Cordier, S. (2016). Chlordecone exposure and adverse effects in French West Indies populations. *Environmental Science and Pollution Research, 23*(1), 3–8.
Nauen, R. (2006). Insecticide mode of action: Return of the ryanodine receptor. *Pest Management Science, 62*(8), 690–692.
Nauen, R., & Bretschneider, T. (2002). New modes of action of insecticides. *Pesticide Outlook, 13*(6), 241–245.
Millar, N. S., & Denholm, I. (2007). Nicotinic acetylcholine receptors: targets for commercially important insecticides. 7(1), 53–66. https://doi.org/10.1007/s10158-006-0040-0
Narahashi, T., Fenster, C. P., Quick, M. W., Lester, R. A. J., Marszalec, W., Aistrup, G. L., Sattelle, D. B., Martin, B. R., Levin, E. D. (2000). Symposium Overview: Mechanism of Action of

Nicotine on Neuronal Acetylcholine Receptors, from Molecule to Behavior. *Toxicological Sciences, 57*(2), 193–202, https://doi.org/10.1093/toxsci/57.2.193
Niwa, R., & Niwa, Y. S. (2014). Enzymes for ecdysteroid biosynthesis: Their biological functions in insects and beyond. *Bioscience, Biotechnology, and Biochemistry, 78*(8), 1283–1292.
Oguh, C. E., Okpaka, C. O., Ubani, C. S., Okekeaji, U., Joseph, P. S., & Amadi, E. U. (2019). Natural pesticides (biopesticides) and uses in pest management – A critical review. *Asian Journal of Biotechnology and Genetic Engineering, 2*(3), 1–18.
Olsen, R. W. (2018). $GABA_A$ receptor: Positive and negative allosteric modulators. *Neuropharmacology, 136*(Pt A), 10–22.
Pan, L., Ren, L. L., Chen, F., Feng, Y. Q., & Luo, Y. Q. (2016). Antifeedant activity of Ginkgo biloba secondary metabolites against Hyphantria cunea larvae: Mechanisms and applications. *PLoS One, 11*, e0155682.
Patočka, J., & Plucar, B. (2003). Pharmacology and toxicology of absinthe. *Journal of Applied Biomedicine, 1*, 199–205.
Praveena, A., & Sanjayan, K. P. (2011). Inhibition of acetylcholinesterase in three insects of economic importance by linalool, a monoterpene phytochemical. In P. Dunston (Ed.), *Insect pest management, a current scenario* (pp. 340–345). Ambrose, Entomology Research Unit, St. Xavier's College.
Proffit, M., Lapeyre, B., Buatois, B., Deng, X., Arnal, P., Gouzerh, F., Carrasco, D., & Hossaert-McKey, M. (2020). Chemical signal is in the blend: Bases of plant-pollinator encounter in a highly specialized interaction. *Scientific Reports, 22*, 10071.
Qasim, M., Islam, W., Ashraf, H. J., Ali, I., & Wang, L. (2020). Saponins in insect pest control. In J. M. Merillon & K. Ramawat (Eds.), *Co-evolution of secondary metabolites. Reference series in phytochemistry* (Vol. 2020, pp. 897–924). Springer Nature Switzerland AG.
Quintavalle, A. (2013). Voltage-gated calcium channels in honey bees: Physiological roles and potential targets for insecticides. *BioSciences Master Reviews, 2013*, 1–11.
Rajashekar, Y., & Shivanandappa, T. (2017). Mode of Action of the Natural Insecticide, Decaleside Involves Sodium Pump Inhibition. *PLoS One, 12*(1), e0170836.
Rajashekar, Y., Gunasekaran, N., & Shivanandappa, T. (2010). Insecticidal activity of the root extract of Decalepis hamiltonii against stored-product insect pests and its application in grain protection. *Journal of Food Science and Technology, 47*(3), 310–314.
Rajashekar, Y., Rao, L. J. M., & Shivanandappa, T. (2012). Decaleside: A new class of natural insecticide targeting tarsal gustatory receptors. *Die Naturwissenschaften, 99*(10), 843–852.
Rajashekar, Y., Raghavendra, A., & Bakthavatsalam, N. (2014). Acetylcholinesterase inhibition by biofumigant (Coumaran) from leaves of *Lantana camara* in stored grain and household insect pests. *BioMed Research International, 2014*, 1–6.
Rattan, R. S. (2010). Mechanism of action of insecticidal secondary metabolites of plant origin. *Crop Protection, 29*(9), 913–920.
Regnault-Roger, C. (2012). Botanicals in pest management. In D. P. Abrol & U. Shankar (Eds.), *Integrated pest management: principles and practice* (pp. 123–124). CPI Group (UK) Ltd.
Reynoso, M. M. N., Lucia, A., Zerba, E. N., & Alzogaray, R. A. (2020). The octopamine receptor is a possible target for eugenol-induced hyperactivity in the blood-sucking bug *Triatoma infestans* (Hemiptera: Reduviidae). *Journal of Medical Entomology, 57*(2), 627–630.
Ribi, W., Sendenb, T. J., Sakellarioub, A., Limayec, A., & Zhang, S. (2008). Imaging honey bee brain anatomy with micro-X-ray-computed tomography. *Journal of Neuroscience Methods, 171*(1), 93–97.
Riddiford, L. M. (2012). How does juvenile hormone control insect metamorphosis and reproduction? *General and Comparative Endocrinology, 179*(3), 477–484.
Rittschof, C. C., & Schirmeier, S. (2017). Insect models of central nervous system energy metabolism and its links to behavior. *Glia, 66*(6), 1160–1175.
Roeder, T. (1999). Octopamine in invertebrates. *Progress in Neurobiology, 59*, 533–561.
Rybak, J. (2012). Chapter 3.1-the digital honey bee brain atlas. In C. G. Galizia, D. Eisenhardt, & M. Giurfa (Eds.), *Honeybee neurobiology and behavior (Chapter 11)* (pp. 125–140).

Samurkas, A., Fan, X., Ma, D., Rajamanikandan, S., Lianyun, L., Li, Y., Ruifang, M., Heng, J., Peng, C., Qingzhi, G., & Zhiguang, Y. (2020). Discovery of potential species-specific green insecticides targeting lepidopteran ryanodine receptor. *Journal of Agricultural and Food Chemistry, 68*(15), 4528–4537.

Sato, K., Pellegrino, M., Nakagawa, T., Nakagawa, T., Vosshall, L. B., & Touhara, K. (2008). Insect olfactory receptors are heteromeric ligand-gated ion channels. *Nature, 452*(7190), 1002–1006.

Sattelle, D. B., Lummis, S. C. R., Wong, J. F. H., Rauh. J. J. (1991). Pharmacology of Insect GABA Receptors. *Neurochemical Research, 16*(3), 363–374.

Sattelle, D. B., Cordova, D., & Cheek, T. R. (2008). Insect ryanodine receptors: Molecular targets for novel pest control chemicals. *Invertebrate Neuroscience, 8*(3), 107–119.

Scialò, F., Fernández-Ayala, D. J., & Sanz, A. (2017). Role of mitochondrial reverse electron transport in ROS signaling: Potential roles in health and disease. *Frontiers in Physiology, 8*, 428.

Shivanandappa, T., & Rajashekar, Y. (2014). Mode of action of plant derived natural insecticides. In D. Singh (Ed.), *Advances in plant biopesticides* (pp. 323–345). Springer.

Smelt, C. L. C., Sanders, V. R., Puinean, A. M., Lansdell, S. J., Goodchild, J., & Millar, N. S. (2021). Agonist and antagonist properties of an insect GABA-gated chloride channel (RDL) are influenced by heterologous expression conditions. *PLoS One, 16*(7), e0254251.

Soderlund, D. M. (2012). Molecular mechanisms of pyrethroid insecticide neurotoxicity: Recent advances. *Archives of Toxicology, 86*(2), 165–181.

Soderlund, D. M., Clark, J. M., Sheets, L. P., Mullin, L. S., Piccirillo, V. J., Sargent, D., Stevens, J. T., & Weiner, M. L. (2002). Mechanisms of pyrethroid neurotoxicity: Implications for cumulative risk assessment. *Toxicology, 171*(1), 3–59.

Souto, A. L., Sylvestre, M., Tölke, E. D., Tavares, J. F., Barbosa-Filho, J. M., & Cebrián-Torrejón, G. (2021). Plant-derived pesticides as an alternative to pest management and sustainable agricultural production: Prospects, applications and challenges. *Molecules, 26*(16), 4835.

Spindler, K. D., Hönl, C., Tremmel, C., Braun, S., Ruff, H., & Spindler-Barth, M. (2009). Ecdysteroid hormone action. *Cellular and Molecular Life Sciences, 24*, 3837–3850.

Sullivan, K. M., Scott, K., Zuker, C. S., & Rubin, G. M. (2000). The ryanodine receptor is essential for larval development in *Drosophila melanogaster*. *Proceedings of the National Academy of Sciences of the United States of America, 97*(11), 5942–5947.

Sun, L. N., Zhang, H. J., Quan, L. F., Yan, W. T., Yue, Q., Li, Y. Y., & Qiu, G. S. (2016). Characterization of the ryanodine receptor gene with a unique 3′-UTR and alternative splice site from the oriental fruit moth. *Journal of Insect Science, 16*(1), 1–9.

Sun, Z., Xue, L., Li, Y., Cui, G., Sun, R., Hu, M., & Zhong, G. (2021). Rotenone-induced necrosis in insect cells via the cytoplasmic membrane damage and mitochondrial dysfunction. *Pesticide Biochemistry and Physiology, 173*, 104801.

Takeshima, H., Nishi, M., Iwabe, N., Miyata, T., Hosoya, T., Masai, I., & Hotta, Y. (1994). Isolation and characterization of a gene for a ryanodine receptor/calcium release channel in *Drosophila melanogaster*. *FEBS Letters, 337*(1), 81–87.

Thany, S. H., Reynier, P., & Lenaers, G. (2013). Neurotoxicité des pesticides – Quel impact sur les maladies neurodégénératives? *Medical Science, 29*(3), 273–278.

Tian, X., Li, Y., Hao, N., Su, X., Du, J., Hua, J., & Tian, X. (2021). The antifeedant, insecticidal and insect growth inhibitory activities of triterpenoid saponins from *Clematis aethusifolia* Turcz against *Plutella xylostella* (L.). *Pest Management Science, 77*(1), 455–463.

Tong, F. (2010). *Investigation of mechanisms of action of monoterpenoid insecticides on insect gamma-aminobutyric acid receptors and nicotinic acetylcholine receptors* (Thesis), 107p.

Tong, F., & Coats, J. R. (2010). Effects of monoterpenoid insecticides on [^{3}H]-TBOB binding in house fly GABA receptor and ^{36}Cl− uptake in American cockroach ventral nerve cord. *Pesticide Biochemistry and Physiology, 98*(3), 317–324.

Toprak, U., Dogan, C., & Hegedus, D. (2021). A comparative perspective on functionally-related, intracellular calcium channels: The insect ryanodine and inositol 1,4,5-trisphosphate receptors. *Biomolecules, 11*(7), 1031.

Truman, J. W. (2019). The evolution of insect metamorphosis. *Current Biology, 29*(23), 1252–1268.

Ujváry, I. (2010). Chapter 3. Pest control agents from natural products. In *Hayes' handbook of pesticide toxicology* (pp. 119–229). Academic Press. https://doi.org/10.1016/b978-0-12-374367-1.00003-3

Vais, H., Atkinson, S., Eldursi, N., Devonshire, A. L., Williamson, M. S., & Usherwood, P. N. (2000). A single amino acid change makes a rat neuronal sodium channel highly sensitive to pyrethroid insecticides. *FEBS Letters, 470*(2), 135–138.

Valbon, W., Andreazza, F., Oliveira, E. E., Liu, F., Feng, B., Hall, M., Klimavicz, J., Coats, J. R., & Donk, K. (2022). Bioallethrin activates specific olfactory sensory neurons and elicits spatial repellency in Aedes aegypti. *Pest Management Science, 78*(2), 438–445.

Voet, D., Voet, J. G., & Domenjoud, L. (2016). Biochimie. In *De boeck supérieur* (Vol. XXIII, 3rd ed., p. 1784). Springer.

Wafford, K. A., Lummis, S. C. R., & Sattelle, D. B. (1989). Block of an insect central nervous system GABA receptor by cyclodiene and cyclohexane insecticides. *Proceedings of the Royal Society B: Biological Sciences, 237*(1286), 53–61.

Wang, K. Y., Jiang, X. Z., Yuan, G. R., Shang, F., & Wang, J. J. (2015). Molecular characterization, mRNA expression and alternative splicing of ryanodine receptor gene in the brown citrus aphid, *Toxoptera citricida* (Kirkaldy). *International Journal of Molecular Sciences, 16*(12), 15220–15234.

Wang, R., Yang, Y., Jing, Y., Segar, S. T., Zhang, Y., Wang, G., Chen, J., Liu, Q. F., Chen, S., Chen, Y., Cruaud, A., Ding, Y. Y., Dunn, D. W., Gao, Q., Gilmartin, P. M., Jiang, K., Kjellberg, F., Li, H. Q., Li, Y. Y., et al. (2021). Molecular mechanisms of mutualistic and antagonistic interactions in a plant-pollinator association. *Nature Ecology & Evolution, 5*(7), 974–986.

Wang, Z., Perumalsamy, H., Wang, X., & Ahn, Y.-J. (2019). Toxicity and possible mechanisms of action of honokiol from Magnolia denudata seeds against four mosquito species. *Scientific Reports, 9*(1). https://doi.org/10.1038/s41598-018-36558-y

Ware, G. W., & Withacre, D. M. (2004). *An introduction to insecticide* (4th ed., pp. 61–69). MeisterPro Information Resources.

Warren, J. T., Petryk, A., Marques, G., Jarcho, M., Parvy, J. P., Dauphin-Villemant, C., O'Connor, M. B., & Gilbert, L. I. (2002). Molecular and biochemical characterization of two P450 enzymes in the ecdysteroidogenic pathway of Drosophila melanogaster. *Proceedings of the National Academy of Sciences of the United States of America, 99*(17), 11043–11048.

Wicher, D., Schäfer, R., Bauernfeind, R., Stensmyr, M. C., Heller, R., Heinemann, S. H., & Hansson, B. S. (2008). Drosophila odorant receptors are both ligand-gated and cyclic-nucleotide-activated cation channels. *Nature, 452*(7190), 1007–1011.

Wigglesworth, V. B. (1945). Transpiration through the cuticle of insects. *The Journal of Experimental Biology, 21*, 97–114.

Xu, X., Bhat, M. B., Nishi, M., Takeshima, H., & Ma, J. (2000). Molecular cloning of cDNA encoding a drosophila ryanodine receptor and functional studies of the carboxyl-terminal calcium release channel. *Biophysical Journal, 78*(3), 1270–1281.

Xu, P., Choo, Y. M., Chen, Z., Zeng, F., Tan, K., Chen, T. Y., Cornel, A. J., Lui, N., & Leal, W. S. (2019). Peripheral, intrareceptor inhibition in mosquito olfaction. *iScience, 19*, 25–38.

Yang, H., Piao, X. M., Zhang, L. X., Song, S. Y., & Xu, Y. H. (2018). Ginsenosides from the stems and leaves of Panax ginseng show antifeedant activity against *Plutella xylostella* (Linnaeus). *Industrial Crops and Products, 124*, 412–417.

Yin, J., Choo, Y. M., Duan, H., & Leal, W. S. (2015). Selectivity of odorant-binding proteins from the southern house mosquito tested against physiologically relevant ligands. *Frontiers in Physiology, 6*, 56.

Yoshizawa, Y., Sato, R., Tsuchihara, K., Ozaki, K., Mita, K., Asaoka, K., & Taniai, K. (2011). Ligand carrier protein genes expressed in larval chemosensory organs of *Bombyx mori*. *Insect Biochemistry and Molecular Biology, 41*, 545–562.

Yuan, G. R., Wang, K. Y., Mou, X., Luo, R. Y., Dou, W., & Wang, J. J. (2017). Molecular cloning, mRNA expression and alternative splicing of a ryanodine receptor gene from the citrus whitefly, *Dialeurodes citri* (Ashmead). *Pesticide Biochemistry and Physiology, 142*, 59–66.

Zeng, F., Xu, P., Tan, K., Zarbin, P. H. G., & Leal, W. S. (2018). Methyl dihydrojasmonate and lilial are the constituents with an "off-label" insect repellence in perfumes. *PLoS One, 13*, e0199386.

Book Links

https://books.google.fr/books?hl=fr&lr=&id=EtqBDwAAQBAJ&oi=fnd&pg=PA83&dq=Plant-derived+Drugs+and+their+Interaction+with+Human+Receptors+chapter+4&ots=G7uYnwwZtE&sig=iS0pBnzCfI2K6xnxUX3Utez24g4#v=onepage&q=Plant-derived%20Drugs20and%20their%20Interaction%20with%20Human%20Receptors%20chapter%204&f=false

Video Links

PhysioPathoPharmaco-Dihydropyridine & Ryanodine Channels – Excitation/Contraction Coupling: https://youtu.be/3Wc7I-H5stQ

Chapter 6
Antimicrobial Use in Animal Food Production

Mary Garvey

Abstract The use of antibiotics in human medicine is being monitored in most developed countries, antimicrobial use in livestock and food production, however, is currently poorly monitored and assessed. It has become increasingly recognized that the widespread application of antimicrobials in food production is contributing to the emergence and proliferation of antimicrobial-resistant species. The presence of clinically relevant multidrug-resistant species in food-producing animals may result in human cases of infectious disease. The push towards intensive livestock production systems to meet food demand has increased antimicrobial use in livestock both metaphylactically and prophylactically as growth-promoting agents. Better monitoring, surveillance, and understanding of the consequences of the uninhibited use of antibiotic agents in veterinary are needed to fully determine its potential impact on animal and human health. To protect human, animal, and environmental health and ensure sustainability, there is a dire need to develop alternative options for use in food production. The impacts of antimicrobial use, the proliferation of resistance, zoonotic disease, and environmental effects are discussed. Adequate disinfection protocols, vaccination programs, and alternative options such as antimicrobial peptides and bacteriophages, which may help alleviate the overuse of antibiotics in food-producing animals, are also discussed.

Keywords Food production · Pathogens · Resistance · Zoonosis · Food safety

1 Introduction

Antimicrobial (AM) use in agriculture and aquaculture has played a central role in food production for decades. Antimicrobials are used in animals metaphylactically, prophylactically, and non-therapeutically as growth promotors and feed proficiency

M. Garvey (✉)
Department of Life Science, Sligo Institute of Technology, Sligo, Ireland

Lir Analytical, Longford, Ireland
e-mail: garvey.mary@itsligo.ie

C. M. Galanakis (ed.), *Biodiversity, Functional Ecosystems and Sustainable Food Production*, https://doi.org/10.1007/978-3-031-07434-9_6

enhancers (Van et al., 2019). Over 200 infectious pathogens are known to infect humans and animals, including Campylobacter, Listeria, Salmonella, E. coli O157, Vibrio, Clostridium, and *Streptococcus* species with zoonotic transmission a constant risk (Hao et al., 2014). The "ESKAPE" pathogens are comprising *Enterococcus faecium, Staphylococcus aureus, Klebsiella pneumoniae, Acinetobacter baumannii, Pseudomonas aeruginosa*, and *Enterobacter* spp. displaying antibiotic and biocidal resistance is also of zoonotic importance (Mulani et al., 2019). There is also crossover in the types of antibiotics used in animals and humans, with penicillin being one of the first antibiotics prescribed for bovine mastitis towards the end of the second world war (Van et al., 2019). Disease prevention in food-producing animals is economically beneficial while also reducing or eliminating food contamination and foodborne outbreaks with pathogenic species. Estimates show that up to 80% of antibiotics produced are used in food-producing animals (Van et al., 2019), with high levels of AB entering the natural environment as agriculture run-off or via aquaculture (Kirchhelle, 2018). As the global pandemic of antimicrobial resistance (AMR) becomes more pronounced and severe, awareness is growing relating to the emergence, dissemination, and impact of AMR on public health. In terms of human health, AMR infectious diseases result in prolonged morbidity and mortality with significant social and economic impacts. Globally, by 2050, it is predicted that 10 million deaths annually will be attributed to AMR with economic costs in the trillions (Tang et al., 2017). While some root causes of AMR are well known, including the overuse and misuse of antibiotics and inadequate antibiotic application, food-producing animals' role is still under debate. Additionally, there is a dire need to promote and establish sustainability in food production to reduce environmental impact while ensuring adequate food supply to meet the growing population. The production and consumption of animal-based products, however, contributes to increased risks of infectious diseases. The impact of using antimicrobials in food-producing animals on human health is a major concern; however, following One Health, the impact on animal health and the environment must also be considered a holistic approach. One Health is based on the belief that the health and well-being of humans, animals, and the environment are interconnected. The One Health approach involves multidisciplinary teams with the overall goals of improving public health, society, animal welfare, and environmental protection while ensuring sustainability (Garcia et al., 2020). In response to this, regulatory bodies including the World Health Organisation (WHO), the World Organisation for Animal Health (OIE), and the Food and Agriculture Organisation (FAO) have implemented criteria and requirements relating to the responsible and appropriate use of AMs in food production, but this varies at a global scale. The WHO Global Action Plan on AMR was published in 2015 detailing 5 strategic objectives aimed at combating AMR via strengthening knowledge, knowledge transfer, and surveillance of antimicrobial use. The WHO has grouped antibiotic classes into categories based on their importance as human therapeutics, groupings of critically important antibiotics (CIA), significant and important applications. The CIA includes high priority therapeutics aminoglycosides, aminopenicillins, and carbapenems; and highest priority cephalosporins (4th and 5th generation), glycopeptides, macrolides, polymyxin, and quinolones (Lekagul et al., 2020). Similarly, the WHO has listed certain pathogens as critical, high, and of medium importance in their threat to public health safety. These pathogens are listed with resistance to certain antibiotic agents, including fluoroquinolones and carbapenems. Similarly, the OIE formed a CIA list of antimicrobials important in veterinary medicine labeled Veterinary Critically Important Antimicrobial Agents (VCIA). The purpose was to use different AB classes for human and animal therapy; there is some overlap, however, such as the macrolides, aminoglycosides, cephalosporins (3rd generation), and polymyxins, which are listed as both WHO CIA and VCIA (Garvey, 2020). In Europe, the new Veterinary Medicines Regulations (EU) 2019/6 requires data on antimicrobial use in food-producing animals to be collected and logged in a consumption database (Martin et al., 2020). The EU Food Safety Authority, EU Centre for Disease Prevention and Control (ECDC), and the European Medicines Agency (EMA) produced Joint Interagency Antimicrobial Consumption and Resistance Analysis (JIACRA) reports. These reports are designed to assess the use of AMs in food-producing animals and the proliferation of AMR in line with One Health (EFSA, 2017). At the current trend, it is expected that antibiotic use in food-producing animals will increase 67% by 2030, with even greater use predicted in China, Brazil, India, South Africa, and Russia (Van et al., 2019). AMR and multidrug resistance (MDR) has led to significant economic and health concerns in both livestock industries and human healthcare, resulting in clinical treatment failure and prolonged morbidity and mortality (Ma et al., 2020). Therefore, it is essential to determine the extent to which antimicrobial use in food production impacts the proliferation of AMR. Furthermore, dissemination and knowledge transfer are essential to inform all persons involved in food production from farm to fork.

2 Antimicrobials in the Food Industry

Antibiotics have traditionally been utilized in food production to control microbial species in livestock, poultry, pig, and fish industries. Antimicrobial agents administered to food-producing animals increases production, protect animal health, and decrease the bacterial contamination of animal products for human consumption. By way of example, virginiamycin reduced the presence of *Clostridium* and *Campylobacter* species, amongst other food-borne pathogens in animals, valinomycin reduced *Clostridium* infection in pigs and piglets by 43%, and neomycin as an animal feed additive reduced *E. coli* O157: H7 in animal faeces (Hao et al., 2014). AB as feed enhancers improve the animal Feed Conversion Ratio (FCR), allowing for greater metabolic efficiency requiring less cropland area for animal feed, decreased manure production, and simultaneous economic benefits for both consumers and producers (Durso & Cook, 2014). ABs' application in crop agriculture and horticulture for the control of phytopathogenic bacteria is also an emerging area of concern (Taylor & Reeder, 2020), as monitoring of crop AB use is currently much lower than that of animal use.

2.1 Agriculture and Aquaculture Use

The use of antimicrobials in agriculture and aquaculture has improved animal health and productivity as they contribute to disease prevention, treatment, management, and control (Hao et al., 2014). Such use of ABs in food production is critical, particularly in intensive livestock situations. It is now recognized that animal husbandry accounts for more antibiotic usage than human therapeutics, with 37% of veterinary antibiotics not used in human therapy (Van et al., 2019). Cross-over occurs with human therapeutics. However, with the use of penicillin and tetracyclines being the most common, and the WHO critically important macrolides, aminoglycosides, polymyxins, and cephalosporins (3rd generation) also implemented in both areas (Table 6.1). The antibiotics implemented in animal agriculture differ by country and are not standardized globally. Antimicrobial growth promotors, including sulfapyridine, streptothricin, and streptomycin, have been added to chicken and pig feed for approximately 50 years with increased production rates (Ma et al., 2020). Chicken feed supplemented with tetracycline and penicillin led to a significant improvement in egg production and hatchability (Chattopadhyay, 2014).

Interestingly, the growth-promoting effect of antibiotics on animals is solely based on observed effects on animal growth; no scientific research has identified the mechanisms by which ABs lead to animal growth (Lekshmi et al., 2017). Growth promoting activity may be related to the animal gut microbiome, leading to improvements in animal nutritional status, reducing chronic infections, and improved vitamin and mineral uptake (Hao et al., 2014). The FDA defines the subtherapeutic addition of ABs to feed as doses of <200 g/ton of feed for <2 weeks and approved the use of growth promoters in animals in 1951. The global food production usage of antibiotics is believed to be 100–200 thousand tons per year (Miranda et al., 2018). Furthermore, estimates show that global antibiotic consumption in animals will increase by 67% by 2030, spanning the chicken, pig, and cattle industries (Van et al., 2019), with China being the largest consumer of ABs in animal husbandry with 52% administered for growth-promoting activity (Ma et al., 2020). The rapid growth in aquaculture and associated disease risks has also contributed to ABs' overall usage in food production. Studies report that 56% of the human population obtains up to 20% of their protein requirements from fish and shellfish sourced from aquaculture (Lulijwa et al., 2019). Many factors such as overcrowding, rapid disease transmission, poor sanitation, and immunosuppressive factors contribute to disease susceptibility in farmed fish with loss of production as a result. ABs are administered to farmed fish predominately via medicated fish feed with large amounts of antibiotics such as oxytetracycline entering the aquatic environment via urine, faces, and uneaten food (Miranda et al., 2018). Antimicrobial consumption in medicated feed and water is higher in pigs than chicken and cattle (Lekagul et al., 2020), with up to 80% of mixed feeds for piglets, veal calves, and poultry containing antibiotic additives in some countries (Kirchhelle, 2018). The benefits of medicated feed are evident in pigs, where10–15% less feed is needed to achieve the animal's desired level of growth (Chattopadhyay, 2014).

Table 6.1 Antimicrobials commonly used in food-producing animals, intended purpose, and additional comments

Antibiotic class	Antibiotic	Food sector	Intended purpose	Comments
Tetracyclines	Chlortetracycline, oxytetracycline, tetracycline, doxycycline	Pig, cattle poultry, plant disease, fish, veal calves	Growth promotors (Granados-Chinchilla & Rodríguez, 2017), therapeutically	WHO Essential, low cost, broad spectrum
Streptogramins	Virginiamycin	Cattle, poultry, rice crops in Japan, fruit trees in the US	Feed additive, growth promotor, (Dzhavakhiya et al., 2016)	Not approved for use in food animals in the EU, WHO Highly important
Macrolides (Van et al., 2019)	Erythromycin, tilmicosin, tylosin tulathromycin,	All food animals, pig, cattle	Metaphylactic, tylosin for growth promotion	WHO critically important, have significant immunomodulatory effects, useful for controlling nematodes (Hao et al., 2014), used in cases where allergy to penicillin is present
Penicillin	Penicillin, amoxicillin, (Lekshmi et al., 2017)	Dairy cows	Growth promotor, therapeutic	WHO critically important, β-lactams, Gram-negative resistance common
Sulfonamides	Sulfadimidine, sulfadiazine, sulfathiazole	Pig, poultry, cattle, fish	Therapeutic, prophylactic	WHO Significant, anti-parasitic activity, one of the first AB used in food animals, some are potentially carcinogenic (Wang et al., 2006)
Fluoroquinolones	Ciprofloxacin, enrofloxacin, danofloxacin	Poultry, cattle, pig	Therapeutic, prophylactic	WHO critically important, detected on barn dust (Schulz et al., 2019)
Aminoglycosides	Gentamicin, neomycin	Prevent early poultry mortality		WHO critically important

(continued)

Table 6.1 (continued)

Antibiotic class	Antibiotic	Food sector	Intended purpose	Comments
Polymyxin	Colistin	Poultry, beef cattle, pig	Growth promotor, prophylactic, pig fattener	WHO important, neurotoxic and nephrotoxic, control swine dysentery, bacterial enteritis; control poultry enteritis
Cephalosporin	Ceftiofur (3rd gen) cefquinome (4th gen)	Cattle, Pig, Poultry	Therapeutic	MRL of 100 μg/kg (Durel et al., 2019), WHO critically important, β-lactams,
Ionophore	Monensin Salinomycin (Lekshmi et al., 2017)		Growth promotor, increase FCR, control of coccidiosis	second most widely used class of antibiotic in US agriculture, not used in human medicine, have favorable effects on rumen fermentation and methane reduction (Hao et al., 2014)

In 2006 the EU forbade the use of ABs for growth promotion purposes in food-producing animals, with the US only eliminating the use of medically necessary AB for growth promotion (Van et al., 2019). Oxytetracycline and chlortetracycline are licensed growth promotors and feed efficiency additives in the US (USFDA, 2020), with tetracycline being prohibited from 2017 (Granados-Chinchilla & Rodríguez, 2017). The WHO, FAO, and EU have implemented maximal residual levels (MRLs) of ABs in animal food products for human consumption, e.g., an MRL of 100 μg/kg for tetracycline, oxytetracycline, and chlortetracycline in milk applies (Aalipour et al., 2015). Prophylactic use of ABs in pigs is not implemented in the fattening period, for example, to prevent exceeding such MRLs at slaughter. The EU legislation, Good Veterinary Practices (GVP), requires withdrawal times following antibiotics administration to food-producing animals such as dairy cows. In developing countries excluding South Africa, however, no legislation is implemented regulating the use of antibiotics in food-producing animals. Studies by Tang et al., assessing the impact of such legislation in food-producing animals and AMR rate in humans, found a positive effect, where AMR was reduced in humans, particularly in persons having direct contact with food-producing animals (Tang et al., 2017). The EU ban on the use of avoparcin in 1997 reduced VRE in poultry, where a reduction in tylosin use decreased erythromycin resistance in Enterobacteriaceae in pigs (Lekshmi et al., 2017).

2.2 Mechanisms of AMR

Antimicrobial agents include antibiotic, antiviral, and antifungal compounds having application in human and animal medicine. In terms of AMR, antibiotics are more commonly considered. However, there is growing recognition of the importance of antifungal and antiviral activity in clinical disease. The mode of action of antibiotics varies depending on the drug itself; β-lactams (penicillin and cephalosporins) interfere with cell wall synthesis, tetracyclines, macrolides, and aminoglycosides disrupt protein synthesis, fluoroquinolones disrupt nucleic acid synthesis, sulphonamides inhibit metabolic pathways. In contrast, polymyxins disrupt the bacterial cell membrane (Savage et al., 2017). Antibiotic resistance mechanisms are intrinsic among certain bacterial species, acquired via spontaneous mutations (de novo) and shared on mobile genetic elements (MBEs) such as plasmids (extrachromosomal DNA) or transposons (DNA segments which can change genomic position) via horizontal gene transfer (conjugation, transformation, and transduction) of antibiotic resistance genes (ARGs) amongst other species. For example, 2 ARGS *sul1* and *sul2* encoding resistance to sulphonamide are carried by transposons and plasmids (Nelson et al., 2019). Resistance mechanisms are resultant from exposure to antimicrobial agents, particularly at sub-therapeutic concentrations with resistance achieved in several ways: mutations modifying gene targets, inhibiting cellular uptake of the antimicrobial, removing the antimicrobial compound, or by deactivating the antibiotic via enzymatic degradation (Savage et al., 2017). Sub-inhibitory concentrations of antibiotics can produce reactive oxygen species (ROS), prompting mutagenesis (Amarasiri et al., 2019). Genetic mutations that confer resistance (to individual AB and entire AB classes) can be retained and transmitted to progeny microbial cells (termed vertical transmission), generating a resistant population following exposure. This selective pressure produces a resistant species with a distinct advantage over its non-resistant counterpart. This selective pressure inflicted by human misuse and overuse of AB has greatly proliferated AMR. For example, mutations that arise in the gyrA and parC genes confer resistance to quinolones, with mutations in the rpsL gene conferring resistance to streptomycin. Importantly, resistance to several different antimicrobials is often linked on the same plasmid, meaning that administering one type of AB can induce resistance to other drug classes, conferring multidrug resistance. Plasmid transmission of MCR genes (Table 6.2) proliferates resistance to colistin with carbapenemases genes present on the same plasmid inducing resistance carbapenems, which are not licensed for food-producing animals. Genes of the IncA/C plasmids isolated from *Salmonella* bestow resistance to aminoglycosides, β-lactams, chloramphenicol, sulfisoxazole, tetracyclines, and trimethoprim (McMillan et al., 2019). The plasmid RP1 first isolated from *P. aeruginosa,* which is transferrable to numerous Gram-negative species, confers resistance to ampicillin, tetracycline, and kanamycin with vertical transmission to daughter cells (Nelson et al., 2019). The first bacterial enzyme capable of antibiotic degradation discovered was an AmpC β-lactamase in *Escherichia coli* affecting the β lactam antibiotics. Extended-spectrum beta-lactamase (ESBL) genes, which

Table 6.2 Outlining zoonotic pathogens associated with human infectious disease, examples of their resistance mechanisms, and biocidal resistance evident

Pathogen	Evident resistance	Resistance mechanisms	Biocidal resistance
Enterobacteriaceae family – *Salmonella, Escherichia coli, Shigella, Proteus, Klebsiella, Enterobacter spp.*	Colistin Carbapenems Pencillins/ cephalosporins MDR	Chromosomal, plasmid mcr-1 gene carbapenem-hydrolyzing β-lactamase genes (Vidovic & Vidovic, 2020), porin OmpC, OmpK36 ES β-lactamases, porin OmpF Efflux pumps AcrAB-TolC, AcrAB, MsdABC/ TolC	Antiseptic resistance via the TolC efflux pump (Impey et al., 2020) Hexachlorophene, benzalkonium chloride conferred by ARGs (Boutarfi et al., 2019), Chlorhexidine, QAC via qacE gene,
Pseudomonas aeruginosa	Colistin Carbapenems MDR	Plasmid mcr-1 gene, Porin OmpD, Efflux pumps MexAB-OprM, MexCD-OprJ, qacE gene, biofilm formation	QACs RND pump mediated, triclosan intrinsic resistance (Amsalu et al., 2020), BAC induces the MexCD-OprJ pump, chlorhexidine
Listeria	Fluoroquinolones, cefotaxime, MDR	FepA, Lde, MdrL, MdrT efflux pumps	Benzalkonium chloride, heavy metals due to the MdrL pump
Campylobacter	MDR, fluoroquinolone, macrolide, tetracycline, aminoglycosides	Efflux pump CmeABC, Thr-86-Ile mutation in *gyrA* and the A2075G mutation in *23S rRNA*, *tetO*, *aphA*, and *aadE* genes, *pTet* plasmid, RE-CmeABC	QACs, chlorhexidine, triclosan and trisodium phosphate (Wales & Davies, 2015)
Acinetobacter	Colistin, β-lactam, Aminoglycosides	Efflux pumps, β-lactamases AMEs, cepA, qacA and qacE genes	Benzalkonium chloride Cetrimide Chlorhexidine gluconate via BRGs (Vijayakumar & Sandle, 2018)
Staphylococcus	Fluoroquinolones, virginiamycin, novobiocin	TetK, QacA, C and B, NorA efflux pumps, plasmid pSK1, MdeA pump,	benzalkonium chloride, QACs via MdeA pump (Schindler & Kaatz, 2016) chlorhexidine gluconate

(continued)

Table 6.2 (continued)

Pathogen	Evident resistance	Resistance mechanisms	Biocidal resistance
Streptococcus	Chloramphenicol, erythromycin, fluoroquinolones, clindamycin	PatAB, SP2073-2075, PmrA, PdrM efflux pumps, ribosomal target site alteration, alteration in antibiotic transport, antibiotic modification	Chlorhexidine (Wales & Davies, 2015)
Clostridium difficile	Aminoglycosides, fluoroquinolones, lincomycin, tetracyclines, erythromycin, clindamycin, penicillins, cephalosporins	CdeA (MATE pump), β-lactamase-like proteins, penicillin-binding protein, MGEs, biofilm formation	Spores resistant to sodium hypochlorite and sodium dichloroisocyanurate (Dyer et al., 2019)
Enterococcus	Norfloxacin, ciprofloxacin, doxycycline, arbekacin, novobiocin, daunorubicin, doxorubicin	EfrAB an ABC pump, qac genes – qacA/B and qacZ	QAC resistance (Vijayakumar & Sandle, 2018)

are plasmid associated and spreadable such as blaTEM, blaCTX-M, blaCMY-2, blaKPC, blaOXA-48, blaPSE, have a broad host range but are largely present in *E. coli* and *Klebsiella spp* (Abrar et al., 2019). Modification of the penicillin-binding protein as typically seen in Gram-positive bacteria including *Staphylococcus aureus* confers resistance to the β lactam antibiotics. The expression of an additional penicillin-binding protein conferrs resistance to methicillin in *S. aureus* (MRSA). Both these resistance mechanisms are termed target alteration or modification. Carbapenem resistance is resultant from carbapenemases enzyme activity. Resistance to aminoglycoside antibiotics is related to the production of aminoglycoside modifying enzymes (AMEs). The WHO lists Carbapenem-resistant P. aeruginosa as a critically important pathogen. It displays intrinsic and acquired resistance to various antibiotics, including aminoglycosides, quinolones, and β-lactams via efflux pumps, porin regulation, enzymatic activity, and potent biofilm formation ability. The chloramphenicol acetyltransferase (CATs) enzymes are known to inactivate the broad-spectrum chloramphenicol therapeutics. Esterases are another hydrolytic enzyme that inactivates macrolide ABs whereas acetyltransferase enzymes modify drugs by attaching acyl groups to the drug moiety, inhibiting the activity of aminoglycosides (Lekshmi et al., 2017). Resistance to tetracycline is achieved via blocking the AB from binding to its target,the 30S ribosome subunitvia the presence of peptides. The reduction of drug permeability is achieved via down-regulating outer membrane proteins (OMPs) such as OmpC, OmpD, which blocks the activity of aminoglycosides, β-lactams, fluoroquinolones and chloramphenicol (Fernández & Hancock, 2012). Efflux pumps have a vital role in cellular homeostasis regulating the uptake of essential nutrients, solutes and expelling waste products and harmful substances (Impey et al., 2020). Intrinsically located or acquired energy-dependent efflux pumps are a major contributor to AMR and MDR having narrow and broad-spectrum activity and promoting biocidal resistance in numerous species (Savage et al., 2017). There are several classes of transporter protein superfamilies involved in this efflux pump system. The ATP requires ATP-binding cassette (ABC) transporters and the passive and secondary active transport systems. Secondary active transport systems utilize ion gradients generated by cellular respiration as the driving energy force. These systems include the resistance-nodulation-cell division (RND) superfamily, the small multidrug resistance (SMR) superfamily, the multidrug and toxic compound extrusion (MATE) superfamily. This group also includes the major facilitator superfamily (MFS) of solute transporters, such as those seen in the Enterobacteriaceae family and *Staphylococcus aureus* (Lekshmi et al., 2017). The main efflux pump in Gram-negative bacteria is the RND AcrAD pump. It is associated with the removal of fluoroquinolones, cephalosporins, tetracyclines, amongst other ABs (Anes et al., 2015) and the AcrAB-TolC pump found in the Enterobacteriaceae family, including *E. coli, K. pneumoniae,* and *Salmonella spp*. The RND TolC efflux pump of the Enterobacteriaceae family confers resistance to numerous antibiotics antiseptics and including β-lactams, chloramphenicol, fluoroquinolones, novobiocin, tetracycline, and macrolides (Impey et al., 2020). Common antiseptics typically have alcohol, hydrogen peroxide, iodine, and trichlorophenols as active ingredients (Nelson et al., 2019). The MexAB-OprM, MexCD-OprJ, MexEF-OprN, and MexXY-OprM pumps of *P. aeruginosa*, CmeABC of *Campylobacter*, and AdeABC of *Acinetobacter baumannii* confer MDR in these species. The emergence of a super efflux pump variant, *RE-CmeABC*, was found to confer MDR in *Campylobacter* as a non-specific efflux pump expelling multiple antibiotics classes and is plasmid-encoded and shared via HGT (Yang et al., 2019). The MFS, SMR, MATE, and ABC superfamily are the efflux families associated with Gram-positive species (Li et al., 2016). The *S. aureus* efflux pumps (e.g., TetK) are found on plasmids with MDR efflux pumps chromosomally encoded. Furthermore, fluoroquinolone resistance in Gram-positive species, including *S. aureus, S. pneumoniae*, the viridans group *Streptococci*, and *Enteroccocci*, is mediated by efflux pumps such as NorA (Poole, 2000). Indeed, the Enterococcus genome possesses 34 potential drug efflux related genes (Li et al., 2016). Biocidal resistance is also being recognized as an issue relating to public health safety, particularly with MDR species. Biocidal resistance is also intrinsic, acquired via gene mutations or HGT of biocide resistance genes (BRGs) such as qac and cepA genes (Vijayakumar & Sandle, 2018). BRGs have been identified in many Gram-positive and Gram-negative species such as the qacE, qacA/B genes common in the *Enterobacteriaceae* family and *Pseudomonas* conferring resistance to quaternary ammonium compounds (QACs) and the plasmid pSK1 of *S. aureus* conferring resistance to antiseptics (Vijayakumar & Sandle, 2018). These qac genes code for efflux pumps active in removing QAC compounds from the bacterial cell following exposure. Furthermore, these genes can be shared via HGT on plasmids to other bacteria and antibiotic resistance genes, explaining why MDR species tend to have biocidal resistance. For every AB there is a complement of genes coding for resistance. For example, there are over 25 ARGs associated with the 3 mechanisms of tetracycline resistance (Durso & Cook, 2014). Additional survival strategies of bacterial species include developing small colony variants (SCV), a phenotype having a slow growth rate rendering ABs less effective. SCVs are often observed in prolonged AMR infections with significant mortality rates (Vidovic & Vidovic, 2020).

2.3 Agri and Aquaculture Promoting AMR and Dissemination

Implementing antimicrobials in food-producing animals for disease management and growth promotion is undoubtedly an area of importance. More and more evidence emerges indicating the impact of such use on animal welfare and the proliferation of AMR in pathogenic species. For example, studies report on the emergence of antibiotic resistance in *Campylobacter* species following fluoroquinolones in the poultry industry, where it is a commensal organism in the gastrointestinal tract (GIT) of birds. Indeed, approximately 94% of *Campylobacter jejuni* isolated from chickens display resistance to one or more classes of antibiotic (Wieczorek et al., 2018). Indeed, *C. jejuni* separated chicken meat was correlated to clinical human isolates in the US (Ma et al., 2020). There is much concern over AMR in *Campylobacter* species as human cases of campylobacteriosis are increasing yearly, with fluoroquinolone resistant *Campylobacter* listed as WHO high priority. The macrolide AB azithromycin is the therapeutic choice for treating *Campylobacter* infections in humans, followed by erythromycin and fluoroquinolones (Yang et al., 2019). The research demonstrated that chickens supplied with tetracycline laden feed possessed tetracycline resistant enteric organisms within 7 days transmitted horizontally on-site to other animals (Levy et al., 1976). Studies monitoring ESBL Enterobacteriaceae in Swedish broilers found that up to 50% of poultry carcasses contained ESBL isolates at slaughter (Boqvist et al., 2018).

Furthermore, the use of virginiamycin in broiler chickens in Denmark in the 1990s was correlated with an increase in *E. faecium* to this AB (Ma et al., 2020) with resistance decreasing following the virginiamycin ban in 1998 (Lekshmi et al., 2017). Studies report on MDR Enterobacteriaceae such as *E. coli* and *Salmonella* isolated from beef carcasses post evisceration and refrigeration with resistant Enterococci isolated continuously from cattle, poultry, and pig carcasses, and often found on fresh meat (Bakhtiary et al., 2016). Studies also identified vancomycin-resistant enterococci (VRE) in meat and faeces of farm personnel (Nelson et al., 2019) with the ban of avoparcin (glycopeptide antibiotic like vancomycin) use in chickens leading to a decrease in VRE isolated from faeces. Dairy and dairy products remain a source of milk microbiota, ARGs and AMR species of zoonotic importance. The isolation of MRSA from milk and milk products is also noteworthy as methicillin is not used to treat *Staphylococcus* mastitis in cattle with staphylococcal enterotoxins (SEs), even a risk to public health safety (Basanisi et al., 2017). In the United States, ceftiofur is often administered to piglets in the swine industry at birth, with males getting a second dose at castration, allowing selective pressure prompting the dissemination of carbapenemase-producing strains such as carbapenem resistance *Enterobacteriaceae* (CRE) (Cheng et al., 2019).

Studies have also demonstrated the transmission of resistance from animal intestines to animal skin. The non-therapeutic tylosin was found in pig intestinal *Enterococcus* isolates and *Staphylococci* skin isolates with erythromycin resistance of gut enterococci shared with dermal *Staphylococcus hyicus* (Marshall & Levy, 2011). The presence of ARG resistomes in the environment from ABs' application in food producing animals promotes the spread of AMR via HGT. Resistance genes including blaCTX-M, blaTEM, blaSHV, blaOXA, tet(A), tet(B), sul1, sul2, and qnr are frequently observed in environmental and livestock areas, where the transmission of ARGs genes across humans, animals, and environment is possible (Kimera et al., 2020). The JIACRA reported in 2015 and 2017 assessing zoonotic pathogens in food producing animals (poultry, pig, beef) showed significant association between the use of fluoroquinolones and resistance in *E. coli*, *C. jejuni* and *Salmonella* and tetracycline in *E. coli* and *Salmonella* in many EU countries (Bennani et al., 2020).

2.4 Externalities of Antimicrobial Use in Food Producing Animals

The consistent use of antimicrobial agents in food producing animals must be considered in terms of its broader impact on the environment, human and long-term animal health/welfare. The short-term effects are undeniable in terms of disease management and increased productivity, and the long-term impact can no longer be ignored. As food production and animal welfare must become more environmentally aware and sustainable, there is a dire need to determine the externalities of this approach to food production. The FAO recognises the impact of climate change on food security as unpredictable climatic events disrupt agriculture, food availability and cause disease outbreaks in human and animal populations (Garcia et al., 2020). As the human population increases, so too will the population of food-producing animals as food requirements need to be met. This increase in livestock production coupled with a global decline in wild animal populations and environmental damage is promoting outbreaks of emerging and re-emerging disease, often of a zoonotic nature (Garvey, 2020). This undeniable interconnectedness of human and animal populations via shared ecosystems (Fig. 6.1) and the impact of human activity on animal and environmental welfare is now a key focus of One Health.

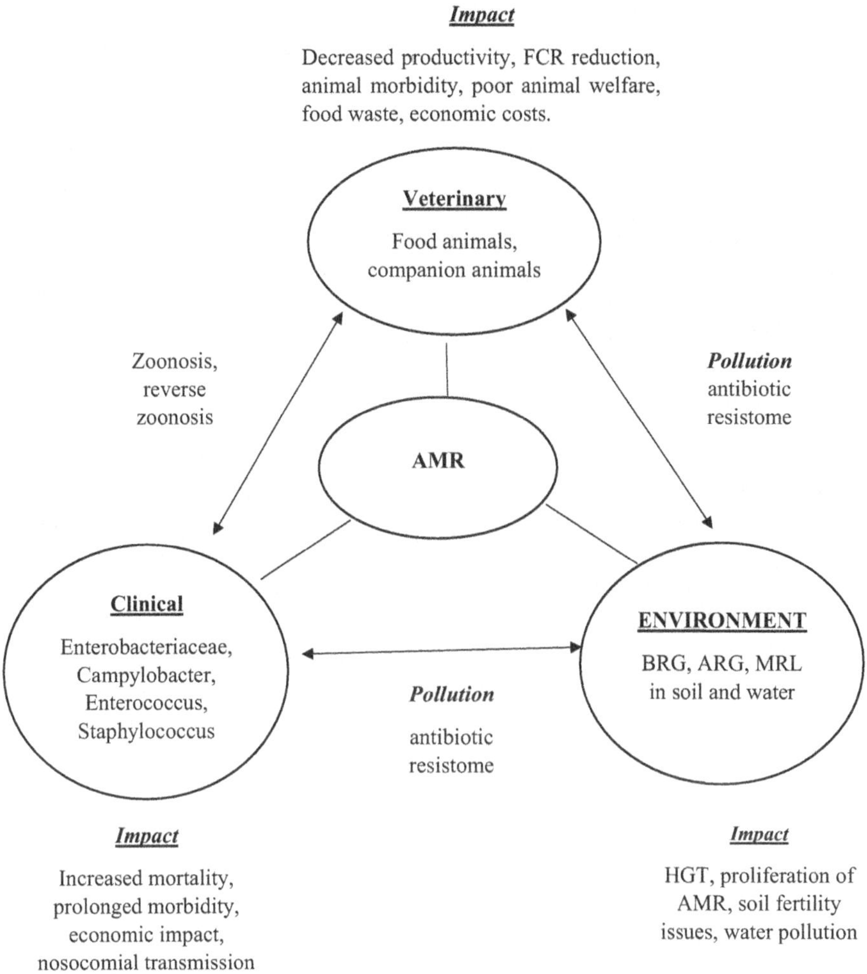

Fig. 6.1 The interplay of clinical therapy, veterinary use and the environmental accumulation of ABs and ARGs

2.4.1 Impacts of AMR

Antibiotics remain one of the most excellent therapeutic aids known to humankind, reducing infectious disease mortality, allowing for invasive surgeries, transplants, and the treatment of less severe infections in humans and animals. With the current antibiotic crisis, however, antibiotics' efficacy is greatly reduced with grim consequences as a result. AMR infections result in prolonged morbidity, extended hospital admissions, treatment costs, and often mortality. The morbidity and mortality rates are particularly increasing in developing countries highly populated due to various AMR infectious diseases. For example, AMR results in treatment failure in 195,763 cases of pneumococcal disease and 2925 child deaths annually in Ethiopia

(Ma et al., 2020). In Europe, MDR ESKAPE pathogens are responsible for approximately 400,000 morbidities and 25,000 mortalities yearly with the US CDC estimating over two million AMR morbidities and 23,000 mortalities yearly in the United States (Prestinaci et al., 2015). Bacteraemia resultant from MRSA has a significantly higher mortality rate than methicillin sensitive *S. aureus* infections with ESBL Enterobacteriaceae also have higher mortality in patients than non ESBL strains (Friedman et al., 2015). The emerging carbapenem-resistant Enterobacteriaceae have a mortality rate of up to 71% with carbapenem-resistant *A. baumannii* bacteraemia having a 14-day mortality of 45% (Nutman et al., 2014). Bacteraemia represents one of the most severe consequences of infection, particularly with Gram-negative *E. coli* responsible for one-third of blood infections globally (Dee et al., 2018) with resistance to 3rd generation cephalosporins further increasing the severity of infection. The presence of underlying conditions and comorbidities contributes to mortality in patients presenting with AMR infectious disease (Bogan et al., 2014). The failing treatment options for MDR Gram-negative pathogens have prompted older less effective and more toxic drugs such as polymyxins. Such antibiotic options pose a risk to the patient in terms of cytotoxicity and organ damage. Polymyxin B and colistin, which are administered for carbapenem-resistant bacterial infections are associated with deteriorating renal function (Friedman et al., 2015), resulting in acute kidney injury. The onset of acute kidney injury during antibiotic therapy is correlated with higher patient mortality rates (Nation et al., 2019). Additionally, prolonged AMR infectious disease may result in iatrogenic disease, as the treatment of AMR infection induces co-morbidities in the patient. Prolonged administration of AB is associated with antibiotic-induced (pseudomembranous) colitis, inflammatory bowel disease (IBD), dysbiosis, obesity, metabolic syndrome, leaky gut syndrome, irritable bowel syndrome (IBS) and autoimmunity (Garvey, 2020). AMR's economic impact is also an essential factor as costs impact at national and international levels, with AMR costing the EU approximately 9 billion euro and the US 20 billion dollars annually (Dadgostar, 2019). Indeed, endemic zoonotic diseases cause illness and mortality in animals and humans and high economic costs in areas where farming is the primary source of income, as seen in developing countries (Belay et al., 2017).

Animal disease and death impose economic losses at the farm level relating to animal treatment and replacement costs. Animal health and productivity can be significantly impacted, leading to a loss of animal products, infertility issues, abortion, and weak offspring (Rahman et al., 2020). Transboundary diseases are a contagious animal disease which typically result in high morbidity and mortality. They are economically devasting for farmers, detrimental to food safety and food security and often zoonotic (Garcia et al., 2020).

2.4.2 Zoonosis – Dissemination to Healthcare

It is well known that >60% of infectious diseases in humans are of zoonotic origin, being transmitted from animals to humans with up to 75% of emerging diseases also zoonotic (Belay et al., 2017). However, the transmission of pathogens from humans to animals termed zooanthroponosis (reverse zoonosis) is also of importance in terms of the overall One Health approach. In terms of zoonosis there are two major routes: (1) the acquisition of pathogens via direct contact with the food-producing animals or their human handlers and (2) the indirect acquisition of pathogens via the food chain or hotspots of high AMR pollution including manure, wastewater and agriculture land (Vidovic & Vidovic, 2020). Once acquired, zoonotic pathogens may spread from human to human (anthroponoses) or from fomites to humans, further extending the infectious disease. There are several bacterial species associated with foodborne zoonosis from animal food production, including MRSA, MDR *Salmonella* sp., CRE, VRE, MDR *Campylobacter, Clostridium sp.* and *Listeria monocytogenes* (Dee et al., 2018). *Campylobacter* remains the most common zoonotic foodborne pathogen in the EU, followed by *Salmonella* with Shiga toxin *E. coli* (STEC) 0157 and *Listeria monocytogenes* frequently resulting in disease outbreaks (Boqvist et al., 2018). Poultry is considered a significant reservoir of quinolone and cephalosporin resistant *Campylobacter* with pork and beef associated with cephalosporin resistant pathogens (Abraham et al., 2020). Research shows that in the United States approximately 9.4 million cases of foodborne disease occur yearly, with 11% attributed to *Salmonella*, 9% to *Campylobacter* with *Salmonella* the primary cause of hospitalization and death (Zaidi et al., 2012). Campylobacteriosis, contracted from contaminated chicken, remains the most common foodborne disease in the EU, with 236,851 confirmed cases in 2014 where cases of foodborne listeriosis increased by 16% in 2014. Studies suggest that Gram-negative bacteria are responsible for approximately 69% of foodborne zoonoses (Abebe et al., 2020). The microbial production of toxins (endotoxins and exotoxins) are also essential factors in foodborne illness and the pathogenesis of bacteria. Toxins produced by STEC strains can result in severe disease in humans, including gastroenteritis, bloody diarrhoea, kidney failure, and fatal haemolytic uremic syndrome with Staphylococcal endotoxins associated with toxic shock syndrome (Pinchuk et al., 2010). The EU Directive 2003/99/EC on zoonosis requires EU member states to monitor and collect information on zoonotic agents and AMR relating to food outbreaks and animal feed for Salmonella *Campylobacter, Listeria monocytogenes*, STEC, *Mycobacterium bovis*, *Brucella*, *Trichinella* and the parasite *Echinococcus*. The 2018 report's findings show an increase in outbreaks associated with eggs and egg products with meat the second biggest contributor to food borne outbreaks where *Salmonella* was the pathogen present (EFSA, 2019). Brucellosis is a widespread zoonotic pathogen resulting in over 500,000 human cases yearly worldwide, via the consumption of unpasteurized dairy products and is considered a neglected zoonotic pathogen by the WHO (Rahman et al., 2020). Tuberculosis is the most important of the bovine zoonotic diseases having severe public health importance and economic impacts in animal production and is caused by *Mycobacterium bovis*, and *M. tuberculosis*. Zoonotic transmission occurs by consuming unpasteurized contaminated milk or via aerosols from coughing infected animals. Tuberculosis is a rare disease in the EU due to the disease control strategy and pasteurisation of milk (EFSA, 2019). Fish are also a significant source of foodborne disease with the most essential pathogens being water borne freshwater bacteria (*Vibrio sp*) and bacteria (*Listeria* and *Salmonella*) present from water pollution, human handling, and transportation (Novoslavskij et al., 2016). There are numerous species associated with fish zoonoses, including *Aeromonas hydrophila, E.coli, Yersinia spp., Brucella spp., Shigella spp., Salmonella spp., Streptococcus iniae, Clostridium botulinum, Listeria* and *Klebsiella spp*., (Haenan et al., 2013). Indeed, in the US fish was the most common foodborne disease outbreak from 2011 to 2014. In developing countries seafood foodborne disease is a significant cause of hospitalisation and death (Ali et al., 2020). The largest outbreak in the US associated with fish resulted in 425 cases in 28 states in 2012 due to the consumption of raw tuna infected with *Salmonella* (Hoffmann et al., 2016).

2.4.3 Environmental Impact

Antibiotics, AMR species and ARGs are present in the environment from agriculture runoff, aquatic fish farms, agri wastewater and manure and clinical wastewater. As AB compounds are only partially metabolised in the body, approximately 30–90% are excreted unchanged in wastewater from animals and humans. Wastewater management systems can only remove 20–80% of pharmaceuticals allowing active pharmaceutical ingredients and their metabolites into the environment (Chen et al., 2021). It is estimated that approximately two-thirds of global rivers, surface waters, groundwater and wetlands are contaminated with micropollutants including antibiotics, pesticides, biocides and industrial chemicals, harming these dynamic ecosystems and their biodiversity (Xie et al., 2020). Hotspots of AMR species and ARGs are present in locations near large animal housings, veterinary clinics, and hospital settings, leading to resistomes in the environment (Kimera et al., 2020). Tetracyclines are the most frequently detected AB in manure and have the highest concentrations, followed by fluoroquinolone, the macrolide tylosin and sulfonamides (Cycoń et al., 2019). This antibiotic pollution is poorly regulated both locally and globally, consequently antibiotics are accumulating in terrestrial, freshwater, and marine environments (Kraemer et al., 2019). Environmental risks of antibiotic pollution include residue accumulation, biodiversity toxicity, selection for antibiotic resistance in species and MDR emergence where soil and water have become reservoirs for AMR (Manyi-Loh et al., 2018). In particular, fish are exposed to a range of ABs at varying concentrations chronically and continuously throughout their life cycle in their natural habitats (Liu et al., 2018).

ABs and ARG affect soil and water microbiota, which are essential for ecological functions, including nutrient cycling, decomposition, and soil fertility. The presence of AB may disrupt this vital balance by promoting the overpopulation of certain species via selective pressure (typically promoting an increase in

Gram-negative bacteria) which impacts microbial ecological roles (Kraemer et al., 2019). Therefore, antibiotics are considered one of the main chemicals affecting aquaculture products together with hydrocarbons, pesticides, hormones, heavy metals and naturally occurring toxins (Varol & Sünbül, 2017). Bioaccumulation of antibiotics is also a risk as plants absorb antibiotic chemicals such as sulfamethoxine in roots and stems (Hao et al., 2014). Studies show ABs' negative effect, including fluoroquinolones, macrolides and tetracyclines on chloroplastic and mitochondrial protein synthesis, plastid replication, morphology, and photosynthesis in plants (Doyle et al., 2020). Studies have also demonstrated that the synergistic effects of antibiotics in the environment include combinations of amoxicillin, erythromycin, levofloxacin, norfloxacin tetracycline were toxic to cyanobacterium and green algae (Xie et al., 2020). Antibiotic degradation in the environment effects their persistence in ecosystems. Hydrolysis of β-lactams is key to the breakdown of penicillin's and cephalosporins whereas photo-degradation contributes to degradation quinolones tetracyclines (Cycoń et al., 2019). Like most ABs, fluoroquinolones are poorly metabolised and so excreted relatively unchanged and remain stable in soil with photodegradation occurring in aquatic environments (Rusch et al., 2018). The presence of different organic compounds in soil also impacts AB degradation. Changes in degradation are seen in manure, biosolids, slurry, sludge, and compost, and soil types (Cycoń et al., 2019) while the presence of microbial species also impacts degradation.

Animals can consume AB compounds from contaminated soil and water allowing for exposure and bioaccumulation. Bioaccumulation of fat-soluble antibiotics or their metabolites occurs in aquatic life. Food producing animals where residues are found in muscles, kidney, liver, fat, milk, and eggs (Manyi-Loh et al., 2018) are associated with numerous health issues, including carcinogenicity and allergic reactions and drug sensitization. In developing countries oxytetracycline, chloramphenicol, and neomycin are still administered to food animals. The adherence to necessary withdrawal periods is also often overlooked meaning humans may consume AB contaminated food products. This has been associated with kidney disease (neomycin), carcinogenicity, mutagenicity, and aplastic anaemia (chloramphenicol) (Wongtavatchai et al., 2004). Studies report on the presence of AB in animal tissues where chloramphenicol and tetracycline residues exceeding the MRL were found in chicken meat and eggs and ciprofloxacin in egg white with oxytetracycline detected in beef tissue exceeding the MRL set out by the WHO (Billah et al., 2015). The presence of different resistance genes in the soil and water environment promotes their combination into the same genetic element, such as integrons, inducing MDR in pathogens, such as MRSA and CRE (Kraemer et al., 2019).

Additionally, resistance genes in the aquatic environment may enter the biosphere where airborne transmission has also been documented. Studies report on the presence of antibiotics on dust in animal housing environments including broiler barns. For example, fluoroquinolones are synthetic compounds used for veterinary application with high stability in biological and environmental systems have been detected on dust particles in broilers, pig and cattle housing units (Schulz et al., 2019).

2.5 Livestock Intensification and Sustainability

Accelerated by rising demand for meat as a food source, urbanisation and market demand, intensive livestock systems highly dependent on antibiotic use have expanded globally. Intensive livestock systems consist of feedlot or housed animals such as cattle, pig, and poultry, which breed and feed for optimal production. These systems have increased global food production efficiency with more food produced at a lower cost per unit (Van Boeckel et al., 2015). Intensive animal farming promotes the emergence and proliferation of AMR species however, due to the physical and genetic closeness of housed animals whereas the consumption of bushmeat and backyard farmed meat increases the risk of disease spillover from wild animals (Espinosa et al., 2020). Horizontal transmission of pathogens occurs rapidly amongst large numbers of animal housed in proximity in intensive systems.

Additionally, breed selection for optimal productivity has promoted genetic similarity within animals in these systems resulting in increased disease susceptibility as all animals are immunologically similar. At present, agriculture occupies approximately half of the worlds land and two thirds of water. This is expected to double to meet the growing human population, increasing the use of fertilisers, pesticides, and antibiotics (Rohr et al., 2019). As economies of scale in agriculture and livestock expand, so too will its impact on the environment. Livestock waste accounts for a large proportion of greenhouse gasses with the production of methane, ammonia, carbon dioxide, nitrate, ethylene acid, methyl mercaptan, hydrogen sulphide, methylamine, and trimethylamine (Hao et al., 2014). Studies show that poorly managed livestock systems can promote eutrophication, overgrazing and soil degradation where 90% of livestock emissions are resultant from ruminant gastric discharges with 10% from manure (Varijakshapanicker et al., 2019). The move towards more significant intensive livestock systems has the benefits of reducing contact between food animals and wild animals and reducing disease transmission. Housed animals, however, have no outdoor access and require feed support to replace foraging crops typically laden with antibiotics to promote growth and prevent disease (Espinosa et al., 2020). This increase demand on food supplies strains natural resources. It contributes to soil erosion, loss of biodiversity, climate change and global environmental pollution, raising sustainable food production and food safety (Garcia et al., 2020).

Additionally, dams and irrigation systems to expand agriculture will ultimately increase the incidence of waterborne and vector borne diseases (Rohr et al., 2019). Indeed, climate change also impacts food production as it affects the availability of grain feed, forage crops, animal health, fecundity, and disease spread where extensive livestock systems of developing countries are under increased pressure (Rust, 2019). Studies suggest that intensive farming systems contribute to diseases such as avian influenza and pig reproductive and respiratory syndrome (Van Boeckel et al., 2015). To ensure sustainability in food production it is essential to consider the ecological impact of food production methods, including intensive farming systems and the negative consequences of excessive AB use. Reports show that livestock systems can contribute to sustainability by acting as greenhouse gas sinks when grassland is converted to pastures of trees, forage and livestock (Feliciano et al., 2018). Land allocation for livestock grazing has lower greenhouse gas emissions than when allocated for crop production and sequesters more carbon per unit area (Varijakshapanicker et al., 2019). Furthermore, 86% of the crop is converted to human food with only 14% of the feed consumed by livestock promoting financial stability and avoiding environmental pollution (Mottet et al., 2017). Mitigating climate change via increased carbon sequestration in soil combined with reduced land degradation and methods to ensure food security such as agroforestry systems which increase agri productivity, soil fertility and protect biodiversity, is key to sustainable food production (Feliciano et al., 2018).

3 Control Measures Reducing AMR

As the issues with antimicrobial use in food producing animals become increasingly evident the need for safe and effective alternatives is clear to ensure animal health and food safety. In this capacity the United States Department of Agriculture (USDA), and the OIE organized the second International Symposium on Alternatives to Antibiotics to assess alternative options and advancements in this area (Seal et al., 2013). Areas focused on included vaccines, microbial-derived products, chemicals, enzymes, and innovative drugs amongst others (Hoelzer et al., 2018). As with any infectious disease prevention is the best means of ensuring public health and animal safety. In terms of animal contagious disease and preventing the proliferating of AMR and the occurrence of foodborne outbreaks, some preventative measures may be used. There are numerous disinfectants on the market for example for farm level and throughout the food chain. Animal vaccination programmes are also implemented to prevent specific infectious disease in animal herds.

3.1 Disinfection and HACCP

Disinfection aims to reduce the viable microbial load on surfaces and ideally, to prevent microbial regrowth. Disinfection is essential in food production but is ineffective where surfaces are not cleaned of interfering substances, particularly in a farm environment with high faecal content. Additionally, biofilm formation on wet surfaces and piping networks acts as a reservoir for difficult pathogens to eradicate. Post-harvest processing by various food industries i.e., such as meat, dairy, and vegetables, has specific microbiota-associated biofilms (Aryal & Muriana, 2019). For example, 2138 cases of foodborne *Salmonella* illnesses were linked to *Salmonella* biofilm-contaminated chicken (Islam et al., 2019). Cleaning is essential followed by adequate disinfection as part of a two-stage cleaning and disinfection (sanitation) program (Holah, 1995). There are numerous disinfectants available in the food industry. Their use depends on efficacy, safety, and the biocide's residual concentration in terms of human health and MRLs in food items. The European Union has set strict regulations on the amount of residue acceptable in food products from animal and non-animal sources. These MRLs dictate the maximum concentration allowable for each compound as outlined in the Commission Regulation (EU) No 37/2010 and Regulation 396/2005. In 2015 the European Commission updated MRLs' list for certain biocides in food and feed of plant and animal origin. Disinfectants approved for use in food production include chlorine compounds, QACs, peracetic acid, hydrogen peroxide, alcohols, surfactants and iodophors (Wirtanen & Salo, 2003). Chlorine-based solutions, however, are the most common and cheapest disinfectants used in food industries (Aryal & Muriana, 2019). Disinfectants' efficacy is affected by interfering substances, typically organic matter, temperature, pH, contact time, and the concentration applied. Gram-negative bacteria tend to be more biocidal resistant than Gram-positive species with fungal species, spores and biofilms demonstrating higher levels of resistance. Efficacy testing of disinfectants to ensure antimicrobial activity are usually determined via suspension tests such as the European standards EN 1276, 1650 and 1656 (amongst others) requiring a 5-log reduction of viable cell numbers within a set number of minutes (Holah, 2014). However, these tests do not mimic the growth conditions found on food production and farm surfaces and so the data is not fully transferable. The EN 13697 is a surface test for disinfectant demonstrating efficacy but does not account for biofilm formation. Quality control systems are typically implemented to monitor cleaning and disinfectants under the Hazard Analysis Critical Control Points (HACCP) system. HACCP is a system that aims to identify, evaluate, and control biological, chemical, and physical hazards in food, ensuring food safety and security at all stages of production. It is essential to prevent pathogen transmission to the carcass or meat at evisceration at animal slaughter in all food industries. Slaughterhouses must ensure that the animal slaughter procedures are following the general requirements of Article 5 of Regulation (EC) No 852/2004 and meet the requirements of the HACCP programme in place. Additionally, good agricultural practice (GAP) and good hygiene practice (GHP) are also implemented to reduce the risk of infectious disease at farm level and foodborne transmission. Recently, studies have reported on the emergence of biocidal resistance in foodborne pathogens, which may be an intrinsic, acquired by chromosomal gene mutation or by acquiring BRGs via plasmids (Vijayakumar & Sandle, 2018). Such biocide resistance may promote AMR and MDR in bacterial species (Boutarfi et al., 2019). For example, studies have demonstrated exposure to chlorhexidine up-regulated vancomycin and daptomycin resistance genes in *E. faecium* (Cheng et al., 2019). Where studies have shown that subtoxic exposure of *P. aeruginosa* and *S. aureus* to QACS and amphoteric promotes resistance in these species (Holah, 2014). The presence of QAC-resistant microbial strains in food processing environments is a risk to public health. To reduce the risk of AMR in food producing animals, the use of antimicrobial agents and AMR species in food animals must be monitored. In the dairy industry, intramammary infections and mastitis are major contributors to animal disease, reduced milk yield, milk wastage and economic difficulties at farm level. As preventative measures, bovine teat dips and disinfectants are used to prevent infection and disease transmission within herds (Fitzpatrick et al., 2021). These dips are formulated containing chlorhexidine, chlorine dioxide, iodine, lactic acid, diamine, or a combination of these. Biocidal resistance to such active ingredients could seriously impact disease prevalence at cow herd level. Safe and effective alternative options to antimicrobial biocides and antibiotics would also greatly impact animal health and food production.

3.2 Vaccination

Prevention of infectious disease via an immunization program may reduce the amount of antibiotics used in food producing animals. Vaccines that prevent infection by inducing the hosts adaptive immune system are undoubtedly a great prophylactic agent in the fight against infectious diseases in humans and animals. Vaccines are one area showing promise as alternatives to antibiotics in food producing animals. Studies demonstrate the use of bacterial and viral vaccines in animals significantly reduced the need for antibiotic therapy (Murphy et al., 2017). For example, in farmed salmon, a vaccination programme against *Aeromonas salmonicida* reduced antibiotic usage in fish farms, where vaccination again *Lawsonia intracellularis* reduced AB use in farmed pigs by up to 80% while improving productivity and weight gains, morbidity and FCR (Hoelzer et al., 2018). Implementing a PCV-2 viral vaccine in Dutch and Canadian pig farms resulted in reduced mortality, increased FCR and reduced AB use (Brockhoff et al., 2009). Likewise, a Belgian pig vaccination programme against viral porcine reproductive and respiratory syndrome (PRRS) resulted in approximately 50% reduction in AB use (Van Looveren et al., 2015). The use of oral vaccination in weaned pigs with live attenuated bacterial vaccines is expected to be useful for reducing enteric diseases in pigs, such as *L. intracellularis* and *S. choleraesuis* infections (Allen et al., 2013). Indeed, the use of vaccines in the pig industry has been ongoing since the 1960s improving pig health and productivity.

Studies also report using a genetically modified live attenuated *Salmonella* strain with induced immunity to 3 pathogenic *Salmonella* species in poultry (Chaudhari et al., 2013). Furthermore, oral administration of the attenuated strain did not induce any morbidity or adverse effects in the poultry (Lee, 2015). Issues arise however, relating to the release of genetically modified species into the environment and the proliferation of such species. There are other types of *Salmonella* immunity acquiring options such as detoxified lipopolysaccharide (LPS), outer membrane proteins (OMPs), and O-polysaccharides however, the attenuated vaccine approach is superior. The Responsible Use of Medicines in Agriculture Alliance (RUMA) is an organisation which represents food production at all stages, aiming to inform on the use of drug therapy and vaccines in food producing animals. RUMA as formulated a set of industrial guidelines for the benefit of vaccines in livestock production, detailing risk assessment and best practice for farmers and veterinarians. The wide use of vaccination programmes in livestock comes with some difficulties, vaccines must be safe, effective, economic, and easily administered to large numbers of animals where biocides are cheaper and easier to use for farmers (Lightowlers et al., 2016).

Additionally, development costs, licensing issues and market value impact developing a commercial vaccines for livestock use (Chambers et al., 2016). However, the development of vaccines for animals has advantages over human vaccines, including lower safety requirements and experimentation in the natural host (Sander et al., 2020). However, vaccination programs are typically costly with limited cross-protection against some pathogens (Economou & Gousia, 2015).

4 Novel Approaches

As we move away from antibiotic use, it is essential to determine alternative options for food production. Additionally, as people increasingly desire less chemically processed foods, there is a demand for more natural food preservation techniques. There is a dire need to source replacement antimicrobial compounds in terms of food safety and security, the emergence and re-emergence of pathogens, sustainability, and environmental impact. As new antibiotic agents' development is not a viable option, there is a need for increased investment and research into implementing alternative options to combat AMR in food producing animals. Ideally such novel antimicrobial substances should have different modes of action to antimicrobial therapeutics to successfully combat AMR pathogenic species. Novel materials for use alone or augmenting current antibiotics which show minimal animal and human toxicity, are economic to produce and do not induce resistance themselves, are desirable.

4.1 Bacteriocins

Bacteriocins are bacterial peptides produced as antibacterial agents as a means of survival and competitive advantage (Meade et al., 2020). The majority of bacteriocins under investigation are produced by Gram-positive bacteria such as lactic acid bacteria, such as colicin produced by Gram-negative species. The mode of action varies depending on the bacteriocin with some having a bactericidal effect, inducing cell lysis, or bacteriostatic preventing cell growth, inhibiting gene expression or protein production (da Silva Sabo et al., 2014). Bacteriocins for food preservation have been ongoing for several years with nisin implemented as a food biopreservative in the dairy industry to inhibit *L. monocytogenes* and *Staphylococcus aureus* and replacing nitrate for the control of *Clostridia* spores (Silva et al., 2018). Similarly, enterocin AS-48 produced by *E. faecalis* species successfully inhibited the growth of *B. cereus, Bacillus macroides, Paenibacillus spp.* and *S. aureus* in fresh vegetables (Ng et al., 2020). Bacteriocins also show promise as therapeutic options in the control of infectious disease in animals and humans. Nisin for example, is useful for the treatment of respiratory disease (pneumonia and pleurisy), meningitis, endocarditis and septicemia in pigs (LeBel et al., 2013) and *E. coli, Staphylococcus* and *Streptococcal* mastitis in dairy cows (Pieterse & Todorov, 2010). Nisin was found to have similar activity to the antibiotic salinomycin for controlling enteric pathogens when feed to broiler hens. Greater body weight in nisin feed hens was also evident (Yang et al., 2014). Lacticin, Macedocin and Garvicin have also demonstrated efficacy against *S. dysgalactiae* and *S. aureus* associated with bovine mastitis (Ng et al., 2020). The Gram-negative produce microcin J25 displays efficacy against MDR strains of *Salmonella, P. aeruginosa, Klebsiella, Acinetobacter* and *E. coli* (Yu et al., 2019). Studies have shown that numerous *E. coli* can produce the bacteriocin colicin, which is active against Shiga toxin-producing *E. coli* present in livestock (Yang et al., 2014).

Issues arise however, as bacterial species may display resistance to bacteriocins such as specific adaptations including the loss of a receptor as seen with pediocin resistance or non-specific transformations altering the cell envelope nisin (O'Connor et al., 2020). This may be overcome by using a bacteriocin combination with varying bacterial toxic effects. Additionally, some bacterial species generate bacteriocins as virulence factors promoting their pathogenicity, cytolysin produced by *Enterococcus* species is one such example which is cytotoxic to mammalian cells (Dicks et al., 2018). Similarly, hemolysins, intermedilysin and streptolysin S produced by *Streptococcus* species promotes their pathogenicity to animal and human cells (Tabata et al., 2019). A clear benefit of bacteriocins relates to their amenability to bioengineering however, bioengineered alternatives display increased potency to select bacterial species and reduced toxicity to animal cells (Meade et al., 2020). Bacteriocins have a narrow spectrum of activity therefore, the identification of the causative agent of infection is required before administration, limiting their efficacy with the aetiology of disease is unknown. However, this type of specificity reduced issues of dysbiosis in the patient as seen with antibiotic therapy, e.g., *C. difficile* associated diarrhoea from antibiotic use (Cotter et al., 2013). Thuricin is a bacteriocin having selective efficacy against *C. difficile* comparable to vancomycin and metronidazole while protecting the gut microbiota (Rea et al., 2010).

Bacteriocins may also be used as combination products, for example in combination with sodium chloride nisin and enterocin had greater antimicrobial efficacy while combining nisin and nitrite delayed botulinal toxin formation, inactivated *Clostridial* spores and inhibited *Listeria* growth in meat (Delesa, 2017). Bacteriocins may also act as animal probiotics protecting gut health and preventing dysbiosis, purified bacteriocins may reduce the pathogen load and preserve the microbiota in chickens, pigs and livestock (Yang et al., 2014). Furthermore, environmentally safe probiotics in aquaculture are approved in the EU where legislation of safety must be adhered to. Issues may arise however when using microorganism as probiotics for example studies have demonstrated the transmission of genes coding for erythromycin resistance from lactic acid bacteria *to L. monocytogenes* (Toomey et al., 2009). The use of bacteriocins instead of bacterial cultures will prevent this and may benefit the animal microbiome's health, boosting the immune system, nutrient use and preventing pathogen colonization (Economou & Gousia, 2015).

4.2 Bacteriophages

The use of bacteriophages or phage's is one area showing potential for controlling bacterial infectious disease. Indeed, phages are often used to control foodborne *Salmonella, E. coli, Listeria* and *Campylobacter* pathogens and their biofilms in food production (Islam et al., 2019). Lytic phage's are suitable for phage therapy as viral species that specifically infect bacteria by injecting their genetic material into the host bacterial cell, replicating and killing the bacterial cell in the process (Garvey, 2020). Phage typically has a narrow bacterial host range making them species specific and, in this way, do not alter the microbiome of the treated animal. Successfully therapy however, requires the identification of the pathogen responsible for disease. Studies report the implementation of lytic phages towards *Salmonella* in poultry and pigs, reduced but not eliminated the target bacterial load present. *Salmonella* is a good target for phage therapy associated with significant disease and production losses in livestock animals and is zoonotic (Gigante & Atterbury, 2019). A cocktail of *Campylobacter* phages was successfully used to treat broiler chickens colonized with *C. jejuni* (Zbikowska et al., 2020) with the commercial product Intralytix successful for treating *C. perfringens* infections. Investigative studies comparing AB efficacy to phage therapy found that phage treatment against *E. coli* in poultry was more effective than the use of chloramphenicol (Xie et al., 2005). Commercial phage preparations ListShieldTM and SalmoFreshTM for the control of *L. monocytogenes* and *S. enterica* are currently Generally Recognized as Safe (GRAS) for use in the food industry (O'Sullivan et al., 2019). Studies also report that phage therapy effectively prevented septicemia and meningitis in calves caused by *E. coli* (Tiwari et al., 2014). A phage cocktail again proved effective against *E. coli* strains associated with mastitis in dairy herds. The commercial phage product, BAFADORR, may be used to eliminate *Pseudomonas* and *Aeromonas* infections in aquaculture. Issues arise however, as phage's do not persist in the animal post eradication of the host bacteria. So repeated treatment is needed, therapy is most effective when administered soon after infection and the animal immune system can neutralise the phage's (Garvey, 2020). Also, phage's can be impacted by pH variations found in the intestinal tract.

Additionally, the efficacy of phage's against intracellular pathogens such as *Mycobacteria* and *Brucella* is currently unknown (Porter et al., 2016). Susceptibility of bacteria to the phage, phage stability, efficacy and bacterial phage resistance need to be assessed and monitored during the phage treatment (Zbikowska et al., 2020). Nevertheless, the UK Department of Health and the Wellcome Trust reported that bacteriophages are among the 10 most promising alternatives to antibiotics at present (Gigante & Atterbury, 2019).

5 Conclusion

With the increasing demand on animal food products to meet the growing human population's needs, concern has also arisen relating to animal welfare, sustainability, and the environmental impact of current food production practices. Changes in farm management and livestock systems are warranted as we move towards more intensive systems globally. To ensure food safety while protecting human, animal, and environmental health, changes need to be made in line with the One Health approach. Such measures must include reducing the use of antimicrobial agents in food production, to reduce antibiotic resistance and the prevalence of resistant disease. This is evident as the treatment options for AMR and MDR infections become increasingly limited with the last-resort antibiotics, including polymyxins and cephalosporins becoming less effective. An acute all-inclusive ban of the use of antibiotics in food producing animals is unrealistic as it will negatively impact animal health, food production and economics. Studies show however, the benefits of such bans as European policy preventing the use of non-therapeutic ABs in food producing animals reduced resistance against clinically relevant antibiotics including vancomycin. Preventative measures including improved cleaning and disinfection in line with HACCP, vaccination programs and the use of alternative antimicrobial agents should protect against disease outbreaks at farm level. While antimicrobial use benefits relating to weight gain and feed efficiency are clear, it must not be valued over the importance of antibiotic clinical therapy. Considering how vital the use of macrolides is in treating human Campylobacteriosis, its application in food production needs to be revised. Useful and timely control measures are essential to limit the emergence and re-emergence of pathogenic species. They include implementing active surveillance systems, legislation, research into novel therapeutics and vaccines at a global level. For this to be successful a joint effort locally, nationally, and globally must be implemented and monitored with transparency in every aspect.

References

Aalipour, F., Mirlohi, M., & Jalali, M. (2015). Dietary exposure to tetracycline residues through milk consumption in Iran. *Journal of Environmental Health Science and Engineering, 13*, 80. https://doi.org/10.1186/s40201-015-0235-6

Abebe, E., Gugsa, G., & Ahmed, M. (2020). Review on major food-borne zoonotic bacterial pathogens. *Journal of Tropical Medicine, 2020*, 4674235. https://doi.org/10.1155/2020/4674235

Abraham, S., Sahibzada, S., Hewson, K., Laird, T., Abraham, R., Pavic, A., Truswell, A., Lee, T., O'Dea, M., & Jordan, D. (2020). Emergence of fluoroquinolone-resistant campylobacter jejuni and campylobacter coli among Australian chickens in the absence of fluoroquinolone use. *Applied and Environmental Microbiology, 86*(8), e02765–e02719. https://doi.org/10.1128/AEM.02765-19

Abrar, S., Ain, N. U., Liaqat, H., Hussain, S., Rasheed, F., & Riaz, S. (2019). Distribution of blaCTX-M, blaTEM, blaSHV and blaOXA genes in Extended-spectrum-β-lactamase-producing Clinical isolates: A three-year multi-center study from Lahore, Pakistan. *Antimicrobial Resistance and Infection Control, 8*, 8.

Ali, A., Parisi, A., Conversano, M., Iannacci, A., D'Emilio, F., Mercurio, V., et al. (2020). Food-borne bacteria associated with seafoods: A brief review. *Journal of Food Quality and Hazards Control, 7*(1), 4–10.

Allen, H. K., Levine, U. Y., Looft, T., Bandrick, M., & Casey, T. A. (2013). Treatment, promotion, commotion: Antibiotic alternatives in food-producing animals. *Trends in Microbiology, 21*(3), 114–119. https://doi.org/10.1016/j.tim.2012.11.001

Amarasiri, M., Sano, D., & Suzuki, S. (2019). Understanding human health risks caused by antibiotic resistant bacteria (ARB) and antibiotic resistance genes (ARG) in water environments: Current knowledge and questions to be answered. *Critical Reviews in Environmental Science and Technology, 50*(19), 2016–2059. https://doi.org/10.1080/10643389.2019.1692611

Amsalu, A., Sapula, S. A., De Barros Lopes, M., Hart, B. J., Nguyen, A. H., Drigo, B., Turnidge, J., Leong, L. E., & Venter, H. (2020). Efflux pump-driven antibiotic and biocide cross-resistance in Pseudomonas aeruginosa isolated from different ecological niches: A case study in the development of multidrug resistance in environmental hotspots. *Microorganisms, 8*, 1647. https://doi.org/10.3390/microorganisms8111647

Anes, J., McCusker, M. P., Fanning, S., & Martins, M. (2015). The ins and outs of RND efflux pumps in Escherichia coli. *Frontiers in Microbiology, 6*, 587. https://doi.org/10.3389/fmicb.2015.00587

Aryal, M., & Muriana, P. M. (2019). Efficacy of commercial sanitizers used in food processing facilities for inactivation of listeria monocytogenes, E. Coli O157:H7, and salmonella biofilms. *Food, 8*(12), 639. https://doi.org/10.3390/foods8120639

Bakhtiary, F., Sayevand, H. R., Remely, M., Hippe, B., Hosseini, H., & Haslberger, A. G. (2016). Evaluation of bacterial contamination sources in meat production line. *Journal of Food Quality, 39*(6), 750–756. https://doi.org/10.1111/jfq.12243

Basanisi, M. G., La Bella, G., Nobili, G., Franconieri, I., & La Salandra, G. (2017). Genotyping of methicillin-resistant Staphylococcus aureus (MRSA) isolated from milk and dairy products in South Italy. *Food Microbiology, 62*, 141–146. https://doi.org/10.1016/j.fm.2016.10.020

Belay, E. D., Kile, J. C., Hall, A. J., Barton-Behravesh, C., Parsons, M. B., Salyer, S., & Walke, H. (2017). Zoonotic disease programs for enhancing global health security. *Emerging Infectious Diseases, 23*(13), S65–S70. https://doi.org/10.3201/eid2313.170544

Bennani, H., Mateus, A., Mays, N., Eastmure, E., Stärk, K., & Häsler, B. (2020). Overview of evidence of antimicrobial use and antimicrobial resistance in the food chain. *Antibiotics (Basel, Switzerland), 9*(2), 49. https://doi.org/10.3390/antibiotics9020049

Billah, M. D. M., Rana, S. M. M., Hossain, M. S., Ahamed, S. K., Banik, S., & Hasan, M. (2015). Ciprofloxacin residue and their impact on biomolecules in eggs of laying hens following oral administration. *International Journal of Food Contamination, 2*, 13. https://doi.org/10.1186/s40550-015-0019-x

Bogan, C., Kaye, K. S., Chopra, T., Chopra, T., Kayakawa, K., & Pogue, J. M. (2014). Outcomes of carbapenem resistant Enterobacteriaceae isolation: Matched analysis. *American Journal of Infection Control, 42*, 612–620.

Boqvist, S., Söderqvist, K., & Vågsholm, I. (2018). Food safety challenges and one health within Europe. *Acta Veterinaria Scandinavica, 60*, 1. https://doi.org/10.1186/s13028-017-0355-3

Boutarfi, Z., Rebiahi, S. A., Morghad, T., Pulido, R. P., Burgos, M. J., Mahdi, F., Lucas, R., & Galvez, A. (2019). Biocide tolerance and antibiotic resistance of Enterobacter spp. isolated from Algerian hospital environment. *Journal of Global Antimicrobial Resistance, 18*, 291–297. https://doi.org/10.1016/j.jgar.2019.04.005

Brockhoff, E., Cunningham, G., & Misutka, C. (2009). A retrospective analysis of a high health commercial pig production system showing improved production and reduced antibiotic use after implementation of a PCV2 vaccination. In *Safe pork conference proceedings*, Quebec, Canada.

Chambers, M. A., Graham, S. P., & La Ragione, R. M. (2016). Challenges in veterinary vaccine development and immunization. *Methods in Molecular Biology, 1404*, 3–35. https://doi.org/10.1007/978-1-4939-3389-1_1

Chattopadhyay, M. K. (2014). Use of antibiotics as feed additives: A burning question. *Frontiers in Microbiology, 5*, 334. https://doi.org/10.3389/fmicb.2014.00334

Chaudhari, A. A., Matsuda, K., & Lee, J. H. (2013). Construction of an attenuated *Salmonella* delivery system harboring genes encoding various virulence factors of avian pathogenic *Escherichia coli* and its potential as a candidate vaccine for chicken colibacillosis. *Avian Diseases, 57*, 88–96.

Chen, Z., Guo, J., Jiang, Y., & Shao, Y. (2021). High concentration and high dose of disinfectants and antibiotics used during the COVID-19 pandemic threaten human health. *Environmental Sciences Europe, 33*, 11. https://doi.org/10.1186/s12302-021-00456-4

Cheng, G., Ning, J., & Ahmed, S. (2019). Selection and dissemination of antimicrobial resistance in Agri-food production. *Antimicrobial Resistance and Infection Control, 8*, 158. https://doi.org/10.1186/s13756-019-0623-2

Cotter, P., Ross, R., & Hill, C. (2013). Bacteriocins — A viable alternative to antibiotics? *Nature Reviews. Microbiology, 11*, 95–105. https://doi.org/10.1038/nrmicro2937

Cycoń, M., Mrozik, A., & Piotrowska-Seget, Z. (2019). Antibiotics in the soil environment-degradation and their impact on microbial activity and diversity. *Frontiers in Microbiology, 10*, 338. https://doi.org/10.3389/fmicb.2019.00338

da Silva Sabo, S., Vitolo, M., González, J. M. D., & De Souza Oliveira, R. P. (2014). Overview of Lactobacillus plantarum as a promising bacteriocin producer among lactic acid bacteria. *Food Research International, 64*, 527–536. https://doi.org/10.1016/j.foodres.2014.07.041

Dadgostar, P. (2019). Antimicrobial resistance: Implications and costs. *Infection and Drug Resistance, 12*, 3903–3910. https://doi.org/10.2147/IDR.S234610

Dee, S., Guzman, J. E., Hanson, D., Garbes, N., Morrison, R., & Amodie, D. (2018). A randomized controlled trial to evaluate performance of pigs raised in antibiotic-free or conventional production systems following challenge with porcine reproductive and respiratory syndrome virus. *PLoS One, 13*(12), e0208430. https://doi.org/10.1371/journal.pone.0208430

Delesa, D. A. (2017). Bacteriocin as an advanced technology in food industry. *International Journal of Advanced Research in Biological Sciences, 4*(12), 178–190. https://doi.org/10.22192/ijarbs.2017.04.12.018

Dicks, L., Dreyer, L., Smith, C., & van Staden, A. D. (2018). A review: The fate of bacteriocins in the human gastro-intestinal tract: Do they cross the gut-blood barrier? *Frontiers in Microbiology, 9*, 2297. https://doi.org/10.3389/fmicb.2018.02297

Doyle, S., Meade, E., Fowley, C., & Garvey, M. (2020). A comprehensive review of current environmental pollutants of pharmaceutical, agricultural and industrial origin. *European Journal of Experimental Biology, 10*(2), 5. https://doi.org/10.36648/2248-9215.10.2.102

Durel, L., Gallina, G., & Pellet, T. (2019). Assessment of ceftiofur residues in cow milk using commercial screening test kits. *Veterinary Record Open, 6*(1), e000329. https://doi.org/10.1136/vetreco-2018-000329

Durso, L. M., & Cook, K. L. (2014). Impacts of antibiotic use in agriculture: What are the benefits and risks? *Current Opinion in Microbiology, 19*, 37–44.

Dyer, C., Hutt, L. P., Burky, B., & Loyleen, T. J. (2019). Biocide resistance and transmission of Clostridium difficile spores spiked onto clinical surfaces from an American health care facility. *Applied and Environmental Microbiology, 85*(17), e01090–e01019. https://doi.org/10.1128/AEM.01090-19

Dzhavakhiya, V., Savushkin, V., Ovchinnikov, A., Glagolev, V., Savelyeva, V., Popova, E., Novak, N., & Glagoleva, E. (2016). Scaling up a virginiamycin production by a high-yield Streptomyces virginiae VKM Ac-2738D strain using adsorbing resin addition and fed-batch fermentation under controlled conditions. *3Biotech, 6*(2), 240. https://doi.org/10.1007/s13205-016-0566-8

Economou, V., & Gousia, P. (2015). Agriculture and food animals as a source of antimicrobial-resistant bacteria. *Infection and Drug Resistance, 8*, 49–61. https://doi.org/10.2147/IDR.S55778

EFSA. (2019). The European Union one health 2018 zoonoses report. *EFSA Journal, 17*(12), e05926. https://doi.org/10.2903/j.efsa.2019.5926

EFSA J. (2017). European Centre for Disease Prevention and Control (ECDC); European Food Safety Authority (EFSA); European Medicines Agency (EMA). ECDC/EFSA/EMA second joint report on the integrated analysis of the consumption of antimicrobial agents and occurrence of antimicrobial resistance in bacteria from humans and food-producing animals. *Joint Interagency Antimicrobial Consumption and Resistance Analysis (JIACRA) Report, 15*(7), e04872. https://doi.org/10.2903/j.efsa.2017.4872. PMID: 32625542; PMCID: PMC7009874.

Espinosa, R., Tago, D., & Treich, N. (2020). Infectious disease and meat production. *Environmental and Resource Economics, 76*, 1019–1044. https://doi.org/10.1007/s10640-020-00484-3

Feliciano, D., Ledo, A., Hillier, J., & Nayak, D. R. (2018). Which agroforestry options give the greatest soil and above ground carbon benefits in different world regions? *Agriculture, Ecosystems & Environment, 254*, 117–129. https://doi.org/10.1016/j.agee.2017.11.032

Fernández, L., & Hancock, R. E. (2012). Adaptive and mutational resistance: Role of porins and efflux pumps in drug resistance. *Clinical Microbiology Reviews, 25*(4), 661–681. https://doi.org/10.1128/CMR.00043-12

Fitzpatrick, S. R., Garvey, M., Flynn, J., O'Brien, B., & Gleeson, D. (2021). The effect of disinfectant ingredients on teat skin bacteria associated with mastitis in Irish dairy herds. *Irish Veterinary Journal, 74*, 1. https://doi.org/10.1186/s13620-020-00179-7

Friedman, N. D., Temkin, E., & Carmeli, Y. (2015). The negative impact of antibiotic resistance. *Clinical Microbiology and Infection, 22*(5), 416–422. https://doi.org/10.1016/j.cmi.2015.12.002

Garcia, S. N., Osburn, B. I., & Jay-Russell, M. T. (2020). One health for food safety, food security, and sustainable food production. *Frontiers in Sustainable Food Systems, 4*. https://doi.org/10.3389/fsufs.2020.00001

Garvey, M. (2020). Bacteriophages and the one health approach to combat multidrug resistance: Is this the way? *Antibiotics, 9*(7), 414. https://doi.org/10.3390/antibiotics9070414

Gigante, A., & Atterbury, R. J. (2019). Veterinary use of bacteriophage therapy in intensively-reared livestock. *Virology Journal, 16*, 155. https://doi.org/10.1186/s12985-019-1260-3

Granados-Chinchilla, F., & Rodríguez, C. (2017). Tetracyclines in food and feedingstuffs: From regulation to analytical methods, bacterial resistance, and environmental and health implications. *Journal of Analytical Methods in Chemistry, 2017*, 1315497. https://doi.org/10.1155/2017/1315497

Haenan, O. L. M., Evans, J. J., & Berthe, F. (2013). Bacterial infections from aquatic species: Potential for and prevention of contact zoonoses. *Revue Scientifique et Technique, 32*, 497–507.

Hao, H., Cheng, G., Iqbal, Z., Ai, X., Hussain, H. I., Huang, L., Dai, M., Wang, Y., Liu, Z., & Yuan, Z. (2014). Benefits and risks of antimicrobial use in food-producing animals. *Frontiers in Microbiology, 5*, 288. https://doi.org/10.3389/fmicb.2014.00288

Hoelzer, K., Bielke, L., Blake, D. P., et al. (2018). Vaccines as alternatives to antibiotics for food producing animals. Part 1: Challenges and needs. *Veterinary Research, 49*, 64. https://doi.org/10.1186/s13567-018-0560-8

Hoffmann, M., Luo, Y., Monday, S. R., Gonzalez-Escalona, N., Ottesen, A. R., Muruvanda, T., Wang, C., Kastanis, G., Keys, C., Janies, D., Senturk, I. F., Catalyurek, U. V., Wang, H., Hammack, T. S., Wolfgang, W. J., Schoonmaker-Bopp, D., Chu, A., Myers, R., Haendiges, J., et al. (2016). Tracing origins of the *Salmonella* bareilly strain causing a food-borne outbreak in the United States. *The Journal of Infectious Diseases, 213*, 502–508.

Holah, J. T. (1995). Disinfection of food production areas. *Revue Scientifique et Technique, 14*(2), 343–363. https://doi.org/10.20506/rst.14.2.850. PMID: 7579635.

Holah, J. T. (2014). *Hygiene in food processing || Cleaning and disinfection practices in food processing* (pp. 259–304). Elsevier. https://doi.org/10.1533/9780857098634.3.259

Impey, R. E., Hawkins, D. A., Sutton, J. M., & Soares da Costa, T. P. (2020). Overcoming intrinsic and acquired resistance mechanisms associated with the cell wall of gram-negative bacteria. *Antibiotics (Basel, Switzerland), 9*(9), 623. https://doi.org/10.3390/antibiotics9090623

Islam, M. S., Zhou, Y., Liang, L., Nime, I., Liu, K., Yan, T., Wang, X., & Li, J. (2019). Application of a phage cocktail for control of salmonella in foods and reducing biofilms. *Viruses, 11*(9), 841. https://doi.org/10.3390/v11090841

Kimera, Z. I., Mshana, S. E., Rweyemamu, M. M., Mboera, L. E. G., & Matee, M. I. N. (2020). Antimicrobial use and resistance in food-producing animals and the environment: An African perspective. *Antimicrobial Resistance & Infection Control, 9*(1), 37. https://doi.org/10.1186/s13756-020-0697-x

Kirchhelle, C. (2018). Pharming animals: A global history of antibiotics in food production (1935–2017). *Palgrave Communications, 4*, 96. https://doi.org/10.1057/s41599-018-0152-2

Kraemer, S. A., Ramachandran, A., & Perron, G. G. (2019). Antibiotic pollution in the environment: From microbial ecology to public policy. *Microorganisms, 7*(6), 180. https://doi.org/10.3390/microorganisms7060180

LeBel, G., Piché, F., Frenette, M., Gottschalk, M., & Grenier, D. (2013). Antimicrobial activity of nisin against the swine pathogen Streptococcus suis and its synergistic interaction with antibiotics. *Peptides, 50*, 19–23. https://doi.org/10.1016/j.peptides.2013.09.014

Lee, J. H. (2015). Protection against salmonella typhimurium, salmonella gallinarum, and salmonella enteritidis infection in layer chickens conferred by a live attenuated salmonella typhimurium strain. *Immune Network, 15*(1), 27–36. https://doi.org/10.4110/in.2015.15.1.27

Lekagul, A., Tangcharoensathien, V., Mills, A., Rushton, J., & Yeung, S. (2020). How antibiotics are used in pig farming: A mixed-methods study of pig farmers, feed mills and veterinarians in Thailand. *BMJ Global Health, 5*(2), e001918. https://doi.org/10.1136/bmjgh-2019-001918

Lekshmi, M., Ammini, P., Kumar, S., & Varela, M. F. (2017). The food production environment and the development of antimicrobial resistance in human pathogens of animal origin. *Microorganisms, 5*(1), 11. https://doi.org/10.3390/microorganisms5010011. PMID: 28335438; PMCID: PMC5374388.

Levy, S. B., FitzGerald, G. B., & Macone, A. B. (1976). Changes in intestinal flora of farm personnel after introduction of a tetracycline-supplemented feed on a farm. *New England Journal of Medicine, 295*(1976), 583–588. https://doi.org/10.1056/NEJM197609092951103

Li, X. Z., Elkins, C. A., & Zgurskaya, H. I. (2016). Antimicrobial drug efflux pumps in other gram-positive bacteria. In *Efflux-mediated antimicrobial resistance in bacteria* (pp. 197–218). Adis. https://doi.org/10.1007/978-3-319-39658-3_8

Lightowlers, M. W., Donadeu, M., Elaiyaraja, M., Maithal, K., Kumar, K. A., Gauci, C. G., et al. (2016). Anamnestic responses in pigs to the Taenia solium TSOL18 vaccine and implications for control strategies. *Parasitology, 143*, 416–420. https://doi.org/10.1017/S0031182016000202

Liu, S., Bekele, T. G., Zhao, H., & Cai., X., Chen, J. (2018). Bioaccumulation and tissue distribution of antibiotics in wild marine fish from Laizhou Bay, North China. *Science of the Total Environment, 631–632*, 1398–1405. https://doi.org/10.1016/j.scitotenv.2018.03.139

Lulijwa, R., Rupia, E. J., & Alfaro, A. C. (2019). Antibiotic use in aquaculture, policies and regulation, health and environmental risks: A review of the top 15 major producers. *Reviews in Aquaculture, 12*(2), 640–663. https://doi.org/10.1111/raq.12344

Ma, F., Xu, S., Tang, Z., Li, Z., & Zhang, L. (2020). Use of antimicrobials in food animals and impact of transmission of antimicrobial resistance on humans. *Biosafety and Health, 3*(1), 32–38. https://doi.org/10.1016/j.bsheal.2020.09.004

Manyi-Loh, C., Mamphweli, S., Meyer, E., & Okoh, A. (2018). Antibiotic use in agriculture and its consequential resistance in environmental sources: Potential public health implications. *Molecules, 23*(4), 795. https://doi.org/10.3390/molecules23040795

Marshall, B. M., & Levy, S. B. (2011). Food animals and antimicrobials: Impacts on human health. *Clinical Microbiology Reviews, 24*(4), 718–733. https://doi.org/10.1128/CMR.00002-11

Martin, H., Manzanilla, E. G., & More, S. J. (2020). Current antimicrobial use in farm animals in the Republic of Ireland. *Irish Veterinary Journal, 73*, 11. https://doi.org/10.1186/s13620-020-00165-z

McMillan, E. A., Gupta, S. K., Williams, L. E., Jové, T., Hiott, L. M., Woodley, T. A., Barrett, J. B., Jackson, C. R., Wasilenko, J. L., Simmons, M., Tillman, G. E., McClelland, M., & Frye, J. G. (2019). Antimicrobial resistance genes, cassettes, and plasmids present in salmonella enterica associated with United States food animals. *Frontiers in Microbiology, 10*, 832. https://doi.org/10.3389/fmicb.2019.00832

Meade, E., Slattery, M. A., & Garvey, M. (2020). Bacteriocins, potent antimicrobial peptides and the fight against multi drug resistant species: Resistance is futile? *Antibiotics (Basel, Switzerland), 9*(1), 32. https://doi.org/10.3390/antibiotics9010032

Miranda, C. D., Godoy, F. A., & Lee, M. R. (2018). Current status of the use of antibiotics and the antimicrobial resistance in the Chilean Salmon farms. *Frontiers in Microbiology, 9*, 1284. https://doi.org/10.3389/fmicb.2018.01284

Mottet, A., de Haan, C., Falcucci, A., Tempio, G., Opio, C., & Gerber, P. (2017). Livestock: On our plates or eating at our table? A new analysis of the feed/food debate. *Global Food Security, 14*, 1–8. https://doi.org/10.1016/j.gfs.2017.01.001

Mulani, M. S., Kamble, E. E., Kumkar, S. N., Tawre, M. S., & Pardesi, K. R. (2019). Emerging strategies to combat ESKAPE pathogens in the era of antimicrobial resistance: A review. *Frontiers in Microbiology, 10*, 539. https://doi.org/10.3389/fmicb.2019.00539

Murphy, D., Ricci, A., Auce, Z., Beechinor, J. G., Bergendahl, H., Breathnach, R., Bureš, J., Da Silva, D., Pedro, J., & Hederová, J. (2017). EMA and EFSA Joint Scientific Opinion on measures to reduce the need to use antimicrobial agents in animal husbandry in the European Union, and the resulting impacts on food safety (RONAFA). *EFSA Journal, 15*, 4666.

Nation, R. L., Rigatto, M. H. P., Falci, D. R., & Zavascki, A. P. (2019). Polymyxin acute kidney injury: Dosing and other strategies to reduce toxicity. *Antibiotics, 8*(1), 24. https://doi.org/10.3390/antibiotics8010024

Nelson, D. W., Moore, J. E., & Rao, J. R. (2019). Antimicrobial resistance (AMR): Significance to food quality and safety. *Food Quality and Safety, 3*(1), 15–22. https://doi.org/10.1093/fqsafe/fyz003

Ng, J. Z., Zarin, M. A., Lee, C. K., & Tan, J. S. (2020). Application of bacteriocins in food preservation and infectious disease treatment for humans and livestock: A review. *RCS Advances, 10*, 38937–38964.

Novoslavskij, A., Terentjeva, M., & Eizenberga, I. (2016). Major foodborne pathogens in fish and fish products: A review. *Annales de Microbiologie, 66*, 1–15. https://doi.org/10.1007/s13213-015-1102-5

Nutman, A., Glick, R., Temkin, E., Hoshen, M., Edgar, R., & Braun, T. (2014). A case control study to identify predictors of 14 day mortality following carbapenem resistant Acinetobacter buamannii bacteraemia. *Clinical Microbiology and Infection, 20*, 01028–01034.

O'Connor, P. M., Kuniyoshi, T. M., Oliveira, R. P. S., Hill, C., Ross, R. P., & Cotter, P. D. (2020). Antimicrobials for food and feed; a bacteriocin perspective. *Current Opinion in Biotechnology, 61*, 160–167. https://doi.org/10.1016/j.copbio.2019.12.023

O'Sullivan, L., Bolton, D., McAuliffe, O., & Coffey, A. (2019). Bacteriophages in food applications: From foe to friend. *Annual Review of Food Science and Technology, 10*, 151–172. https://doi.org/10.1146/annurev-food-032818-121747

Pieterse, R., & Todorov, S. D. (2010). Bacteriocins: Exploring alternatives to antibiotics in mastitis treatment. *Brazilian Journal of Microbiology, 41*(3), 542–562. https://doi.org/10.1590/S1517-83822010000300003

Pinchuk, I. V., Beswick, E. J., & Reyes, V. E. (2010). Staphylococcal enterotoxins. *Toxins, 2*(8), 2177–2197. https://doi.org/10.3390/toxins2082177

Poole, K. (2000). Efflux-mediated resistance to fluoroquinolones in gram-positive bacteria and the mycobacteria. *Antimicrobial Agents and Chemotherapy, 44*(10), 2595–2599. https://doi.org/10.1128/AAC.44.10.2595-2599.2000

Porter, J., Anderson, J., Carter, L., Donjacour, E., & Paros, M. (2016). In vitro evaluation of a novel bacteriophage cocktail as a preventative for bovine coliform mastitis. *Journal of Dairy Science, 99*(3), 2053–2062. https://doi.org/10.3168/jds.2015-9748

Prestinaci, F., Pezzotti, P., & Pantosti, A. (2015). Antimicrobial resistance: A global multifaceted phenomenon. *Pathogens and Global Health, 109*(7), 309–318. https://doi.org/10.1179/2047773215Y.0000000030

Rahman, M. T., Sobur, M. A., Islam, M. S., Ievy, S., Hossain, M. J., Zowalaty, E., Mohamed, E., Rahman, A. M. M. T., & Ashour, H. M. (2020). Zoonotic diseases: Etiology, impact, and control. *Microorganisms, 8*(9), 1405. https://doi.org/10.3390/microorganisms8091405

Rea, M. C., Sit, C. S., Clayton, E., O'Connor, P. M., Whittal, R. M., Zheng, J., Vederas, J. C., Ross, R. P., & Hill, C. (2010). Thuricin CD, a posttranslationally modified bacteriocin with a narrow spectrum of activity against Clostridium difficile. *Proceedings of the National Academy of Sciences, 107*(20), 9352–9357. https://doi.org/10.1073/pnas.0913554107

Rohr, J. R., Barrett, C. B., Civitello, D. J., Craft, M. E., Delius, B., Deleo, G. A., Hudson, P. J., Jouanard, N., Nguyen, K. H., Ostfeld, R. S., Remais, J. V., Riveau, G., Sokolow, S. H., & Tilman, D. (2019). Emerging human infectious diseases and the links to global food production. *Nature Sustainability, 2*, 445–456. https://doi.org/10.1038/s41893-019-0293-3

Rusch, M., Spielmeyer, A., Zorn, H., & Hamscher, G. (2018). Biotransformation of ciprofloxacin by Xylaria longipes: Structure elucidation and residual antibacterial activity of metabolites. *Applied Microbiology and Biotechnology, 102*, 8573–8584. https://doi.org/10.1007/s00253-018-9231-y

Rust, J. M. (2019). The impact of climate change on extensive and intensive livestock production systems. *Animal Frontiers, 9*(1), 20–25. https://doi.org/10.1093/af/vfy028

Sander, V. A., Sánchez López, E. F., Mendoza Morales, L., Ramos Duarte, V. A., Corigliano, M. G., & Clemente, M. (2020). Use of veterinary vaccines for livestock as a strategy to control foodborne parasitic diseases. *Frontiers in Cellular and Infection Microbiology, 10*, 288. https://doi.org/10.3389/fcimb.2020.00288

Savage, M., Mead, E., Slattery, M. A., & Garvy, M. (2017). Antibiotic resistance: An important issue for public health safety. *Annals of Microbiology Research, 1*(1), 26–30.

Schindler, B. D., & Kaatz, G. W. (2016). Multidrug efflux pumps of gram-positive bacteria. *Drug Resistance Updates, 27*, 1–3. https://doi.org/10.1016/j.drup.2016.04.003

Schulz, J., Kemper, N., Hartung, J., Janusch, F., Mohring, S. A. I., & Hamscher, G. (2019). Analysis of fluoroquinolones in dusts from intensive livestock farming and the co-occurrence of fluoroquinolone-resistant Escherichia coli. *Scientific Reports, 9*(1), 5117. https://doi.org/10.1038/s41598-019-41528-z

Seal, B. S., Lillehoj, H. S., Donovan, D. M., & Gay, C. G. (2013). Alternatives to antibiotics: A symposium on the challenges and solutions for animal production. *Animal Health Research Reviews, 14*, 78–87.

Silva, C. C. G., Silva, S. P. M., & Ribeiro, S. C. (2018). Application of bacteriocins and protective cultures in dairy food preservation. *Frontiers in Microbiology, 9*, 594. https://doi.org/10.3389/fmicb.2018.00594

Tabata, A., Yamada, T., Ohtani, H., Ohkura, K., Tomoyasu, T., & Nagamune, H. (2019). β-Hemolytic Streptococcus anginosus subsp. anginosus causes streptolysin S-dependent cytotoxicity to human cell culture lines in vitro. *Journal of Oral Microbiology, 11*(1), 1609839. https://doi.org/10.1080/20002297.2019.1609839

Tang, K. L., Caffrey, N. P., Nobrega, D. B., Cork, S. C., Ronksley, P. E., Berkema, H. W., Polachek, A. J., Ganshorn, H., Sharma, N., & Kellner, J. D. (2017). Restricting the use of antibiotics in food-producing animals and its associations with antibiotic resistance in food-producing animals and human beings: A systematic review and meta-analysis. *The Lancet Planetary Health, 1*, e316–e327.

Taylor, P., & Reeder, R. (2020). Antibiotic use on crops in low and middle-income countries based on recommendations made by agricultural advisors. *CABI Agriculture and Bioscience, 1*(1), 1. https://doi.org/10.1186/s43170-020-00001-y

Tiwari, R., Dhama, K., Chakraborty, S., Kumar, A., Rahal, A., & Kapoor, S. (2014). Bacteriophage therapy for safeguarding animal and human health: A review. *Pakistan Journal of Biological Sciences, 17*, 301–315. https://doi.org/10.3923/pjbs.2014.301.315

Toomey, N., Monaghan, A., Fanning, S., & Bolton, D. J. (2009). Assessment of antimicrobial resistance transfer between lactic acid bacteria and potential foodborne pathogens using in vitro methods and mating in a food matrix. *Foodborne Pathogens and Disease, 6*(8), 925–933.

United States Food and Drug Administration (USFDA). (2020). Animal drugs, feeds, and related products. Part 558. *New animal drugs for use in animal feeds*. http://www.accessdata.fda.gov/scripts/cdrh/cfdocs/cfcfr/CFRSearch.cfm?CFRPart=558

Van Boeckel, T. P., Brower, C., Gilbert, M., Grenfell, B. T., Levin, S. A., Robinson, T. P., Teillant, A., & Laxminarayan, R. (2015). Global trends in antimicrobial use in food animals. *Proceedings of the National Academy of Sciences, 112*(18), 5649–5654. https://doi.org/10.1073/pnas.1503141112

Van Looveren, F., De Jonghe, E., Maass, P., & De Backer, P. (2015). Reduction of antibiotic use after implementation of Ingelvac® PRRS MLV piglet vaccination in a Belgian wean to finish farm. In *Safe pork conference proceedings*, Porto, Portugal.

Van, T. T. H., Yidana, Z., Smooker, P. M., & Coloe, P. J. (2019). Antibiotic use in food animals in the world with focus on Africa: Pluses and minuses. *Journal of Global Antimicrobial Resistance, 20*, 170–177. https://doi.org/10.1016/j.jgar.2019.07.031

Varijakshapanicker, P., Mckune, S., Miller, L., Hendrickx, S., Balehegn, M., Dahl, G. E., & Adesogan, A. T. (2019). Sustainable livestock systems to improve human health, nutrition, and economic status. *Animal Frontiers, 9*(4), 39–50. https://doi.org/10.1093/af/vfz041

Varol, M., & Sünbül, M. R. (2017). Organochlorine pesticide, antibiotic and heavy metal residues in mussel, crayfish and fish species from a reservoir on the Euphrates River, Turkey. *Environmental Pollution, 230*, 311–319. https://doi.org/10.1016/j.envpol.2017.06.066

Vidovic, N., & Vidovic, S. (2020). Antimicrobial resistance and food animals: Influence of livestock environment on the emergence and dissemination of antimicrobial resistance. *Antibiotics (Basel, Switzerland), 9*(2), 52. https://doi.org/10.3390/antibiotics9020052

Vijayakumar, R., & Sandle, T. (2018). A review on biocides reduced susceptibility due to plasmid-borne antiseptic resistant genes – Special notes on pharmaceutical environmental isolates. *Journal of Applied Microbiology, 126*(4), 1011–1022. https://doi.org/10.1111/jam.14118

Wales, A. D., & Davies, R. H. (2015). Co-selection of resistance to antibiotics, biocides and heavy metals, and its relevance to foodborne pathogens. *Antibiotics, 4*(4), 567–604. https://doi.org/10.3390/antibiotics4040567

Wang, S., Zhang, H. Y., Wang, L., Duan, Z. J., & Kennedy, I. (2006). Analysis of sulphonamide residues in edible animal products: A review. *Food Additives and Contaminants, 23*(4), 362–384. https://doi.org/10.1080/02652030500499359

Wieczorek, K., Wołkowicz, T., & Osek, J. (2018). Antimicrobial resistance and virulence-associated traits of Campylobacter jejuni isolated from poultry food chain and humans with diarrhea. *Frontiers in Microbiology, 9*, 1508. https://doi.org/10.3389/fmicb.2018.01508

Wirtanen, G., & Salo, S. (2003). Disinfection in food processing – Efficacy testing of disinfectants. *Reviews in Environmental Science and Biotechnology, 2*(2–4), 293–306. https://doi.org/10.1023/b:resb.0000040471.15700.03

Wongtavatchai, J., McLean, I. G., Ramos, F., & Arnold, D. (2004). *WHO food additives series 53: Chloramphenicol JECFA (WHO: Joint FAO/WHO Expert Committee on Food Additives), IPCS (International Programme on Chemical Safety), INCHEM* (pp. 7–85). WHO.

Xie, H., Zhuang, X., Kong, J., Ma, G., & Zhang, H. (2005). Bacteriophage Esc-A is an efficient therapy for Escherichia coli 3-1 caused diarrhea in chickens. *The Journal of General and Applied Microbiology, 51*, 159–163.

Xie, H., Du, J., & Chen, J. (2020). Concerted efforts are needed to control and mitigate antibiotic pollution in coastal waters of China. *Antibiotics, 9*(2), 88. https://doi.org/10.3390/antibiotics9020088

Yang, S., Lin, C. H., Sung, C. T., & Fang, J. Y. (2014). Antibacterial activities of bacteriocins: Application in foods and pharmaceuticals. *Frontiers in Microbiology, 5*, 241. https://doi.org/10.3389/fmicb.2014.00241

Yang, Y., Feye, K. M., Shi, Z., Pavlidis, H. O., Kogut, M., J Ashworth, A., & Ricke, S. C. (2019). A historical review on antibiotic resistance of foodborne campylobacter. *Frontiers in Microbiology, 10*, 1509. https://doi.org/10.3389/fmicb.2019.01509

Yu, H., Li, N., Zeng, X., Liu, L., Wang, Y., & Wang, G. (2019). A comprehensive antimicrobial activity evaluation of the recombinant microcin J25 against the foodborne pathogens *Salmonella* and *E. coli* O157:H7 by using a matrix of conditions. *Frontiers in Microbiology, 10*, 1954. https://doi.org/10.3389/fmicb.2019.01954

Zaidi, B. M., Campos, F. D., Estrada-Garcia, T., Gutierrez, F., Leon, M., Chim, R., & Calva, J. J. (2012). Burden and transmission of zoonotic foodborne disease in a rural community in Mexico. *Clinical Infectious Diseases, 55*(1), 51–60. https://doi.org/10.1093/cid/cis300

Zbikowska, K., Michalczuk, M., & Dolka, B. (2020). Review the use of bacteriophages in the poultry industry. *Animals, 10*(5), 872. https://doi.org/10.3390/ani10050872

Chapter 7
Impacts of Environment-Friendly Unit Operations on the Functional Properties of Bee Pollen

Aydin Kilic

Abstract This section comprehensively discuss the functional properties of bee pollen of some environmental-friendly novel unit operations. In this context, fluidized bed assisted cold drying (LTHVfb), microwave (MW), freezing (FD), vacuum (VD), infrared (IR) and microwave assisted vacuum drying (MW-VD) are investigated as new environment-friendly unit operations. The protective properties of these unit operations on the raw material properties during processing, storage and kitchen applications were investigated. In addition, many specific bioactive properties of pollen bioactive components such as antimicrobial, antioxidant and anticarcinogenic properties are discussed. In addition, the pollen production chain, sustainability in this life cycle and the environment-friendly features of these new applications on sustainability have been revealed. According to the literature data obtained, it is explained that novel unit operations contribute to sustainability with their environmentally friendly features as well as their protective effects on bioactive components in drying and production process.

Keywords Sustainability · Environment-friendly · Novel unit operations · Pollen · Functional food · Bioactive component

Abbreviations

bb	brine basis
bw	brewing
Cp	cold press
CP	control point
da	drying air
db	dry bases

A. Kilic (✉)
Department of Gastronomy Culinary Arts, Faculty of Tourism, University of Recep Tayyip Erdogan, Rize, Turkey
e-mail: aydin.kilic@erdogan.edu.tr

C. M. Galanakis (ed.), *Biodiversity, Functional Ecosystems and Sustainable Food Production*, https://doi.org/10.1007/978-3-031-07434-9_7

d_p	diameter of amorf particles
Dp	sample particle diameter (m)
EGG	epigallocatechin gallate
EKG	epicatechin gallat
fb	fluidized bed
FD	freeze-drying
g	gravity constant
GAE	gallic acid equivalent.
GHG	greenhouse gas
IR	Infrared drying
L	height of the fluidized bed (m)
ls	liquid smoking
$LTHV_{fb}$	fluidized bed assisted low-temperature high velocity
MWD	microwave drying
nd	not determined
PCP	Process control point
Reh	Reynolds number
Rep	Reynolds particle number
RH	relative humidity
RM	raw material
RSA	radical scavenging activities
SD	standard deviation
T	temperature (°C)
t	time, h
TPC	total phenol content
TPh	Total phenolics
U	superficial bed velocity, m/s
u_{mf}	superficial fluidization velocity
VD	vacuum drying
Vp	particle volume, m
W	moisture content, %
wb	wet bases
wt	withering
ΔE	lowest total color difference
ΔP	the pressure difference, Pa
E	void fraction
ε_{mf}	fluidization voidage
μ	viscosity of fluidized material, kg/(m-s)
ρ	density of fluidized mater kg/m^3
ρh	density of material
ϕ_S	sphericity

1 Introduction

Pollen, the male reproductive cell of flowers, can be consumed as a raw or dried food form (Kilic & Oztan, 2013; Bogdanov, 2004; De-Melo et al., 2016; Campos et al., 2008). Although such a definition is made for flower pollen, bee pollen, which is used as human food or for medicinal purposes in various ways, can be defined as a unique bee product consisting of a mixture of flower nectar, pollen, and bee saliva (Kilic, 2020a; Kilic, 2022; De-Melo & de Almeida-Muradian, 2017; Ulusoy & Kolayli, 2014). While bees collect pollen as a food source, they pollinate flowering plants. Bee pollen is collected from flowers and a complex substance containing beeswax, honey, and bee secretion. It is also used for the anti-inflammatory, local analgesic, and immune stimulant (Kilic, 2020a; Komosinska-Vassev et al., 2015; Dias et al., 2016; Dias et al., 2013; Feas et al., 2012).

In the content of bee pollen, there are many vitamins, protein, amino acids, phenolic compounds, some enzymes, antioxidants, beta carotene, some steroids, selenium, lectin, cysteine, magnesium, calcium, B1, B2, C, and E vitamins. Bee pollen can be consumed in dried or raw product form as human food. There is approximately 20–30% water in bee pollen as a perishable material (De-Melo et al., 2016). The water content in the pollen must be below 10% for microbial stability (Bogdanov, 2004; De-Melo et al., 2016; Campos et al., 2008; de Melo & de Almeida-Muradian, 2011; Kanar & Mazı, 2019).

In addition, it has been found that the flavonoids in the bee pollen are species for every plant-specific flora (Tomas- Lorente et al., 1992; Campos et al., 2016, 2002, 1997). Raw pollen is usually stored by the drying or freezing process. Drying processes can be applied to hot, cold, or open-air sun drying (Kieliszek et al., 2018; Kanar & Mazı, 2019; Barajas et al., 2012). To preserve the bioactive content and prevent Maillard components, novel applications should be developed (Kilic et al., 2014; Kilic, 2020a; Kanar & Mazı, 2019; Collin et al., 1995).

At the same time, in addition to the general food components found in bee pollen, bee pollen, which has functional properties due to its food fiber and some essential fatty acids, can be considered as a bioactive food or food additive (Kilic, 2020a). Bee pollen contains 7.4% water, 6 lipids, 20 proteins, and 2.2 minerals on average (Kilic, 2020a; Almeida-Muradian et al., 2005). On the other hand, the bioactivity of the product may vary depending on the flower variety, type, and environmental factors (Kilic, 2020a). According to the numerical data obtained from numerous studies, it can be said that it contains more than 250 bioactive components like caffeic, chlorogenic, Gallic, myricetin, ferulic, coumaric acid, kaempferol, galangin (Kilic, 2020a; Komosinska-Vassev et al., 2015; Başdogan et al., 2019; Bell et al., 1983). The drying process can affect the sensitive bioactive components of raw pollen (Kanar & Mazı, 2019; Dominguez- Valhondo et al., 2011). In a study, after the hot drying process, vitamin E, β-carotene, and tocopherol contents of bioactive components decreased significantly (15–19%). In the other study, the tocopherol content of raw material decreased using microwave completely (Kanar & Mazı, 2019; de Melo & de Almeida-Muradian, 2011). Heat treatment of foods such as pollen causes the emergence of some carcinogenic components such as Hydroxymethylfurfural (HMF), which are risky in terms of food safety due to the presence of sugar. Many experimental studies carried out in this context confirm this result. Therefore, it is essential to process the product using low temperatures. On the other hand, it was found that cold drying applications were the best preservation applications for the bee pollen (Dominguez- Valhondo et al., 2011; De-Melo et al., 2016; Kilic, 2020a; Kanar & Mazı, 2019; Dias et al., 2016).

The biological value of the material, which is known to contain more than 250 bioactive components, may vary depending not only on the type and variety of flowers included in it but also on the environmental parameters of the environment in which it is produced (Komosinska-Vassev et al., 2015; Kilic, 2020a; Başdogan et al., 2019; Bell et al., 1983). Bee pollen has some functional characteristics for the many bioactive contents (gallic, myricetin, chlorogenic, kaempferol, caffeic, coumaric acid, ferulic, galanin, and quartset) are antiallergic, antibiotic, antidiarrheic, and antimicrobiologic, anti-carcinogenic, radical scavenger, antimicrobiologic, and antioxidant activities. Bee pollen, which contains many bioactive components (myricetin, kaempferol, gallic, caffeic, chlorogenic, coumaric acid, ferulic, galangin, and quartzet), has many functions such as antiallergic, antibiotic, diarrheal and antimicrobial, anti-carcinogenic, radical scavenger, antimicrobial and antioxidant activity (Kilic, 2020a; Ranieri et al., 2019; Almaraz-Abarca et al., 2007).

In addition, experimental studies have shown that taking into the body without storage or processing provides better bioactive characteristics against gastrointestinal, cardiovascular, reproductive, and skin problems (Song et al., 2020). Although bee pollen is consumed mostly fresh, it can also be consumed by drying with various methods and after a prolonged storage period. However, depending on the drying and storage conditions of the raw material, significant losses occur in the phenol content, which highlights the total quality and especially its functional properties (Kilic, 2020a; Anjos et al., 2019; Campos et al., 2008).

Bee pollen can have a shelf life of different times depending on the amount of moisture in its content. Therefore, it can be evaluated in various standards according to the moisture content in its content and evaluated with different criteria according to its moisture content. It has been determined that although the bioactive components are damaged to a great extent in the pollen where the microwave drying method is applied, these losses can be tolerated when used under vacuum (Campos et al., 2008; Başdogan et al., 2019; Kilic, 2020a; De-Melo et al., 2016).

In order to preserve the stability of the sensitive components in the raw material, it is essential to dry them at low temperatures and not be exposed to light in an open environment. Many studies have also numerically demonstrated the bioactive components of pollen exposed to light and heat application, where significant losses occur (Kilic, 2020a;).

Although many traditional drying methods are applied in the food industry today, new drying applications such as low temperature and high speed (LTHV) drying and fluidized bed-supported LTHV stand out, especially with their protective effects on the bioactive components of the raw material (Kilic, 2020a). For example, many components in pollen, known as bioactive, can be preserved at higher rates by cold drying application or using a fluidized bed-assisted cold drying method (Kilic, 2009, 2017, 2020a). The fluidized bed drying technique is known as the fluid behavior of solid particles depending on the passage of compressed air between them. Fluidized bed drying is preferred in many industrial applications, as it has many production advantages such as quality, cost, and efficiency (Kilic, 2020a).

Studies have shown that these losses that may occur in sensitive bioactive components will be reduced to a minimum with cold drying air, fluidized bed, and fully controlled closed lightless systems (Kilic, 2020a; Kosuke et al., 2006). However, if the drying time is long in cold weather conditions, this negative situation can be neglected to protect the critical components for health (Kilic, 2017, 2020a Raw pollen has approximately 30% water content (db), and the following drying process, it is decreased to 10> (db) around. The vitamin C of dry pollen had considerable variations depending on the drying air temperature (40–60 °C, 2 m/s). It was stated that the sensory qualities and color values of the products obtained by drying at the lowest drying temperature gave the best results. In contrast, the best temperature to be applied was 40 °C. Experimental studies have determined that drying temperature also affects oil stability in pollen (Isik et al., 2019; Kilic, 2020a; Song et al., 2020).

The primary purpose of this chapter is to discuss the new findings of the novel technologies applied in drying bee pollen, known as functional food, and to determine the specific preservative properties of these new applications on the bioactive components of the raw material. On the other hand, another primary purpose is to examine in detail the LTHVfb drying system, known as a new drying application, and to determine the preservative effects of the method on the bioactive components during drying of bee pollen. In addition, this section discusses the new findings of the bioactive components of pollen, known as functional food. Finally, it discusses the stability of these components with bioactive characteristics after novel technological applications like as LTHVfb.

2 General Considerations About Pollen

Pollen is formed due to the combination of pollen, enzymes, and nectar collected from many flowers with the saliva of worker honey bees. It varies significantly according to plants' diversity that bees can collect in pollen production and according to the climatic conditions and the activity of the bees, which are greatly affected by these conditions. However, the color change in amorphous pollen may vary according to biodiversity and climatic conditions. (Nainu et al., 2021; Al-Yousef et al., 2020). Bee bread, which has an overall consumption as another bee product, can also be called fermented bee pollen, defined as a mixture of pollen and honeybee saliva. Pollen is also a nutrient that bees use to feed their larvae. Although bee pollen is used as a food source for bees, it has also been a source of food widely used by humans since ancient times. Physical, chemical, sensory, and biological quality characteristics of raw pollen can vary greatly depending on many environmental vectors such as humidity, pressure, temperature, plant diversity, pressure, humidity, and wind used by bees during production (Kilic, 2020a; Silici, 2014; Nainu et al., 2021). The most common bee plant species as bee pollen sources known in the literature are *Trifolium spp., Thyme spp., Zea spp., Rosmarinus spp., Crataegus spp., Salix spp., Cystus spp., Castanea spp., Acacia spp., Aesculus spp., Eucalyptus spp., Brassica spp., Centaurea spp., Citrus spp., Lavandula spp., Malus spp., Papaver spp., Pinus spp., Prosopis spp., Prunus spp., Tara spp., and Schisandra spp.* (Kilic, 2020a; Abid et al., 1990; Song et al., 2020).

2.1 Bioactive Characteristics of Bee Pollen

Pollen, which is commercially consumed in raw, granule, powder, capsule, and pellet, is an essential supplementary supplement that can provide nutritional balance. However, the raw materials bio quality or biochemical characteristics have considerable variations depending on many environmental factors such as plant flora in the ecosystem, precipitation, temperature, air pressure, and humidity (Kilic, 2020a; Kanar & Mazı, 2019; Kalaycıoğlu et al., 2017; Kostić et al., 2020).

Bee pollen is also defined as the most perfect food in the world because it contains all nutritional components in a balanced way (Ares et al., 2018; Kilic, 2022; Kieliszek et al., 2018). Due to the antimicrobial, antioxidant, and ligand properties of polyphenols and carotenoids, in addition to many bioactive food components in its content, it has specific bioactivity for complementary medical purposes against heart diseases, cancer, liver disorders, many specific ailments, therapeutic characteristics, and many allergic reactions. The most basic components whose bioactivity has been confirmed are phytosterols, phospholipids, essential fatty acids, fitometabolitlerinin (flavonoidler, fenoller, karotenoidler, and vitaminler, etc.) (Li et al., 2018; Kostić et al., 2020; Antonelli et al., 2019).

Bee pollen includes β-carotene (as vitamin A), ascorbic acid (as vitamin C), thiamine (as vitamin B1), riboflavin (as vitamin B2), niacin (as vitamin B3), pantothenic acid (as vitamin B5), pyridoxine (as vitamin B6), biotin (as vitamin B7), folic acid (as vitamin B9), para-aminobenzoic acid (as vitamin B10), salicylic acid (as vitamin B11), cobalamin (as vitamin B12) and tocopherol (as vitamin E) such as bioactive components (Ang et al., 2008). In addition, components such as *rutin, apigenin, epicatechin, cristine, phytoalexin, isoquercetin, vanillic, chlorogenic, gallic, ferulic, cumaric, caffeic, quercetin, luteolin, myricetin, and resveratrol* can be counted as other active compounds in bee pollen. Although bee pollen contains many phenolics as bioactive components, especially flavonoids have greater importance due to their solid bioactive characteristics. These components also have a much greater extent with their powerful antioxidant and antiradical properties. Simple phenolics and polyphenols form phenolic substances. Flavonoids with high bioactive properties such as catechins flavanones, proanthocyanidins, anthocyanidins, and flavonols are also classified as phenolics (De-Melo & de Almeida-Muradian, 2017; Kilic, 2020a;

Ranieri et al., 2019; Almaraz-Abarca et al., 2007; Silva et al., 2006; Kaškonienėa et al., 2020).

The composition and amounts of phenolic compounds in pollen vary greatly depending on environmental conditions, especially plant flora (Kaškonienėa et al., 2020).

It is known that bee pollen is consumed as an alternative medical practice against ailments such as anemia, colitis, ulcers, influenza, colds, allergies, enteritis, and anti-cholesterol (Ulusoy & Kolaylı, 2014).

2.2 *The Presentation and Consumption of Bee Pollen*

It is stated that daily pollen consumption should be a maximum of 40 g for an adult individual. Should be between Since it has a capsule that resists enzymatic or chemical digestion, it is thought that the evaluation rate of its consumption can be increased from 4% to 80% by grinding it in the form of bee bread, which can be called fermented pollen instead of direct consumption, for maximum metabolic evaluation (Li et al., 2018; Kostić et al., 2020; Antonelli et al., 2019; Kieliszek et al., 2018; Franchi et al., 1997).

Bee pollen may not be a beautiful food product or food additive in terms of its sensory qualities for a consumer. Therefore, the consumption of bee pollen, which is highly recommended, especially in terms of child health and nutrition, may need to be more attractive for children and babies. In this context, gastronomy and culinary arts must find a way to make the presentation of the product visually, emotionally, or texturally more appealing. In order to make bee pollen more attractive with additives, several traditional or new forms of presentation have recently emerged with various foods or food additives. In this context, some suggestions regarding the production of bee pollen in the literature can be listed as follows;

1. Bee pollen with yogurt,
2. Bee pollen paste with honey,
3. Green spring,
4. Bee pollen oats,
5. Bee pollen, ice cream, etc.

It is stated that bee pollen has low digestibility (10–15%) and bioavailability in human metabolism due to chemical resistance (Kaškonienėa et al., 2020; Denisow & Denisow-Pietrzyk, 2016). On the other hand, it is stated that bee bread, which can be described as a fermented pollen product, has higher digestible phenol content and therefore has a much higher bioavailability and therefore bioactivity. In this context, it can be thought that a similar effect may occur by mixing pollen with honey and yogurt and resting it in an oxygen-free environment. The density, structural properties, and sensory qualities of the bioactive compounds that will be formed due to the mixing of yogurt containing lactic acid bacteria, bee pollen, and honey and keeping them in an oxygen-free environment may be the target of a new study (Kaškonienėa et al., 2020; Barene et al., 2014; Vásquez & Olofsson, 2009; Karakaya & Kilic, 1994). It has been determined that bee pollen mixed with fermented yogurt has antioxidant capacity, higher polyphenol content, and better sensory qualities (Karabagias et al., 2018; Zlatev et al., 2018; Kostić et al., 2020). According to the data obtained from the experimental studies, the addition of pollen to yogurt had positive effects on the traditional texture of yogurt. The sensory properties and chemical properties of traditional yogurt enriched with pollen or vice versa, pollen enriched with yogurt differ depending on the plant diversity of the existing flora (Khider et al., 2013; Atallah, 2016; Atallah & Morsy, 2017; Kostić et al., 2020). As in the odor formation of honey, some plants can create the dominant fragrance in the shape of the pollen's smell. It is known that plants such as thyme, thyme, and chestnut usually reveal the prevalent odor in honey, as well as these plants, can be dominant in the formation of odor in pollen. However, in order to ensure the appearance of the desired fragrance in the pollen, it can be assured that the bio flora can be directed naturally. Thus, the plant odor expected to be dominant can be brought to the fore in the natural flora.

2.3 *Drying of Bee Pollen*

In the production process of bee pollen, generally, there may be steps such as pre-cleaning, freezing, thawing, drying, airing, final cleaning, packaging, and storage. However, experimental data obtained from numerous scientific studies, aromatic properties, color, smell, vitamins, proteins, and oils in the biomaterial suffer some quality losses due to processing at high temperatures. Some losses may occur, especially in free essential amino acids and polyunsaturated fatty acids. On the other hand, drying processes with the low-temperature application can reduce these losses that may occur in such quality characteristics (Kilic, 2020a; Kilic, 2020b; Barreto et al., 2005; Abid et al., 1990; Song et al., 2020).

There are many drying applications to be used for bee pollen dryings like hot drying, vacuum drying, lyophilization, and microwave drying. In addition, depending on the drying methods and parameters applied, significant losses occur mainly in the bioactive components of pollen and, accordingly, in its functional properties (Kilic, 2020a; Ranieri et al., 2019; Kayacan et al., 2018).

In addition, there are some novels drying applications in food literature like HAD, MWD, LTHVfb, VD, IR, and FD. It is stated that these new drying methods provide protective advantages on sensitive bioactive food ingredients, unlike traditional methods such as open-air drying and hot drying. It is noted that the main reason for this protective feature is the drying parameters applied in these new drying methods.

2.4 *General Considerations About Novel Fluidized Bed Assisted Cold Drying (LTHVfb)*

Low drying temperatures applied in cold drying may have some disadvantages due to its long drying time and higher cost compared to the classically applied fluid bed drying technique. However, it has been determined that it is a protective method that should be preferred for drying precious bioactive foods such as bee pollen with high functional properties such as the following. Furthermore, while the fluidized bed drying application ensures homogeneous, better quality, safe and fast drying of the product, performing all these processes at low temperatures also provides significant advantages in terms of the stability of components with bioactive properties (Kilic, 2020a; 2021; Raveendran et al., 2019). Figure 7.1 gives the diagram of the fluidization of raw bee pollen.

In experimental studies, a minimum drying air velocity of 1 m/s is required to fluidize raw pollen samples. There are many equations for calculating the fluidization rate (Kilic, 2020a). On the other hand, the LTHV drying method can be used

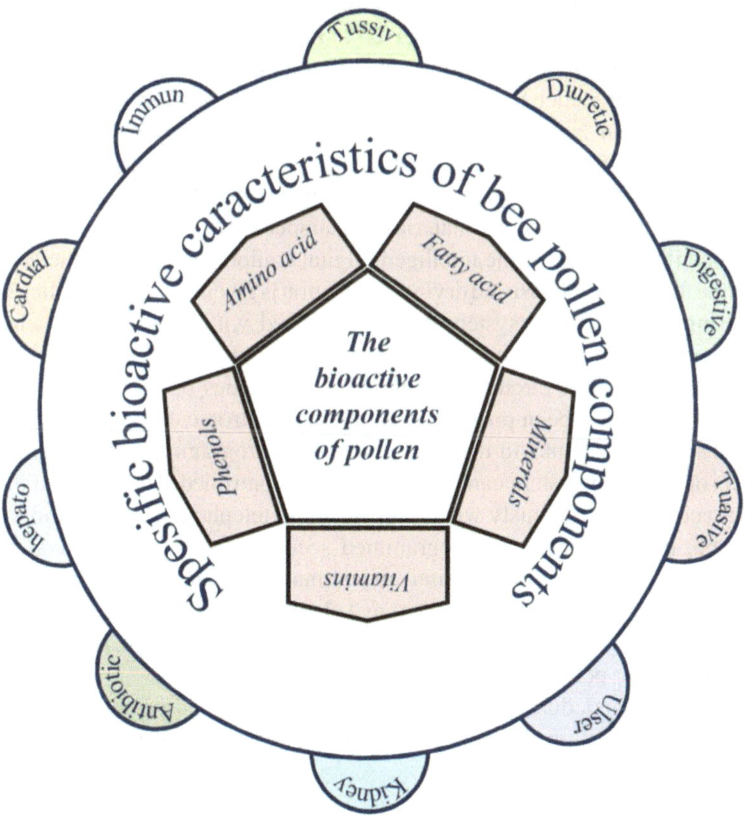

Fig. 7.1 The general bioactive characteristics of bee pollen components

between 2–4 m/s as $LTHV_{fb}$ application. If the lowest air velocity can be set equal to the pressure drop to ensure fluidization, it can be determined according to the following equation;

$$\frac{150\mu u_{mf}}{\varnothing_s^2 d_p^2}\frac{1-\varepsilon_{mf}}{\varepsilon_{mf}^3}+\frac{1,75\rho u_{mf}^2}{\varnothing_s d_p}\frac{1}{\varepsilon_{mf}^3}=g\left(\rho h-\rho\right) \tag{7.1}$$

In here g presents the gravity constant; ρ presents the density of the fluid mater; d_p presents the diameter of nominal amorf particles; ρh presents the density of material; μ presents the viscosity of fluidized material; ε_{mf} presents the fluidization voidage; u_{mf} presents the superficial fluidization velocity, and finally ϕ_S presents (1–4.4 mm) the sphericity of the raw pollen (Cobzaru & Inglezakis, 2015).

The physical characteristics of pollen were identified as;

$\varepsilon = 0,39$
$dp = 0.04\ m$
$\rho h = 376\ k/m^3$

The detailed schematic layout of the fluidized bed-assisted LTHV drying system applied at refrigerator temperature is given in Fig. 7.2.

Three basic units, namely cooling, conditioning and drying, are designed in the fluid bed dryer unit, which is a new drying method at refrigerator temperature (Kilic, 2020a). While thin layer drying was applied in the traditional LTHV system, a cylindrical drying tray was added to the existing system for fluid bed application. The depth of the fluid bed drying tray is recommended as 5 m/s for pollen, depending on the flight distance of the material and airspeed (Kilic, 2020a).

The detailed diagram of the intelligent digital equipment made for the dryer unit used in the fluidized bed-assisted drying application is given in Fig. 7.3 (Kilic, 2020a).

Each unit of the dryer system can be equipped with experimental sensors to determine and record environmental parameters, and all this equipment is connected to an electronic circuit. Depending on the values obtained from the dryer unit, the software has been programmed into the electronic circuit for the system's automatic control. Thanks to the electronic circuit programmed, the parameters in the units of the dryer system can be automatically controlled. The obtained data can also be recorded continuously with computers and telephones' contribution. At the same time, the system can be programmed according to the parameters, and the work's functions can be converted into an automatic system. For example, while the fluidized bed can be obtained with an air velocity of 6 m/s at the beginning, this value can be programmed to decrease according to the change in the product's weight in the later periods of drying (Kilic, 2020a).

On the other hand, during fluidized bed drying, pressure drops during drying can be determined based on the minimum fluidization rate using the Ergun equation (Trahana et al., 2014).

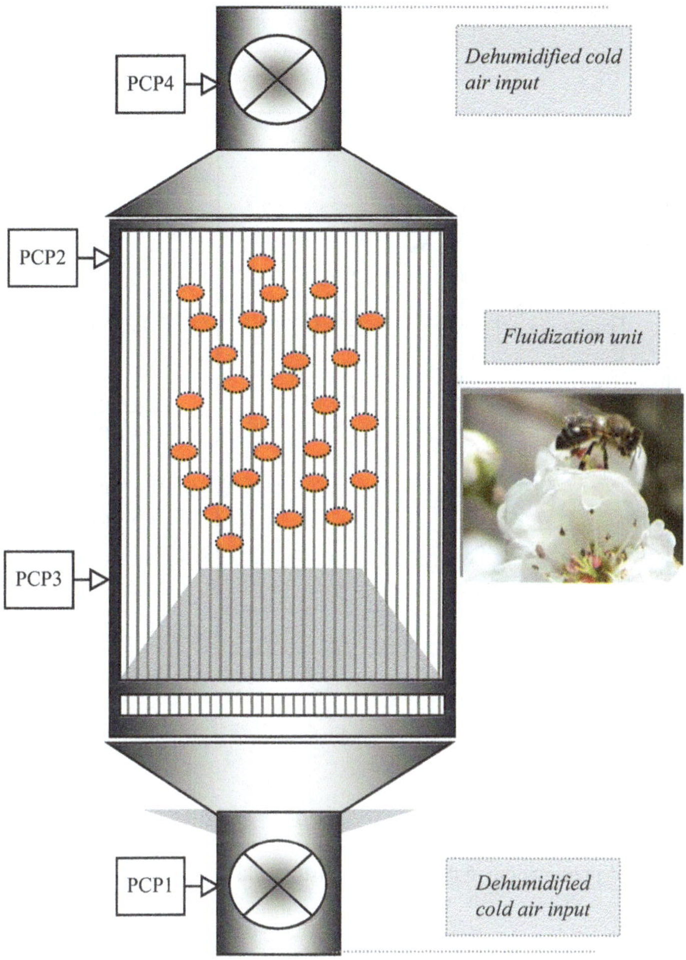

Fig. 7.2 The diagram of fluidization of raw bee pollen (*Modified from Kilic, 2020*)

$$\frac{\Delta P}{L} = A\frac{(1-\varepsilon)^2}{\varepsilon^3}\frac{\mu U}{D_p^2} + B\frac{1-\varepsilon}{\varepsilon^3}\frac{\rho U^2}{D_p} \tag{7.2}$$

In where A is 150 and B is 1.75 as a coefficient, D_p is the particle diameter (m), L is fluidized bed layer height (m), ρ is density, kg/m^3, Pa is μ viscosity, ε is the void fraction, kg/(m-s), U is superficial velocity, ΔP present pressure drop.

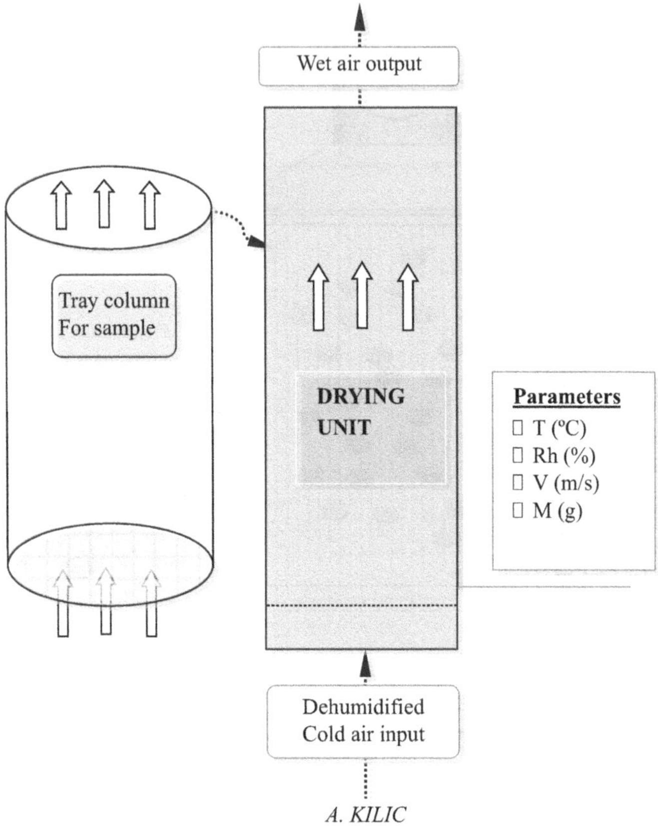

Fig. 7.3 Detailed presentations fluidized bed assisted cold drying system (LTHVfb) (*Modified from Kilic 2020*)

2.5 *Low Temperature and LTHVfb Drying Characteristics of Bee Pollen*

Since bee pollen is a sensitive food with moisture content, some traditional processes are applied during its long-term storage. Although standard preservation methods such as freezing, drying, or lyophilization are used, it is stated that these applied methods have adverse effects on the functional characteristics and, therefore, on the bioactivity of the bioactive components (Kilic, 2020a; Kostić et al., 2020; Kanar & Mazı, 2019; Conte et al., 2016). Therefore, some experimental analyzes were carried out to determine the moisture, protein, lipid, ash, carbohydrate, and ascorbic acid content of bee pollen before and after drying in an LTHVfb tower drying system. The moisture value determined in the dried products after the drying process was found to be ~6% (wb). On the other hand, a_w value of dried pollen should be approximately 0.28 at the most. Figure 7.4 shows the general bioactive characteristics of bee pollen components (Kilic, 2020a; Mărgăoan et al., 2019). To

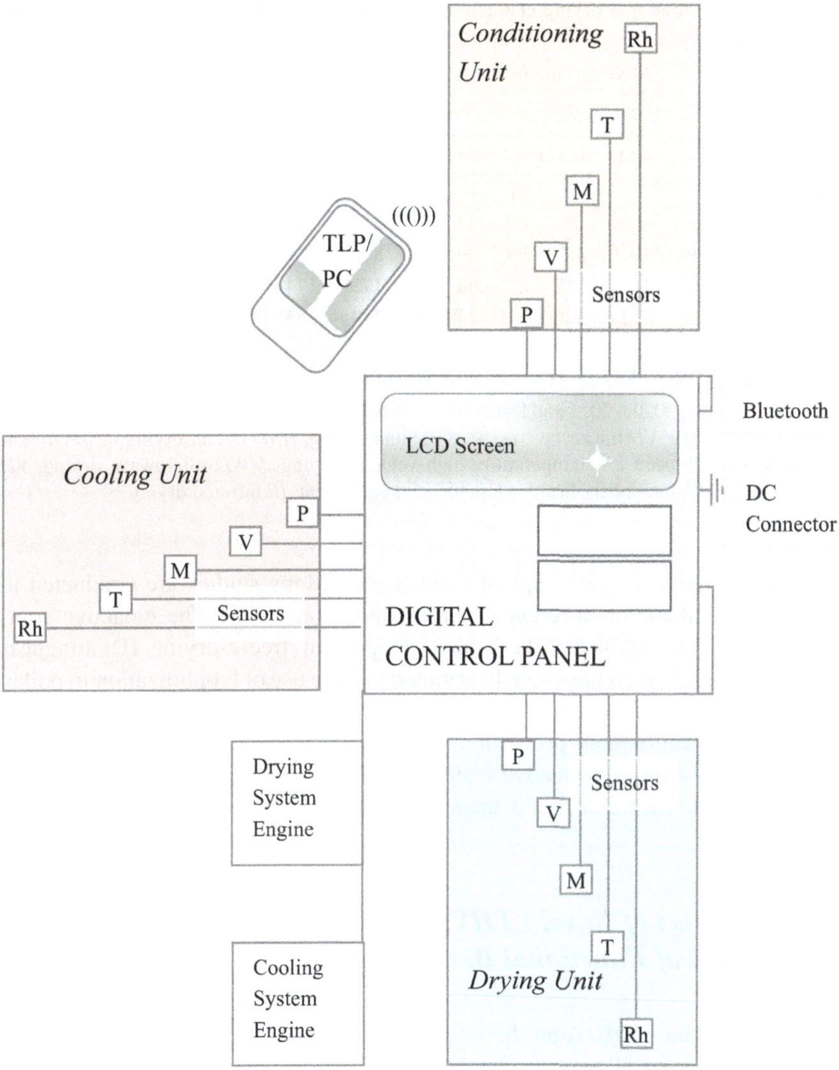

Fig. 7.4 The detailed diagram of the intelligent digital equipment made for the dryer unit (Rh; relative humidity sensor; M: weight sensor; T: thermal sensor; V: Velocity sensor; P: Air pressure sensor)

preserve the bioactive content and prevent Maillard components, novel applications should be developed (Kilic et al., 2014; Kilic, 2020a; Kanar & Mazı, 2019; Collin et al., 1995). Table 7.1 gives drying parameters for the different traditional and novel drying applications.

As novel drying applications, the LTHVfb, VD, MWD, MWVD, and IR drying methods were applied to dry bee pollen. Kayacan et al. (2018) detected that the phenolic and flavonoid amounts in experimental samples in which vacuum drying,

Table 7.1 Drying time and drying characteristics of bee pollen depending on different drying applications (w.b)

Experimental parameters	Novel drying methods									
	RM	HAD		LTHVfb		VD	MWD	MW-VD	Storage GAE g^{-1}	IR
Temperaturer (°C)	–	35	65	4	40	35	300 W	300 W	–	–
Time (h)	–	29	4.5	34	7	257	24 min	24 min	–	–
RSA (mg TEAC/g)	10.3	3.3	2.8	nd	nd	8.5	9.2	9.0	–	–
Vitamine E (µg/g)	162	117	132	nd	nd	123	174	125	–	–
Vitamine C (µg/g)	451	426	202	430	400	174	253	245	–	370–200
TPh	14.42	9.34	11.1	nd	nd	12.51	14.2	9.9	37	–

Modified from Kilic, 2020a, 2021 and Kanar & Mazı, 2019; Anjos et al., 2019; Kayacan et al., 2018
RM raw material, *MW-VD* microwave-assisted vacuum drying, *HAD* hot air drying, *LTHVfb* fluidised inadequate assisted low-temperature, high-velocity drying, *MWD* microwave drying, *VD* vacuum drying, *TPh* total phenolics, *GAE* gallic acid equivalent, *IR* infrared drying

a new drying method, were applied were higher. Many studies are conducted to determine the effects of different drying methods on pollen. The bioactive compounds were more stable in the pollen samples of freeze-drying (Dominguez-Valhondo et al., 2011). It has been determined that the use of lyophilization in pollen drying leads to fewer losses than hot drying in terms of preserving polyphenols, flavonoids, and antioxidant properties (Dias et al., 2016). In addition, it has been determined that it has a protective feature on bioactive compounds in pollen samples dried by microwave drying (Canale et al., 2016; Conte et al., 2016).

2.6 The Effect of Novel LTHVfb Drying on the Bioactive Content of Functional Bee Pollen

In a study by Kilic (2020a) on the effects of the fluidized bed-supported cold air drying on the functional compounds of the raw material, the impact on ascorbic acid was investigated. According to the experimental data obtained, it has been revealed that the method known as the LTHVfb method, which is a new method, has a protective effect on ascorbic acid. In this context, although they stated that temperatures of 4–45 °C could be applied, they indicated that they obtained the best result from the lowest temperature application. Before the drying process, the Vitamin C value for bee pollen is identified as 4.29 (% w.b.). The ascorbic acid amount of bee pollen dried with a cold air-assisted fluidized bed drying system is found at 4.28 (% w.b.) for LTHVfb dried pollen (04 °C). The amount of ascorbic acid (Vitamine C) of raw bee pollen dried with a cold air-assisted fluidized bed drying method is found 4.28 (% w.b.) for LTHVfb dried pollen at 04 °C and dried pollen at 40 °C. Studies have shown that drying bee pollen at temperatures higher than 40 °C causes chemical and sensory changes (Kilic, 2020a; Kostić et al., 2020; Barajas et al., 2012; Szczêsna, 2006a, b).

The experimental results demonstrate the protective effect of LTHVfb administration on vitamin C. Depending on the LTHVfb drying applied at +4 °C, the vitamin C value of the product was determined as 4.28. In contrast, this value was 4.04 in the product where 40 °C LTHVfb was used. Furthermore, it has been determined from experimental studies that the protein composition, functional properties, and amount of raw honey bee pollen produced from a single flower ecosystem depend on the flower source of the pollen (Kilic, 2020a; 2021; Barajas et al., 2012; Szczêsna, 2006a, b).

According to the results obtained from the new studies, the * value increases while the b* value decreases depending on the drying temperature. However, it was determined that the drying process applied at a low temperature had protective effects on the color and structure of the biological product (Kilic & Oztan, 2013; Kilic, 2021; Song et al., 2020; Bogdanov, 2004). Table 7.2 gives low-temperature drying characteristics of some bioactive components of bee pollen.

Barajas et al. (2012) revealed that although drying at 45 °C takes a short time, the carotene and vitamin C values of raw pollen dried at the lower temperature of 35 °C were higher, as shown in Table 7.2.

In short, similar to the LTHVfb method, which we defined as a new method, it was determined that vitamin C, known for its bioactive property, was highly preserved in this drying application, which was carried out at low temperature. All these results reveal the protective effects of the new LTHVfb method, which is a common temperature application, on bioactive components. In this context, although the same studies have revealed the protective effects of cold drying and fluidized bed applications on functional components, it has been shown that it does not matter if the geographical or regional areas are different.

Ranieri et al. (2019) studied the FD method as a novel drying application, and the technique was adequate to the proline and amino acid in chestnut pollen collected by the honeybee. In addition, It has been stated that infrared drying, which is one of the new drying techniques, is a new technology that can be preferred over traditional applications with its fast, homogeneous, economic, and protective features (Sadin et al., 2014; Isik et al., 2019).

Table 7.2 Low-temperature drying characteristics of some bioactive components of bee pollen (Kilic, 2020a; 2021; Barajas et al., 2012)

Environmental origin	Groups (Bee pollen)	Vitamin C (mg/100 g)	Carotene (mg/g)
1. Ecosystem	Raw pollen	40.22	0.77
	Dry pollen 35 °C	31.75	0.78
	Dry pollen 45 °C	27.35	0.51
2. Ecosystem	Raw pollen	40.37	0.21
	Dry pollen 35 °C	32.79	0.22
	Dry pollen 45 °C	28.75	0.17

2.7 Environmental Friendly Strategies for the Product Life Cycle for Bee Pollen

In the bee products and bee pollen production chain, which is similar to the food production chain, the environmental resources required for the bee and the supply of these resources, the care of the bees, milking, drying of the pollen, packaging, retail marketing, and consumption consist of many steps. The leading actor of the whole process is the honey bee in these steps. Therefore, there are both environmental effects and the effects from the production process to the environment.

Many political strategies have been developed to ensure sustainability in the food production chain. In this context, energy sources and usage, sustainability in food production depending on environmental effects are considered. In this context, it is possible to provide sustainable production with technological applications with minimum environmental impacts, depending on sustainable energy sources and sustainable environmental strategies. In this context, the main things to do are;

- Identifying new technological strategies by re-evaluating old practices depending on novel technological applications.
- Introducing new environmental strategies at macro and micro scale,
- To present new short, medium, and long-term environmental strategies,
- To determine new strategies for the use of sustainable energy resources,
- To consider the strategies mentioned above in the steps in the production chain of bee and bee products (Kilic et al., 2010, 2009).

As a result, the implementation of environmentally friendly production processes with less environmental impact should be emphasized by applying the sustainable practices developed for the food sector in the beekeeping sector as well. With the implementation of these strategies, an environment-friendly production policy based on sustainable resources will be realized (Kilic et al., 2009). In the bee products and bee pollen production chain, which is similar to the food production chain, the environmental resources required for the bee and the supply of these resources consist of many steps such as the care of the bees, milking, drying of the pollen, packaging, retail marketing, and consumption. The leading actor of the entire production process is the honey bee in these steps. In this context, it is necessary to determine both environmental and environmental effects in the production process and develop the required policies and strategies to reduce these mutual adverse effects. These policies should aim to implement environmentally friendly agricultural practices that prevent fossil fuels, pesticides, and chemical fertilizers. Figure 7.5 shows the bee life cycle and environmental interactions (Kilic et al., 2010). Figure 7.5 shows the bee pollen production life cycle and environmental interaction parameters of the bee ecosystem.

Bees are among the most essential biodiversity and sustainable primary production actors, providing pollination between plants in ecosystems. Bees are the most critical creatures that ensure the reproduction of more than half of the plants in nature by pollinating. For this reason, it is the most critical link of the chain in the

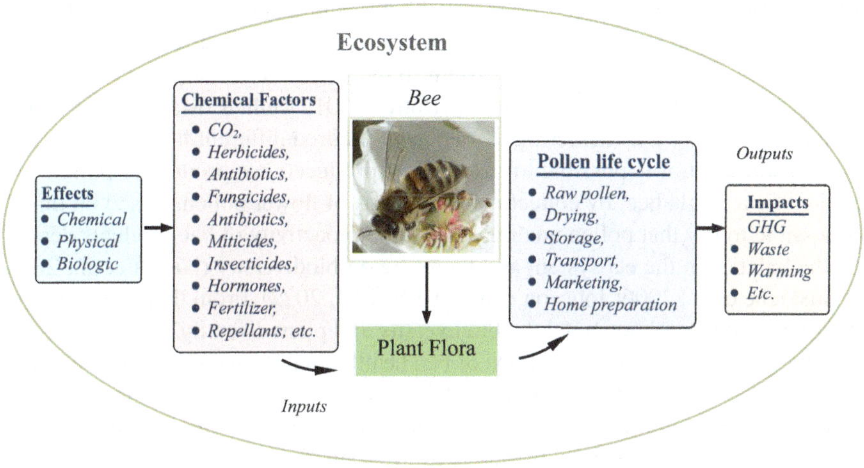

Fig. 7.5 Bee pollen production life cycle and environmental interaction parameters in the ecosystem (Kilic et al., 2010)

ecosystem and the most critical element of a sustainable food production chain. For this reason, it is the most crucial element of the ecosystem that must be protected and supported for sustainable biodiversity and, accordingly, pollen production. It is stated that the population of pollinating bees is at risk due to transport between ecosystems, industrial chemicals, agricultural, pharmaceuticals, light, sound pollution, and similar environmental wastes. Accordingly, the biodiversity, pollen production, and the bees' quality of life fed with pollen have decreased. In the ecosystem, the main factors that pose a risk to the life of bees can be grouped under three different headings as chemical, physical or biological. Plant diversity supports sustainability in the secondary consumer bee population (Di Noi et al., 2021; Rortais et al., 2017; Bloch et al., 2017; Klein et al., 2007; Thomann et al., 2013; Kösoğlu et al., 2021). In order to ensure sustainability in pollen production, first of all, the diversity of plant resources in the ecosystem and the bee population must be sustainable. Nutritional stress and illness due to malnutrition are the two main factors affecting honey bee health. In this context, plant diversity and the stability of genetic diversity should also be preserved. In order to ensure sustainability, biological sustainability needs to protect the genetic variations in the plant and animal resources that constitute the biological resources in the ecosystem (Bonhevi & Jorda, 1997; Dolezal & Toth, 2018; Kösoğlu et al., 2021).

Each bee product (Bee bread, pollen, milk, propolis, gel) is a vital link in the sustainability of bees' life cycle in the honeycomb. However, we cannot consider one of these rings alone. In the pollen production process, the ecosystem where the bee uses the flowers to produce the raw material is damaged under various environmental effects. These are generally polluted atmosphere, groundwater, rainwater under the influence of industry, pesticides, hormones, fertilizers, solid wastes can also be found in the environment. Such pollutants in the environment can be found on the plant flora, the source of bee pollen, or accumulated in its biological structure. According to the data obtained from the studies, bees carry the pollen they collect from the flowers in the ecosystem to the honeycombs on average 15 times per day (Nisbet et al., 2013; Conti & Botre, 2001; Celli & Maccagnani, 2003; Rashed et al., 2009). Considering that about a hundred different flowers are visited in each transport, it is possible to say that the produced bee product or pollen was created by a single bee, by collecting 1500 kinds of flower pollen. Based on this, it is possible to say that pollen quality, safety, and bioactivity of bee products depend on the balance in the ecosystem and therefore on biodiversity (Nisbet et al., 2013; Staniskiene et al., 2006; Johnson et al., 2010; Kilic, 2022). From this point of view, it will be understood how vital the biodiversity and plant diversity in the ecosystem is in the production process in terms of bee health, quality, and safety of bee pollen (Nisbet et al., 2013; Conti et al., 2007; Brodschneider & Crailsheim, 2010). Environmental pollution and global warming lead to consequences that directly affect the bee activity and the bioactivity of raw products produced by the bee and the consumer.

All these bioaccumulations can threaten public health by affecting both the bee and the products produced by the bee. However, the bee pollen production cycle may not only be affected by environmental risks but may also reveal some effects that may pose a risk to the environment. These risks can occur in raw pollen production, drying, storage, transport, marketing, retail consumption steps. Therefore, it is crucial to develop new environmentally friendly production strategies in this context. Environmentally friendly practices and sustainable resources should support all these steps with low environmental impact. Other bee products such as bee honey and pollen are directly affected by the chemical and physical vectors exposed to the plant flora it collects. It can also carry these effects to the products it produces.

On the other hand, environmental effects may occur at different rates depending on the energy sources and technological processes used in the whole process. Environmental pollution and then the impact of global warming can occur on a local, regional or international scale. Technologies and energy sources with minimum environmental emissions should be used in this context. Environmentally friendly practices and sustainable resources should support all these steps with environmental impact. Other bee products such as bee honey and pollen are directly affected by the chemical and physical vectors exposed to the plant flora it collects. It can also carry these effects to the products it produces.

Although bee pollen production appears to have few industrial applications, it is relatively energy-intensive and process-intensive. The process is not only how the bee collects the pollen, contains it in the honeycomb, and turns it into ready-to-eat food. The goals of protecting the plant diversity and plant safety in the ecosystem, moving them to the most suitable natural ecosystems, or creating these ecosystems artificially appear as costly investments that require energy use. Again in this context, the establishment, maintenance, renewal, periodic transportation, and off-season maintenance of production honeycombs are processes that require energy use. In this sense, selecting sustainable energy sources and environmentally friendly practices is essential (Kilic et al., 2010; Ziesemer, 2007; Kendall et al., 2013).

Although there is CO_2 emission to the environment due to energy use in the production process of all food products, low environmental impacts occur both in the production process based on traditional methods and in sustainable production processes where renewable energy sources are used (Kendall et al., 2013; Kilic et al., 2010). In these production processes, the environmental effects of each product vary significantly according to the nature of the product, the production method, and the characteristics of the energy sources used. Environment-friendly production technologies have a lower environmental impact due to the renewable energy sources they use and the least usage rate. In this sense, it has been reported that the LTHVfb method and similar low-temperature applications have low environmental impacts as an environmentally friendly application. However, depending on storage, it has been stated that cooling applications constitute half of the energy consumption in total production (Kilic et al., 2010, 2009; Pimentel & Pimentel, 1996; Dalsgaard & Abbotts, 2000).

The deterioration of the existing balance in the ecosystem by unnatural effects affects water resources and primarily the plant flora known as the primary producer, depending on the substance cycle. The deterioration of the plant flora, which is the primary source of pollen production, impairs bees' quality of life. The main change that this change in the ecosystem will bring about in the plant flora will appear as a decrease in plant diversity. Subsequently, bioaccumulation and mutagenic effects will occur due to the accumulation of toxic compounds in biodiversity plant sources. These effects will affect not only the plants known as primary producers but also the bees that use the plants as a source of pollen, and finally, the quality and safety of the bee pollen produced by the bees.

In order to reduce the deterioration of this balance in the ecosystem, which directly affects quality and safety in food production processes, it is essential to eliminate the environmental effects of industrial applications as much as possible (Kilic et al., 2009, 2010).

2.8 *Environmentally Friendly Characteristics of New Technologies Applied in Bee Pollen Production*

LTHVfb, Drying is a traditional, natural, and environmentally friendly basic process to minimize microbial and biochemical activity by reducing moisture (Kilic et al., 2010). Although there are many industrial drying techniques, LTHVfb drying application, a new drying technique that we recommend in pollen drying, is also recommended as an environmentally friendly technology. It protects food quality and has no environmental warming effect and emission (Kilic et al., 2010). Therefore, both LTHV and LTHVfb are recommended in drying technologies as an environmentally friendly application.

MW, Although the positive effects of microwave drying on food quality are known, it may not be considered an environmentally friendly technology due to its high electricity use and environmental impact. However, inevitably, this negative point of view will not be correct if electricity is supplied from renewable sources. In this context, it would not be wrong to call this technology an environmentally friendly technology in sustainable energy policies where electricity production is based on renewable energy sources.

FD, drying is also recommended as an environmentally friendly technology, as it preserves food quality and does not have environmental warming effects and emissions, as is the case with LTHV and LTHVfb as a drying application aimed at drying resources with high economic value or withering biomaterial (Kilic et al., 2010).

2.9 Conclusions

In this study, bioactive properties of functional bee pollen, new unit operations used in pollen drying, preservative properties of new unit operations on sensitive pollen bioactive content, sustainability in pollen production, and environmentally friendly features of new basic processes are discussed. In this context, this chapter presents the functional and bioactive functions of bee pollen comprehensively. It examines how some new protective unit processes have been applied to minimize quality losses during applications. In this context, it has been understood that new unit operations such as fluidized bed assisted cold drying (LTHVfb), microwave (MW), freezing (FD), vacuum (VD), infrared (IR), and microwave-assisted vacuum drying (MW-VD) are used in pollen drying. On the other hand, it has been demonstrated by literature studies that these new basic processes provide significant advantages in drying. The protective effects on raw material properties were evaluated during these unit operations, processing, storage, and kitchen applications. In addition, many specific bioactive properties of bee pollen components such as antimicrobial, antioxidant, and anti-carcinogenic properties are discussed. The protective effects of these new unit operations on bioactive components such as vitamin A (Carotenoids), vitamin B, vitamin C, vitamin E, fatty acids, amino acids, and phenolics are discussed. It was found that all these new studies have critical protective effects on the product's functional properties. However, as a critical study, the application method of the new LTHVfb drying method, which is known to provide significant advantages in terms of drying quality and product safety, and its protective effects on bioactive components, are discussed in more detail.

Finally, environmentally friendly LTHVfb, MW, FD, VD, IR, or MW-VD new drying methods have been shown to have specific protective effects on the stability of the bioactive components of pollen with their environmentally friendly properties (Kilic, 2020a; Kilic, 2021).

In summary, the advantages of new drying methods can be listed as follows;

1. Homogeneous drying,
2. Fast drying,
3. Protective effect on product quality,

4. Protective effect on bioactive components,
5. Economic application,
6. Safe drying,
7. Environmentally friendly practices,

References

Abid, M., Gibert, R., & Laguerie, C. (1990). An experimental and theoretical analysis of the mechanisms of heat and mass transfer during the drying of corn grains in a fluidized bed. *International Chemical Engineering, 30*(4), 632–642.

Almaraz-Abarca, N., Graça-Campos, M., Ávila-Reyes, J. A., Naranjo-Jiménez, N., Herrera-Corral, J., & González-Valdez, L. S. (2007). Antioxidant activity of polyphenolicextract of monofloral honeybee-collected pollen from mesquite (Prosopisjuliflora, Leguminosae). *Journal of Food Composition and Analysis, 20*, 119–124.

Almeida-Muradian, L. B., Pamplona, L. C., Coimbra, S., & Barth, O. M. (2005). Chemical composition and botanical evaluation of dried bee pollen pellets. *Journal of Food Composition and Analysis, 18*(1), 105–111.

Al-Yousef, H. M., Amina, M., Alqahtani, A. S., Alqahtani, M. S., Malik, A., Hatshan, M. R., Siddiqui, M. R. H., Khan, M., Shaik, M. R., Ola, M. S., & Syed, R. (2020). Pollen bee aqueous extract-based synthesis of silver nanoparticles and evaluation of their anti-cancer and anti-bacterial activities. *PRO, 8*, 524.

Ang, C. D., Alviar, M. J. M., Dans, A. L., Bautista-Velez, G. G. P., Villaruz-Sulit, M. V. C., Tan, J. J., Co, H. U., Bautista, M. R. M., & Roxas, A. A. (2008). Vitamin B for treating peripheral neuropathy. *Cochrane Database of Systematic Reviews, 16*(3), CD004573.

Anjos, O., Paula, V., Delgado, T., & Estevinho, L. (2019). Influence of the storage conditions on the quality of bee pollen. *Zemdirbyste- Agriculture, 106*(1), 87–94.

Antonelli, M., Donelli, D., & Firenzuoli, F. (2019). Therapeutic efficacy of orally administered pollen for nonallergic diseases: An umbrella review. *Phytotherapy Research, 33*(11), 2938–2947.

Ares, A. M., Valverde, S., Bernal, J. L., Nozal, M. J., & Bernal, J. (2018). Extraction and determination of bioactive compounds from bee pollen. *Journal of Pharmaceutical and Biomedical Analysis, 147*, 110–124.

Atallah, A. A. (2016). The production of bio-yoghurt with probiotic bacteria, royal jelly and bee-collected pollen grains. *International Journal of Food Sciences and Nutrition., 6*, 510.

Atallah, A. A., & Morsy, K. M. (2017). Effect of incorporating royal jelly and bee-collected pollen grains on texture and microstructure profile of probiotic yoghurt. *Journal of Food Processing Technology, 8*, 693.

Barajas, J., Cortes-Rodriguez, M., & Rodriguez-Sandoval, E. (2012). Effect of temperature on the drying process of bee pollen from two zones of Colombia. *Journal of Food Process Engineering, 35*, 134–148.

Barene, I., Daberte, I., & Siksna, S. (2014). Investigation of bee bread and development of its dosage forms. *Theory and Practice in Medicine., 21*(1), 16–22.

Barreto, L. M. R. C., Funari, S. R. C., & Orsi, R. O. (2005). Pólen apícola: perfil da produção no Brasil. In *Annals of Congresso De Apicultura Del Mercosur 1* (p. 20). Sociedad de Apicultores Uruguaya.

Başdogan, G., Sağdıç, O., Daştan, T., Düz, G., & Acar, S. (2019). Farklı Bölgelerden Toplanan Arı Polenlerinin Fizikokimyasal Özelliklerive Şeker Profillerinin Belirlenmesi. *Avrupa Bilim ve Teknoloji Dergisi, 15*, 627–631.

Bell, R. R., Thornber, E. J., Seet, J. L. L., Groves, M. T., Ho, N. P., & Bell, D. T. (1983). Composition and protein quality of honey bee collected pollen of Eucalyptus marginata and Eucalyptus calophylla. *Journal of Nutrition, 113*(12), 2479–2484.

Bloch, G., Bar-Shai, N., Cytter, Y., & Green, R. (2017). Time is honey: Circadian clocks of bees and flowers and how their interactions may influence ecological communities. *Philosophical Transactions of the Royal Society B, 372*, 20160256.

Bogdanov, S. (2004). Quality and standards of pollen and beeswax. *Apiacta, 38*, 334–341.

Bonhevi, S. J., & Jorda, E. J. (1997). Nutrient composition and microbiological quality of honey-bee- collected pollen in Spain. *Journal of Agricultural and Food Chemistry, 45*(3), 725–732.

Brodschneider, R., & Crailsheim, K. (2010). Nutrition and health in honey bees. *Apidologie, 41*, 278–294.

Campos, M., Markham, K. R., Mitchell, K. A., & Pro enca da Cunha, A. (1997). An approach to the characterization of bee pollens via their flavonoid/ phenolic profiles. *Phytochemical Analysis, 8*, 181–185.

Campos, M. G., Webby, R. F., & Markham, K. R. (2002). The unique occurrence of the flavone aglycone tricetin in Myrtaceae pollen. *Zeitschrift fur Naturforschung. C, 57*(9–10), 943–946.

Campos, M. G. R., Bogdanov, S., Almeida-Muradian, L. B., Szczesna, T., Mancebo, Y., Frigerio, C., & Ferreira, F. (2008). Pollen composition and standardization of analytical methods. *Journal of Apicultural Research and Bee World., 47*(47), 156–163.

Campos, M. G., Olena, L., & Anjos, O. (2016). Chemical composition of bee pollen. In S. M. Cardoso et al. (Eds.), *Chemistry, biology and potential applications of honeybee plant-derived products* (pp. 67–88). Bentham eBooks.

Canale, A., Benelli, G., Castagna, A., Sgherri, C., Poli, P., Serra, A., Mele, M., Ranieri, A., Signorini, F., Bientinesi, M., & Nicolella, C. (2016). Microwave-assisted drying for the conservation of honeybee pollen. *Materials (Basel)., 9*(5), 363.

Celli, G., & Maccagnani, B. (2003). Honey bees as bioindicators of environmental pollution. *Bulletin of Insectology, 56*(1), 137–139.

Cobzaru, C., & Inglezakis, V. (2015). Ion exchange. In T. Steve (Ed.), *Progress in filtration and separation* (Vol. 10, pp. 425–498). Academic, is an imprint of Elsevier, USA.

Collin, S., Vanhavre, T., Odart, E., & Bouseta, A. (1995). Heat treatment of pollens: Impact on their volatile flavor constituents. *Journal of Agricultural and Food Chemistry, 43*, 444–448.

Conte, G., Benelli, G., Serra, A., Signorini, F., Bientinesi, M., Nicolella, C., Mele, M., & Canale, A. (2016). Lipid characterization of chestnut and willow honey bee collected pollen: Impact of freeze-drying and microwave-assisted drying. *Journal of Agricultural and Food Chemistry., 55*, 12–19.

Conti, M. E., & Botre, F. (2001). Honeybees and their products as potential bioindicators of heavymetals contamination. *Environmental Monitoring and Assessment., 69*, 267–282.

Conti, M. E., Stripeikis, J., Campanella, L., Cucina, D., & Tudino, M. B. (2007). Characterization of Italian honeys (Marche Region) on the basis of their mineral content and some typical quality parameters. *Chemistry Central Journal., 1*, 14–15.

Dalsgaard, H., & Abbotts, A. W. (2000). Improving energy efficiency. In B. Mattsson & U. Sonnesson (Eds.), *Environmentally-friendly food processing* (p. 33431). North America by CRC Press LLC.

de Melo, I. L. P., & de Almeida-Muradian, L. B. (2011). Comparison of methodologies for moisture determination on dried bee pollen samples. *Ciência e Tecnologia de Alimentos, 31*, 194–197.

De-Melo, A. A. M., & de Almeida-Muradian, L. B. (2017). Chemical composition of bee pollen. In J. M. Alvarez-Suarez (Ed.), *Bee products – chemical & biological properties, (*pp. 10–300). Springer press, USA.

De-Melo, A. A. M., Estevinho, M. L. M., Sattler, F. J. A. G., Souza, B. R., Freitas, A. S., Barth, O. M., & Almeida-Muradian, L. B. (2016). Effect of processing conditions on characteristics of dehydrated bee-pollen and correlation between quality parameters. *LWT - Food Science and Technology., 65*, 808–815.

Denisow, B., & Denisow-Pietrzyk, M. (2016). Biological and therapeutic properties of bee pollen: A review. *Journal of the Science of Food and Agriculture, 96*(13), 4303–4309.

Di Noi, A., Casini, S., Campani, T., Cai, G., & Caliani, I. (2021). Review on sublethal effects of environmental contaminants in honey bees (*Apis mellifera*), knowledge gaps and future perspectives. *International Journal of Environmental Research and Public Health, 18*, 1863.

Dias, T. R., Tomas, G., Teixeira, N. F., Alves, M. G., Oliveira, P. F., & Silva, B. M. (2013). White tea (*Camellia sinensis* (L.)): Antioxidant properties and beneficial health effects. *International Journal of Food Science, Nutrition and Dietetics, 2*(2), 19–26.
Dias, L. G., Tolentino, G., Pascoal, A., & Estevinho, L. M. (2016). Effect of processing conditions on the bioactive compounds and biological properties of bee pollen. *Journal of Apicultural Research, 55*(2), 357–365.
Dolezal, A. G., & Toth, A. L. (2018). Feed backs between nutrition and disease in honey bee health. *Current Opinion in Insect Science, 26*, 114–119.
Dominguez- Valhondo, D., Gil, D. B., Hernandez, M. T., & Gonzalez-Gomez, D. (2011). Influence of the commercial processing and floral origin on bioactive and nutritional properties of honeybee-collected pollen. *International Journal of Food Science and Technology, 46*, 2204–2211.
Feas, X., Vazquez-Tato, M. P., Estevinho, L., Seijas, J. A., & Iglesias, A. (2012). Organic bee pollen: Botanical origin, nutritional value, bioactive compounds, antioxidant activity, and microbiological quality. *Molecules, 17*(7), 77.
Franchi, G. G., Franchi, G., Corti, P., & Pompella, A. (1997). Microspectrophotometric evaluation of digestibility of pollen grains. *Plant Foods for Human Nutrition, 50*(2), 115–126.
Isik, A., Ozdemir, M., & Doymaz, I. (2019). Infrared drying of bee pollen: Effects and impacts on food components. *Czech Journal of Food Sciences, 37*(1), 69–74.
Johnson, R. M., Ellis, M. D., Mullin, C. A., & Frazier, M. (2010). Pesticides and honey bee toxicity. *Apidologie, 41*, 312–331.
Kalaycıoğlu, Z., Kaygusuz, H., Döker, S., Kolaylı, S., & Erim, F. B. (2017). Characterization of Turkish honeybee pollens by principal component analysis based on their individual organic acids, sugars, minerals, and antioxidant activities. *LWT - Food Science and Technology, 84*, 402–408.
Kanar, Y., & Mazı, B. G. (2019). HMF formation, diastase activity, and proline content changes in bee pollen dried by different drying methods. *LWT - Food Science and Technology, 113*, 108273.
Karabagias, I. K., Karabagias, V. K., Gatzias, I., & Riganakos, K. A. (2018). Bio-functional properties of bee pollen: The case of "bee pollen yoghurt". *Coatings, 8*, 423.
Karakaya, M., & Kilic, A. (1994). The effect of yogurt bacteria (Lactobacillus bulgaricus, Streptococcus thermophilus) on the fermentation of sausage. *Journal of Food, 19*(2), 97–101.
Kaškonienėa, V., Adaškevičiūtė, V., Kaškonas, P., Mickienė, R., & Maruška, A. (2020). Antimicrobial and antioxidant activities of natural and fermented bee polen. *Food Bioscience, 34*, 100532.
Kayacan, S., Sagdic, O., & Doymaz, I. (2018). Effects of hot-air and vacuum drying on drying kinetics, bioactive compounds, and color of bee pollen. *Journal of Food Measurement and Characterization, 12*, 1274–1283.
Kendall, A., Yuan, J., & Brodt, S. B. (2013). Carbon footprint and air emissions inventories for US honeyproduction: Case studies. *The International Journal of Life Cycle Assessment, 18*, 392–400.
Khider, M., Elbanna, K., Mahmoud, A., & Owayss, A. A. (2013). Egyptian honeybee pollen as antimicrobial, antioxidant agents, and dietary food supplements. *Food Science and Biotechnology, 22*, 1–9.
Kieliszek, M., Piwowarek, K., Kot, A. M., Blazejak, S., Chlebowska-Smigiel, A., & Wolska, I. (2018). Pollen and bee bread as new health-oriented products: A review. *Trends in Food Science and Technology, 71*, 170–180.
Kilic A. (2009). Low temperature and high velocity (LTHV) application in drying: Characteristics and effects on the fish quality. *Journal of Food Engineering, 91*(1) 173–182.
Kilic, A. (2017). Mathematical modeling of low temperature high velocity (LTHV) drying in Foods. *Journal of Food Process Engineering, 40*(2) e12378.
Kilic, A. (2020a). Low temperature and high velocity assisted fluidized bed drying characteristics of bee pollen as bioactive food. *Journal of Food Process Engineering, 43*(8), e13439.
Kilic, A. (2020b). The cooking sensitivity of ω-3 fatty acids in fish, Nutrition Research Virtual Conference, E-Abstract Book, Roma

Kilic, A. (2021). Novel cold process applications for the preservation of bioactive components of several natural functional foods. In Atta-Ur-Rahman (Ed.), *Studies in natural products chemistry* (Bioactive natural products) (Vol. 74). Elsevier. In press.

Kilic, A. (2022). Novel cold process applications for the preservation of bioactive components of several natural functional foods. In Atta-Ur-Rahman (Ed.), *Studies in natural products chemistry* (Bioactive natural products) (Vol. 74). Elsevier. In press.

Kilic, A., & Oztan, A. (2013). Effect of ascorbic acid utilization on cold-smoked fish quality (*Oncorhynchus mykiss*) during process and storage. *Food Science Technology Research, 19*, 823–831.

Kilic, A., Midilli, A., & Dincer, I. (2009). A strategic program to reduce greenhouse gas emissions from food industry. In İ. Dincer, A. Hepbasli, A. Midilli, & T. H. Karakoc (Eds.), *Global Warming* (pp. 197–209). Springer, Boston, MA.

Kilic, A., Midilli, A., & Dincer, I. (2010). A novel fish-drying technique for better environment, quality and sustainability. *International Journal of Global Warming, 2*, 262–278.

Kilic, A., Küçük, H., & Midilli, A. (2014). Environmental friendly food smoking technologies. In I. Dincer, A. Midilli, & H. Kucuk (Eds.), *Progress in sustainable energy technologies: Creating sustainable development* (pp. 557–576). Springer.

Klein, A. M., Vaissiere, B. E., Cane, J. H., Steffan-Dewenter, I., Cunningham, S. A., Kremen, C., & Tscharntke, T. (2007). Importance of pollinators in changing landscapes for world crops. *Proceedings of the Royal Society B: Biological Sciences, 274*, 303–313.

Komosinska-Vassev, K., Olczyk, P., Kafmierczak, J., Mencner, L., & Krystyna Olczyk, K. (2015). Bee pollen: Chemical composition and therapeutic application. *Evidence-based Complementary and Alternative Medicine, 297425*, 1–6.

Kösoğlu, M., İvgin Tunca, R., Topal, E., Yücel, B., Balkanska, R., & Tavlı Yıldırır, Z. (2021). Arıcılıkta Sürdürülebilirlik Mümkün Mü? *MAS Journal of Applied Sciences, 6*(3), 610–623.

Kostić, A. Ž., Milinčić, D. D., Barać, M. B., Ali Shariati, M., Tešić, Ž. L., & Pešić, M. B. (2020). The application of pollen as a functional food and feed ingredient- the present and perspectives. *Biomolecules, 10*, 84.

Kosuke, N., Li, Y., Jin, Z., Fukumuro, M., Ando, Y., & Akaishi, A. (2006). Low-temperature desiccant-based food drying system with airflow and temperature control. *Journal of Food Engineering., 75*, 71–77.

Li, Q.-Q., Wang, K., Marcucci, M. C., Sawaya, A. C. H. F., Hu, L., Xue, X. F., Wu, L. M., & Hu, F. L. (2018). Nutrient-rich bee pollen: A treasure trove of active natural metabolites. *Journal of Functional Foods., 49*, 472–484.

Mărgăoan, R., Strant, M., Varad, A., Erkan, T., Yüce, B., Cornea-Cipcigan, M., Campos, M. G., & Vodnar, D. C. (2019). Review; bee collected pollen and bee bread: Bioactive constituents and health benefits. *Antioxidants, 8*, 568.

Nainu, F., Masyita, A., Bahar, M. A., Raihan, M., Prova, S. R., Mitra, S., Emran, T. B., & Simal-Gandara, J. (2021). Pharmaceutical prospects of bee products: Special focus on anticancer, antibacterial, antiviral, and antiparasitic properties. *Antibiotics, 10*, 822.

Nisbet, C., Güler, A., Gül Yarım, F., Cenesiz, S., & Ardalı, Y. (2013). Relationship between environmental and flora change with mineral content of honey bee products. *Turkish Journal of Biochemistry, 38*(4), 494–498.

Pimentel, D., & Pimentel, M. (1996). Energy use in livestock production. In D. Pimentel & M. Pimentel (Eds.), *Food, energy and society* (pp. 77–84). University Press of Colorado.

Ranieri, A., Benelli, G., Castagna, A., Sgherri, C., Signorini, F., Bientinesi, M., Nicolella, C., & Canale, A. (2019). Freeze-drying duration influences the amino acid and rutin content in honeybee-collected chestnut pollen. *Saudi Journal of Biological Sciences, 26*, 252–255.

Rashed, M. N., El-haty, M. T. A., & Mohamed, S. M. (2009). Bee honey as environmental indicator for pollution with heavy metals. *Toxicological & Environmental Chemistry, 91*(3), 389–403.

Raveendran, K., Jayarathna, W. A. R., Amarasınghe, A. D. U. S., & Botheju, W. S. (2019). The effect of shrinkage on fluidized bed drying of orthodox broken type tea. *Chemical Industry and Chemical Engineering Quarterly, 25*(3), 299–307.

Rortais, A., Arnold, G., Dorne, J.-L., More, S. J., Sperandio, G., Streissl, F., Szentes, C., & Verdonck, F. (2017). Risk assessment of pesticides and other stressors in bees: Principles, data gaps and perspectives from the European Food Safety Authority. *Science of the Total Environment, 587–588*, 524–537.

Sadin, R., Chegini, G. R., & Sadin, H. (2014). The effect of temperature and slice thickness on drying kinetics tomato in the infrared dryer. *Heat and Mass Transfer, 50*, 501–507.

Silici, S. (2014). Bee pollen and bee bread. *Uludag Bee Journal, 14*(2), 99–105.

Silva, T. M. S., Camara, C. A., Silva Lins, A. C., Maria Barbosa-Filho, J., Silva, E. M. S., Freitas, B. M., & Santos, F. A. R. (2006). Chemical composition and free radical scavenging activity of pollen loads from stingless bee Melipona subnitida Ducke. *Journal of Food Composition and Analysis., 19*, 507–511.

Song, X.-D., Mujumdar, A. S., Law, C. L., Fang, X. M., Peng, W. J., Deng, L. Z., Wang, J., & Xiao, H. W. (2020). Effect of drying air temperature on drying kinetics, color, carotenoid content, antioxidant capacity, and oxidation of fat for lotus pollen. *Drying Technology, 38*(9), 1–15.

Staniskiene, B., Matusevicius, P., & Budreckiene, R. (2006). Honey as an indicator of environmental pollution. *Environmental Research, Engineering and Management, 2*(36), 53–58.

Szczêsna, T. (2006a). Protein content and amino acid composition of bee collected pollen from selected botanical origins. *Journal of Apicultural Science, 50*(2), 81–90.

Szczêsna, T. (2006b). Long-chain fatty acids composition of honeybee- collected pollen. *Journal of Apicultural Science, 50*(2), 65–79.

Thomann, M., Imbert, E., Devaux, C., & Cheptou, P. O. (2013). Flowering plants under global pollinator decline. *Trends in Plant Science, 18*, 353–359.

Tomas- Lorente, F., Garcia-Grav, M. M., Nieto, J. L., & Tomás-Barberán, F. A. (1992). Flavonoids from Cis tus ladanifer bee pollen. *Phytochemistry, 31*, 2027–2029.

Trahana, J., Graziani, A., Yogi Goswamia, D., Stefanakosa, E., Jotshia, C., & Goel, N. (2014). Evaluation of pressure drop and particle sphericity for an air-rock bed thermal energy storage system. *Energy Procedia, 57*, 633–642.

Ulusoy, E., & Kolayli, S. (2014). Phenolic composition and antioxidant properties of anzer bee pollen. *Journal of Food Biochemistry, 38*, 73–82.

Vásquez, A., & Olofsson, T. C. (2009). The lactic acid bacteria involved in the production of bee pollen and bee bread. *Journal of Apicultural Research and Bee World, 48*(3), 189–195.

Ziesemer, J. (2007). Energy use. In *Organic food systems* (pp. 3–24). Natural Resources Management and Environment Department Food and Agriculture Organization of the United Nations, FAO.

Zlatev, Z., Taneva, I., Baycheva, S., & Petev, M. A. (2018). Comparative analysis of physicochemical indicators and sensory characteristics of yogurt with added honey and bee-collected pollen. *Bulgarian Journal of Agricultural Science, 24*, 132–144.

Chapter 8
Microbiome Applications for Sustainable Food Systems

Monica Trif, Alexandru Vasile Rusu, M. Pilar Francino, Gabriel Delgado, and Jose Ángel Rufián-Henares

Abstract The present chapter, which combines theoretical and experimental knowledge, is divided into two parts. The first part provides critical insights to allow scientists to generate discoveries across microbiome applications for sustainable food systems. It offers a broad view of research of interest to early and experienced scientists, as well as an understanding of the role of microbiomes as vital ecosystems and inter-relations among microbiomes across food chains. In the second part of the chapter, the reuse of spent coffee grounds to increase the resilience of agro-food systems is described as an example of a successful application of a microbiome-related intervention.

Keywords Microbiome · Food chain · Vital ecosystem · Spent coffee grounds

---

M. Trif (✉)
Centre for Innovative Process Engineering GmbH, Food Research Department, Stuhr, Germany

A. V. Rusu (✉)
Biozoon Food Innovations GmbH, Strategic Research Department, Bremerhaven, Germany
e-mail: rusu@biozoon.de

M. P. Francino
Unitat Mixta d'Investigació en Genòmica i Salut, Fundació per al Foment de la Investigació Sanitària i Biomèdica de la Comunitat Valenciana (FISABIO-Salut Pública)/Institut de Biologia Integrativa de Sistemes (Universitat de València), València, Spain

CIBER en Epidemiología y Salud Pública, Madrid, Spain

G. Delgado
Departamento de Edafología y Química Agrícola, Universidad de Granada, Granada, Spain

J. Á. Rufián-Henares (✉)
Departamento de Nutrición y Bromatología, Instituto de Nutrición y Tecnología de los Alimentos, Centro de Investigación Biomédica, Universidad de Granada, Granada, Spain

Instituto de Investigación Biosanitaria ibs.GRANADA, Universidad de Granada, Granada, Spain
e-mail: jarufian@ugr.es

C. M. Galanakis (ed.), *Biodiversity, Functional Ecosystems and Sustainable Food Production*, https://doi.org/10.1007/978-3-031-07434-9_8

1 Introduction

The global human population is expected to increase from 7.8 billion in 2020 to 8.6 billion by mid-2030, 9.8 billion by mid-2050, and 11.2 billion by 2100, according to World Population Prospects 2020. Therefore, food supply and equal access to food will be two of the significant issues and challenges in the near future (Godfray & Garnett, 2014; Funabashi, 2018; Calicioglu et al., 2019). The environmental degradation is a major problem that threatens future food production growth (Sundström et al., 2014). Integrated landscape management for the delivery of ecosystem services is a priority research area that needs to be well linked to the other priority research areas discussed in this chapter: understanding the factors and effects of land use change, improving water management and water use efficiency at multiple levels, and improving soil health, nutrient management and use (FAO, 2015; Hatfield & Dold, 2019). Multidisciplinary research, in turn, is essential to test policies and technologies that have been proposed to improve agricultural production and livelihoods, to create a knowledge base for global planning, and to foster cross-border collaboration (FAO, 2010; Liu & Li, 2017).

The promotion of better use of biodiversity in sustainable agriculture and sustainable food systems requires the integration of research activities on the topics of food, nutrition, health and diseases with research on genetic resources, control, guidance and institutions, and on socio-ecological interactions (Zimmerer & de Haan, 2017). The initiative innovation-driven of companies globally is to work to protect the health and safety of people and the environment and to ensure the supply of safe and affordable food (European Commission, 2020). Therefore, there is an increasing amount of discussion about how plant protection products affect human health and the environment (Frische et al., 2018; Official Journal of the European Union (OJEU), 2008; NAP, 2014). Farms often face numerous challenges. These include extreme weather conditions, weed infestation, insect pests and plant diseases. Pesticides are an important tool for farmers to meet these challenges and produce enough food on the existing arable land (Popp et al., 2013). In this way, the surrounding areas are less polluted and biodiversity is promoted in these areas. In keeping with the vision of "health for all, hunger for none", by merging all forces can be guarantee a resilient food system and a safe promotion, and a sustainable agriculture (Brodt et al., 2011). Globally authorities are responsible for the approval of plant protection products having a common goal: to protect human health and the environment with the help of risk assessments and permits (Dietrich et al., 2016).

In general a commitment to sustainable agriculture and food systems is based on the obligation to achieve a balance between production and protection (FAO, 2014; Caron et al., 2018). But how do be feed a growing world population without overloading the planet? By working together with authorities, farmers, scientists, partners and other representatives of civil society to make agriculture part of the solution when it comes to climate change would be a rational initiative (GFFA, 2019).

Microbes represent unexploited tools to increase the competitiveness and sustainability of food systems, including productivity, quality, and safety (BMBF and

BMEL, 2015). Currently, the focus is on identifying innovative solutions derived from microbiomes (Zdolec et al., 2018; Lindgren et al., 2018).

A microbiome is the entirety of microorganisms that inhabit a specific habitat – including bacteria, archaea, fungi, unicellular organisms, and viruses (Boon et al., 2014). The terms microbiome and microbiota are often used synonymously. However, the microbiome includes the entire genetic information of the microorganisms in a living space (Checinska Sielaff et al., 2019).

Research focus on intestinal microbiomes (totality of all microorganisms colonizing the intestine) as well as environmental microbiomes (totality of all environmentally relevant habitats colonizing microorganisms). The study of the microbiome is an increasingly important research area that provides connections between the health of the environment, animals, and humans (Foo et al., 2017; Trinh et al., 2018). Research has focussed on the influence of environmental changes on food chain systems – including those from plants to ruminants (Simon et al., 2019). Microbial relationships at the various interfaces of the food chain may modify the exposures to heavy metals, pesticides, and secondary plant substances, affecting the health of the different links of the food chain and the whole system (Singh et al., 2017). Research on the cascade of health effects related to the food chain is a new approach in the health research field. Researchers recognize that it is essential to understand better the impacts of environmental chemicals on the health of food chains (Thompson & Darwish, 2019).

In addition to harmful pathogens, there are also many beneficial microbes, present in and around plants, especially in the soil near the roots (Olanrewaju et al., 2019). Microorganisms are invisible, but should not be underestimated because they can affect plant health, soil quality, and essential ecosystem functions (Mueller & Sachs, 2015; Nanjundappa et al., 2019).

The importance of microbes to us and our environment is still insufficiently understood (de Vries et al., 2018). Understanding the role of the microbiome is becoming increasingly important in plant research (Compant et al., 2019). Studies in recent years show that microbes play a far more significant role for plants than they have long been thought to do. They can influence the genetic structure of plants, their health, and also affect the interaction with other plants (Turner et al., 2013).

Today the focus is on holistic approaches to include microorganisms in current studies (Kumar et al., 2019). Both the plants and the microbes emit signaling molecules, creating a chemical communication level (Shagas et al., 2018). With the help of microorganisms, nutrients are made available, which can then be absorbed by the plants. The interaction among plant-microbiome is dynamic and can be conditioned by both sides and influenced to their advantage (Orozco-Mosqueda et al., 2018).

Probably the best-known example of a nutrient exchange is the mutually beneficial (mutualistic) symbiosis of nodule bacteria (rhizobia) and plants such as legumes (Zgadzaj et al., 2016). The soil bacteria are attracted by the released root substances and penetrate the root hairs of the plants, which after infection, develop the characteristic root nodules to which the bacteria owe their name (McNear Jr., 2013).

Nodule bacteria can bind elemental nitrogen from the air and make it available to the plant (Wolińska et al., 2017).

However, the bacteria can only fix nitrogen under low-oxygen conditions (Wagner, 2011), since the enzyme required for this is inactivated even at low oxygen concentrations. The plant creates the concentration of oxygen needed in the root nodules and also provides the bacteria with the necessary nutrients (Ma et al., 2020). The plant receives nitrogen from the bacteria, which it cannot take from the air itself (Morgan & Connolly, 2013). This is an excellent benefit for legumes, as it allows them to grow on low-nitrogen soil and thus have an important location advantage (Mus et al., 2016).

In addition to nitrogen, which is essential for growth (Mahmud et al., 2020), bacteria can supply the plant with other nutrients and even form phytohormones or precursors (Kudoyarova et al., 2019). Plants can actively influence the microbiome of the rhizosphere around their roots. Even in the same soil, species-specific microbial communities settle around the different plant species that grow in it. The plant defense strategies also influence the rhizosphere (Pérez-Jaramillo et al., 2016; Lu et al., 2018; de Faria et al., 2020).

Microbiome-based agricultural product are one of the fastest-growing sectors in agronomy (Busby et al., 2017; Sessitsch et al., 2018) with a Compound Annual Growth Rate (CAGR) of 15–18% and a predicted value of over 10 billion U.S. dollars by 2025 (Dunham Trimmer L.L.C., 2017).

2 Microbiomes as Vital Ecosystems

Microorganisms make up 70% of the biomass on our planet (Cavicchioli et al., 2019). Archaea, bacteria, yeasts, and other eukaryotic unicellular organisms are the oldest and most diversified forms of life. Microbes have existed on Earth for more than three billion years. Due to their short generations, they are masters of adaptation (Cavalier-Smith et al., 2006). They have enormously diverse metabolic functions and have conquered practically every living space in the course of evolution, whether in the oceans, on the Earth's crust, or in the air (San Roman & Wagner, 2018).

Microbes made the Earth habitable for other living beings in the first place (Mann, 2018). In addition, they colonize every multicellular living being. Every multicellular organism has developed its mix of microbial tenants during its evolution (Rivera-Yoshida et al., 2018). In humans, all microbes found in the gut represent the gut microbiome (Thursby & Juge, 2017). These microorganisms take on essential functions in digestion and defense against disease (Belkaid & Hand, 2014; Wu & Wu, 2012). In plants, such a vital microbiome exists in the soil: microorganisms, especially bacteria that occur in the ground near the roots, ensure that the plant remains healthy (Jacoby et al., 2017; Deng et al., 2019; Hassani et al., 2018). They represent a kind of external immune system.

2.1 A New Look at Complex Communities

It is becoming increasingly clear what a vital role microbial ecosystems play for their hosts (Laforest-Lapointe & Arrieta, 2018). The developmental biologist Thomas Bosch from Kiel University speaks of an evolutionary partnership: "The fascinating image has prevailed that organisms are always multi-organismic. They form a complex community called a metaorganism." (Deines & Bosch, 2016; Bang et al., 2018).

There has been another change of perspective in microbiology: for a long time, the primary interest was in pathogenic microbes (Ayres, 2016). However, these only make up a fraction of the microbial diversity. Most bacteria, fungi, and protozoa, on the other hand, are benign and live in symbiosis with their hosts (D'Abramo & Neumeyer, 2020; Sachidanand et al., 2019).

In humans, microbes help digestion, shape the immune system, and take on other protective functions (Rusu et al., 2020a). In the case of cows, they ensure that the ruminants can break down grass into its components and use them (Mamuad et al., 2019; Matthews et al., 2019; Cui et al., 2019).

In the soil, the microcosm is responsible, among other things, for the decomposition of dead biomass, for the digestion of nutrients, and the fixation of nitrogen (Frąc et al., 2018). In addition to the interaction between microbes and host organisms, there are close relationships between the often highly specialized species of a microbiome – cooperative and competing – based on the need for nutrients on the one hand and the metabolic products made from them on the other (Longford et al., 2019).

Microbiomes are by no means to be viewed in isolation (Timm et al., 2020). In the agricultural context in particular, there is a constant mixing of animal, plant and soil-borne microbiomes. Parts of the plant microbiome get into the body of ruminants when they are eaten; conversely, part of the animal microbiome is carried into the soil through the excrement (van Veelen et al., 2020; Ottman et al., 2019). It is therefore important to clarify to what extent cross-habitat management strategies are necessary and to be developed (Moss et al., 2020; Prakash et al., 2020).

There is a worldwide unique opportunity to investigate the role of microbiomes in relation to different production systems (animal and vegetable, and human). The rapid methodological developments, especially in the field of "omics" technologies, also offer the opportunity for the first time not only to describe microbial communities and to establish correlations to abiotic and biotic factors, but also to understand mechanisms that can be used for sustainable agricultural and food production (Berg et al., 2020; Crandall et al., 2020). Due to the close contact between fundamental application-related research, possible new processes can be quickly transferred into practice via basic field tests (including organic farming) to operational application. Economic assessments can also be carried out (e.g. production and resource economics of agricultural holdings) (Guth & Smędzik-Ambroży, 2020). In particular, there would be strong synergies with the chairs for microbiology, microbial ecology and biofunctionality of food.

2.2 Microbiomes in Plants and Soil

Billions of microorganisms live in soil, on the roots of plants, and inside plants and on their surface. They form complex communities in which individual fungal and bacterial species compete and perform various functions (Souza et al., 2015).

Today it is known that the microbiome interacts with plants in a variety of ways (Rodriguez et al., 2019). Many microorganisms contribute to the nutrient supply. Well-known examples are the root nodule bacteria mentioned above, which enter into a symbiosis with legumes such as beans, peas, or lupine and supply the plant with nitrogen – in exchange for sugar, an interplay with mutual benefit (Peoples et al., 2009; Liu et al., 2018, 2011).

How the microbiome is composed depends, among other things, on the interaction between the plant and the microorganisms (Nelson, 2018). Plants produce a large number of chemical signaling substances, which attract certain selected microorganisms. Such signaling substances also play an essential role in the defense against pathogens or predators (Guerrieri et al., 2019).

The soil microbiome in the roots may play the most significant role among the various plant-associated microbiomes. It is estimated that there are up to 50,000 different species in one gram of soil (Raynaud & Nunan, 2014). While the aboveground microbes play a role primarily in the interaction with plant pathogens, the underground microbes have an influence on nutrient uptake and nutrient cycles in the soil. The soil microbiome is also an essential store of CO_2 (Schloter et al., 2018).

Around 20,000 different bacterial strains colonize the model plant thale cress (*Arabidopsis thaliana*) (Gan et al., 2011). Scientists have been able to isolate around half of them and analyze 432 in more detail. It turns out that the microbial communities on leaves and roots are 50% identical. A majority of the microorganisms should, therefore, have their origin in the soil microbiome (Hassani et al., 2018; Compant et al., 2019).

Research into the plant microbiome has grown in importance in recent years (Berg et al., 2016). Scientists today are focusing on studying the microbiomes of various crops for sustainable agriculture and food production (Blum et al., 2019), aiming of determine their functions and the interactions that exist between the microbiome, the plant, and the environment. The main goal is to influence and optimize the microbiome of essential crops in such a way that the use of synthetic fertilizers and pesticides will be reduced or completely superfluous in the future (Busby et al., 2017). If these adapted microbiomes are spread on the field, the use of pesticides can possibly be dispensed with. The function of the microbiome concerning the supply of nutrients in the soil, the productivity of the plant, and tolerance to abiotic stress are also being examined (Hunter, 2016). It is hoped that the application of optimized microbiomes will increase productivity and reduce the use of mineral fertilizers (Qiu et al., 2019; Trivedi et al., 2017; Arif et al., 2020). The overall vision is that, along with the seeds of the plants best suited to the respective soils and cultivation methods, farmers will also receive those soil microorganisms that optimally promote growth. About the composition of the microbiome, researchers can already state that no solution is optimal for every plant and all environmental conditions (Velásquez et al., 2018).

Plant researchers have already tried to understand many individual aspects of the plant-microbiome association. The Alpine rockcress, for example, selects which fungi colonize its roots and thus promotes the symbiosis with a specific fungus. In this way, it is ensured that the plant can grow well, even in soils that are poor in phosphorus, by supporting the absorption of this element. It is conceivable that this fungus could also affect other plants if it succeeds in securing a place in their microbiome (van der Heijden et al., 2016).

The plant microbiome also has an effect beyond cultivation. Wine connoisseurs have long spoken of the "terroir" when they mean the influence of the soil on the taste of a wine. In fact, not only mineral factors but also the composition of the soil microbiome play a central role, as has meanwhile been shown (Compant et al., 2019; Cordovez et al., 2019). For instance, climate change alters wine aroma in this way.

It is also known that monocultures in agriculture have a substantial impact on the soil microbiome. In nature, there is a closed nutrient cycle between plants and soil. In intensively used soils, it is not only the harvest that removes nutrients from the cycle. These are also washed out or outgassed. As a result, the diversity of the microbiome is reduced, which usually benefits the microorganisms that are harmful to the plant (O'Banion et al., 2020).

The reduction of biodiversity is problematic because plants assemble their "own" microbiome by secreting certain metabolic products in the root space (Pascale et al., 2020). In this way, they promote the growth of certain groups of bacteria, which are conducive to plant development. In fact, this coexistence goes so far that plants take up certain soil bacteria in their seeds to lay the basis for a favorable microbiome for their offspring (Flandroy et al., 2018).

Microbiomes pervade all living agrarian domains. Agriculture without microbiomes is unimaginable. Research into the functionality of microbial metagenomes across different agricultural production systems and scale levels holds great innovative potential. The soil microbiome not only contributes to sustainable plant growth, but also controls other important ecosystem services, such as the carbon storage function of soils or the potential of soils to break down or fix pollutants and thus protect groundwater. In addition, soil microorganisms and their activity are also significantly involved in the formation of climate-relevant trace gases in the soil. The importance of microbial polysaccharides for the formation of structures in the soil and thus for the stability of soils and the resilience to erosion has recently become increasingly clear.

The microbiome of livestock has a significant impact on animal health and thus also the yield and quality of animal foods such as meat, milk, eggs, etc. Especially with regard to the discussion about the use of drugs in animal husbandry (e.g. antibiotics), the current one has opened up The role of the intestinal microbiome in protecting against infectious diseases is a highly topical field of research. However, the microbiome of farm animals is also of particular importance as a substance converter in the digestive tract. All farm animals use the capabilities of the intestinal microbiome to break down indigestible components of the feed into absorbable nutrients. Only with the help of the ruminal microbiome are ruminants, for example, able to digest high-fiber feed to a large extent and thus transform plant biomass into high-quality animal products without food competition with humans.

2.3 Interactions among Microorganisms in the Soil and Plant Microbiomes

Soil is a living, natural raw material that determines the fertility and sustainability of the earth through the activity of a great diversity of microorganisms. A lot of basic research is still necessary to examine individual microorganisms for their effect on the plant and to understand the genetic basis. In addition, it is also required to understand the interplay between the many different microorganisms and the interplay with the plant. The viability of the soil microorganisms is highly dependent on the available energy and nutrient sources. Among the nutrients, one of the most important is nitrogen, which is mainly brought into the soil by plants. The vegetable nitrogen is mainly contained in proteins, which make up 2–5% of the dry weight, since the concentration of freely available amino acids is normally 100 times lower.

Recently, scientists at the Max Planck Institute for Plant Breeding Research in Cologne have found that a particular fungus (*Colletotrichum tofieldiae*) can help the plant to absorb phosphate. Phosphorus is an essential nutrient for the plant, and better absorption of the phosphorus present in the soil could reduce the use of fertilizers. The next step will now be to check whether the increased phosphate uptake also works in the field when many other microorganisms come into play. The effect may then be canceled or reversed, as many different species of this fungus are known to be plant pathogens.

The thale cress (*Arabidopsis thaliana*) ensures a balance between certain bacteria and fungi in its root area to optimize its growth. In the roots themselves, however, bacteria dominate and displace fungi and false fungi. If certain bacteria are missing, the plant cannot defend itself sufficiently against certain fungi and dies. In addition to displacement, the so-called induced systemic resistance also has another effect: certain microorganisms trigger a plant reaction through their presence, which is used to defend against other microorganisms. With an increasing understanding of how the microbiomes in the soil and plants are composed and interact, research is setting out to influence these microbiomes to optimize agricultural yields in a specific manner.

But how easily humans can destroy the functions of the microbiome is shown by the example of tomatoes. There, usually beneficial bacteria on the leaves keep the harmful ones at bay by successfully competing for nutrients. However, if the plant is heavily fertilized and there is an excess of nutrients, this mechanism fails, and the plant becomes ill more often. This phenomenon also affects the shelf life of potatoes and the weight loss of papayas after harvesting.

2.4 *Microbiomes and Microbiota, Inter-Relations Between microbiomes across Food Chains*

Food chains represent the flow of energy and matter. Simply, they are the relationships between living beings based on the principle of eating and being eaten. At the beginning of every food chain, there are mostly plants and some bacteria. They build complex organic compounds from the inorganic substances found in water, soil, and air, using the energy of sunlight. This process is called photosynthesis.

The primary consumers are the herbivores, and the secondary consumers, the carnivores. The food chain is closed by decomposing organisms, the destructors. These are bacteria, fungi, and many soil-living organisms that break down the dead organic substances and ultimately produce the inorganic substances that the plants need to rebuild the organic compounds. In species-rich communities, numerous food chains are linked to form food webs.

In addition to energy and nutrients, numerous pollutants are passed on in food chains from green plants to consumers in the ecosystem concerned. In this process, depending on the substance properties, the contaminants in the course of the food chain can be gradually enriched at each stage so that the end consumer ingests quantities of the toxic and is chronically or fatally damaged. Permanent poisons and heavy metals in the fatty tissue of living beings and their organs reach exceptionally high rates of accumulation because they are not biodegradable or are challenging to break down. Humans are at the end of the food chain and are, therefore, most at risk (Oltmanns et al., 2020).

Plant and soil microbiota has an important role for sustainable intensification of agricultural and food production (Cassán & Díaz-Zorita, 2016).

3 Environmental Microbiomes as a Resource for Biotechnology

Microbiomes are valuable for their genetic and metabolomic information alone. This gives researchers access to the blueprints of genes and proteins of organisms that they cannot successfully cultivate in the laboratory. All they have to do is analyze the complete genetic information – the metagenome – in water, soil, or a tissue sample (Lawson et al., 2019). A few years ago, this would have been technically impossible due to the small number of individual genomes within the samples. Modern high-throughput sequencing and other omics technologies have now created the basis for this approach.

3.1 *Importance of the Microbiome for Humans*

Bacteria, viruses, and fungi – billions of microorganisms live in symbiosis in and on people. Each person carries an estimated 100 trillion microorganisms that colonize almost all internal and external surfaces of the body. As many microbes as human body cells live with us on the skin, in the mouth, in body cavities – most of them in the gastrointestinal tract. The various microbiomes in the human body are complex and highly specialized ecosystems that influence the immune system, counteract toxins and pathogens and support metabolic processes. After all, around 30% of the metabolic products in the blood are of microbial origin. It is now clear: the microbiome is of inestimable value for our health (Kemp et al., 2020).

By far, the most densely populated habitat of the human body is the intestine, where microorganisms belonging to hundreds of different species account for up to 1.5 kilograms of biomass (Berg et al., 2020). Microorganisms are endowed with multiple different genomes, unlike human body cells. While human cells contain around 23,000 genes, microorganisms communicate with each other and with human cells in various ways, using 3.3 million genes. The sum of microbial genes gives the intestinal bacteria an immense metabolic capacity, and many scientists now regard the intestinal microbiome as an independent organ, the health of which goes far beyond supporting digestion.

The human microbiome develops in the first 3–4 years of life. An important infuencing factor is the birth process in which the mostly germ-free embryo comes into contact with microorganisms for the first time. Every human has a very individual microbiome. It's like a fingerprint. While humans only differ from one another by 0.1 percent in terms of genetic make-up, our microbiome is 60–70% different. The microbiome generally remains stable until old age, although a permanent change in lifestyle can strongly modify it. This happens, for example, in the case of a change in diet, but also regular consumption of nicotine, medication, or even sweeteners can impact the composition of the microbiome.

Throughout human evolution, key innovations, such as the introduction of agriculture, the industrial revolution, the discovery of antibiotics, or the advent of heavily processed foods have had a profound impact on human health and disease. The major lifestyle changes that occurred during the last century had beneficial effects in controlling infectious diseases, but they likely also had a leading role in the increased prevalence of atopic, inflammatory and autoimmune diseases. Current understanding of these negative effects proposes a significant role for the microbiome alterations that have been associated with significant dietary and lifestyle changes (Rusu et al., 2020b). The "hygiene hypothesis" (Strachan, 1989) stated that lack of exposure to microbes during early life was responsible for the increased prevalence of allergy and asthma in westernized populations (Forstner & Rusu, 2015; Trif et al., 2019). Today, this hypothesis has been reformulated as the "microbiome hypothesis" or "old friends hypothesis", stating that alterations in microbiome development are responsible for numerous diseases related to immune and/or metabolic imbalance (Noverr & Huffnagle, 2005; Rook & Brunet, 2005). Indeed, there is growing evidence that early microbiome-host interactions are critical in determining lifelong health or disease (Wold, 1998; Borre et al., 2014; Mulligan & Friedman, 2017; Vallès & Francino, 2018). Notably, given the close feedback between the microbiota and immune system development, an inadequate microbial colonization can produce imbalances among the different types of T cells. In turn, the immune deregulation resulting from such imbalances can provoke pathological outcomes such as atopy or autoimmune disease (Bennett et al., 2018).

In fact, there is probably almost no human disease that is not related to the microbiome, even if causality has not yet been proven. But for some conditions, we already know that a changed microbiome can trigger them. Scientists are investigating this with the help of gnotobiotic – i.e., initially germ-free, then specifically colonized – mice. By giving individual microbes or combinations thereof, they determine to what extent a disease develops, how it changes, whether it is more or less severe. There are already adequate studies for causal relationships with the microbiome in tumor diseases, obesity, depression, asthma, or conditions of the cardiovascular and immune systems. This knowledge is currently being channeled towards a variety of innovative medical therapies that target the microbiome in order to promote health and treat or decrease disease risk.

Unfortunately, microbiome diversity is being lost due to the restricted exposure to microbes in westernized societies. Recently, the microbiome of the feces, mouth, and skin of the Yanomami Amerindians, who live in isolation from Western society, has been analyzed (Clemente et al., 2015). The microbiome that differed the most between the Yanomami and the U.S. residents examined was that of the skin, which showed a significantly higher diversity in the former. While the skin microbiome in Western societies is dominated by staphylococci, no single group of bacteria dominated the Yanomami skin microbiome. Even bacteria previously assigned to soil bacteria were enriched in their skin microbiome. This seems easy to explain, as they live in closer contact with nature and wear less clothing than people in Western societies who spend most of their lives indoors, washing away the dirt after having been outside (Singh & Trivedi, 2017). Fecal microbiota diversity was also substantially increased in the Yanomami and it actually was higher than in other analyzed populations in South America and Africa that are currently undergoing a process of transculturation. In comparison to U.S. subjects, the Yanomami fecal microbiota showed higher and lower relative abundances of *Prevotella* and *Bacteroides*, respectively, similarly to what has been observed in African hunter-gatherers (Ponter et al., 2012).

Diet is undoubtedly one of the main factors shaping the gut microbiota, as has been shown by multiple studies comparing populations with widely different dietary habits (De Filippo et al., 2010; Shankar et al., 2017), including those living in the same society, such as vegans and omnivores in an urban U.S. environment (Wu et al., 2016). The typical high-fat, high-sugar Western diet is obesogenic and drives the gut microbiota towards decreased diversity. In contrast, high-fibre diets are associated with a healthy, well-structured microbiota characterized by a slow release of nutritional energy, protecting against inflammation and non-infectious colonic diseases (De Filippo et al., 2010; Meybeck & Gitz, 2017). Moreover, not only food composition but also cooking methods play an important role in the modulation of the gut microbiota, due to chemical changes in foodstuffs during the cooking process (Pérez-Burillo et al., 2018).

3.2 Capturing the Microbial World with New Technologies

It is estimated that around a trillion species of microbes live in the world. A fraction of it has been described so far. One of the reasons for this is that it is often impossible to cultivate individual species in the laboratory.

Technological advances in molecular biology and bioinformatics, however, have made a previously largely invisible world visible. The ability to examine and analyze microbiomes precisely has only become possible thanks to new technologies. In the past, it was only possible to detect those microorganisms that could be cultivated on nutrient media. Still, today the entire microbiome can be recorded with the help of D.N.A. sequencers that work at high throughput and the latest microscopic methods with which individual species can be made visible. With the use of sequencing techniques and other omics technologies, genetic information can be sequenced and compared with enormous databases for analysis.

In a soil sample or a stool sample, the genetic fingerprint of countless microorganisms can be determined. The analysis of the genetic information allows an inventory of the species of an individual microbiome. The vast genetic diversity is also a treasure trove of enzymes and other metabolic products that biotechnologists want to unlock.

The investigation of the key organisms that control the nutrient cycles (carbon and nitrogen) is possible by investigating the functional and structural composition of the environmental microbiomes (totality of all environmentally relevant habitats colonizing microorganisms) community. Soil metaproteomics enables the description of the microbial composition. Plant production in the twenty-first century should ensure a high quality food supply. Need less fertilizer and chemical-synthetic pesticides and reduce the discharge of harmful substances and greenhouse gases (EPA, 2010). Modern crop production must also take into account the species loss it causes. And pursue the goal of high biodiversity.

In addition, agrochemical residues present in food may also affect the gut microbiota (Mussatto et al., 2011; Claus et al., 2016). The engineering of environmental microbiomes may soon replace toxic agrochemicals (Wu et al., 2019; Hutchins et al., 2019; Sessitsch et al., 2018).

But how can be reduce the use of pesticides to a minimum? A reduction in the use of pesticides can be achieved by redesigning cultivation systems, using ecological principles and using digitalization and new technologies.

How can we better understand the plant health promoting effects of the plant and soil microbiome and use them systematically? The dynamics examination of the plant and soil microbiome with regard to its beneficial properties in useful plants and apply this knowledge to crop rotation, fertilization, soil cultivation and breeding can be done.

How can we preserve soils and their functions in agriculture and the landscape in the long term? The key solution is to investigate the effect of soil cultivation to improve the soil structure and structure as well as the stabilization of the organic soil substance and develop specific, site-specific cultivation strategies.

Overall, the process of identifying the scientific possibilities for sustainable and healthy nutrition and food systems needs to be more ambitious. Living within planetary boundaries (including those for nutrients, water and climate) and a healthy population require new approaches to food systems. It is necessary to develop research, teaching and innovation, and to mobilize these resources sufficiently to work with policy makers and other stakeholders. Besides, there is a need to promote nutrition-conscious agriculture to provide healthy and sustainable nutrition with associated problems in resource use and food prices.

Special attention has been given to the potential role of food waste products for ameliorating the microbiome of agricultural soils. Besides, various pesticides and additives have been shown to be harmful to an intact human microbiome (Jin et al., 2017).

An optimization of the microbiome in such a way that fertilizers and pesticides are superfluous in the field is under current development. It is still not known very much about the complex interplay between microorganisms and plants. In humans, the microorganisms make up the gut microbiome, take on important functions in digestion and defense against disease. In plants, for example, one finds such a vital microbiome in the soil: microorganisms, especially bacteria that occur in the soil near the roots, ensure that the plant remains healthy. They represent a kind of external immune system.

Scientists today are working to study the microbiomes and their composition of various crops. The aim is to find out which functions the microorganisms take on and which interactions exist between the microbiome, the plant and the environment. In the future, one direction would be to develop very specific microbiomes for individual crops such as wheat, rice or potatoes, which specifically protect the respective plants from pests and diseases. If these adapted microbiomes are applied to the field, the use of pesticides can possibly be dispensed with.

The microbiome interacts with plants in a variety of ways (Schirawski & Perlin, 2018). Among the myriad of microorganisms there are those that are harmful to the plant, but also many that protect the plant from diseases and pests or help it cope with stress. Others are important to the growth of the plant. Many microorganisms contribute to the nutrient supply.

The function of the microbiome with regard to the supply of nutrients in the soil, the productivity of the plant and the tolerance to abiotic stress are also examined. It is hoped that the application of optimized microbiomes will, for example, increase productivity and reduce the use of mineral fertilizers (Singh & Trivedi, 2017).

Microbial research has evolved that it is no longer just the breakdown of a compound by a single microorganism that is examined, but rather the breakdown of several compounds or mixtures of substances by microbial communities. In fact, these communities are the natural organizational form of microbial life, which, due to their high diversity, are able to cope with a wide range of extreme ecological habitats.

The amino acids can be taken up directly via several membrane transport systems, while proteins must first be depolymerized by extracellular proteases before uptake, which is the rate-limiting step in the breakdown of organic matter from plants. The resulting, simple, organic nitrogen compounds can be readily used by microorganisms as a source of energy and nutrients. The cycle of inorganic nitrogen, such as nitrification and denitrification, has been researched and described in more detail, whereas the cycle of organic nitrogen in soil is not yet fully understood. In contrast, the degradation and assimilation of organic carbon by soil microorganisms has already been demonstrated several times

For instance, the main chemical compounds present in spent coffee ground extracts are considered a source of nutrients for communities of microorganisms. Various studies have explored the use of spent coffee grounds to serve as a substrate for multiple species of microorganisms, and they have been proven to promote a healthier microbial community (Massaya et al., 2019; Pérez-Burillo et al., 2019; Vítězová et al., 2019).

4 Spent Coffee Grounds Composition, Global Production, and Potential Environmental Problems

Composted or fresh S.C.G. could be a used as an organic amendment (Cruz et al., 2014) in many many agricultural soils with a very low carbon content, decreasing then their susceptibility to erosion (ICO, 2020). In this sense, it has been proposed (COP21, November 30 to December 11, 2015) to increase the levels of organic carbon in soils (S.O.C.) 0.4% per year. The amendments with S.C.G. suppose an addition of organic carbon and other compounds (such as different mineral elements, phenolic compounds or melanoidins), which will impact on soil properties (nutrients, structure, and microbiota), and in turn will affect plant growth and its nutritional value. These points will be discussed in the subsequent sections.

4.1 Effect of Spent Coffee Grounds on Soil Organic Matter, Hydrophobicity, and Structure

The quality of agricultural soils is steadily being deteriorated, mainly in the Mediterranean basin (Rodríguez-Entrena et al., 2014). In order to increase chemical, physical and biological properties of the organic matter found in solis, it is imperative a deep study on how to use biowastes (like S.C.G.) as organic amendments. For example, some authors (Aranda et al., 2016) described a adverse effects like alterations in soil-water dynamics or a decrease in physical activity.

As stated above, a low organic matter content is a severe problem for soil structure formation and stabilization (Bronick & Lal, 2005). Thus, the reduction of soil organic matter content involves soil physical degradation, which leads to a drop in water and air infiltration or an increase in surface water run-off and soil compaction (Lal, 2015). Physical soil properties as bulk density, plant-available water, aggregate size distribution, and stability, as well as organic matter content, are commonly used to evaluate soil physical quality (Dexter, 2004; Reynolds et al., 2008). In this sense, the effect of S.C.G. on soil physical properties is not widely studied, since the majority of reports are only focused on the chemical and physicochemical properties of S.C.G. (Cruz et al., 2012, 2014, 2015a, b; Yamane et al., 2014; Hardgrove & Livesley, 2016; Cervera-Mata et al., 2017). However, Hardgrove and Livesley (2016) reported that the addition of S.C.G. increased water retention in the soil, and Cervera-Mata et al. (2017) confirmed this, also noting a decrease in bulk density.

Organic matter composition and quality is influenced by organic amendments. For example, municipal solid waste compost, sewage sludges and animal manures have humic-like properties (Senesi & Plaza, 2007). Other authors (Orlova et al., 2019) described the modification of organic matter in soil by addition of biochar. However, it is almost unknown the influence of S.C.G. on organic matter. The adition of S.C.G. to poultry manure or olive mill wastewater sludge made them highly compostable (Hachicha et al., 2012). Humic substances also increased by the addition of cow dung and S.C.G. (Zhang & Sun, 2017) since lignocellulose degradation enzimes increased the concentrations of low molecular weight compounds (more stable compost obtained). Very recently, Comino et al. (2020) demonstrated in microcosm conditions the effects of S.C.G. on quantity-quality of soil organic matter. They found that S.C.G. increased all organic matter fractions, especially the levels of the more labile ones, with a 600–700% increase in water-extractable carbon. S.C.G. also increased humic acids and fulvic acids around 200%. Still, the functionality of humic acids was affected by a reduction of the functional groups with more recalcitrant and stable character. In addition, the degree of incorporation of S.C.G. into the structure of the soil and the interaction between soil and S.C.G. particles (Fig. 8.1) affected carbon retention under stable forms, increasing carbon stabilization. In this sense, it has been estimated (Cervera-Mata et al., 2017) that the addition of 2.5% of S.C.G. to soil would mean the retention of 5299 kg of organic carbon/ha in the form of humus, representing a decrease of 19,426 kg of CO_2 (greenhouse gas) emitted into the atmosphere.

Hydrophobicity (appart of less stable and labile organic matter) is the worst side-efect of inadequately composted by-products and raw bio-residues, like S.C.G. (Murthy & Madhava-Naidu, 2012; Ballesteros et al., 2014; Alves et al., 2017; Comino et al., 2017). Hydrophobic compounds comes from the decomposition of vegetation (de Blas et al., 2010), microorganisms (Schaumann et al., 2007) or no humificated raw organic matter (Comino et al., 2017). Soil hydrophobicity have an adverse effect on different soil water properties such as infiltration capacity (Mohawesh et al., 2014), water availability for the plant (Doerr & Thomas, 2000) or

Fig. 8.1 S.E.M. image of an S.C.G. particle embedded in the surface of a soil aggregate. The S.C.G. particle is partially covered with mineral ones

hydraulic conductivity. In addition, hydrophobicity is the responsible to a decreased accessibility of microorganisms to organic matter, decreasing the decomposition rate of organic matter (Goebel et al., 2005). Hydrophobicity also improves soil aggregates stability by occudling organic matter (Lützow et al., 2006). All these effects of soil hydrophobicity are crucial to maintain soil organic matter, allowing the soil being a carbon sink (Chagas et al., 2018).

In this sense, it has previously been determined that S.C.G. are a highly hydrophobic material, probably due to their high lipid content (Gross et al., 1997). Cervera-Mata et al. (2020a; Ćosović et al., 2010) evaluated the effects of S.C.G. on the hydrophobicity, soil organic matter, and structure of two Mediterranean agricultural soils. Two S.C.G. doses (2.5 and 10%) in two different soils and two incubation times (30 and 60 days) were studied in vitro (Fig. 8.2). These authors found that S.C.G. addition increased water drop penetration time, contact angle, and surface free energy components. A larger amount of labile organic matter and lower humus quality index were related with soil hydrophobicity by means of UV-vis and infrared spectroscopy (Zanella et al., 2018). Nevertheless, hydrophobicity increase improved soil physical quality: water retention, saturated hydraulic conductivity, a high aggregate stability index, low bulk density and porosity. A notable decrease on available water content was the worst effect related with hydrophobicity increase. Figure 8.1 (Scanning electron microscopy (S.E.M.) images) shows a more considerable incorporation of S.C.G. particles in soil, which in turn is assited by smectitic clays and carbonates. Figure 8.3 depicts calcium carbonate biomineralization

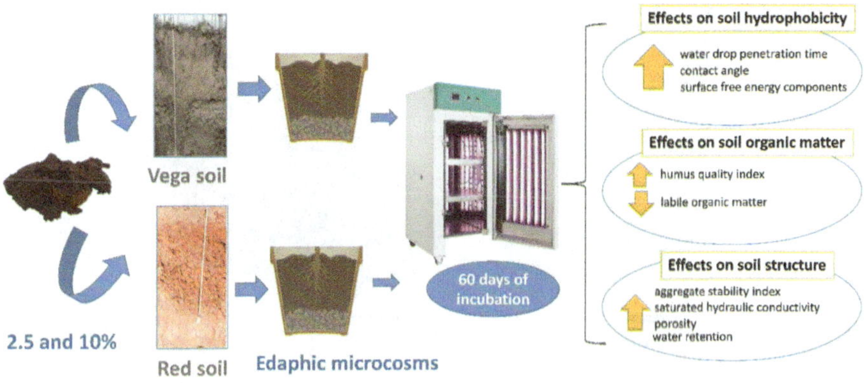

Fig. 8.2 Experimental design for the determination of the S.C.G. effects on soil hydrophobicity, structure and organic matter

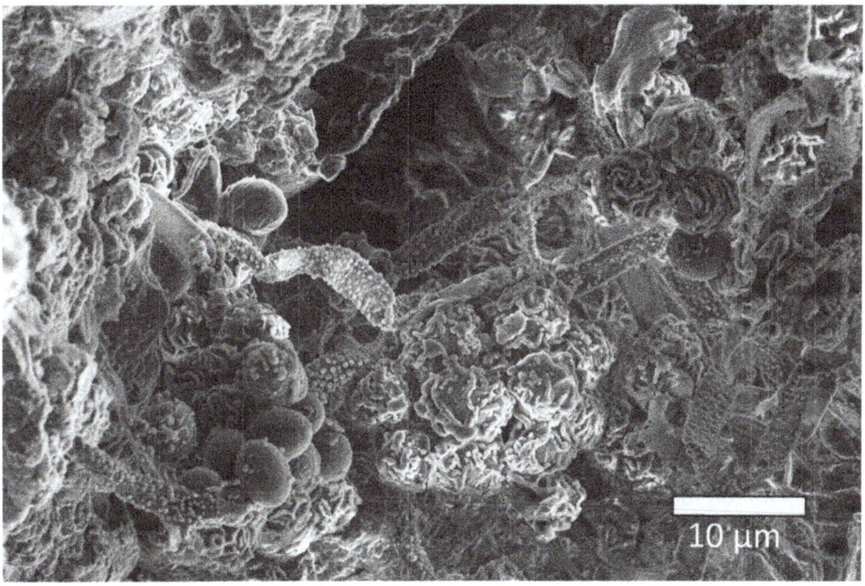

Fig. 8.3 S.E.M. image showing the development of fungal hyphae on a S.C.G. particle embedded in the surface of a soil aggregate

mediated by fungus, which could be an exciting and little-studied process of inorganic carbon fixation related to S.C.G. incorporation into soil.

Other recent literature (Cervera-Mata et al., 2019a) reflects the short-term effects of S.C.G. on soil physical properties of two Mediterranean agricultural soils (Calcisol and Luvisol) at 60 and 240 mg/ha doses, with two incubation times (30 and 60 days) and two modalities: with lettuce seedlings and without lettuce seedlings (*Lactuca sativa* var. longifolia). These authors found that S.C.G. addition

Fig. 8.4 Stereomicroscopic images of soil microaggregates. (**a**) before S.C.G. amendment. (**b**) after S.C.G. amendment

increased water retention and decreased bulk density and plant-available water content. S.C.G. addition also increased the percentage of macroaggregates and their structural stability. The stereomicroscope images showed that structural aggregates were rounded, with a structure hierarchized in smaller sums (Fig. 8.4b) compared to the original soil structure without S.C.G. (Fig. 8.4a). The incorporation of S.C.G. particles could happen in intra-ped cracks, and S.C.G. particles acted as cements. Finally, as indicated above, S.C.G. particles interacted with mineral particles and thus integrated into the soil structure (Fig. 8.1).

4.2 *Effect of Spent Coffee Grounds on Soil Microbiota*

Soil microbial diversity, abundance and metabolism is deeply affected by the addition of organic amendments (Wang et al., 2013). Therefore, the analysis of soil microbiota could be an useful indicator of soil quality (Giacometti et al., 2013). There's no too much scientific information about the effects of S.C.G. addition on soil microbiota; however, several studies point out to the phytotoxic effect of S.C.G. addition to horticulture soils (Hardgrove & Livesley, 2016; Cervera-Mata et al., 2017) or soils contaminated with heavy metals (Kim et al., 2014). Such phitotoxicity is though to be caused by chelating compounds like melanoidins, polyphenols or caffeine, among others (Jiménez-Zamora et al., 2015). For example, Cervera-Mata et al. (2017) demonstrated that the use of S.C.G. stimulates microbial activity in a concentration-dependent manner due to increased soil respiration rates, concurrently with high fungal hyphae densities (Fig. 8.3).

Very recently, Vela-Cano et al. (2019) studied the effect of S.C.G. on soil microbiota at two concentrations (2.5 and 10%) in two different soils (vega and red soil). These authors found that S.C.G. addition increased species richness (with a maximum diversity at 2.5% of S.C.G. added) under a stable microbial community structure. In general, the phyla abundance profiles over the time of the experiment were mostly unchanged. The thermophilic and radiotolerant *Rubrobacter* was overall the most abundant genus observed in both soils, although its abundance decreased with S.C.G. percentage. In addition, the genera *Pseudomonas*, *Clostridium*, *Caulobacter*, *Variovorax*, *Rhizobium*, *Ohtaekwangia*, and *Phenylobacterium* also increased by S.C.G. %. Since high concentrations of phenolic compounds were incorporated into soils via S.C.G. addition, potential polyphenol degraders (such as *Caulobacter*, *Pseudomonas*, *Achromobacter,* and *Rhodococcus*) increased their relative abundances after 30 days of S.C.G. addition (Fig. 8.5a). Finally, relevant genera

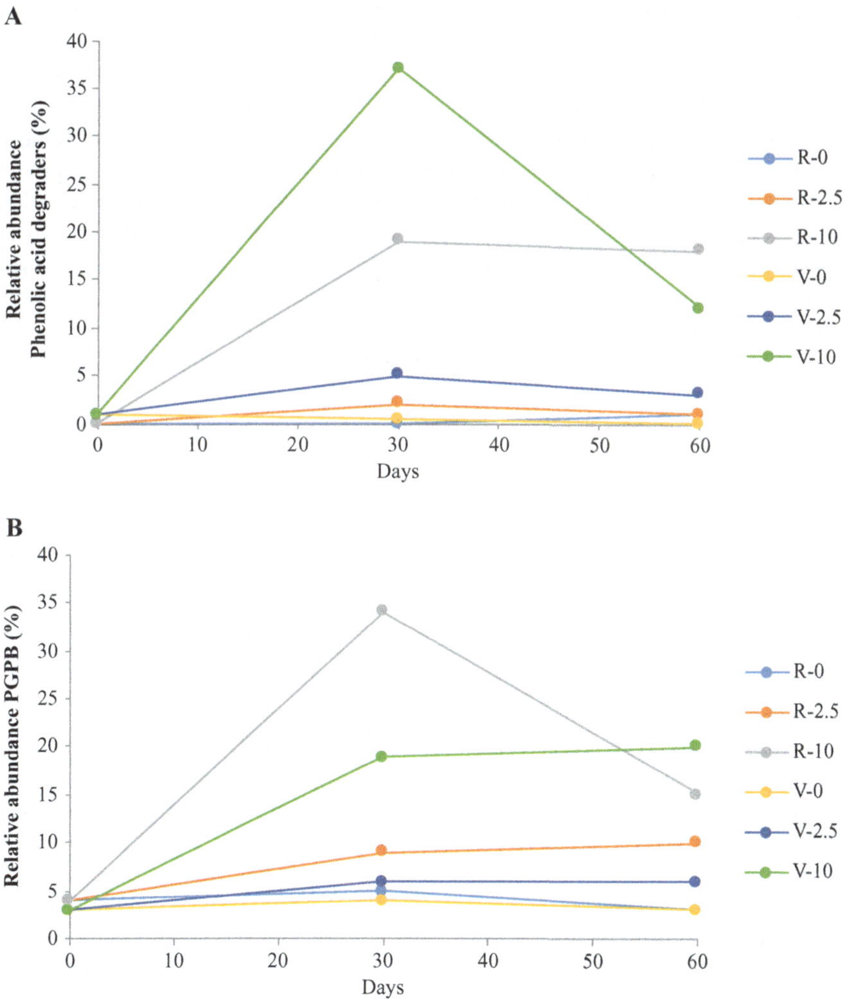

Fig. 8.5 Effect of S.C.G. on soil microbiota. (**a**) phenolic acids degraders. (**b**) potential plant growth promoting bacteria (P.G.P.B.)

belonging to the potential plant growth-promoting bacteria (P.G.P.B.) were also correlated with S.C.G. % and time (Fig. 8.5b).

4.3 Effect of Spent Coffee Grounds on Plant Growth

Lettuce cultivar has been used in many papers related with S.C.G., due to the high worldwide lettuce consumption, steadily expanding in the last decades (Heimler et al., 2012). Lettuce is an important vegetable for human nutrition (Sofo et al., 2016) due to the presence of essential elements (such as Na, K, Mg, among others) which are important enzymatic cofactors for the human metabolism. Cruz et al. (2014), (2015a) were the first authors to study S.C.G. reuse as an organic amendment to improve the mineral content of lettuces. A decrease on mineral levels (Ca, Cu, Fe, Mg, Mn, Na, P) was observed after the addition of fresh S.C.G. to letucces cultivars; this effect was related with mineral retention by the presence of chelating compounds in the coffee matrix. However, low doses of composted S.C.G. improved mineral concentration (i.e. K, Mg, Mn and Na) in lettuces due to enhanced plant uptake related to caffeine degradation (Cruz et al., 2014). S.C.G. directly composted in soil at high concentrations (20–30%) for 4 months (Cruz et al., 2015a) gave rise the mineral content of lettuces compared to those grown with a lower S.C.G. percentage. Finally, Morikawa and Saigusa (2011) described in brown rice the increase on Fe, Mn and Zn levels by means of top-dressing application of Fe- and Zn-enriched S.C.G.

Cervera-Mata et al. (2017) found that the addition of fresh S.C.G. to the soil, at doses of 2.5 and 10%, w/w, increased available K, P, Cu, Zn, and Fe in lettuce while limiting their growth at the same time. In order to overcome the effects of S.C.G. on lettuce growth, these authors tested the concomitant development of commercial inorganic fertilizer (N.P.K., 15:15:15) with S.C.G. (Cervera-Mata, Navarro-Alarcón, et al., 2019b) in lettuces grouped in 5 categories (baby, cultivated in agricultural soils with low or high percentages of S.C.G., without S.C.G. and with N.P.K.). Lettuces developed with only S.C.G. had higher levels of several essential (V, Fe, Co, V, Mn, and Zn) and toxic elements (Al and As) without reaching their toxicological limits. Such effect was not observed for those plants amended with S.C.G. and N.P.K. (Fig. 8.6). Very recently, Cervera-Mata et al. (2020b) studied the effect of different S.C.G.-based amendments on lettuce growth and mineral content (Cu, Fe and Zn). The improvent of agronomic properties of Mediterranean soils was obtained by using S.C.G. as biochars pyrolyzed at 400 and 270 °C, hydrolyzed, composted, vermicomposted, washed with water or ethanol or even as fresh S.C.G. These amendments were also compared against the common use of a N.P.K. fertilizer. Both hydrochar obtained at 400 °C and vermicompost didn't limited lettuces growth, although they decreased their mineral content. It has been suggested (Jiménez-Zamora et al., 2015) that such behavior could be related with the thermal destruction of chelating compounds like melanoiidins or polyphenols. Contrary, the mineral content increased with those treatments that did not wholly

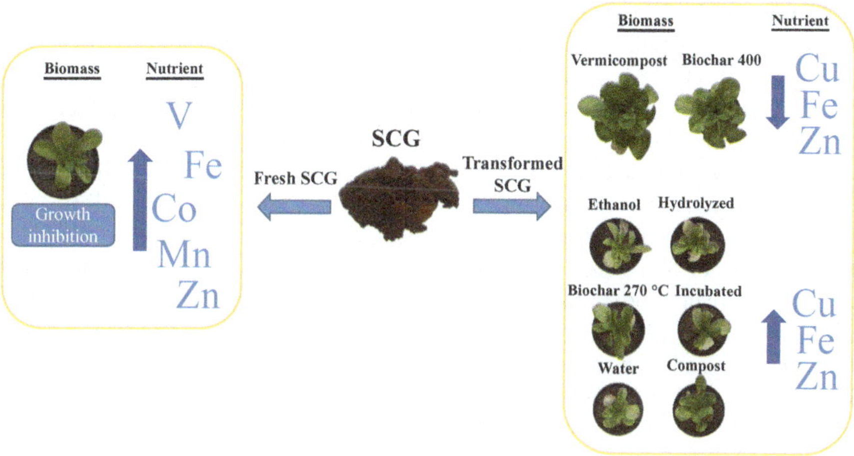

Fig. 8.6 Effect of raw and transformed S.C.G. on lettuce biomass and mineral content

eliminate chelating compounds. Finally, it was found that N.P.K. fertilizer increased lettuces biomass but decreased the micronutrient content (Chrysargyris et al., 2020).

5 Conclusions

Microorganisms influence not only each other and plants but also soil and thus the entire ecosystem and food chains. The analysis of the dynamic interactions between microbes and plants can help better to understand plants and their complex relationships in the ecosystem. The findings could also be used to secure and even improve the yields of our crops.

The microbiome plays a decisive role both as a "substance converter" in the breakdown and build-up of biomass and as an important factor in the area of health, immunity and resistance of crops and animals to pathogenic factors. Sustainable agriculture and food productions are simply not possible without the integration of the microbiome. Modern management strategies must therefore include the microflora of agricultural systems and their services as core elements of robust and efficient plant and animal production in their concepts. This is all the more true because global change (climate change and intensified agriculture) means that sustainable options for action are urgently required. Microorganisms in the root space are responsible for the conversion of nutrients and their transport into the plant, and are therefore crucial for the plant yield (Simmons et al., 2018). In addition, bacteria and fungi protect the plants directly from infestation by phytopathogenic organisms by means of "biocontrol" or induce a kind of "immune response" in the plant, which makes the plant more resistant to pathogens. Ultimately, microorganisms contribute to an improved resistance of plants to biotic and abiotic stressors. It is not surprising that in the field of "plant breeding", there has recently been a targeted attempt to influence the structure and function of the plant-associated microbiome.

The use of S.C.G. as an organic amendment could provide multiple solutions for the agricultural and environmental problems existing today. This practice facilitates the reuse of such bio-waste, increases soil organic carbon and hydrophobicity, and decreases CO_2 emissions into the atmosphere. At the same time, S.C.G. addition activates the microbial activity in the soil, increasing soil chemical fertility as an alternative to inorganic fertilizers, which could, in some cases, improve the nutritional quality of foods such as lettuces. Therefore, there is a transparent interconnection between food production and food-waste recycling, allowing for more sustainable crop practices, improving ecosystem functions (including the soil microbiome) and consumer health.FundingThis work was supported from the European Union's Horizon 2020 research and innovation programme under grant agreement No. 816303 (STANCE4HEALTH), and project AGL2014-53895-R from the Spanish Ministry of Economy and Competitiveness and by the European Regional Development Fund (FEDER).

References

Alves, R. C., Rodrigues, F., Nunes, M. A., Vinha, A. F., & Oliveira, M. P. (2017). State of the art in coffee processing by-products. In C. M. Galanakis (Ed.), *Handbook of coffee processing by-products: Sustainable applications* (pp. 1–22). Academic Press.

Aranda, V., Calero, J., Plaza, I., & Ontiveros-Ortega, A. (2016). Long-term effects of olive mill pomace co-compost on wettability and soil quality in olive groves. *Geoderma, 267*, 185–195.

Arif, I., Batool, M., & Schenk, P. M. (2020). Plant microbiome engineering: Expected benefits for improved crop growth and resilience. *Trends in Biotechnology*.

Ayres, J. S. (2016). Cooperative microbial tolerance behaviors in host-microbiota mutualism. *Cell, 165*(6), 1323–1331.

Ballesteros, L. F., Teixeira, J. A., & Mussatto, S. I. (2014). Chemical, functional, and structural properties of spent coffee grounds and coffee silver skin. *Food and Bioprocess Technology, 7*, 3493–3503.

Bang, C., Dagan, T., Deines, P., et al. (2018). Metaorganisms in extreme environments: Do microbes play a role in organismal adaptation? *Zoology (Jena, Germany), 127*, 1–19.

Belkaid, Y., & Hand, T. W. (2014). Role of the microbiota in immunity and inflammation. *Cell, 157*(1), 121–141.

Bennett, J. M., Reeves, G., Billman, G. E., & Sturmberg, J. P. (2018). Inflammation-nature's way to efficiently respond to all types of challenges: Implications for understanding and managing "the epidemic" of chronic diseases. *Frontiers in Medicine, 5*, 316.

Berg, G., Rybakova, D., Grube, M., & Köberl, M. (2016). The plant microbiome explored: Implications for experimental botany. *Journal of Experimental Botany, 67*(4), 995–1002.

Berg, G., Rybakova, D., Fischer, D., et al. (2020). Microbiome definition re-visited: Old concepts and new challenges. *Microbiome, 8*, 103.

Blum, W., Zechmeister-Boltenstern, S., & Keiblinger, K. M. (2019). Does soil contribute to the human gut microbiome? *Microorganisms, 7*(9), 287.

Boon, E., Meehan, C. J., Whidden, C., Wong, D. H., Langille, M. G., & Beiko, R. G. (2014). Interactions in the microbiome: Communities of organisms and communities of genes. *FEMS Microbiology Reviews, 38*(1), 90–118.

Borre, Y. E., O'Keeffe, G. W., Clarke, G., Stanton, C., Dinan, T. G., & Cryan, J. F. (2014). Microbiota and neurodevelopmental windows: Implications for brain disorders. *Trends in Molecular Medicine, 20*(9), 509–518.

Brodt, S., Six, J., Feenstra, G., Ingels, C., & Campbell, D. (2011). Sustainable agriculture. *Nature Education Knowledge, 3*(10), 1.

Bronick, C. J., & Lal, R. (2005). Soil structure and management: A review. *Geoderma, 124*, 3–22.

Busby, P. E., Soman, C., Wagner, M. R., Friesen, M. L., Kremer, J., Bennett, A., Morsy, M., Eisen, J. A., Leach, J. E., & Dangl, J. L. (2017). Research priorities for harnessing plant microbiomes in sustainable agriculture. *PLoS Biology, 15*(3), e2001793.

Calicioglu, O., Flammini, A., Bracco, S., Bellù, L., & Sims, R. (2019). The future challenges of food and agriculture: An integrated analysis of trends and solutions. *Sustainability, 11*, 222.

Caron, P., Ferrero y de Loma-Osorio, G., Nabarro, D., et al. (2018). Food systems for sustainable development: Proposals for a profound four-part transformation. *Agronomy for Sustainable Development, 38*, 41.

Cassán, F. D., & Díaz-Zorita, M. (2016). Azosprillum sp. in current agriculture: From the laboratory to the field. *Soil Biology and Biochemistry, 103*, 117–130.

Cavalier-Smith, T., Brasier, M., & Embley, T. M. (2006). Introduction: How and when did microbes change the world? *Philosophical Transactions of the Royal Society of London. Series B, Biological Sciences, 361*(1470), 845–850.

Cavicchioli, R., Ripple, W. J., Timmis, K. N., et al. (2019). Scientists' warning to humanity: Microorganisms and climate change. *Nature Reviews. Microbiology, 17*, 569–586.

Cervera-Mata, A., Pastoriza, S., Rufián-Henares, J. A., Martín-García, J. M., & Delgado, G. (2017). Impact of spent coffee grounds as organic amendment on soil fertility and lettuce growth in two Mediterranean agricultural soils. *Archives of Agronomy and Soil Science, 64*, 790–804.

Cervera-Mata, A., Martín-García, J., Delgado, R., Sánchez-Marañón, M., & Delgado, G. (2019a). Short-term effects of spent coffee grounds on the physical properties of two Mediterranean agricultural soils. *International Agrophysics, 33*, 205–216.

Cervera-Mata, A., Navarro-Alarcón, M., Delgado, G., Pastoriza, S., Montilla-Gómez, J., Llopis, J., Sánchez-González, C., & Rufián-Henares, J. A. (2019b). Spent coffee grounds improve the nutritional value in elements of lettuce (Lactuca sativa L.) and are an ecological alternative to inorganic fertilizers. *Food Chemistry, 282*, 1–8.

Cervera-Mata, A., Aranda, V., Ontiveros-Ortega, A., Comino, F., Martín-García, J., Vela-Cano, M., & Delgado, G. (2020a). Hydrophobicity and surface free energy to assess spent coffee grounds as soil amendment. Relationships with soil quality. *Catena*. In press.

Cervera-Mata, A., Navarro-Alarcón, M., Rufián-Henares, J. Á., Pastoriza, S., Montilla-Gómez, J., & Delgado, G. (2020b). Phytotoxicity and chelating capacity of spent coffee grounds: Two contrasting faces in its use as soil organic amendment. *Science of the Total Environment, 717*, 137247.

Chagas, F. O., Pessotti, R. C., Caraballo-Rodríguez, A. M., & Pupo, M. T. (2018). Chemical signaling involved in plant-microbe interactions. *Chemical Society Reviews, 47*(5), 1652–1704.

Checinska Sielaff, A., Urbaniak, C., Mohan, G. B. M., et al. (2019). Characterization of the total and viable bacterial and fungal communities associated with the International Space Station surfaces. *Microbiome, 7*, 50.

Chrysargyris, A., Antoniou, O., Xylia, P. *et al.* (2020). The use of spent coffee grounds in growing media for the production of *Brassica* seedlings in nurseries. Environmental Science and Pollution Research

Claus, S. P., Guillou, H., & Ellero-Simatos, S. (2016). The gut microbiota: a major player in the toxicity of environmental pollutants? *npj Biofilms Microbiomes, 2.*

Clemente, J. C., Pehrsson, E. C., Blaser, M. J., Sandhu, K., Gao, Z., Wang, B., Magris, M., Hidalgo, G., Contreras, M., Noya-Alarcón, Ó., Lander, O., McDonald, J., Cox, M., Walter, J., Oh, P. L., Ruiz, J. F., Rodriguez, S., Shen, N., Song, S. J., et al. (2015). The microbiome of uncontacted Amerindians. *Science Advances, 1*, e1500183.

Comino, F., Aranda, V., Domínguez-Vidal, A., & Ayora-Cañada, M. J. (2017). Thermal destruction of organic waste hydrophobicity for agricultural soils application. *Journal of Environmental Management, 202*, 94–105.

Comino, F., Cervera-Mata, A., Aranda, V., Martín-García, J. M., & Delgado, G. (2020). Short-term impact of spent coffee grounds over soil organic matter composition and stability in two contrasted Mediterranean agricultural soils. *Journal of Soils Sediments, 20*, 1182–1198.

Compant, S., Samad, A., Faist, H., & Sessitsch, A. (2019). A review on the plant microbiome: Ecology, functions, and emerging trends in microbial application. *Journal of Advanced Research, 19*, 29–37.

Cordovez, V., Dini-Andreote, F., Carrión, V. J., & Raaijmakers, J. M. (2019). Ecology and evolution of plant microbiomes. *Annual Review of Microbiology, 73*, 69–88.

Ćosović, B., Vojvodić, V., Bošković, N., Plavšić, M., & Lee, C. (2010). Characterization of natural and synthetic humic substances (melanoidins) by chemical composition and adsorption measurements. *Organic Geochemistry, 41*, 200–205.

Crandall, S. G., Gold, K. M., Jiménez-Gasco, M., Filgueiras, C. C., & Willett, D. S. (2020). A multiomics approach to solving problems in plant disease ecology. *PLoS One, 15*(9), e0237975.

Cruz, R., Baptista, P., Cunha, S., Pereira, J. A., & Casal, S. (2012). Carotenoids of lettuce (*Lactuca sativa L.*) grown on soil enriched with spent coffee grounds. *Molecules, 17*, 1535–1547.

Cruz, R., Morais, S., Mendes, E., Pereira, J. A., Baptista, P., & Casal, S. (2014). Improvement of vegetables elemental quality by espresso coffee residues. *Food Chemistry, 148*, 294–299.

Cruz, R., Mendes, E., Torrinha, Á., Morais, S., Alberto Pereira, J., Baptista, P., & Casal, S. (2015a). Revalorization of spent coffee residues by a direct agronomic approach. *Food Research International, 73*, 190–196.

Cruz, S., Sc, C., & Cordovil, S. (2015b). Espresso coffee residues as a nitrogen amendment for small-scale vegetable. *Journal of the Science of Food and Agriculture, 95*, 3059–3066.

Cui, K., Qi, M., Wang, S., et al. (2019). Dietary energy and protein levels influenced the growth performance, ruminal morphology and fermentation and microbial diversity of lambs. *Scientific Reports, 9*, 16612.

D'Abramo, F., & Neumeyer, S. (2020). A historical and political epistemology of microbes. *Centaurus, 62*(2), 321–330.

de Blas, E., Rodríguez-Alleres, M., & Almendros, G. (2010). Speciation of lipid and humic fractions in soils under pine and eucalyptus forest in northwest Spain and its effect on water repellency. *Geoderma, 155*, 242–248.

de Faria, M. R., Costa, L. S. A. S., Chiaramonte, J. B., et al. (2020). The rhizosphere microbiome: Functions, dynamics, and role in plant protection. *Tropical Plant Pathology*.

De Filippo, C., Cavalieri, D., Di Paola, M., Ramazzotti, M., Poullet, J. B., Massart, S., Collini, S., Pieraccini, G., & Lionetti, P. (2010). Impact of diet in shaping gut microbiota revealed by a comparative study in children from Europe and rural Africa. *Proceedings of the National Academy of Sciences of the United States of America, 107*(33), 14691–14696.

de Vries, F. T., Griffiths, R. I., Bailey, M., et al. (2018). Soil bacterial networks are less stable under drought than fungal networks. *Nature Communications, 9*, 3033.

Deines, P., & Bosch, T. C. (2016). Transitioning from microbiome composition to microbial community interactions: The potential of the metaorganism *Hydra* as an xperimental model. *Frontiers in Microbiology, 7*, 1610.

Deng, S., Wipf, H. M., Pierroz, G., et al. (2019). A plant growth-promoting microbial soil amendment dynamically alters the strawberry root bacterial microbiome. *Scientific Reports, 9*, 17677.

Dexter, A. R. (2004). Soil physical quality part I. theory, effects of soil texture, density, and organic matter, and effects on root growth. *Geoderma, 120*, 201–214.

Dietrich, T., Del Carmen Villaran Velasco, M., Echeverría, P. J., Pop, B., & Rusu, A. (2016). Crop and plant biomass as valuable material for BBB. Alternatives for valorization of green wastes. In *Biotransformation of agricultural waste and by-products: The food, feed, fibre, fuel (4F) economy*.

Doerr, S. H., & Thomas, A. D. (2000). The role of soil moisture in controlling water repellency: New evidence from forest soils in Portugal. *Journal of Hydrology, 231-232*, 134–147.

Dunham Trimmer (LLC). (2017). *Biological control global market overview*. http://wrir4.ucdavis.edu/events/2017_SLR_Meeting/Presentations/GeneralPresentations/1%20Trimmer%20-%20Global%20Biocontrol%20Market%202017.pdf. Accessed 07 Aug 2020.

Environmental Protection Agency (EPA). (2010). *Inventory of US U.S. greenhouse gas emissions and sinks: 1998–2008. 15 April 2010*. EPA. 430-R-10-006. Washington (DC).

European Commission. Communication from the Commission to the European Parliament, the Council, the European Economic and Social Committee and the Committee of the Regions A Farm to Fork Strategy for a fair, healthy and environmentally-friendly food system, COM/2020/381 final. https://eur-lex.europa.eu/legal-content/EN/TXT/?qid=1590404602495&uri=CELEX%3A52020DC0381. Accessed 17 Oct 2020.

FAO. (2010). Sustainable diets and biodiversity. Directions and solutions for policy, research and action. In *Proceedings of the international scientific symposium: Biodiversity and sustainable diets: United against hunger, 3–5 November 2010*. Food and Agriculture Organization of the United Nations (FAO).

FAO. (2014). *SAFA sustainability assessment of food and agriculture systems*. Guideline. Version 3.0. http://www.fao.org/3/a-i3957e.pdf

FAO. (2015). *Healthy soils are the basis for healthy food production*. Food and Agriculture Organization of the United Nations [FAO]. pp. 1–4. http://www.fao.org/documents/card/en/c/645883cd-ba28-4b16-a7b8-34babbb3c505/. Accessed 17 Oct 2020.

Federal Ministry of Education and Research (BMBF) and Federal Ministry of Food and Agriculture (BMEL). (2015). *Bioeconomy in Germany opportunities for a bio-based and sustainable future*. https://www.bmbf.de/upload_filestore/pub/Biooekonomie_in_Deutschland_Eng.pdf

Flandroy, L., Poutahidis, T., Berg, G., et al. (2018). The impact of human activities and lifestyles on the interlinked microbiota and health of humans and of ecosystems. *Science of the Total Environment, 627*, 1018–1038.

Foo, J. L., Ling, H., Lee, Y. S., & Chang, M. W. (2017). Microbiome engineering: Current applications and its future. *Biotechnology Journal, 12*(3). https://doi.org/10.1002/biot.201600099

Forstner, S., & Rusu, A. (2015). *Development of personalised food for the nutrition of elderly consumers. Know your food: Food ethics and innovation* (pp. 24–27). Wageningen Academic Publishers.

Frąc, M., Hannula, S. E., Bełka, M., & Jędryczka, M. (2018). Fungal biodiversity and their role in soil health. *Frontiers in Microbiology, 9*, 707.

Frische, T., Egerer, S., Matezki, S., et al. (2018). 5-Point programme for sustainable plant protection. *Environmental Sciences Europe, 30*, 8.

Funabashi, M. (2018). Human augmentation of ecosystems: Objectives for food production and science by 2045. *Npj Science of Food, 2*, 16.

Gan, X., Stegle, O., Behr, J., et al. (2011). Multiple reference genomes and transcriptomes for *Arabidopsis thaliana*. *Nature, 477*, 419–423.

Giacometti, C., Demyanb, M. S., Cavania, L., Marzadoria, C., Ciavatta, C., & Kandelerc, E. (2013). Chemical and microbiological soil quality indicators and their potential to differentiate fertilization regimes in temperate agroecosystems. *Applied Soil Ecology, 64*, 32–48.

Global Forum for Food and Agriculture (GFFA). (2019). *Agriculture goes digital – Smart solutions for future farming*. https://www.gffa-berlin.de/wp-content/uploads/2019/10/ergebnisbroschuere_2019-EN.pdf

Godfray, H. C., & Garnett, T. (2014). Food security and sustainable intensification. *Philosophical transactions of the Royal Society of London. Series B, Biological Sciences, 369*(1639), 20120273.

Goebel, M., Bachmann, J., Woche, S. K., & Fischer, W. R. (2005). Soil wettability, aggregate stability, and the decomposition of soil organic matter. *Geoderma, 128*, 80–93.

Gross, G., Jaccaud, E., & Hugget, A. C. (1997). Analysis of the content of the diterpenes cafestol and kahweol in coffee brews. *Food and Chemical Toxicology, 35*, 547–554.

Guerrieri, A., Dong, L., & Bouwmeester, H. J. (2019). Role and exploitation of underground chemical signaling in plants. *Pest Management Science, 75*(9), 2455–2463.
Guth, M., & Smędzik-Ambroży, K. (2020). Economic resources versus the efficiency of different types of agricultural production in regions of the European union. *Economic Research-Ekonomska Istraživanja, 33*(1), 1036–1051.
Hachicha, R., Rekik, O., Hachicha, S., Ferchichi, M., Woodward, S., Moncef, N., Cegarra, J., & Mechichi, T. (2012). Chemosphere co-composting of spent coffee ground with olive mill wastewater sludge and poultry manure and effect of Trametes versicolor inoculation on the compost maturity. *Chemosphere, 88*, 677–682.
Hardgrove, S. J., & Livesley, S. J. (2016). Applying spent coffee grounds directly to urban agriculture soils greatly reduces plant growth. *Urban Forestry & Urban Greening, 18*, 1–8.
Hassani, M. A., Durán, P., & Hacquard, S. (2018). Microbial interactions within the plant holobiont. *Microbiome, 6*(1), 58.
Hatfield, J. L., & Dold, C. (2019). Water-use efficiency: Advances and challenges in a changing climate. *Frontiers in Plant Science, 10*, 103.
Heimler, D., Vignolini, P., Arfaioli, P., Isolani, L., & Romani, A. (2012). Conventional, organic and biodynamic farming: Differences in polyphenol content and antioxidant activity of Batavia lettuce. *Journal of the Science of Food and Agriculture, 92*, 551–556.
Hunter, P. (2016). Plant microbiomes and sustainable agriculture: Deciphering the plant microbiome and its role in nutrient supply and plant immunity has great potential to reduce the use of fertilizers and biocides in agriculture. *EMBO Reports, 17*(12), 1696–1699.
Hutchins, D. A., Jansson, J. K., Remais, J. V., Rich, V. I., Singh, B. K., & Trivedi, P. (2019). Climate change microbiology – problems and perspectives. *Nature Reviews. Microbiology, 17*, 391–396.
ICO. (2020). *Historical data on the global coffee trade*. International Coffee Organization; [accesed 2020 August 09]. http://www.ico.org/new_historical.asp?section=Statistics
Jacoby, R., Peukert, M., Succurro, A., Koprivova, A., & Kopriva, S. (2017). The role of soil microorganisms in plant mineral nutrition-current knowledge and future directions. *Frontiers in Plant Science, 8*, 1617.
Jiménez-Zamora, A., Pastoriza, S., & Rufián-Henares, J. A. (2015). Revalorization of coffee by-products. Prebiotic, antimicrobial and antioxidant properties. LWT – food Sci. *Technology, 61*, 12–18.
Jin, Y., Wu, S., Zeng, Z., & Fu, Z. (2017). Effects of environmental pollutants on gut microbiota. *Environmental Pollution, 222*, 1–9.
Kemp, L., Adam, L., Boehm, C. R., Breitling, R., Casagrande, R., Dando, M., Djikeng, A., Evans, N. G., Hammond, R., Hills, K., Holt, L. A., Kuiken, T., Markotić, A., Millett, P., Napier, J. A., Nelson, C., ÓhÉigeartaigh, S. S., Osbourn, A., Palmer, M., et al. (2020). Bioengineering horizon scan 2020. *Elife, 9*, e54489.
Kim, M. S., Min, H. G., Koo, N., Park, J., Lee, S. H., Bak, G. I., & Kim, J. G. (2014). The effectiveness of spent coffee grounds and its biochar on the amelioration of heavy metals-contaminated water and soil using chemical and biological assessments. *Journal of Environmental Management, 146*, 124–130.
Kudoyarova, G., Arkhipova, T., Korshunova, T., Bakaeva, M., Loginov, O., & Dodd, I. C. (2019). Phytohormone mediation of interactions between plants and non-symbiotic growth promoting bacteria under edaphic stresses. *Frontiers in Plant Science, 10*, 1368.
Kumar, M., Ji, B., Zengler, K., et al. (2019). Modelling approaches for studying the microbiome. *Nature Microbiology, 4*, 1253–1267.
Laforest-Lapointe, I., & Arrieta, M. C. (2018). Microbial eukaryotes: a missing link in gut microbiome studies. *mSystems, 3*(2), e00201–e00217.
Lal, R. (2015). Restoring soil quality to mitigate soil degradation. *Sustainability, 7*, 5875–5895.
Lawson, C. E., Harcombe, W. R., Hatzenpichler, R., Lindemann, S. R., Löffler, F. E., O'Malley, M. A., García Martín, H., Pfleger, B. F., Raskin, L., Venturelli, O. S., Weissbrodt, D. G.,

Noguera, D. R., & McMahon, K. D. (2019). Common principles and best practices for engineering microbiomes. *Nature Reviews. Microbiology, 17*(12), 725–741.

Lindgren, E., Harris, F., Dangour, A. D., et al. (2018). Sustainable food systems—a health perspective. *Sustainability Science, 13*, 1505–1517.

Liu, Y., & Li, Y. (2017). Revitalize the world's countryside. *Nature, 548*, 275–277.

Liu, W., Baddeley, J., & Watson, C. (2011). Models of biological nitrogen fixation of legumes. Areview. *Agronomy for Sustainable Development, Springer Verlag/EDP Sciences/INRA, 31*(1), 155–172.

Liu, A., Contador, C. A., Fan, K., & Lam, H. M. (2018). Interaction and regulation of carbon, nitrogen, and phosphorus metabolisms in root nodules of legumes. *Frontiers in Plant Science, 9*, 1860.

Longford, S. R., Campbell, A. H., Nielsen, S., et al. (2019). Interactions within the microbiome alter microbial interactions with host chemical defences and affect disease in a marine holobiont. *Scientific Reports, 9*, 1363.

Lu, T., Ke, M., Lavoie, M., et al. (2018). Rhizosphere microorganisms can influence the timing of plant flowering. *Microbiome, 6*, 231.

Lützow, M. V., Kögel-Knabner, I., Ekschmitt, K., Matzner, E., Guggenberger, G., Marschner, B., & Flessa, H. (2006). Stabilization of organic matter in temperate soils: Mechanisms and their relevance under different soil conditions – a review. *European Journal of Soil Science, 57*, 426–445.

Ma, H., Pineda, A., Hannula, E. S., Kielak, A. M., Setyarinia, S. N., & Bezemer, T. M. (2020). Steering root microbiomes of a commercial horticultural crop with plant-soil feedbacks. *Applied Soil Ecology, 150*, 103468.

Mahmud, K., Makaju, S., Ibrahim, R., & Missaoui, A. (2020). Current progress in nitrogen fixing plants and microbiome research. *Plants, 9*, 97.

Mamuad, L. L., Kim, S. H., Biswas, A. A., et al. (2019). Rumen fermentation and microbial community composition influenced by live *Enterococcus faecium* supplementation. *AMB Express, 9*, 123.

Mann, A. (2018). Inner workings: Hunting for microbial life throughout the solar system. *Proceedings of the National Academy of Sciences of the United States of America, 115*(45), 11348–11350.

Massaya, J., Prates Pereira, A., Mills-Lamptey, B., Benjamin, J., & Chuck, C. (2019). Conceptualization of a spent coffee grounds biorefinery: A review of existing valorisation approaches. *Food and Bioproducts Processing, 118*, 149–166.

Matthews, C., Crispie, F., Lewis, E., Reid, M., O'Toole, P. W., & Cotter, P. D. (2019). The rumen microbiome: a crucial consideration when optimising milk and meat production and nitrogen utilisation efficiency. *Gut Microbes, 10*(2), 115–132.

McNear, D. H., Jr. (2013). The rhizosphere – roots, soil and everything in between. *Nature Education Knowledge, 4*(3), 1.

Meybeck, A., & Gitz, V. (2017). Sustainable diets within sustainable food systems. *The Proceedings of the Nutrition Society, 76*, 1–11.

Mohawesh, O., Mahmoud, M., Janssen, M., & Lennartz, B. (2014). Effect of irrigation with olive mill wastewater on soil hydraulic and solute transport properties. *International journal of Environmental Science and Technology, 11*, 927–934.

Morgan, J. B., & Connolly, E. L. (2013). Plant-soil interactions: Nutrient uptake. *Nature Education Knowledge, 4*(8), 2.

Morikawa, C. K., & Saigusa, M. (2011). Recycling coffee grounds and tea leaf wastes to improve the yield and mineral content of grains of paddy rice. *Journal of the Science of Food and Agriculture, 91*, 2108–2111.

Moss, E. L., Maghini, D. G., & Bhatt, A. S. (2020). Complete, closed bacterial genomes from microbiomes using nanopore sequencing. *Nature Biotechnology, 38*, 701–707.

Mueller, U. G., & Sachs, J. L. (2015). Engineering Microbiomes to Improve Plant and Animal Health. *Trends in Microbiology, 23*(10), 606–617.

Mulligan, C. M., & Friedman, J. E. (2017). Maternal modifiers of the infant gut microbiota: Metabolic consequences. *The Journal of Endocrinology, 235*(1), R1–R12.

Murthy, P. S., & Madhava-Naidu, M. (2012). Sustainable management of coffee industry by-products and value addition—A review. *Resources, Conservation and Recycling, 66*, 45–58.

Mus, F., Crook, M. B., Garcia, K., et al. (2016). Symbiotic nitrogen fixation and the challenges to its extension to nonlegumes. *Applied and Environmental Microbiology, 82*(13), 3698–3710.

Mussatto, S. I., Machado, E. M. S., Martins, S., & Teixeira, J. A. (2011). Production, composition, and application of coffee and its industrial residues. *Food and Bioprocess Technology, 4*, 661–672.

Nanjundappa, A., Bagyaraj, D. J., Saxena, A. K., et al. (2019). Interaction between arbuscular mycorrhizal fungi and *Bacillus* spp. in soil enhancing growth of crop plants. *Fungal Biology and Biotechnology, 6*, 2.

National Action Plan (NAP) to reduce the use of plant protection products stems from Directive 2009/128/EC of the European Parliament and of the Council of 21October 2009 establishing a framework for Community action to achieve the sustainable use of pesticides, which was transposed into national law by the Law of 19 December 2014 on plant protection products. https://ec.europa.eu/food/sites/food/files/plant/docs/pesticides_sup_nap_lux-rev_en.pdf

Nelson, E. B. (2018). The seed microbiome: Origins, interactions, and impacts. *Plant and Soil, 422*, 7–34.

Noverr, M. C., & Huffnagle, G. B. (2005). The 'microflora hypothesis' of allergic diseases. *Clinical and Experimental Allergy, 35*(12), 1511–1520.

O'Banion, B. S., O'Neal, L., Alexandre, G., & Lebeis, S. L. (2020). Bridging the gap between single-strain and community-level plant-microbe chemical interactions. *Molecular Plant-Microbe Interactions, 33*(2), 124–134.

Official Journal of the European Union (OJEU). (2008). *Directive 2008/98/EC of the European Parliament and of the Council of 19 November 2008 on waste and repealing certain Directives*. 22.11.2008. L312/30.

Olanrewaju, O. S., Ayangbenro, A. S., Glick, B. R., et al. (2019). Plant health: Feedback effect of root exudates-rhizobiome interactions. *Applied Microbiology and Biotechnology, 103*, 1155–1166.

Oltmanns, J., Licht, O., Bohlen, M.-L., Schwarz, M., Escher, S. E., Silano, V., MacLeod, M., Noteborn, H. P. J. M., Kass, G. E. N., & Merten, C. (2020). Potential emerging chemical risks in the food chain associated with substances registered under REACH. *Environmental Science: Processes Impacts, 22*, 105–120.

Orlova, N., Abakumov, E., Orlova, E., Yakkonen, K., & Shahnazarova, V. (2019). Soil organic matter alteration under biochar amendment: Study in the incubation experiment on the Podzol soils of the Leningrad region (Russia). *Journal Soils Sediments, 19*, 2708–2716.

Orozco-Mosqueda, M. D. C., Rocha-Granados, M. D. C., Glick, B. R., & Santoyo, G. (2018). Microbiome engineering to improve biocontrol and plant growth-promoting mechanisms. *Microbiological Research, 208*, 25–31.

Ottman, N., Ruokolainen, L., Suomalainen, A., Sinkko, H., Karisola, P., Lehtimäki, J., Lehto, M., Hanski, I., Alenius, H., & Fyhrquist, N. (2019). Soil exposure modifies the gut microbiota and supports immune tolerance in a mouse model. *Journal of Allergy Clinical Immunology, 143*(3), 1198–1206.e12.

Pascale, A., Proietti, S., Pantelides, I. S., & Stringlis, I. A. (2020). Modulation of the root microbiome by plant molecules: The basis for targeted disease suppression and plant growth promotion. *Frontiers in Plant Science, 10*, 1741.

Peoples, M. B., Brockwell, J., Herridge, D. F., et al. (2009). The contributions of nitrogen-fixing crop legumes to the productivity of agricultural systems. *Symbiosis, 48*, 1–17.

Pérez-Burillo, S., Pastoriza, S., Jiménez-Hernández, N., D'Auria, G., Francino, M. P., & Rufián-Henares, J. A. (2018). Effect of food thermal processing on the composition of the gut microbiota. *Journal of Agricultural and Food Chemistry, 66*(43), 11500–11509.

Pérez-Burillo, S., Pastoriza, S., Fernández-Arteaga, A., Luzón, G., Jiménez-Hernández, N., D'Auria, G., Francino, M. P., & Rufián-Henares, J. Á. (2019). Spent coffee grounds extract, rich in mannooligosaccharides, promotes a healthier gut microbial community in a dose-dependent manner. *Journal of Agricultural and Food Chemistry, 67*(9), 2500–2509.

Pérez-Jaramillo, J. E., Mendes, R., & Raaijmakers, J. M. (2016). Impact of plant domestication on rhizosphere microbiome assembly and functions. *Plant Molecular Biology, 90*, 635–644.

Ponter, H., Raichlen, D. A., Wood, B. M., Mabulla, A. Z. P., Racette, S. B., et al. (2012). Hunter-gatherer energetics and human obesity. *PLoS One, 7*(7), e40503.

Popp, J., Pető, K., & Nagy, J. (2013). Pesticide productivity and food security. A review. *Agronomy for Sustainable Development, 33*, 243–255.

Prakash, O., Nimonkar, Y., & Desai, D. A. (2020). Recent overview of microbes and microbiome preservation. *Indian Journal of Microbiology, 60*, 297–309.

Qiu, Z., Egidi, E., Liu, H., Kaur, S., & Singh, B. K. (2019). New frontiers in agriculture productivity: Optimised microbial inoculants and in situ microbiome engineering. *Biotechnology Advances, 37*(6), 107371.

Raynaud, X., & Nunan, N. (2014). Spatial ecology of bacteria at the microscale in soil. *PLoS One, 9*(1), e87217.

Reynolds, W. D., Drury, C. F., Yang, X. M., & Tan, C. S. (2008). Optimal soil physical quality inferred through structural regression and parameter interactions. *Geoderma, 146*, 466–474.

Rivera-Yoshida, N., Arias Del Angel, J. A., & Benítez, M. (2018). Microbial multicellular development: Mechanical forces in action. *Current Opinion in Genetics & Development, 51*, 37–45.

Rodriguez, P. A., Rothballer, M., Chowdhury, S. P., Nussbaumer, T., Gutjahr, C., & Falter-Braun, P. (2019). Systems biology of plant-microbiome interactions. *Molecular Plant, 12*(6), 804–821.

Rodríguez-Entrena, M., Espinosa-Goded, M., & Barreiro-Hurlé, J. (2014). The role of ancillary benefits on the value of agricultural soils carbon sequestration programmes: Evidence from a latent class approach to Andalusian olive groves. *Ecological Economics, 99*, 63–73.

Rook, G. A., & Brunet, L. R. (2005). Microbes, immunoregulation, and the gut. *Gut, 54*(3), 317–320.

Rusu, A., Alvarez, P., Schwarze, B., Trif, A.-K., & M. (2020a). The impact of Candida spp. on physiological alterations in gut microbiota. Candida elimination diet therapy. In T. Askun & J. S. M. Tondolo (Eds.), *Candidiasis*. IntechOpen. ISBN: 978-1-83881-112-9.

Rusu, A., Randriambelonoro, M., Perrin, C., et al. (2020b). Aspects influencing food intake and approaches towards personalising nutrition in the elderly. *Population Ageing, 13*, 239–256.

Sachidanand, B., Mitra, N. G., Kumar, V., Roy, R., & Mishra, B. B. (2019). Soil as a huge laboratory for microorganisms. *Agricultural Research & Technology: Open Access Journal, 22*(4), 556205.

San Roman, M., & Wagner, A. (2018). An enormous potential for niche construction through bacterial cross-feeding in a homogeneous environment. *PLoS Computational Biology, 14*(7), e1006340.

Schaumann, G. E., Braun, B., Kirchner, D., Rotard, W., Szewzyk, U., & Grohmann, E. (2007). Influence of biofilms on the water repellency of urban soil samples. *Hydrological Processes, 21*, 2276–2284.

Schirawski, J., & Perlin, M. H. (2018). Plant-microbe interaction 2017-the good, the bad and the diverse. *International Journal of Molecular Sciences, 19*(5), 1374.

Schloter, M., Nannipieri, P., Sørensen, S. J., et al. (2018). Microbial indicators for soil quality. *Biology and Fertility of Soils, 54*, 1–10.

Senesi, N., & Plaza, C. (2007). Role of humification processes in recycling organic wastes of various nature and sources as soil amendments. *Clean – Soil, Air, Water, 35*, 26–41.

Sessitsch, A., Brader, G., Pfaffenbichler, N., Gusenbauer, D., & Mitter, B. (2018). The contribution of plant microbiota to economy growth. *Microbial Biotechnology, 11*, 801–805.

Shankar, V., Gouda, M., Moncivaiz, J., Gordon, A., Reo, N. V., Hussein, L., & Paliy, O. (2017). Differences in gut metabolites and microbial composition and functions between Egyptian and U.S. children are consistent with their diets. *mSystems, 2*, e00169-16.

Simmons, T., Caddell, D. F., Deng, S., & Coleman-Derr, D. (2018). Exploring the root microbiome: Extracting bacterial community data from the soil, rhizosphere, and root endosphere. *Journal of Visualized Experiments : JoVE, 135*, 57561.

Simon, J.-C., Marchesi, J. R., Mougel, C., & Selosse, M. A. (2019). Host-microbiota interactions: From holobiont theory to analysis. *Microbiome, 7*, 5.

Singh, B. K., & Trivedi, P. (2017). Microbiome and the future for food and nutrient security. *Microbial Biotechnology, 10*(1), 50–53.

Singh, N., Gupta, V. K., Kumar, A., & Sharma, B. (2017). Synergistic effects of heavy metals and pesticides in living systems. *Frontiers in Chemistry, 5*, 70.

Sofo, A., Lundegårdh, B., Mårtensson, A., Manfra, M., Pepe, G., Sommella, E., De Nisco, M., Tenore, G. C., Campliglia, P., & Scopa, A. (2016). Different agronomic and fertilization systems affect polyphenolic profile, antioxidant capacity and mineral composition of lettuce. *Scientia Horticulturae, 204*, 106–115.

Souza, R. d., Ambrosini, A., & Passaglia, L. M. (2015). Plant growth-promoting bacteria as inoculants in agricultural soils. *Genetics and Molecular Biology, 38*(4), 401–419.

Strachan, D. P. (1989). Hay fever, hygiene, and household size. *BMJ, 299*(6710), 1259–1260.

Sundström, J. F., Albihn, A., Boqvist, S., et al. (2014). Future threats to agricultural food production posed by environmental degradation, climate change, and animal and plant diseases – a risk analysis in three economic and climate settings. *Food Security, 6*, 201–215.

Thompson, L. A., & Darwish, W. S. (2019). Environmental chemical contaminants in food: Review of a global problem. *Journal of Toxicology, 2345283.*

Thursby, E., & Juge, N. (2017). Introduction to the human gut microbiota. *The Biochemical Journal, 474*(11), 1823–1836.

Timm, C. M., Loomis, K., Stone, W., et al. (2020). Isolation and characterization of diverse microbial representatives from the human skin microbiome. *Microbiome, 8*, 58.

Trif, M., Vodnar, D. C., Mitrea, L., Rusu, A. V., & Socol, C. T. (2019). Design and development of oleoresins rich in carotenoids coated microbeads. *Coatings, 9*, 235.

Trinh, P., Zaneveld, J. R., Safranek, S., & Rabinowitz, P. M. (2018). One health relationships between human, animal, and environmental microbiomes: A mini-review. *Frontiers in Public Health, 6*, 235. https://doi.org/10.3389/fpubh.2018.00235

Trivedi, P., Schenk, P. M., Wallenstein, M. D., & Singh, B. K. (2017). Tiny microbes, big yields: Enhancing food crop production with biological solutions. *Microbial Biotechnology, 10*(5), 999–1003.

Turner, T. R., James, E. K., & Poole, P. S. (2013). The plant microbiome. *Genome Biology, 14*(6), 209.

Vallès, Y., & Francino, M. P. (2018). Air pollution, early life microbiome, and development. *Current Environmental Health Reports, 5*(4), 512–521.

van der Heijden, M. G., de Bruin, S., Luckerhoff, L., van Logtestijn, R. S., & Schlaeppi, K. (2016). A widespread plant-fungal-bacterial symbiosis promotes plant biodiversity, plant nutrition and seedling recruitment. *The ISME Journal, 10*(2), 389–399.

van Veelen, H. P. J., Falcão Salles, J., Matson, K. D., et al. (2020). Microbial environment shapes immune function and cloacal microbiota dynamics in zebra finches *Taeniopygia guttata*. *Animal Microbiome, 2*, 21.

Vela-Cano, M., Cervera-Mata, A., Purswani, J., Pozo, C., Delgado, G., & González-López, J. (2019). Bacterial community structure of two Mediterranean agricultural soils amended with spent coffee grounds. *Applied Soil Ecology, 137*, 12–20.

Velásquez, A. C., Castroverde, C., & He, S. Y. (2018). Plant-pathogen warfare under changing climate conditions. *Current Biology : CB, 28*(10), R619–R634.

Vítězová, M., Jančiková, S., Dordević, D., Vítěz, T., Elbl, J., Hanišáková, N., Jampílek, J., & Kushkevych, I. (2019). The possibility of using spent coffee grounds to improve wastewater treatment due to respiration activity of microorganisms. *Applied Sciences, 9*, 3155.

Wagner, S. C. (2011). Biological nitrogen fixation. *Nature Education Knowledge, 3*(10), 15.

Wang, M., Sahu, A. K., Rusten, B., & Park, C. (2013). Anaerobic co-digestion of microalgae Chlorella sp. and waste activated sludge. *Bioresource Technology, 142*, 585–590.

Wold, A. E. (1998). The hygiene hypothesis revised: Is the rising frequency of allergy due to changes in the intestinal flora? *Allergy, 53*(46 Suppl), 20–25.

Wolińska, A., Kuźniar, A., Zielenkiewicz, U., et al. (2017). Metagenomic analysis of some potential nitrogen-fixing bacteria in arable soils at different formation processes. *Microbial Ecology, 73*, 162–176.

Wu, H. J., & Wu, E. (2012). The role of gut microbiota in immune homeostasis and autoimmunity. *Gut Microbes, 3*(1), 4–14.

Wu, G. D., Compher, C., Chen, E. Z., et al. (2016). Comparative metabolomics in vegans and omnivoresreveal constraints on diet-dependent gut microbiotametabolite production. *Gut, 65*, 63–72.

Wu, L., Ning, D., Zhang, B., Li, Y., Zhang, P., Shan, X., et al. (2019). Global diversity and biogeography of bacterial communities in wastewater treatment plants. *Nature Microbiology, 4*, 1183–1195.

Yamane, K., Kono, M., Fukunaga, T., Iwai, K., & Sekine, R. (2014). Field evaluation of coffee grounds application for crop growth enhancement, weed control, and soil improvement. *Plant Production Science, 17*, 93–102.

Zanella, A., Bolzonella, C., Lowenfels, J., Ponge, J.-F., Bouché, M., Saha, D., Kukal, S. S., Fritz, I., Savory, A., & Blouin, M. (2018). Humusica 2, article 19: Techno humus systems and global change – Conservation agriculture and 4/1000 proposal. *Applied Soil Ecology, 122*(2), 271–296. https://doi.org/10.1016/j.apsoil.2017.10.036

Zdolec, N., Lorenzo, J. M., & Ray, R. C. (2018). Use of microbes for improving food safety and quality. *BioMed Research International, 2018*, 3902698.

Zgadzaj, R., Garrido-Oter, R., Jensen, D. B., Koprivova, A., Schulze-Lefert, P., & Radutoiu, S. (2016). Root nodule symbiosis in Lotus japonicus drives the establishment of distinctive rhizosphere, root, and nodule bacterial communities. *Proceedings of the National Academy of Sciences of the United States of America, 113*(49), E7996–E8005.

Zhang, L., & Sun, X. (2017). Using cow dung and spent coffee grounds to enhance the two-stage co- composting of green waste. *Bioresource Technology, 245*, 152–161.

Zimmerer, K. S., & de Haan, S. (2017). Agrobiodiversity and a sustainable food future. *Nature Plants, 3*, 17047.

Chapter 9
Healthier and Sustainable Food Systems: Integrating Underutilised Crops in a 'Theory of Change Approach'

Elisabete Pinto, Helena Ferreira, Carla S. Santos, Marta Nunes da Silva, David Styles, Paola Migliorini, Georgia Ntatsi, Anestis Karkanis, Marie-Fleur Brémaud, Yann de Mey, Miranda Meuwissen, Janos-Istvan Petrusan, Sergiy Smetana, Beatriz Silva, Lina Maja Marie Krenz, Daniel Pleissner, Adriano Profeta, Marko Debeljak, Aneta Ivanovska, Bálint Balázs, Diego Rubiales, Cathy Hawes, Pietro P. M. Iannetta, and Marta W. Vasconcelos

Abstract Increasingly, consumers are paying attention to healthier food diets, "healthy" food attributes (such as "freshness", "naturalness" and "nutritional value"), and the overall sustainability of production and processing methods. Other significant trends include a growing demand for regional and locally produced/supplied and less processed food. To meet these demands, food production and processing need to evolve to preserve the raw material and natural food properties while ensuring such sustenance is healthy, tasty, and sustainable. In parallel, it is necessary to understand the influence of consumers' practices in maintaining the beneficial food attributes from purchasing to consumption. The whole supply chain must be resilient, fair, diverse, transparent, and economically balanced to make different food systems sustainable. This chapter focuses on the role of dynamic value chains using biodiverse, underutilised crops to

E. Pinto
Universidade Católica Portuguesa, CBQF – Centro de Biotecnologia e Química Fina – Laboratório Associado, Escola Superior de Biotecnologia, Porto, Portugal

EPIUnit – Instituto de Saúde Pública, Universidade do Porto, Porto, Portugal

H. Ferreira · C. S. Santos · M. N. da Silva · M. W. Vasconcelos (✉)
Universidade Católica Portuguesa, CBQF – Centro de Biotecnologia e Química Fina – Laboratório Associado, Escola Superior de Biotecnologia, Porto, Portugal
e-mail: mvasconcelos@ucp.pt

D. Styles
School of Chemical and Biological Sciences, University of Galway, Galway, Ireland

Bernal Institute, School of Engineering, University of Limerick, Limerick, Ireland

P. Migliorini
University of Gastronomic Science, Bra, CN, Italy

C. M. Galanakis (ed.), *Biodiversity, Functional Ecosystems and Sustainable Food Production*, https://doi.org/10.1007/978-3-031-07434-9_9

improve food system resilience and deliver foods with good nutritional and health properties while ensuring low environmental impacts, and resilient ecosystem functions.

Keywords Nutrition · Sustainability · Underutilised crops · Value chains

1 Introduction

There are about 50,000 edible plants on the Earth; however, current food systems are concentrated on only three: wheat, maize, and rice (Khoury et al., 2014). These species provide more than 50% of the plant-based calories consumed by the world's population and occupy 40% of the world's arable land. The lack of agricultural

G. Ntatsi
Laboratory of Vegetable Production, Department of Crop Science, Agricultural University of Athens, Athens, Greece

A. Karkanis
Department of Agriculture Crop Production and Rural Environment, University of Thessaly, Volos, Greece

M.-F. Brémaud · Y. de Mey · M. Meuwissen
Business Economics Group, Wageningen University & Research, Wageningen, the Netherlands

J.-I. Petrusan · S. Smetana · A. Profeta
German Institute of Food Technologies – DIL e.V, Quakenbrück, Germany

B. Silva
Universidade Católica Portuguesa, CBQF – Centro de Biotecnologia e Química Fina – Laboratório Associado, Escola Superior de Biotecnologia, Porto, Portugal

German Institute of Food Technologies – DIL e.V, Quakenbrück, Germany

L. M. M. Krenz
Institute for Food and Environmental Research (ILU), Bad Belzig, Germany

D. Pleissner
Institute for Food and Environmental Research (ILU), Bad Belzig, Germany

Leuphana University of Lueneburg, Institute for Sustainable Chemistry, Lueneburg, Germany

M. Debeljak · A. Ivanovska
Department of Knowledge Technologies, Jozef Stefan Institute, Ljubljana, Slovenia

B. Balázs
Environmental Social Science Research Group (ESSRG), Budapest, Hungary

D. Rubiales
Institute for Sustainable Agriculture, CSIC, Córdoba, Spain

C. Hawes · P. P. M. Iannetta
Ecological Sciences, James Hutton Institute, Dundee, Scotland, UK

diversity has severe consequences on biodiversity and global environmental sustainability, namely soil degradation and higher global emissions (FAO, 2010a). Therefore, it is vital to stimulate the cultivation of less common species, known by "underutilised crops" (UCs) and enhance awareness of, plus improve where necessary, their nutritional and environmental proprieties. Here we describe how this can be achieved via integrating UCs in value chains while realising their benefits using a "system function approach".

Agri-food systems comprise actors and activities involved in the production, processing, distribution, consumption, and disposal of food products (FAO, 2021). By 2050, food demand is projected to increase by 60% relative to 2005. However, this projection is highly sensitive to, among other things, consumption patterns (diets), distribution, and levels of food waste (FAO, 2018b; Hunter et al., 2017). By 2067 the population is expected to reach 10.4 billion, with Africa and Asia accounting for 81% of this growth (Britt et al., 2018). Meanwhile, there is increasing pressure on agriculture and the broader land sector to deliver food, feed, fibre, fuel, bio-based materials, and ecosystem services – including nature-based solutions to climate change (Huppmann et al., 2018). Rapid cuts in greenhouse gas (GHG) emissions, alongside adaptation to a changing climate, are essential to maintain food system viability, let alone sustainability. Food systems account for 21–37% of global anthropogenic GHG emissions (Masson-Delmotte et al., 2019). On current trajectories, these emissions alone (excluding other industry, transport, and building sources) could exceed Paris Agreement targets for climate stabilisation (Clark et al., 2020). Similarly, food systems are key drivers of "Planetary Boundaries" exceedances across land use, biodiversity loss, and nutrient cycling (Steffen et al., 2015). Livestock production dominates many of these impacts (Foley et al., 2011; Rogelj et al., 2018; Vermeulen et al., 2012; Eshel et al., 2014; FAO, 2018a; Steinfeld et al., 2006). Up to now, food system intensification has been highly successful at delivering more output per unit of land (Burney et al., 2010). This has helped reducing GHG emission intensity per unit of output and sustain increasing levels of consumption to the point where a larger share of global population is obese rather than under-nourished (Benton & Bailey, 2019). To meet increasing food demands, focus during since the 60's was on intensification of agricultural systems, characterized by low crop diversity and large use of chemical inputs. Together with the implementation of low-input agronomic practices, crop diversification is highlighted today as a key issue for future sustainable development of agroecosystems valorising natural and cultivated biodiversity for agricultural purposes (Stagnari et al., 2017), resulting in greater ecosystem services and resilience (Springmann et al., 2018). Going beyond the recent focus on efficiency to deliver more food, fibre, and fuel at a dramatically lower aggregate environmental cost and with resilience is a massive challenge – necessitating transformative change beyond the incremental improvement of business-as-usual (Fanzo et al., 2021). The effective transformation will require integration of demand-side measures (e.g., reduced consumption of livestock products) alongside reconfiguration of value chains. This, in turn, will deliver both food and value-added more fairly, changing primary production to provide food and a plethora of ecosystem services, while preserving large areas of land for nature-based solutions (Fanzo et al., 2021; IPCC, 2019; Willett et al., 2019).

Sustainable diets can be defined as diets with a low environmental impact that contribute to food and nutrition security, and health in the present and future generations. They are "*protective and respectful of biodiversity and ecosystems, culturally acceptable, accessible, economically fair and affordable; nutritionally adequate, safe and healthy; while optimising natural and human resources*" (FAO, 2010a). Consumers are increasingly aware of the impact of their choices and behaviours on environmental sustainability. Accordingly, they are looking for demonstrations of sustainability commitment from food industries along the entire product life cycle. These include the interest in the origin of ingredients, higher demand for locally sourced products, and clean labels. For instance, food trends for 2022 are in line with those observed in the last years, looking ahead to an increase in the demand for plant-based and alternative proteins, better ingredients, and functional foods and beverages (Mintel, 2021).

This chapter explores the prospects of better integration of UCs in value chains and presents how their multiple benefits can be realised in a systems function approach. We show the current status of underutilised crops from cultivation to various uses and benefits. Then we turn to challenges in the value chain from farm to fork. Finally, we introduce policies that could improve investment in underutilised crops.

2 The Role of Underutilised Crops

2.1 What Are Underutilised Crops?

Staple crops currently dominate agricultural production and global food supplies. Diets around the globe are becoming more standardised, relying on very few crops or crop varieties– together with high intakes of sugar and oil. With staple crops' technological and policy investments, these new dietary habits have led to the neglect of about 7000 plant species documented as human food (FAO, 2010a). These neglected, underutilised, minor, or orphan crops are locally adapted to challenging environments, domesticated by local communities, and require lower inputs than staple crops, but are characterised by having low agricultural production and no perceived economic importance in advanced economies (Ebert, 2014). There are many possible definitions for a UC, and taking into account the FAO's work on this topic, these can be defined as '*a neglected, but valuable species, landrace, variety, or cultivar that has limited current use in a given geographic, social, and economic context and that holds great promise to diversify agricultural systems, create resilient agroecosystems, diversify diets, and create economically viable dynamic value chains (for feed, food, and non-food uses)*'.

UCs are therefore considered key to sustainable food security in the future, as scientists have been discussing their role in climate change mitigation and potential for exploitation for years (Padulosi et al., 2011). Given their more rustic nature, they contribute to agroecological resilience through system diversification and are an important component of the culture and diets of specific regions of the world (Massawe et al., 2016). However, they are underexploited and under-conserved resources, with minimal research and investment in their development (FAO, 2010b). For example, UC seed conservation is mainly made by gene banks, complemented by family or local seed networks (Padulosi et al., 2011). Current factors hindering the broader incorporation of UCs in food systems have been reviewed. They include their lower productivity and yield potential, lack of trait improvement and processing technology, limited market availability, lower cooking quality, and lack of knowledge both at the cultivation and at the consumption level (Bekkering & Tian, 2019; Hunter et al., 2019; Saini et al., 2021).

General examples of UCs include millets, roots and tubers, pulse crops, fruits and vegetables, and tree nuts (FAO, 2010b). These are traditionally grown for food, fibre, fodder, oil, or medicinal value (Ebert, 2014). Although these potential uses are acknowledged, further exploitation is yet to be developed due to these crops' semi-domesticated and neglected nature (Murthy & Bapat, 2020).

Millets, such as pearl millet (*Pennisetum glaucum*), foxtail millet (*Setaria italica*), barnyard millet (*Echinochloa* spp.), little millet (*Panicum miliar*), kodo millet (*Paspalum scrobiculatum*), finger millet (*Eleusine coracana*), are small-grained C4 cereals with a rich nutritional profile and many of them resilient to drought and high temperatures (Saini et al., 2021).

There are also many C3 underutilized cereals emmer (*Triticum dicoccum*), einkorn (*T. monococcum*), spelt (*T. spelta*), or rye (*Secale cereale*). Other UCs such as amaranth (*Amaranthus* spp.), buckwheat (*Fagopyrum esculentum*), chia (*Salvia hispanica*), or quinoa (*Chenopodium quinoa*), are considered pseudocereals because despite having seeds resembling the cereal grains, do not belong to Poaceae. The other large family containing many UCs is the Fabaceae, including many species of interest for forage or pasture, but also many others whose seeds are valued for food and feed. These species are known as grain legumes or pulses, and include good examples of warm-season pulses like winged bean (*Psophocarpus tetragonolobus*), horse gram (*Macrotyloma uniflorum*), lima bean (*Phaseolus lunatus*), hyacinth bean (*Lablab purpureus*), cowpea (*Vigna unguiculata*), mung bean (*Vigna radiata*) or barbara groundnut (*Vigna subterranea*), and of cool-season ones like faba bean (*Vicia faba*), grass pea (*Lathyrus sativus*) or lupin (*Lupinus albus*), among many others (Rubiales et al., 2021).

Underutilised roots and tubers are represented mainly by taro (*Colocasia esculenta*), yam (*Dioscorea* sp.), ulluco (*Ullucus tuberosus*), yautia (*Xanthosoma sagittifolium*), arrowroot (*Maranta arundinaceae*), and giant swamp taro (*Cyrtosperma paeonifolius*), and sweet potato (*Ipomoea batatas*) (FAO, 2010b; Li et al., 2020b). These crops are essential food on small-holder farms in marginal rural areas where they can be the primary source of nutrients during periods of food scarcity (Siddique et al., 2021).

Fruits and vegetables, such as wild melon (*Citrullus lanatus*), wild mustard (*Sinapis arvensis*), jackfruit (*Artocarpus heterophyllus*), mangosteen (*Garcinia mangostana*), African eggplant (*Solanum aethiopicum*), or grape (*Vitis* spp.) (FAO, 2010b; Massawe et al., 2016) are essential foods to support small-scale farmers and serve as the primary source of nourishment in underdeveloped countries (Siddique et al., 2021). Hence, these underutilised horticultural crops are vital in supporting nutritional security and avoiding malnutrition and hidden hunger issues with respect to the lack of specific vitamins and micronutrients (Nandal & Bhardwaj, 2014). Unfortunately, most tree nuts are also considered underutilised, as little or no research has been dedicated to their development. Some examples include cashew nut (*Anacardium occidentale*), Brazil nut (*Bertholletia excelsa*), chestnut (*Castanea sativa*), acorns (*Quercus nuts*), and tiger nut (*Cyperus esculentus*) (Asare et al., 2020; Murthy & Bapat, 2020). Their importance has been put forward in the latest diet recommendations due to their high contents in bioactive compounds and their biological activity (Willett et al., 2019). The harvested area and production levels of some examples of UCs mentioned above are listed in Table 9.1, although is still challenging to trace UCs production numbers on a worldwide scale. However, for the crops where numbers are available, increasing UCs agricultural production needs further investment in technology and research. This could be focused on their beneficial impacts and resilience in semi-arid and arid areas and their adaptation to different climatic scenarios. Additionally, when looking at the countries with higher levels of UCs production (Table 9.1), food systems benefit from UCs incorporation and adaptation.

2.2 Preservation of Biodiversity

Biodiversity and ecosystems are the very foundations of human existence and contribute to human well-being in three fundamental ways: through the production of goods (food, fibres, water, air, medicines, and recreational spaces); provision of services (cultural, religious, aesthetic, and spiritual); and the processes that balance and regulate the above (pollination, prevention of soil erosion, microclimate control and nutrient cycling and transfer) (Buiatti et al., 2010). Cultural and natural biodiversity that include thousands of UCs are the basis of agrobiodiversity, which is preserved, alike nutrition and health, by traditional farming practices and cultural identities. These practices also make long-term sustainable use of natural resources and the environment, increasing productivity and ensuring food security and sovereignty (Buiatti et al., 2010). Unfortunately, however, biodiversity and agrobiodiversity are in a state of decline worldwide, and with it, the inclusion of UCs in local agri-food systems. The key factors contributing to the loss of biodiversity include unsustainable farming, fishing, and forest practices which lead, among other things, to natural resource consumption, habitat loss and fragmentation, soil deterioration, water and atmospheric pollution, and genetic pollution (MEA, 2005). Moreover, global climate change threatens biodiversity by altering habitats and modifying the

Table 9.1 World harvested area, production, and main country and production levels of millets and some examples of roots and tubers, pulse crops, fruits, vegetables, and tree nuts in 2020, and of the staple crops rice, wheat, and maize (FAOSTAT, 2022)

Underutilised crop	Type	World harvested area (ha)	World production (tonnes)	Main country and production (tonnes)
Millets	Cereals	32,117,837	30,463,642	India: 12,490,000
Taro (*Colocasia esculenta*)	Roots and tubers	1,809,485	12,838,664	Nigeria: 3,205,317
Yams (*Dioscorea* sp.)	Roots and tubers	8,831,037	74,827,234	Nigeria: 50,052,977
Yautia (*Xanthosoma sagittifolium*)	Roots and tubers	32,020	398,290	Cuba: 101,618
Sweet potato (*Ipomoea batatas*)	Roots and tubers	7,400,472	89,487,835	China: 48,949,495
Faba bean (*Vicia faba*)	Legumes	2,671,497	5,669,185	China: 1,723,598
Cowpea (*Vigna unguiculata*)	Legumes	15,056,435	8,901,644	Nigeria: 3,647,115
Barbara groundnut (*Vigna subterranea*)	Legumes	354,619	230,619	Burkina Faso: 57,428
Grapes (*Vitis* spp.)	Fruits	6,950,930	78,034,332	China: 14,769,088
Mangoes, mangosteens, guavas	Fruits	5,522,933	54,831,104	India: 24,748,000
Mustard seed	Vegetables	619,495	540,454	Nepal: 214,055
Brazil nut (*Bertholletia excelsa*)	Tree nuts	11,561	69,658	Brazil: 33,118
Chestnut (*Castanea sativa*)	Tree nuts	582,545	2,321,780	China: 1,743,354
Cashew nut (*Anacardium occidentale*)	Tree nuts	7,101,967	4,180,990	Côte d'Ivoire: 848,700
Rice	Staple crop	164,192,164	756,743,722	China: 211,860,000
Wheat	Staple crop	219,006,893	760,925,831	China: 134,250,000
Maize	Staple crop	201,983,645	1,162,352,997	USA: 360,251,560

equilibria of crucial species. At the same time, the narrow spectrum of products traded from agriculture, forestry, and fisheries make ecosystems increasingly vulnerable (FAO, 2019a).

When considering genetic resources for food and agriculture (GRFA), we refer to crop diversity created by man (FAO, 1999). It underpins agriculture's productivity, resilience, and adaptive capacity and is an integral part of people's cultural identity (IAAKSTD, 2009). Given that they supply most of the food for human consumption, they are fundamental for creating sustainable agriculture and food safety. Yet we are losing them at an alarming rate. Since agriculture began to develop about 15,000 years ago, it is estimated that around 10,000 species have been used for human food. Currently, no more than 120 cultivated species provide 90% of the human food supplied by plants. Only four plant species (potatoes, rice, maize, and wheat) and three animal species (cattle, pigs, and chickens) provide more than 50% of all human food. The degree of biodiversity in agroecosystems depends on: (i) variety of vegetation inside and around the agroecosystem; (ii) length of different crops; (iii) intensity of management; and (iv) degree of isolation from wild vegetation (Altieri, 1999).

The best way to conserve traditionally cultivated plants, including UCs, and raised animals, is to utilise them. Two distinctive conservation methods of UCs can be identified as *in-situ* (and on farms) and *ex-situ* practices. The first is carried out in conditions that allow a natural and continuous evolution and co-adaptation through cultivation or breeding. In contrast, the second entails protecting endangered species and genetic resources (plant varieties and animal breeds) outside their natural habitat, for example, by preserving seeds in a germplasm bank. All conservation measures should be planned and implemented on a scale determined by ecological and social criteria, focusing on densely populated areas, and protected natural areas. It is interesting to note that the leading cause of the loss of Genetic Resources for Food and Agriculture (GRFA) would appear to be underutilisation as opposed to overexploitation. Given the high interdependency of countries on GRFA, international cooperation in this area is not an option but a must. This cooperation has led to intergovernmental negotiations and the adoption of the legally binding International Treaty for Plant Genetic Resources for Food and Agriculture (ITPGRFA). The objectives of the ITPGRFA (FAO, 2001) are the conservation and sustainable use of all plant genetic resources for food and agriculture and the fair and equitable sharing of the benefits arising out of their use, in harmony with the Convention on Biological Diversity, for sustainable agriculture and food security.

Wider cultivation of UCs would help to deliver on the pledge to diversify agricultural systems, create resilient agroecosystems, diversify diets, and develop economically viable dynamic value chains for feed, food, and non-food uses (Fig. 9.1) (Bavec et al., 2017; Gregory et al., 2019). Therefore, characterising their nutritional and health attributes is essential to promote their wider adoption amongst populations.

2.3 Health and Nutritional Benefits of UCs

Recent data suggests that ending world hunger and malnutrition in all its forms is becoming increasingly more challenging, particularly exacerbated by the COVID-19 pandemic (Lowe, 2021). Indeed, the latest international reports highlight concerns regarding the world's ability to meet the sustainable development agenda by 2030, including the Zero Hunger target (FAO, 2021). Around 118 million more people were facing hunger in 2020 compared to 2019, representing 768 million undernourished people worldwide (FAO, 2021). Limited access to a healthy, balanced, diverse, and nutritious diet, particularly in low- and middle-income countries, contributes to this scenario (Li et al., 2020a, b). In 2020, nearly 1 in 3 people, around 2.37 billion people in the world, did not have access to adequate food, 40% of which or almost

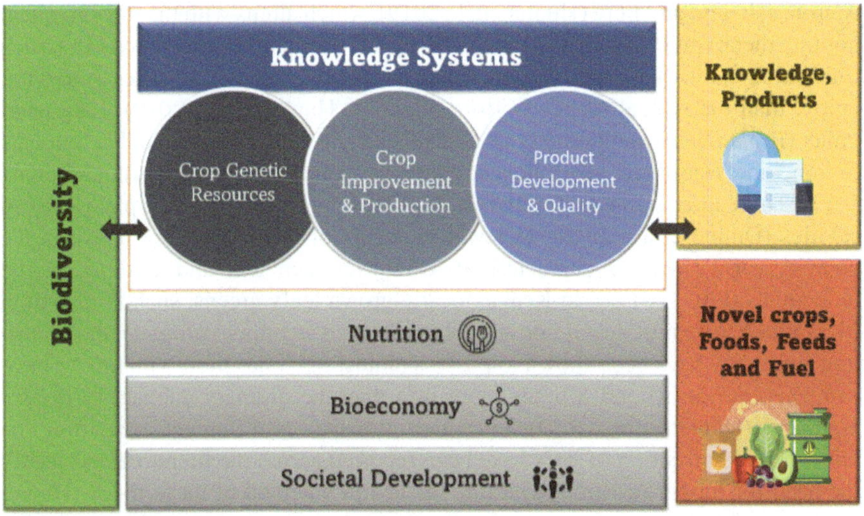

Fig. 9.1 The role of underutilised crops (UCs) in the preservation of biodiversity. (Modified from Gregory et al., 2019)

928 million, faced severe levels of food insecurity (FAO, 2021). Simultaneously, overweight and obesity rates continuously spread worldwide as technological developments push societies towards more sedentary lifestyles with easier access to highly caloric but nutritionally poor and highly processed foods and beverages (Popkin et al., 2020). Hence, it is estimated that multiple micronutrient deficiencies, frequently underlying a "hidden hunger" phenomenon, affect more than 2 billion people globally (von Grebmer et al., 2014). For instance, of the 7 billion world's population, more than 1.6 billion and more than 200 million people suffer from iron deficiency and vitamin A insufficiency, respectively (Li et al., 2020a, b). Such nutrient deficits can not only impair human health but may ultimately hamper socioeconomic development (Ibeanu et al., 2020).

Although a shortage in food supply to adequately feed the growing world's population has become a critical reality, it is advocated that current food systems should shift from the quantity-oriented mindset to a more quality-focused approach. This will allow exploration of new and more sustainable means to nourish populations (Hunter et al., 2019). The investment in UCs seems required to restore sustainable agriculture practices and address the global food challenges (Li & Siddique, 2020). These crops represent a local, affordable, sustainable, and culturally acceptable way to improve diversity in food supply systems and, therefore, access to nutrient-dense foods. Yet, many countries fail to recognise their rightful value (Hunter et al., 2019). Among various socio-economic and political reasons, the over-reliance on more profitable high-yielding monocultures has caused the marginalisation of minor crops, including primarily wild or semi-domesticated crops (Li et al., 2020a, b). However, data suggests that these crops often provide greater levels of essential nutrients in comparison to current major staple crops, including vitamin C, vitamin

A, iron, calcium, and fibre (Hunter et al., 2019), which aligns with the most frequent limiting dietary micronutrients (Lowe, 2021). For example, an indigenous Brazilian fruit, camu-camu (*Myrciaria dubia*), has been found to contain 40 times more vitamin C than the typical orange (*Citrus sinensis*) (Hunter et al., 2019). Like pearl millet (*P. glaucum*), traditional crops in Pakistan and Nepal possess higher amounts of iron, zinc, riboflavin, and folic acid than rice, maize, and wheat (Adhikari et al., 2017). These three staple crops provide more than 50% of the world's plant-derived calories (Dulloo et al., 2016). Noteworthy, many of these crops are native in the poorest world's regions characterised by nutrient-deficient and health-impaired individuals (FAO, 2019a). In Kenya, locally grown leafy greens, such as amaranth (*A. dubius*), were introduced into school meals in an attempt to mitigate undernutrition since they have been shown to possess almost 3.5 times more vitamin A (beta-carotene equivalent) and 6 times more iron than the ordinary cabbage (*Brassica oleracea*) (Hunter et al., 2019). Inter and intra-species differences regarding the nutritional composition of UCs have been reported across the literature and justify the need for further research regarding the health potential of these foods (Hunter et al., 2019). Also, challenges are still present when considering the processing associated with including UCs in food product development.

2.4 Integration of UCs in Food production and Processing

It is vital to provide populations with diverse and nurturing foods to keep them healthy without damaging the environment (Willett et al., 2019). As the global population keeps growing, it presses for the intensification of the current food system, causing environmental impact to increase beyond sustainable levels (Poore & Nemecek, 2018). The highest impact of food production comes from raw materials sourcing (Poore & Nemecek, 2018; Willett et al., 2019). Therefore, many environmental problems can be easily traced back to this point: from deforestation to the desertification of arable land to lixiviation, loss in biodiversity and others (Mentis, 2020; Zhao et al., 2015).

Farming has a considerable environmental impact that could be lowered. Intensification leads to higher yields per given area and higher resource consumption (fuel, irrigation, fertilisers, and pesticides). Sustainable farming can help lower the impact from these inputs (Commission on Sustainable Agriculture and Climate Change, 2012) and growing quality plant proteins tackles many different sustainability points (e.g., agrobiodiversity and potentially avoiding high-impact animal-protein production). As mentioned above, presently, only a few crops are responsible for almost 50% of global food intake (FAO, 2018b). This becomes a problem when the repetitive growth of the same cultures reduces soil biodiversity and depletes nutrients beyond natural replenishment rates (Zhao et al., 2015). This then leads to an increasing need for synthetic fertilisers that cause additional damage to the environment, as explained previously. Increasing the consumption of different proteins is often suggested to diversify diets. Multiple studies show these alternative proteins and developed products have a lower impact (Smetana et al., 2015, 2021; Tello et al., 2021). Increasing the production and consumption of UCs can also decrease the environmental strain of the repetitive growth of crops.

Crop replacement is not easy since their cultivation is adapted to the different areas where they are grown. One possibility that can help with this is the revalorization of crops adapted to those areas, but that lost competitiveness. These crops, however, may need to be improved to increase their competitiveness and lead to agricultural diversification and reduction of risks (e.g., pest attacks that can destroy entire cultures; Popp et al., 2013). Another possibility could be the introduction on new UCs, their adaptation needing testing and probably further improvement. Still, the use of these crops can lead to novel product development, and this, in return, revalorize the crops.

The development of these novel products should consider consumer trends with more fresh-like attributes and long shelf-life (Palou et al., 2020). Due to this, implementation of novel technologies such as pulsed electric fields, high-pressure processing, and high-pressure homogenisation or ultrasound, can provide an interesting starting point. All these technologies can provide potential solutions to the pressing challenge of sustainability (Matthews et al., 2019). The plant protein industry shows many advantages and strengths over the animal protein industry, as shown by Petrusán et al. (2016). These advantages of plant proteins could underpin the success of integrating UC-derived products through marketing and certification strategies that support their broader commercialisation. As a result of the increasing demand of local vegetable protein food, both in traditional uses and in novel processed food business (Cusworth et al., 2021) legume cultivation is speedily recovering.

2.5 *Consumers, Cultural Barriers, and Leverages*

Most of the research to increase the consumption of UCs has been from the supply side (Cheng et al., 2017; Dawson et al., 2009; Mayes et al., 2012). These have focused on highlighting their nutritional and environmental properties to justify the additional effort in improving the characteristics of those crops (e.g., yields, agronomic properties, environmental impact). Understanding consumer knowledge, acceptance, and preferences for UCs are essential in enhancing their consumption levels to increase micronutrient intake.

A barrier to higher adoption rates of UCs as a staple food is the rise of convenience foods and modern consumption patterns. In particular, consumers in developing countries are increasingly abandoning the traditional diets that these crops are part of and are replacing them with western diets (Cordain et al., 2005). Likewise, in industrialized countries, many recipes and products have fallen into oblivion in the last century, partly due to the change in direction to a society where meat is the dominant food (Holm & Møhl, 2000). For example, in Germany, the consumption of legumes decreased from 20.7 kg in 1850 to 3.0 kg per capita and year in 2017

(BLE, n.d.; Teuteberg, 2006). In this context, many legumes varieties became extinct or have been forgotten, e.g., lentils from Swabian Alb (Reif et al., 2021). Moreover, consuming such traditional UCs often requires know-how, i.e., how to prepare and cook them (sometimes depending on the stage of maturity), and perhaps even knowing which cultivar (variety) is more desirable for a particular use. These knowledge gaps often render UCs non-competitive against well-known and globally consumed staples such as rice, wheat, maize, soybean, and potatoes.

There are several approaches to increase the consumption of underutilised crops. One way is to convey knowledge about such food products and their preparation. Activities like the Bavarian specialty database (https://www.spezialitaetenland-bayern.de/spezialitaeten) or the Slow Food 'Ark of Taste' (https://www.fondazioneslowfood.com/it/arca-del-gusto-slow-food) try to preserve and to promote the knowledge and to create consumer awareness. Through such measures, consumer preferences for traditional specialties (Profeta et al., 2007) and authentic foods (Wirsig et al., n.d.) can be addressed and triggered. Furthermore, consumers are becoming increasingly conscious of their food basket's health and nutritional profile (Profeta, 2019). The tendency is to avoid chemicals and synthetic foods and preference for nutrition through foods that bring "natural" attributes. In this context, many UCs have advantages compared to staple foods, as outlined in the chapters before. In this situation, marketing communication to the final consumer should highlight UCs' special health and environmental characteristics.

Looking at new ways of incorporating UCs into consumers' diets requires creativity. UCs could easily fit into a modern lifestyle by adding value and creating ready-made convenience products. Finished convenience products do not need consumers to prepare or cook the corresponding UC. Thus, by finding novel and innovative methods to organise, sell and consume UCs, consumers can discover more diverse ways to enjoy this nutritional and culturally relevant food source. Also, due to the dominant role of taste in consumers' purchase decisions, there is the need to bring UCs closer to consumers' preferences. In this way, the value chain for UCs will get sustainable economic, environmental, food security, and nutritional benefits.

2.6 The Role of Markets, Labelling, and Certification

Many UCs are locally popular crops, are nutritionally superior, they generate income, are resistant to drought, they conserve natural resources, are tasty and delicious, are necessary for climate adaptation, and often have long culinary traditions. Still, they continue to be marginalised by research and undervalued by development (Eyzaguirre et al., 1999). Most importantly, market factors are responsible for rendering these crops underutilised; consequently, UCs become unable to meet the global market requirements, industrialised agri-techniques, and uniformity standards. Similarly, the policy is also often divisive, even "food discriminatory", and this explains why UCs are undervalued and underinvested (Chishakwe, 2008). Furthermore, increasing their value for more comprehensive production and commerce depends on research-intensive activities. These crops are mostly not suitable for cultivation or cannot meet uniformity standards, often due to genetic erosion. UCs are niche-specific, versatile, and differ substantially from mainstream crop value chains. Breeding programs, seed multiplication, collective actions of value chain actors are indispensable for market development (Stamp et al., 2012). Increasing the consumption of UCs requires not only systemic demand-markets development, but more expansive capacity building in the value chain. Any attempt at commercialising UCs requires demand expansion, increased supply, marketing channels efficiency, and a supply control mechanism (Gruere et al., 2008). Scarano et al. (2021) identified several research-intensive factors that could raise awareness of and fully realise the benefits of UCs. These include research on the genetic traits linked to the climate adaptation; characterisation of main nutrient classes and their biosynthesis pathways; quantification and characterisation of the main antimetabolic factors/antinutrients; and understanding biological activities in the prevention of human diseases. Finally, any research on UCs needs to benefit from the full participation in exploration and action learning of value chain actors in a participatory setting (Vernooy, 2021). In sum, value addition would be a high potential for UCs in a diverse and sustainable food system only if more significant investment in research and development becomes more available.

Consumers are increasingly interested in local, traditional, or sustainably produced fruits, vegetables, or arable crops. This provides an excellent premise to label such products to make consumers aware of unique product qualities, taste, shape, and colour (Wirsig et al., 2011). However, there is no label for UCs in the food market. Nonetheless, at least in the European Union, there are different food quality labels as, e.g., PDO (Protected Designation of Origin), PGI (Protected Geographical Indication), and TSG (Traditional Specialties Guaranteed), or Protected Mountain Products. These allow covering aspects of crop diversity or seed origin (Benner et al., 2008; Profeta et al., 2006). According to this scheme, many underutilised food products, e.g., Bamberger Hörnla or Alho da Graciosa (Berbereia, 2015), are protected and profit from marketing campaigns promoting the EU quality system. A recent case study from Germany shows the positive effects of this official labelling scheme for such products (Chilla et al., 2020). Since the EU regulation even allows applications for PDO, PGI, and TSG from third countries, there is a legal labelling framework that nearly all countries can use. Despite this, the existing regulation was not specially developed for underutilised groups. Such a long and complicated application process is too great a task for small-producer groups. In the next revision of the EU regulation, the unique requirements and needs of producers of UCs should be considered to improve their access to the existing scheme.

2.7 The Non-food Uses for Underutilised Crops

Crops are most likely underutilised when their potential is unknown, or their availability is not sufficient to establish an economically feasible utilisation. This is particularly challenging in rural areas where long distances need to be bridged to harvest, treat, and utilise biomass. Nevertheless, there are economic opportunities, and most UCs possess a high potential to serve as a source of food and non-food products. A combination of both uses may foster the cultivation of UCs. As with UCs considered for food use, the prospect of UCs for non-food use can be assessed based on the biochemical components such as lipid, carbohydrate, and protein contents. Depending on the composition, tailor-made harvesting and utilisation approaches can be developed, allowing the implementation of a biorefinery and the generation of products and services even in rural areas.

An example of a successful new crop is late-harvested grass, usually cut in autumn at natural conservation areas. While fresh grass has been considered as feedstock in green biorefineries or as feed, late-harvested grass utilisation is still at the early stage. The biochemical composition is the reason for the different utilisation intensities of fresh and late-harvested grasses. On a dry matter basis (w/w), grass can contain 20–30% cellulose, 15–25% hemicellulose, 3–10% lignin, 6–25% protein, 1–2.5% fat, 1–2.5% starch, and 5–20% ash (Grass, 2004). Fresh grass is rich in proteins and is easier to digest. Contrarily, matured grass contains less metabolisable energy, for instance, due to a reduced degradable protein content (Bovolenta et al., 2008; Waramit et al., 2012; Boob et al., 2019; Koidou et al., 2019) and reduced nutrient contents such as P, N, and K (Bokdam & Wallis de Vries, 1992; Mládek et al., 2011; Schlegel et al., 2016; Boob et al., 2019; Koidou et al., 2019).

Even though the protein content is comparably low, protein extraction can be worthwhile. About 30–60% (w/w) of the original protein can be recovered by mechanical pressing or alkaline extraction (Bals et al., 2012; Hermansen et al., 2017). The highest value arises from the fibres present in the grass. After mechanical pressing, up to 95% (w/w) of the fibres remain in the press cake (O'Keeffe et al., 2011) and can be used as a feedstock in pulp and paper production (Finell, 2003), for biocomposites (Biowert, 2021), or building materials (King et al., 2013).

Although late-harvested grass has been investigated as a substrate for combustion (Tonn et al., 2010; Lewandowski et al., 2003), pyrolysis (Wilson et al., 2013; Mos et al., 2013) or as lignocellulosic feedstock in fermentation (Dien et al., 2018; Jungers et al., 2013) a biorefinery that operates purely on late-harvested grass is currently not working. As mentioned above, the challenges are the availability of biomass as late-harvested grass appears once, maximum twice per year. However, the availability of biomass and the services that the biomass delivers during its cultivation stage should be considered. Late-harvested grass is vital to conserving biodiversity and storing carbon in the soil. Thus, the use of late-harvested grass can be an example where ecosystem services are preserved, and the potential of the biomass is simultaneously utilised.

2.8 Environmental Benefits of UCs

Modern crop varieties deliver reliable and high yields, but the widespread adoption of monocultures in intensive agriculture often leads to environmental depletion and higher chemical inputs. Most of the cereal crops that dominate global production, such as wheat, rice, and maize, require an increased water supply and have low adaptive resilience to water shortage, raising concerns about their suitability to under the forecasted scenarios of more frequent and severe droughts (Mueller et al., 2012). Pesticides and herbicides target harmful organisms that can harm or compete with crops. Still, they can also reach animals and plants beyond the seemingly restricted area of their application. For example, several pesticides are harmful to bees and other insects, limiting their ability to pollinate crops and other plants (Uhl & Brühl, 2019). Phytosanitary products also impair soil microorganisms involved in carbon and nitrogen cycling, contributing to climate change. Highly disturbed soils with low microbial biodiversity quickly lose carbon to waterways and the atmosphere, propelling the accumulation of greenhouse gases in the atmosphere (Lazcano et al., 2021). This will ultimately lead to warming temperatures and extreme weather events, further impairing plant and soil communities and favouring the adaptation of invasive species that disrupt native ecosystems (Diffenbaugh et al., 2008). Intensive agriculture is also largely reliant on nitrogen fertilisers, which can run off into waterways, decrease the available oxygen in the water and cause eutrophication of both fresh and saltwater ecosystems, making them uninhabitable for aquatic organisms (Huang et al., 2017).

In the forthcoming decades, food systems are estimated to have an increasing environmental impact by intensifying global ecological pressures and destabilising key ecosystem processes, fostering climate change (Springmann et al., 2018). On the other hand, climate change will also pose challenges to ecosystems worldwide, as plants will have to endure in drier, saltier soils (Onyekachi et al., 2019). UCs are typically native to the environments in which they are grown, thus requiring fewer external and economic inputs than conventional crops. They can show adaptation to dryland cropping systems, high water use efficiency, and short growing seasons while delivering similar yields to major cereal crops (Karunaratne et al., 2015). They can also prosper in harsh environments and poor soils by fixing carbon from the atmosphere and nitrogen in the ground, offering opportunities for nutrient use efficiency and lowering global GHG emissions (Mabhaudhi et al., 2019). Developing powerful sustainable and bio-based agronomic strategies for crop nutrition, irrigation, soil fertility, and stress tolerance could allow a significant reduction in the use of chemical fertilisers and water for agriculture (Karkanis et al., 2018; Karavidas et al., 2022). They can also improve environmental resilience and quality of crops (Rivero et al., 2022; Dubey et al., 2020). Moreover, integrating neglected landraces, ecotypes, and varieties with increased nutrient use efficiency, water use efficiency, and stress tolerance into such farming systems could help in this direction (Dwivedi et al., 2016; Rivero et al., 2022). Specifically, UCs can contribute to environmental resilience and in mitigating climate change by the following means:

(a) Delivering tolerance to drought, salt, and toxic metals stress, as is the case of several cultivars of tomato, chickpea, barley, rice, wheat, and sunflower that possess specific genes involved in abiotic-stress tolerance (Mammadov et al., 2018 and references therein; Kumar Rai et al., 2021);
(b) Improving water use efficiency, as they can grow as a dryland crop without supplemental irrigation (e.g., millets), as well as by improving water quality (e.g., winged bean); (Kamel et al., 2018 and references therein);
(c) Fostering biodiversity and beneficial wild animals, thus promoting resilience against pests and diseases as part of integrated pest management, as detailed above (Villegas-Fernández et al., 2011; Sardana et al., 2017; Mammadov et al., 2018 and references therein);
(d) Decreasing the need for inputs and supporting natural carbon and nitrogen cycles, particularly concerning legumes that promote the accumulation of nitrogen in the soil while capturing carbon from the atmosphere (Mabhaudhi et al., 2019 and references therein);
(e) Reducing the high environmental impact of large-scale food and feed production and consumption worldwide by creating shorter value chains and decreasing transportation burdens (Weinberger & Swai, 2006; Will, 2008; Imathiu, 2021).

The exploitation of UCs as part of a holistic transformation of food systems plays a pivotal role in environmental sustainability (Haddad et al., 2016). Table 9.2 showcases the environmental and ecosystem services provided by distinct UCs that can lever the security of the global food supply while ensuring the sustainable use of environmental resources. Figure 9.2 illustrates the multiple benefits of UCs that go beyond the farm level.

2.9 Genetics and Breeding of UCs

Being minor crops, there has been a lag in the overall genetic improvement of UCs due to limited investment compared to major crops. Applicable breeding methods are the same that could be used for any crop, from classical selection, to genomic assisted-breeding, being the availability of resources and the targets what makes the difference. Breeding more adapted and productive cultivars, thus meeting producer and consumer needs, enables a wider adoption in the value chain. However, when the surfaces are limited, the return of the breeding activity is not sufficient to support strong breeding programs. The agroecological transition requires not only greater UCs cultivation but also different cultivation approches, such as intercropping, organic, etc., each one requiring specific breeding strategies. Greater adaptation to low input conditions will be a leading priority in UC breeding, particularly

Table 9.2 Benefits of UCs in environmental stress resilience and supporting ecosystem services

Crop	Environmental stress resilience and ecosystem services	References
Asparagus (*Asparagus officinalis*)	Resilience to alkaline and saline soils, including in dry regions.	Shannon and Grieve (1999)
Bambara groundnut (*Vigna subterranean*)	Higher pod yield than groundnut under limited water supply, possessing All three drought tolerance mechanisms—Avoidance, escape, and tolerance.	Linnemann and Azam-Ali (1993) Collinson et al. (1996) Collinson et al. (1997)
Barley landraces (*Hordeum vulgare, Hordeum maritimum*)	Adaptation to high temperatures, drought, and salinity stress through the temporal accumulation of specific metabolites (e.g., proline).	Lakew et al. (2011) Ferchichi et al. (2018)
Christ's thorn jujube (*Ziziphus spina-christi*)	Leaves can serve as forage to animals under open grazing conditions. Root architecture supports sand dunes and other unstable soils. Heat and drought tolerance and suitability for growing in areas with little annual rainfall. Moderate tolerance to salinity and has been suggested for revegetation of moderately degraded saline lands-	Rao et al. (2014) and references therein
Citron watermelon (*Citrullus lanatus* var. *citroides*)	Through the accelerated transition from vegetative growth to reproductive growth, drought tolerance and avoidance.	Mandizvo et al. (2021) and references therein
Common bean (*Phaseolus vulgaris*)	Resilience to elevated atmospheric CO_2. Ability to accumulate nitrogen in the soil, improve soil quality, and require fewer fertiliser inputs.	Soares et al. (2019) Wilker et al. (2019)
Cotton landraces (*Gossypium somalense, G. barbadense, G. hirsutum, G. darwinii, G. longicalyx*)	Tolerance to insects, nematodes, and diseases (e.g., *Pseudatomoscelis seriatus, Rotylenchulus reniformis*, bacterial blight, leaf curl virus). Resilience to drought, salinity, and heat.	Mammadov et al. (2018)

(continued)

Table 9.2 (continued)

Crop	Environmental stress resilience and ecosystem services	References
Cowpea (*Vigna unguiculata*)	High grain yield and water use efficiency under semi-arid conditions through no-till farming. It improved cereal yield by increasing nitrogen use efficiency when in rotation with cowpea. Intercropping cowpea and amaranth improves phosphatase activity providing inorganic P in both species under limited NPK fertilisation, and intercropping sorghum with cowpea is recommended for semi-and arid environments due to higher water-use efficiency. Genetic markers associated with salt tolerance in cowpea could support improved tolerance to salinity.	Bationo et al. (2002) Chimonyo et al. (2016) Freitas et al. (2019) Ravelombola et al. (2018) Mndzebele et al. (2020)
Cruciferous vegetables (Brassicaceae family)	Higher salt tolerance and potential for saline regions than more common leafy greens. Resilience to drought.	Rao and Shahid (2016) Mafakheri and Kordrostami (2020)
Drumstick tree (Moringaceae family)	It can be planted as a windbreak or living fence and support climbers such as beans, black pepper, and yams. Leaves and twigs can be used as forage for livestock. Ground seeds can purify drinking water and flocculate contaminants. Potential for biofuel production.	Jahn et al. (1986) Azam et al. (2005) Fahey (2005) Ebert (2014), and references therein
Faba bean (*Vicia faba*)	Yield and nitrogen fixation are improved under elevated atmospheric CO_2 and terminal drought. Resistance to rust and chocolate spot.	Villegas-Fernández et al. (2011) Parvin et al. (2019a, b)
Fruit trees	Sequestration of atmospheric CO_2. Soil restoration and biodiversity conservation. Habitat provision for pollinating insects. Water transport over terrestrial surfaces.	Jansen et al. (2020) and references therein
Grasspea	Rusticity, adaptation to marginal environments, N fixing, nutricious seeds	Vaz Patto and Rubiales (2014) Rubiales et al. (2020)
Hyacinth bean (*Lablab purpureus*)	Displays drought tolerance and high water-use efficiency. Nitrogen-fixing activity.	Morris (2009) Naeem et al. (2009) Vidigal et al. (2018)

(continued)

Table 9.2 (continued)

Crop	Environmental stress resilience and ecosystem services	References
Jack bean (*Canavalia ensiformis*)	Able to grow in poor and acidic soils. Tolerant to drought, heat, and pest attacks. When intercropped with maize, it improves nitrogen fixation and weed control.	Dantas et al. (2019) Popoola et al. (2019) and references therein
Kersting's groundnut (*Macrotyloma geocarpum*)	Resilience to drought, pests, and diseases.	Obasi and Ezedinma (1991) Foyer et al. (2016)
Leafy greens (*Amaranthus* spp., *Bidens pilosa*, *Cleome gynandra*, *Corchorus olitorius*, *Cucurbitaceae* spp., *Ipomoea batatas*)	It can be produced with relatively low inputs. High water use efficiency under high temperatures and high radiation intensity.	Kuo et al. (1992) Slabbert and van den Heever (2006) Wang and Ebert (2013) Ebert (2014), and references therein Shayanowako et al. (2021)
Lentil (*Lens culinaris*)	Adaptation to multiple environments Resilience to lentil rust caused by *Uromyces viciae-fabae*. Adaptation to elevated atmospheric CO_2 and acute heat. Nitrogen fixation through the formation of root nodules containing symbiotic bacteria.	Wright et al. (2021) Bourgault et al. (2017) Parihar et al. (2018) Parvin et al. (2019a, b)
Lima beans (*Phaseolus lunatus*)	Resistant to viral and rust diseases and insect pests. Tolerance to drought and aluminium and manganese toxicity.	Ballhorn et al. (2009) Azeke et al. (2011)
Maize relatives (*Zea nicaraguensis*, *Z. diploperennis*, *Z. mays ssp. mexicana*)	Tolerance to insects, diseases, and weeds (e.g., *Cotesia marginiventris*, *Meteorus laphygmae*, gray leaf spot, *Striga hermonthica*). Resilience to drought, acid soil and aluminium, salinity, and waterlogging.	Mammadov et al. (2018) and references therein
Millets (*Eleusine coracana*, *Panicum miliaceum*, *Setaria italica*)	High water-use efficiency and resilience to low water availability and high temperatures. Potential as a rotational crop for wheat-based dryland farming by preserving soil moisture. Contributes to controlling winter annual grass weeds, pests and diseases. Tolerance to herbicides applied to corn. Potential for biofuel production.	Meldrum et al. (2016) Habiyaremye et al. (2017) Nielsen and Vigil (2017), Das et al. (2019) and references therein
Mungbean (*Vigna radiata* var. *radiata*)	Improves soil properties and provides additional nitrogen to subsequent crops (e.g., rice). Decreased pest and disease pressure.	Weinberger (2003)

(continued)

Table 9.2 (continued)

Crop	Environmental stress resilience and ecosystem services	References
Quinoa (*Chenopodium quinoa*)	It is a facultative halophyte, able to cope with high salinity levels.	Rao et al. (2014)
Rice relatives (*Oryza meridionalis*, *O. officinalis*, *O. rufipogon*, *O. Nivara*, *O. glaberrima*)	Tolerance to insects and diseases (e.g., *Nilaparvata lugens*. Rice blast, bacterial blight, grassy stunt virus, tungro disease). Resilience to drought, heat, cold, acid soils, aluminium, and salinity.	Thanh et al. (2006) Ndjiondjop et al. (2010) Mammadov et al. (2018) and references therein
Safflower (*Carthamus tinctorius*)	Displays high yield under high salinity. Tolerance drought and ability to adapt to hot and dry climates. The root system can access subsoil water at high depths.	Oelke et al. (1992) Fraj et al. (2013)
Salicornia dolichostachya	High tolerance to salinisation of agricultural land.	Katschnig et al. (2013)
Sorghum (*Sorghum bicolor*)	High water-use efficiency and adaptation to water stress can improve cereal productivity under water scarcity.	Hadebe et al. (2017)
Soybean relatives (*Glycine max, G. soja, G. argyrea, and G. pescadrensis, G. canescens, G. tabacina, and G. tomentella*)	Tolerance to cyst nematodes and fungi (e.g., *Heterodera glycines*, *Phakopsora pachyrhizi*, *Sclerotinia sclerotiorum*), Resilience to drought and salinity. Adaptation to elevated atmospheric CO_2 through greater harvest index and short stature.	Bishop et al. (2015) Mammadov et al. (2018) Soares et al. (2019)
Sword bean (*Canavalia gladiata*)	Tolerance to pests and diseases. Potential for lead (Pb) phytoremediation. When used for intercropping, it improves nitrogen fixation and provision	Ekanayake et al. (2003) Souza et al. (2013) Oyelakin and Olaniyi (2019)
Wheat ancestors (*Triticum secale, Triticum dicoccoides*)	Higher tolerance to salinity to drought than domesticated varieties.	Budak et al. (2013) Fraj et al. (2014)
Vetches (*Vicia sativa, V. narbonensis, V. articulata, V. ervilia* and other *Vicia* spp.)	Forage legumes providing also environmental services, including also nutritious seeds often used for birds feeding and even in traditional food uses	Rubiales and Flores (2020)
Winged bean (*Psophocarpus tetragonolobus*)	Highly resistant to biotic and abiotic stresses Seeds exhibit tolerance to storage pests. Suitable for cultivation in poor soils through nitrogen fixation. It can be used as a natural coagulant and improve water quality.	Keatinge et al. (2010) Ebert (2014), and references therein Kamel et al. (2018)

for organic systems. The global change and increasing instability of the climate pose additional challenges to breeders, emphasizing a need for greater nutrient use efficieny and greater tolerance to major abiotic stresses (Rubiales et al., 2021). The

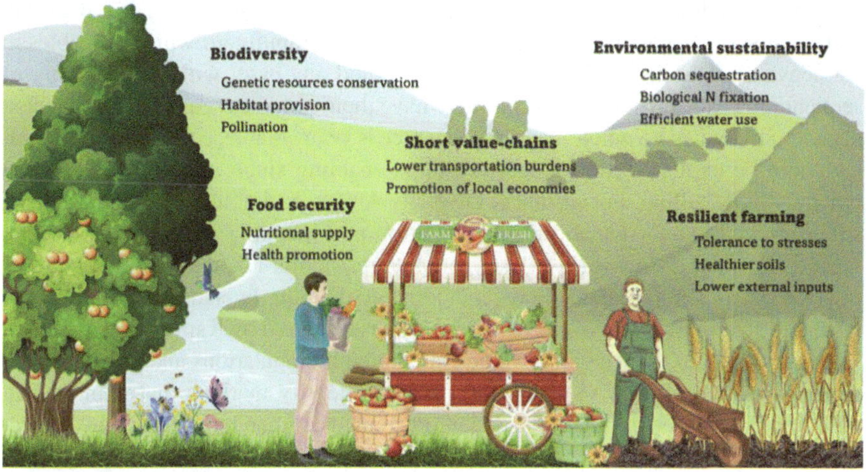

Fig. 9.2 Illustration of the diversity of benefits delivered by underutilised crops (UCs), from the farm to the final consumer

need to improve pest and disease resistance will be increasingly critical with the mandatory decrease in pesticide uses and with the expected effects of climatic change on the geographic distribution and frequency of epidemic outbreaks (Skendžić et al., 2021). Also, consumer preferences are affecting breeding priorities in terms of quality, that used to focus mainly on improving protein yield and reducing "undesirable" compounds contents, currently demanding increasing attention to important sensory or processing traits (Vaz Patto et al., 2015; Mecha et al., 2021).

Breeding relies on genetic diversity, and, for this, collection and conservation of genetic resourses is crucial. The breeding of elite cultivars of any crop tends to focus on selected germplams, progressively reducing the genetic diversity in the given species used in agriculture. This would be easily remedied by pre-breeding, with infusion of genetic diversity coming from landraces, ecotypes, or wild relative. However, UCs breeders have to cope with ever-increasing quantitative target traits with modest budgets, being often forced to focus on short-term breeding goals, preventing the needed exploitation of valuable germplasm that would require lengthy pre-breeding (Dwivedi et al., 2016; Rubiales et al., 2015). There are already excellent global collections in which wild and cultivated (e.g., landraces, old varieties) accessions of most crop species, including most UCs, are effectively stored, multiplied, and shared (EURISCO, 2022; GENESYS, 2022). However, a real limitation for effective use in breeding is the insufficient characterization (phenotypic and genotypic data) of these stored accessions.

Despite the modest investment made on UCs, significant advances were made in biotechnology and genomics over the last two decades, with funded initiatives and web resources available (Gregory et al., 2019; Jamnadass et al., 2020; Rubiales et al., 2021). This offers great opportunities to adapt to UCs advanced tools already used form major crops, such as whole-genome and transcriptome sequencing, genomic selection, genome editing and speed breeding (Kamenya et al., 2021). In fact, the list of UCs with their genomes sequenced is rapidly growing (see tables in Kamenya et al., 2021; Rubiales et al., 2021), and any case, with the dropping of sequencing costs, most of UCs will likely have their genomes sequenced in the next decade. Still, when no whole-genome secuence is yet available, comparative genomics could be exploited alongside other tools enabling single nucleotide polymorphisms (SNP) calling. For instance, Diversity Arrays Technology (DArT) has been successfully used for genetic characterization and mapping in many UCs. More recently DArT-sequencing (DArT-seq) or other restriction-associated DNA sequencing (RADseq) genotyping methods, including genotyping-by-sequencing, are being used for rapid marker discovery in many UCs (as reviewed by Kamenya et al., 2021; Rubiales et al., 2021). Mapping studies in biparental populations and genome-wide association studies (GWAS) are being used to identify markers that explain trait variation in a chosen population. Also, monogenic traits can be exploited by marker assisted selection. However, most agronomically important traits are polygenic, thus genomic selection could help to incorporate small-effect loci into prediction equations. Genomic selection has potential for UC breeding, enhancing selection efficiency once prediction equations are available (Annicchiarico et al., 2020). To develop these prediction equations not only Next Generation Sequencing (NGS) genotyping data are needed, but, most importantly, good phenotypic data. Field phenotyping remains a bottleneck for crop genetic improvement. Therefore, affordable low-cost phenotyping tools are needed to decrease the cost of field evaluations (Araus et al., 2018).

2.10 Agronomic Challenges of UC Cultivation

Growing UCs sometimes comes with agronomic challenges. As for any crop, UCs' demand for nutrients is not constant during the growth period, as nutrient availability is affected by environmental factors such as soil type and climate (Havlin, 2020). Therefore, to better utilise UCs, the supply and demand of fertiliser can be synchronised by fine-tuning its application to the needs of such crops, and thus the input be significantly reduced without compromising yield (Shah & Wu, 2019; Gatsios et al., 2021a, b). The loss of nutrients from the soil can also be appreciably reduced by the use of new intelligent fertilisers, such as nano-fertilisers, slow-release fertilisers, fertilisers enriched with nitrification inhibitors, compost, and microbial biostimulants such as arbuscular mycorrhiza fungi (AMF) and plant growth-promoting rhizobacteria (PGPR) (Mejias et al., 2021; Rouphael & Colla, 2020a, b; Ghafoor et al., 2021; Alonso-Ayuso et al., 2016; Cristofano et al., 2021; Sabatino et al., 2020). Applying such integrated nutrient management (INM) strategies in UC cultivation could enhance nutrient use efficiency (Shah & Wu, 2019.). Similarly, grafting onto nitrogen-efficient rootstocks can also lead to reduced nitrogen application (Liang et al., 2021). Some UCs could also be used as rootstocks, the wild relatives of cultivated crops (Razi et al., 2021). Introducing these crops to innovative farming practices spanning from the agroecological (integrated, organic, conservation) to high controlled technology (soilless culture, vertical farming) could improve their performance. Importantly, it could also lead to a measurable increase in farm income (Savvas & Gruda, 2018; Gatsios et al., 2021a, b; van Delden et al., 2021).

Organic crop production is facing the challenge of the yield gap due to nitrogen shortage availability at critical growth stages (Ponisio et al., 2015; Birkhofer et al., 2016). Identifying elite and UC genotypes suitable for low-input farming systems may also reduce the yield gap (Ntatsi et al., 2018a, b; Anastasi et al., 2019; Ronga et al., 2021). Taking also into consideration that organic farming relies on the inclusion of legumes as green manure, or in the rotation, due to the contribution of significant quantities of atmospheric nitrogen (N_2) (Gatsios et al., 2019, 2021a, c), the need to use legumes with high biological nitrogen fixation (BNF) ability are imperative for enhancing nitrogen inputs to the soil, thereby improving crop yield (Ntatsi et al., 2018a, b).

In addition, due to climate change, choosing the appropriate tillage system is extremely important. The adoption of conservation tillage systems (e.g., reduced tillage or no-tillage) can make a significant contribution to the reduction of greenhouse gas emissions due to a decrease in fuel consumption and lowered soil mineralisation rates (Stošić et al., 2021). Also, several studies show that conservation tillage systems improve soil properties (e.g., soil organic matter and water storage) and increases crop yields (Li et al., 2020a, b; Dong et al., 2021). Thus, for all the reasons mentioned above, it is essential to evaluate the effects of tillage systems on the growth and yield of UCs.

Another important limiting factor in UC cultivation is their competition with weeds and the lack of registered herbicides integrated into weed management programs. Thus, weed control is mainly based on hand hoeing and mechanical equipment. As in other "minor crops," these species should be planted in rows at distances to allow natural weed control (Karkanis et al., 2022). An appropriate design of the crop rotation system can also make a significant contribution to weed management (Kanatas, 2020; Shahzad et al., 2021). Ideally, this should be done using a 'Theory of Change' approach where the system's long-term and robust (stable) functional capacities determine the degree to which a system is resilient. All these obstacles and opportunities pave the way for developing new agri-food systems, including UCs. There is a need to implement a 'Theory of Change' approach where food system actors are included in the process of problem identification and solving, using true multi-actor approaches. The views and knowledge from breeders, farmers, chefs, consumers, food retailers, scientists, food/non-food industry and civil society in general need to be integrated to strengthen the evidence base of UCs multiple dimensions of value. This 'Theory of Change' approach for UCs will help also to identify the governance and policy frameworks needed for effective implementation of UCs in food and non-food value chains and ensure that agrobiodiversity is used sustainably.

3 Integrating UCs Using a Theory of Change Approach

Resilience can be defined as the maintenance of system functions in the face of stress from biotic or abiotic perturbations, whether gradual or sudden. The functional capacities of an ecological system are determined by interactions between biotic and abiotic components and the influence of specific pedoclimate, biogeography, land-use or -management approaches, socio-economic- and -technical aspects on the resulting ecosystem processes. Systems comprising a balance of functions tend to be more stable due to internal regulation of specific essential processes, or "system-function indicators", such as primary production, nutrient-, carbon- and water cycling, etc. These system-function indicators should be selected and monitored at the relevant spatial scale, such as field, farm, catchment, or bioregion. These system function indicators can also serve as a measure of system resilience where acceptable upper- and lower-thresholds can be defined.

3.1 Defining Better Farming System Functions with UCs

What constitutes a well-functioning and resilient farming system depends upon the ecological interactions at field- farm- and catchment-scales. Better farming operations can be defined as those which maintain a balance of all the essential farmed habitat functions such that they maintain stable levels over time, in response to shocks, and with minimal inputs from outside the system – since external inputs present dependencies, and therefore a risk.

(a) **System functional indicators** can be divided into biotic, abiotic, and socioeconomic categories. Biotic indicators include crop productivity and yield qualities, non-crop vegetational diversity, and the diversity and functional composition of trophic groups of microbes and invertebrates. These are organisms responsible for ecological processes needed to maintain system functions of soil and water quality, nutrient cycling, primary productivity, pollination, and the trophic and competitive interactions driving population regulation. Abiotic indicators relate to soil physical structure and environmental pollution (greenhouse gas emissions, leaching, and erosion). Socioeconomic indicators include cost-benefit analysis at the farm business scale and social aspects (employment, countryside access, etc.) beyond the farm gate, depending on the system boundaries (Hawes et al., 2009, 2016)

(b) **Stability** is defined here as a fluctuation within the upper- and lower limits or thresholds, which will vary depending on the environment and desired system states. Resilience is then determined by the system's capacity (farm) to keep within these thresholds over time and is the speed at which the system returns to a stable state following a disturbance. Resilience is strongly influenced by diversity and by a system feature called "functional redundancy" or "compensating complementarity". The similarity in functional role between species allows those functions to be maintained in the face of species extinctions (Ehrlich & Ehrlich, 1981), i.e., where numerous species possess a specific ecosystem function, the loss of one or few can be compensated by the others present in the system. In this way, system function is not compromised by such loss(es). Sufficient diversity accommodates functional redundancy and is an insurance measure for protection against shocks, as may occur due to management or climate (Yachi & Loreau, 1999).

(c) **Minimal inputs** should be the defining feature of well-functioning production systems that are semi "closed" (Hadavi & Ghazijahani, 2018), i.e., reduce reliance on external inputs by enhancing resource use efficiency and introducing nature-based solutions, minimising pollution and diversity loss, and so maintain stable functioning. However, fully closed systems are not entirely possible at the field-farm scale since harvested material must be removed for consumption. Therefore, offtake or loss from the system must be replaced to maintain stable states. Consequently, maintaining productivity (offtake) demands renewable and sustainably (and preferably locally) sourced inputs. Suppose the offtake is consumed locally and sourced from the same region. In that case, the system could be considered "closed" within a more comprehensive spatial boundary (i.e., bioregion) than the literal confines of the farm-scale management unit. Furthermore, reducing reliance on external inputs requires that resource use efficiency is optimised. This can be accomplished through agronomy (e.g., precision fertiliser placement in time and space precise targeting of crop protection chemicals through forecasting and mapping technology), plant diversity (e.g., niche complementarity giving rise to complete utilisation of inputs and selection of varieties to optimise resource capture and pest and disease resistance), soil biophysical function (e.g., microbial and invertebrate communities for nutrient turnover and optimal rooting for uptake efficiency), and non-crop biodiversity (e.g., alternative resources to support pollinator and natural enemy populations). Finally, interventions such as minimum tillage, cover cropping, riparian buffers, field margins, and fertiliser injection can be used to help "close the loop" by minimising inputs losses through erosion, leaching, and GHG emissions.

3.2 Implementing the System-Functions Approach

In the contexts defined above, management of production systems for the needs, or "health" of the environment, society, and economics requires optimisation across system functions. There will be inevitable trade-offs, at least in the short term, e.g., productivity/profit versus diversity/ecological functions, until the long-term benefits of more sustainable approaches can be achieved. As such, decisions need to be agreed on what system functions (health states) are desired or/and are to be prioritised. It is then necessary that: key indicators are identified for the desired functions; that upper- and lower-thresholds of acceptability in these functions are determined;

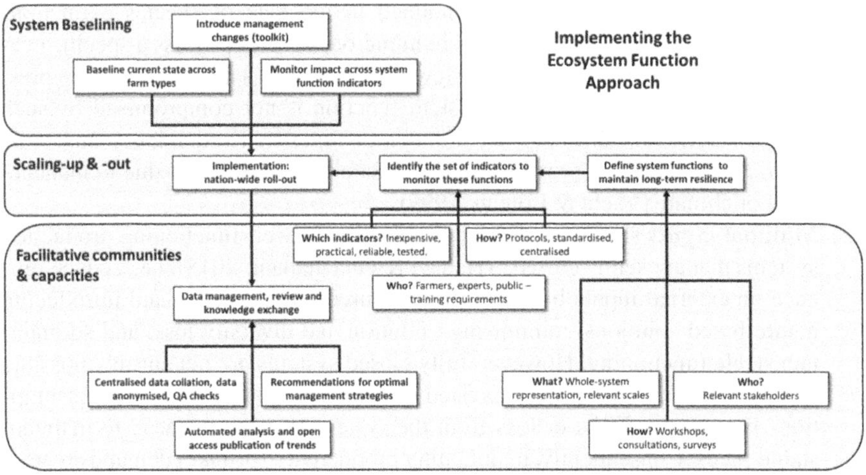

Fig. 9.3 A schematic flow-diagram illustrating the main steps involved in implementing the Ecosystem Function Approach. The approach is socioecological and demands the involvement of cooperative communities of stakeholders from across the value chain at appropriate spatial scales – from "system baselining" to "scaling-up and –out" of the approach. The pivotal importance of the interactive and cooperative socio-ecological approach is highlighted by the facilitative communities and capacities necessary to underpin the success of the process

and that they can be practically applied at different scales (e.g., field-farm-catchment-region-national). Finally, questions are raised at each step, and decisions need to be made for successful design and implementation, as illustrated in the conceptual model (Fig. 9.3).

There are various national and global environmental impact accounting tools, especially concerning GHG emissions. However, such inventories present data at national levels. Moreover, they do not dissect the detail of landscape structures and land-uses at levels related with confidence to ecosystem functions. So, the Ecosystem Function Approach has not yet been achieved for conventional farming systems, let alone those using less common agronomic strategies or underutilised crops.

Even with an agreed indicator set and using accredited, open, and transparent monitoring- and accounting- strategies, we still face the challenge of how monitoring approaches can be effectively taken up? Successful implementation of the system function approach requires an objective assessment of impact through accurate baselining and subsequent monitoring of the effect of any change in management intervention. Monitoring needs to be sensitive enough to detect trends over time. Land managers can ensure (and prove) that their interventions result in a move in the right direction towards the set target. Traditionally, agri-environment schemes have been incentivised through payments based on implementing a specified management intervention (length of hedgerow planted, area of cover crop sown, etc.). Still, these schemes suffer from a lack of evidence for any subsequent ecosystem function benefit. They frequently fall short of their original goal (biodiversity gain, species conservation, etc.) (Hawes et al., 2016). An alternative in the form of outcomes-based monitoring allows a proper assessment of impact, the opportunity for iterative development of improved management, and incentive payment is based on the extent to which the goals have been met. However, this approach requires indicator monitoring protocols that are quick, inexpensive, and easy to carry out by the land managers themselves while providing sufficiently accurate data that can detect trends in the right (or wrong) direction. Some examples are currently being piloted by the James Hutton Institute and NatureScot with farmers across Scotland and Ireland, focusing on biodiversity conservation for specific habitats and species. There is a need to test the quality and objectivity of data collected and then extend these specific protocols to more generally applicable assessments of farmland system functioning.

Research and innovation to realise validated approaches of system function accounting are rare and generally restricted to field-farm scales, which are also experimental sites, rather than in commercial farm settings. Across Europe, there are only a minimal number of long-term experimental platforms, mainly in arable stages, and livestock and perennial orchard systems, offering well-developed frameworks of indicators to the main system elements. However, such long-term farm-monitoring platforms appear absent for even major cropped systems and critical commercial species, such as tomatoes (Quesada et al., 2019; Tran et al., 2021). Also, there are no known life cycle analysis (LCA) studies of such platforms' ex-farmgate impacts (or functions). Additionally, it may be that a typical synthesis of system function indicators could be achieved even from the existing platforms, however limited. It may be possible to identify that sub-set of system-function indicators that are relatively easy for farmers to carry out themselves. Automated, high throughput technological solutions for monitoring system function – such as satellite imagery, other remote methods, and molecular diagnostics – also have the potential to bolster farmer-led data collection. Nonetheless, these will require a significant increase in research and development support before ground-truth testing and subsequent roll out.

3.3 *Novel AI Methods for Integrating UCs in Sustainable Food production Systems*

The transition to sustainable agriculture with UCs requires simultaneously considering the questions "How much food needs to be produced?" and "How will this food be produced?". So far, the agri-food sector has failed to address this challenge comprehensively and successfully. A promising approach to this challenge is the introduction of sustainable agriculture (Piñeiro et al., 2020; Rocchi et al., 2020). It is becoming increasingly clear that the transition to sustainable agriculture is impossible without using modern information technologies and artificial intelligence (AI) methods. With their help, the discovery of synergistic links between environmental conditions, biodiversity, and food production has been dramatically accelerated, enabling the adoption of sustainable agriculture. Cropland is no longer considered a basic input for food production, but a complex dynamic agri-ecosystem managed based on cognitive approaches. This means constantly monitoring its condition and maintaining a stable balance between "how much" and "how" by flexible management decisions. Artificial intelligence has become a new tool with which agriculture successfully introduces new principles and criteria for sustainable food production (Liakos et al., 2018).

Artificial intelligence is the computer science of complex dynamic systems that help extract information from large amounts of data, research already carried out, and experts' experience and knowledge (e.g., agronomists, pedologists, entomologists) (Russel & Norvig, 2021). The information gathered in this way is integrated into knowledge structures that help us understand, predict, and manage complex dynamic systems such as sustainable food production. This type of research approach allows us to acquire new knowledge very quickly and design scenarios for an efficient transition to sustainable multifunctional agriculture. Artificial intelligence, therefore, plays a critical role in the development of modern decision support systems for sustainable food production (Zhai et al., 2020).

One such system that illustrates the use of artificial intelligence for assessment of the sustainability of agri-food chains, including legumes as the target UC is the PATHFINDER (http://pathfinder.ijs.si/) Decision Support System (DSS) (Fig. 9.4). The system assesses the sustainability and its pillars (environmental, economic, social) of both the individual links and the chain as a whole. If the user wants to improve sustainability, the DSS finds and suggests changes to enhance sustainability or its unique sustainability pillars of the whole agri-food chain. With the help of artificial intelligence methods, a system like this can be further developed and upgraded to consider dynamic agri-food chains that would introduce, promote and strengthen the role of UCs in the agri-food chain.

Artificial intelligence is a very effective new tool to build advanced decision support systems that enable qualitative and quantitative breakthroughs in agriculture.

Fig. 9.4 Landing page of the PATHFINDER web-based DSS (http://pathfinder.ijs.si/)

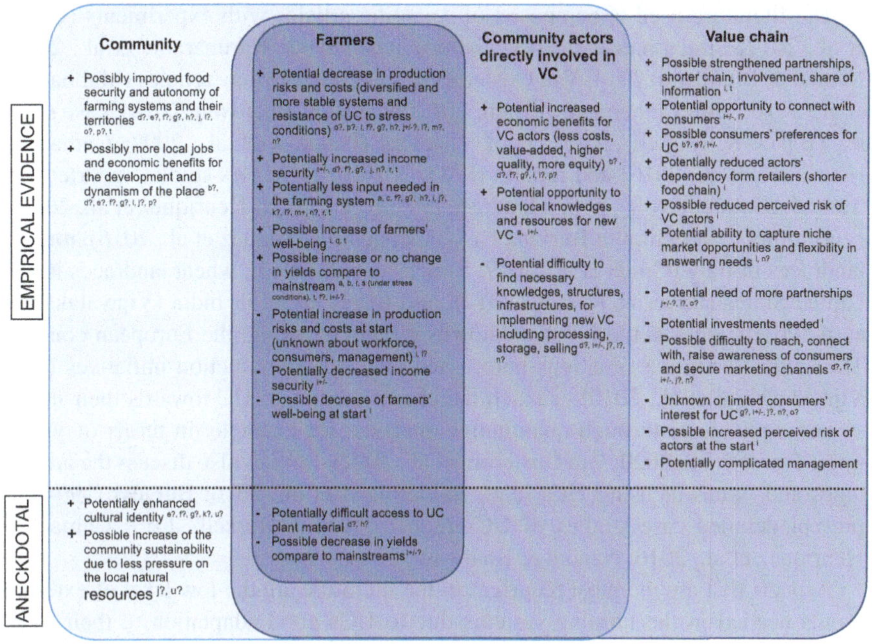

Fig. 9.5 Challenges (+) and opportunities (−) of UCs found in the literature: aspects of underutilized crops with empirical evidence described in the paper itself are above the dashed line, while aspects without empirical evidence are below. References referring to potential effects of UC in a hypothetical manner are marked with (?). [a] Guida et al., 2017; [b] Siracusa et al., 2013; [c] Galmes et al. 2011; [d] Padulosi et al., 2002; [e] Padulosi et al. 2013; [f] FAO, 2010b; [g] Padulosi et al., 1999; [h] Altieri & Merrick, 1987; [i] (Baker & Russell, 2017; [j] Baldermann et al., 2016; [k] Burgess, 1994; [l] Camacho-Henriquez et al., 2016; [m] Karunaratne et al., 2015; [n] Mabhaudhi et al., 2016; [o] Murevanhema & Jideani, 2013; [p] Nandal and Bhardwaj, 2014; [q] Nganga, 2014; [r] Traoré et al., 2020; [s] Van Oosterom et al., 2002; [t] Vijayalakshmi et al., 2010; [u] Will, 2008

With its help, we can make responsible decisions about measures to achieve the Sustainable Development Goals in general and sustainable food production in particular.

3.4 Social and Economic Considerations

In this section, the various opportunities and challenges of UCs are discussed from a socio-economic perspective. Figure 9.5 presents an overview of findings from the literature at the farm level, the potential impacts for the local community, and value chain aspects. We conceptualise farmers as both community and value chain actors; the value chain is embedded within the community but goes beyond (e.g., remote consumers). The community includes both actors directly involved in the value chain and indirectly impacted members.

The literature used is composed of scientific articles with experiments (Guida et al., 2017; Siracusa et al., 2013; Galmes et al., 2011; Karunaratne et al., 2015; Nandal & Bhardwaj, 2014; Van Oosterom et al., 2002), review articles (Mabhaudhi et al., 2016; Murevanhema & Jideani, 2013; Nandal & Bhardwaj, 2014), case studies (Baker & Russell, 2017), book chapters (Padulosi et al., 2002; Camacho-Henriquez et al., 2016) and reports (FAO, 2010b). The cases studies reported are from all continents (e.g., potato landrace in Peru (Camacho-Henriquez et al., 2016), pearl millet or sorghum in Burkina Faso (Camacho-Henriquez et al., 2016), tomato landraces in Italy (Guida et al., 2017; Siracusa et al., 2013), wheat landraces in the United States (Baker & Russell, 2017), and finger millet in India (Vijayalakshmi et al., 2010), etc.). Yet, a lot of unknowns remain regarding the European context. The studies compare situations before and after UC introduction initiatives (e.g., Vijayalakshmi et al., 2010) or reach the characteristics of UC towards their mainstream equivalents through quantitative analysis, for example, in terms of yields (e.g., Traoré et al., 2020; Van Oosterom et al., 2002). Studies also discuss the advantages and issues of using UCs (e.g., Baldermann et al., 2016; Burgess, 1994) or present detailed case studies of UCs (e.g., (Baker & Russell, 2017; Camacho-Henriquez et al., 2016; Nandal & Bhardwaj, 2014).

Aspects that are the most recurrent in the literature are the low level of external inputs needed in the farming systems due to UCs good adaptation to their local context (Guida et al., 2017; Galmes et al., 2011; FAO, 2010b; Padulosi et al., 1999; Altieri & Merrick, 1987; (Baker & Russell, 2017; Baldermann et al., 2016; Burgess et al., 1994; Camacho-Henriquez et al., 2016; Karunaratne et al., 2015; Mabhaudhi et al., 2016; Traoré et al., 2020; Vijayalakshmi et al., 2010). There are, however, difficulties in implementing and managing local value chains that can be competitive with mainstream crops, such as the need to access plant materials (Padulosi et al., 2002; Altieri & Merrick, 1987), to connect with other VC actors (e.g., consumers) (Padulosi et al., 2002; FAO, 2010b; Baker & Russell, 2017; Baldermann et al., 2016; Mabhaudhi et al., 2016), to create partnerships (Camacho-Henriquez et al., 2016; Murevanhema et al., 2013), and to find necessary structures, infrastructures, funds, knowledge, etc. (Padulosi et al., 2002; Baker & Russell, 2017; Baldermann et al., 2016).

3.4.1 Scotland as a Socio-economic Case Study

The Scottish socioeconomic paradigm and its impact on the Scottish arable system are particular, with barley occupying around two-thirds of cultivated arable land annually. Moreover, this barley is apportioned 2:1 for animal feed and malting markets, respectively (Scottish Government, 2021). In particular, the high demand for Scottish whisky, a heritage product, facilitates economic security mainly via a global trade which accounts for 75% of the total value of Scottish drink *and* food exports (21% of all of the UK), and high tax revenue for the government which is are currently estimated at £5.5bn in Gross Value Added (GVA) (Scottish Whisky Association, 2021), and this is doubled when beer and other spirits are taken into consideration. Nevertheless: *how can crop systems be diversified using UCs while maintaining the commercial success of the whisky sector?*

An exemplary Scottish farm (www.arbikie.com) took the approach of developing a short value chain which they termed their 'field to bottle' approach, and which encompasses five key elements (attractive location, traditional ethos, master craftsmen, small scale, and very close proximity of crop-production and distilling-elements), and the offer of products whose provenance and environmental credentials are fully traceable. Arbikie developed rye as a forgotten Scottish crop to diversify their cropped system and reintroduce Scottish Whisky made from rye after a 150-year absence. Arbikie's approach is now allied to the use of intercropping, the use of heritage barley types, and adoption of under-cultivated crops, including pea (*Pisum sativum*), used to produce the world's first climate-positive gin and vodka (Lienhardt et al., 2019a, b) known as the Nàdar Collection (Arbikie Distillery, 2021). Other smaller production units have advertised their products along with similar principles, advertising their products based on their whole value chain (*e.g.*, Nc'nean Distillery, 2021).

A defining feature here is that sustainability matters have emerged as the language of modern marketing. Additionally, the increasing awareness among consumers of "greenwashing" (Chen & Chang, 2013) has meant that the importance of any sustainability-related marketing claim should be evidenced. This requires open access to and transparent data about the claims made and value chain operations and processes (Beulens et al., 2005). Additionally, independent agencies should validate evidence using methods and procedures that are also approved, accredited, or certified.

Given the complex nature of environmental sustainability and resilience assessments and the need to adopt the Ecosystem Function Approach (described in this chapter – Sect. 3.3), producers and processors forming business partnerships with specialists, including research and technology organisations, are becoming commonplace. It is these research organisations that develop and offer state-of-the-art methodologies, which can be exploited to evidence new unique selling points. These are critical in a competitive marketplace, and in one where consumers have become very well informed. One solution is to account for ecosystem service functions alongside environmental impact assessments using LCA – since evidence of reduced impact does not necessarily inform on improved ecosystem functions (Koellner et al., 2013).

4 Programmes, Policies, and Research to Promote the Inclusion of UCs in Agro Food Systems

Public and private policies can directly influence the adoption of UCs (Table 9.3). Several International Fund for Agricultural Development (IFAD)-funded projects of Bioversity International confirmed UCs' livelihood benefits to poor people in numerous countries, including Bolivia, Peru, Guatemala, Mali, Nepal, and India

Table 9.3 Policies and areas of state interventions to support UCs/recommendations for policymakers

Consumption policy: increase demand	Production policy: increase supply
Consumption taxes on ultra-processed staple food Tax exemptions on healthy and sustainable foods from UCs Local food procurement to purchase UC-foods in public institutions Promotions and marketing campaigns to increase demand for UC-foods Food literacy programs to increase consumers' taste and health motivation for UCs Informative and educational program on UCs Promotions in schools and retail food stores for UCs Introduce UCs into the curricula Labels for UCs Create UCs farmers markets Create technology hubs to foster innovation and facilitate the adoption UCs	Define UCs in agricultural policies Fund UC-oriented agricultural research and development Make available UCs germplasm for breeding and multiplication Inventory and situ conservation of UCs Preservation of knowledge on their medicinal, cosmetic, nutritional, cultural values Reduce subsidies for non-UCs Capacity building of independent extension service programs for UCs Provide long-term, low cost-financing for UCs Tax incentives to UC-farmers Foster farmer-to-farmer knowledge exchange and technology transfer Increase incentives (cross-compliance programs) for UC-farmers Mainstream UCs in agricultural marketing policies

Based on: Chishakwe (2008), Bioversity International, and IFAD (2021)

(Padulosi et al., 2013). Their Holistic Value Chain Approach created participatory interventions at different value chain stages to overcome barriers. Public food procurement (e.g., school, hospital food programs) could shape diets by offering healthy and nutritious food for students purchased from local producers. Agrobiodiversity conservation programs that link UC farmer groups with public food procurement proved effective. For example, the Bioversity International program ('Linking agrobiodiversity value chains, climate adaptation, and nutrition', and 'Empowering the poor to manage risk') targeted the promotion of UCs in African countries funded by the IFAD and the European Commission. Finally, thousands of followers can reach a broader public via food champions and influencers. For example, when Crops for the Future launched the Forgotten Foods Network, they partnered with Prince Charles of Wales.

In Europe, perhaps the most common means by which UC could be supported is *via* the Common Agricultural Policy (CAP). The CAP has several functions, including increasing productivity while stabilising markets by avoiding the over-production of dominant crops and crop products, protecting income for farmers, ensuring food availability, and the affordability of food for consumers (EC, 2021). Ensuring sufficient levels of crop diversity is not the main aim of the CAP. Nevertheless, it is possible that the production of specific UCs could be encouraged *via* direct financial support under either CAP Pillar 1- or Pillar 2-payments. While provided by the EU, this money is administered by national or federal governments. While Pillar 1 payments relate to the area of land owned, payment is made on the basis that additional 'cross compliant' criteria are met. Such cross-compliance criteria can include attaining specific standards, often referred to as "Greening," as these encompass protection measures for the environment and biodiversity – through this ambition also aims to ensure production levels are maintained and even increased (Erjavec et al., 2015). Among interventions intended to aid the environment, there is: maintaining permanent grassland; maintaining a not (necessarily) cropped "ecological focus area" (EFA), of at least 7% of the total farm area; and crop diversification, which is defined as cultivating more than 2 crops when the area which can be cropped is >10 ha, or 3 crops if >30 ha. So, a farmer need not cultivate many crop species to qualify for payment, and the current diversification standard (3 crop minimum), highlights that holistic crop rotations are largely an ideal. There is, in fact, a "crop sequence" whose composition is determined by the demand of dominant markets and less by the protection of either crop from disease or the production environment from degradation. There is no substantial accommodation of high crop diversity or UCs. Though favourable markets exist, neither is there a restriction to using underutilised species under the 2 or 3 crop-minimum rules. Also, it is possible that where a crop also fulfils environmental or biodiversity protection goals, a crop can qualify payment as an EFA too. This service has been (controversially) acknowledged for grain legumes like fava beans. No synthetic fertiliser has been applied and based on its ability to provide a resource to pollinating- and beneficial insects. However, a different reality is that while the upscaling of the cultivation and consumption of legumes is required, these are common crops which domestic EU market has elected to import. While grain legumes are under-cultivated (in Europe), they qualify as UC species.

The fact is that there are no specific means by which government schemes support crop diversification *via* the use of UCs and that the use of UCs is mainly realised in short-value chains (Will, 2008). These, as mentioned before, are often cultivated by small-holder farmers, and utilised by relatively small processing units operated by artisans. As such, UCs are a bastion for maintaining and developing regional food cultures and ensuring food security among the neediest in many parts of the less-industrialised world (Massawe et al., 2016). In Europe, the farmers who most commonly grow underutilised crops own a land area that is too small (<5 ha) for the production unit to qualify for income protection via the CAP. Nevertheless, several Non-Governmental Organisations and community-led groups support underutilised crops, including Crops for the Future, La Via Campesina (the International Peasants' Movement), and the Permaculture Association, as well as community seed banks (Let's Liberate Diversity) and Slow Food. The EU agricultural policy did not identify and define UCs per se but considers them context dependent. Therefore, UCs are not the target of any CAP policies specifically. UCs can still be embedded into the current CAP, but the current governance system marginalises them. Only recently, the Farm-to-Fork Strategy and the Biodiversity Strategy, as part of the EU Green Deal Roadmap, started to focus on the sustainability of cropping systems, which creates room for valuing UCs.

Therefore, the commercial success of UCs is often achieved by the entrepreneurship of individuals who recognise the potential of underutilised crops in strong existing markets for products already accepted by consumers. The markets of

UC-based products also usually involve highlighting historic food cultures, forgotten heritage, and any additional attractive environmental, nutritional, and organoleptic attributes.

5 Concluding Remarks: The Critical Importance of a 'Theory of Change' Approach for the Promotion of UCs

As showcased above, UCs provide multiple nutritional, health, environmental, social, and economic benefits that go well beyond the farm level and contribute to agri-food system resilience. However, a 'Theory of Change' Approach is needed for UCs to be successfully included in sustainable agri-food systems. This approach recognises the need for an unprecedented degree of multi-actor strategies, whole-of-society engagement, and transformative actions. It is being developed and implemented, e.g., in the European H2020 project RADIANT (Realising Dynamic Value Chains for Underutilised Crops). It recognises the need to: (i) support multilateral learning among farmers, breeders, chefs, food retailers, scientists, representatives of food/non-food industry, and civil society; (ii) strengthen the evidence base in multiple dimensions of UC value (agronomic, environmental, economic and resilience), that also go beyond farm level (nutritional, techno-functional, health) and devise tools that integrate and showcase the potential benefits for adoption into new marketing schemes; (iii) identify the governance and policy frameworks needed for effective implementation of UCs in food and non-food value chains; and (iv) ensure that agrobiodiversity is used sustainably to meet people's needs and that agrobiodiversity promoting actions are supported by enabling conditions (educational, financial, technological, and capacity) that effectively get UCs to farmers fields and consumers' tables.

Implementation of the 'Theory of Change' Approach needs to be allied to equally facilitative socio-ecological frameworks or communities to ensure that farmers, other land managers, and citisens more generally co-develop and co-deliver interventions. These social networks and partnerships will need to seek agreed system function targets in a manner that avoids polarisation, adopting a flexible and non-prescriptive approach to land management – based on the rate or degree of change from the baseline, rather than absolute values. This will also help ensure that data-gathers and -users are receptive to the approaches and accept downstream data management, – analyses, and syntheses. Whatever procedures are undertaken, the transition from the dominant paradigm of conventional, high-input, intensive farming to more integrated, regenerative approaches supported by healthy ecosystem function and UC inclusion exposes farmers to risk. Direct and indirect costs are likely to be incurred before longer-term efficiency gains and system function improvements materialise. Although financial remuneration for undertaking the transition is required, it can be expected that such an incentive would not be necessary for the long term.

Acknowledgements This research was supported by the European Union's Horizon 2020 Research and Innovation Programme through the project "Realising Dynamic Value Chains for Underutilised Crops" (RADIANT), Grant Agreement number 101000622. The authors would also like to thank the scientific collaboration under the FCT project UIDB/50016/2020. in. The James Hutton Institute (CH and PPMI) are supported by the "Rural and Environmental Science and Analytical Services" (RESAS), a Division of the Scottish Government.

Conflicts of Interest The authors declare no conflict of interest.

References

Adhikari, L., Hussain, A., & Rasul, G. (2017). Tapping the potential of neglected and underutilized food crops for sustainable nutrition security in the mountains of Pakistan and Nepal. *Sustainability, 9*(2), 291. https://doi.org/10.3390/su9020291

Alonso-Ayuso, M., Gabriel, J. L., & Quemada, M. (2016). Nitrogen use efficiency and residual effect of fertilizers with nitrification inhibitors. *European Journal of Agronomy, 80*, 1–8. https://doi.org/10.1016/j.eja.2016.06.008

Altieri, M. A. (1999). The ecological role of biodiversity in agroecosystems. *Agriculture, Ecosystems and Environment, 74*(1), 19–31.

Altieri, M. A., Merrick, L., (1987). In Situ Conservation of Crop Genetic Resources through Maintenance of Traditional Farming Systems, *Economic Botany 41*(1): 86–96.

Anastasi, U., Corinzia, S. A., Cosentino, S. L., et al. (2019). Performances of durum wheat varieties under conventional and no-chemical input management systems in a semiarid Mediterranean environment. *Agronomy, 9*, 788. https://doi.org/10.3390/agronomy9120788

Annicchiarico, P., Nazzicari, N., Laouar, M., et al. (2020). Development and proof-of-concept application of genome-enabled selection for pea grain yield under severe terminal drought. *International Journal of Molecular Sciences, 21*, 2414. https://doi.org/10.3390/ijms21072414

Araus, J. L., Kefauver, S. C., Zaman-Allah, M., et al. (2018). Translating high-throughput phenotyping into genetic gain. *Trends in Plant Science, 23*, 451–466. https://doi.org/10.1016/j.tplants.2018.02.001

Arbikie Distillery. (2021). The Nadar collection. https://arbikie.com/pages/nadar-collection. Accessed 31 Jan 2022.

Asare, P. A., Kpankpari, R., Adu, M. O., et al. (2020). Phenotypic charcaterization of tiger nuts (Cyperus esculentus L.) from major growing areas in Ghana. *Scientific World Journal, 2020*, 7232591. https://doi.org/10.1155/2020/7232591

Azam, M. M., Waris, A., & Nahar, N. M. (2005). Properties and potential of fatty acid methyl esters of some non-traditional seed oils for use as biodiesel in India. *Biomass & Bioenergy, 29*, 293–302.

Azeke, M. A., Elsanhoty, R. M., Egielewa, S. J., et al. (2011). The effect of germination on the phytase activity, phytate and total phosphorus contents of some Nigerian-grown grain legumes. *Journal of the Science of Food and Agriculture, 91*(1), 75–79.

Baker, B. P., & Russell, J. A. (2017). Capturing a value-added niche market: Articulation of local organic grain. *American Journal of Agricultural Economics, 99*, 532–545.

Baldermann, S., Blagojević, L., Frede, K., et al. (2016). Are neglected plants the food for the future? *Critical Reviews in Plant Sciences, 35*(2), 106–119.

Ballhorn, D. J., Kautz, S., Heil, M., et al. (2009). Cyanogenesis of wild lima bean (*Phaseolus lunatus* L.) is an efficient direct defence in nature. *PLoS One, 4*(5), e5450.

Bals, B. D., Dale, B. E., & Balan, V. (2012). Recovery of leaf protein for animal feed and high-value uses. In C. Bergeron, D. J. Carrier, & S. Ramaswamy (Eds.), *Biorefinery co-products* (pp. 179–197). Wiley.

Bationo, A., Ntare, B., Tarawali, S., et al. (2002). Soil fertility management and cowpea production in the semiarid tropics. In *Challenges and opportunities for enhancing sustainable cowpea production* (pp. 301–318). IITA.

Bavec, F., Lisec, U., & Bavec, M. (2017). Importance of underutilized field crops for increasing functional biodiversity. In B. Şen & O. Grillo (Eds.), *Selected studies in biodiversity*. IntechOpen. https://doi.org/10.5772/intechopen.70472

Bekkering, C. S., & Tian, L. (2019). Thinking outside of the cereal box: Breeding underutilized (pseudo) cereals for improved human nutrition. *Frontiers in Genetics, 10*, 1289. https://doi.org/10.3389/fgene.2019.01289

Benner, E., Profeta, A., Wirsig, A., et al. (2008). Übergangsregelung Zum Herkunftsschutz Bei Agrarprodukten Und Lebensmitteln Aus Dem Blickwinkel Der Transaktions- Und Der Informationsökonomie. https://doi.org/10.22004/AG.ECON.52644

Benton, T. G., & Bailey, R. (2019). The paradox of productivity: Agricultural productivity promotes food system inefficiency. *Global Sustainability, 2*, e6. https://doi.org/10.1017/SUS.2019.3

Berbereia, V. L. D. (2015). Promoção do uso de alimentos promotores de saúde na dieta de grupos socioeconómicos desfavorecidos. https://repositorio.uac.pt/handle/10400.3/3551

Beulens, A. J., Broens, D. F., Folstar, P., et al. (2005). Food safety and transparency in food chains and networks relationships and challenges. *Food Control, 16*, 481–486.

Bioversity International and IFAD. (2021). *How to do. Mainstreaming NUS in national policy for nutritionoutcomes*. IFAD. https://www.ifad.org/en/web/knowledge/-/how-to-do-note-mainstreaming-nus-in-national-policy-for-nutrition-outcomes

Biowert (2021) Products. https://biowert.com/products. Accessed 8 Nov 2021.

Birkhofer, K., Smith, H. G., & Rundlöf, M. (2016). Environmental impacts of organic farming. In *eLS*. Wiley. https://doi.org/10.1002/9780470015902.a0026341

Bishop, K. R., Betzelberger, A. M., Long, S. P., et al. (2015). Is there potential to adapt soybean (Glycine max Merr.) to future [CO_2]? An analysis of the yield response of 18 genotypes in free-air CO_2 enrichment. *Plant, Cell & Environment, 38*, 1765–1774.

BLE. (n.d.). Pro-Kopf-Konsum von Hülsenfrüchten in Deutschland bis 2016/2017 | Statista. https://de.statista.com/statistik/daten/studie/175416/umfrage/pro-kopf-verbrauch-von-huelsenfruechten-in-deutschland-seit-1935/. Accessed 21 Jan 2022.

Bokdam, J., & Wallis de Vries, M. F. (1992). Forage quality as a limiting factor for cattle grazing in isolated Dutch nature reserves. *Conservation Biology, 6*, 399–408.

Boob, M., Elsaesser, M., Thumm, U., et al. (2019). Harvest time determines quality and usability of biomass from lowland hay meadows. *Agriculture, 9*(9), 198.

Bourgault, M., Brand, J., Tausz-Posch, S., et al. (2017). Yield, growth and grain nitrogen response to elevated CO_2 in six lentil (*Lens culinaris*) cultivars grown under Free Air CO2 Enrichment (FACE) in a semi-arid environment. *European Journal of Agronomy, 87*, 50–58. https://doi.org/10.1016/j.eja.2017.05.003

Bovolenta, S., Spanghero, M., Dovier, S., et al. (2008). Chemical composition and net energy content of alpine pasture species during the grazing season. *Animal Feed Science and Technology, 140*, 164–177.

Britt, J. H., Cushman, R. A., Dechow, C. D., et al. (2018). Invited review: Learning from the future – A vision for dairy farms and cows in 2067. *Journal of Dairy Science, 101*(5), 3722–3741. https://doi.org/10.3168/jds.2017-14025

Budak, H., Kantar, M., & Kurtoglu, K. Y. (2013). Drought tolerance in modern and wild wheat. *The Scientific World Journal, 2013*, 548246. https://doi.org/10.1155/2013/548246

Buiatti, M., Esquinas-Alcázar, J., Lazzerini, G., et al. (2010). Biodiversity and ecosystems. In *AAVV. Food policy and sustainability*. SlowFood.

Burgess, M. A. (1994). Cultural responsibility in the preservation of local economic plant resources. *Biodiversity and Conservation, 3*, 126–136.

Burney, J. A., Davis, S. J., & Lobell, D. B. (2010). *Greenhouse gas mitigation by agricultural intensification*. https://doi.org/10.1073/pnas.0914216107

Camacho-Henriquez, A., Kraemer, F., Galluzzi, G., et al. (2016). Decentralized collaborative plant breeding for utilization and conservation of neglected and underutilized crop genetic resources. In *Advances in plant breeding strategies: Breeding, biotechnology and molecular tools* (Vol. 1, pp. 25–61).

Chen, Y. S., & Chang, C. H. (2013). Greenwash and green trust: The mediation effects of green consumer confusion and green perceived risk. *Journal of Business Ethics, 114*, 489–500.

Cheng, A., Mayes, S., Dalle, G., et al. (2017). Diversifying crops for food and nutrition security – A case of teff. *Biological Reviews, 92*(1), 188–198. https://doi.org/10.1111/BRV.12225

Chilla, T., Fink, B., Balling, R., et al. (2020). The EU food label 'protected geographical indication': Economic implications and their spatial dimension. *Sustainability, 12*(14), 5503. https://doi.org/10.3390/SU12145503

Chishakwe, N. E. (2008). The role of policy in the conservation and extended use of underutilised plant species: A cross-national policy analysis. N. 31 p. ISBN: 978-92-9043-782-6.

Clark, M. A., Domingo, N. G. G., Colgan, K., et al. (2020). Global food system emissions could preclude achieving the 1.5° and 2°C climate change targets. *Science, 370*(6517), 705–708. https://doi.org/10.1126/SCIENCE.ABA7357/SUPPL_FILE/ABA7357_DATAS3.ZIP

Collinson, S. T., Azam-Ali, S. N., Chavula, K. M., et al. (1996). Growth, development and yield of bambara groundnut (*Vigna subterranea*) in response to soil moisture. *Journal of Agricultural Science, 126*, 307–318.

Collinson, S. T., Clawson, E. J., Azam-Ali, S. N., et al. (1997). Effect of soil moisture deficits on the water relations of bambara groundnut (*Vigna subterranea* L. Verdc.). *Journal of Experimental Botany, 48*, 877–884.

Commission on Sustainable Agriculture and Climate Change (2012). Achieving food security in the face of climate change. Final report from the Commission on Sustainable Agriculture and Climate Change Commission Secretariat. www.ccafs.cgiar.org/commission

Cordain, L., Eaton, S. B., Sebastian, A., et al. (2005). Origins and evolution of the Western diet: Health implications for the 21st century. *The American Journal of Clinical Nutrition, 81*(2), 341–354. https://doi.org/10.1093/AJCN.81.2.341

Cristofano, F., El-Nakhel, C., & Rouphael, Y. (2021). Biostimulant substances for sustainable agriculture: Origin, operating mechanisms and effects on cucurbits, leafy greens, and nightshade vegetables species. *Biomolecules, 11*, 1103. https://doi.org/10.3390/biom11081103

Cusworth, G., Garnett, T., & Lorimer, J. (2021). Legume dreams: The contested futures of sustainable plant-based food systems in Europe. *Global Environmental Change, 69*, 102321. https://doi.org/10.1016/j.gloenvcha.2021.102321

Dantas, E. F., Freitas, A. D., Lyra, M. D., et al. (2019). Biological fixation, transfer and balance of nitrogen in passion fruit (Passiflora edulis Sims) orchard intercropped with different green manure crops. *Australian Journal of Crop Science, 13*(03), 465–471.

Das, S., Khound, R., Santra, M., et al. (2019). Beyond bird feed: Proso millet for human health and environment. *Agriculture, 9*(3), 64. https://doi.org/10.3390/agriculture9030064

Dawson, I. K., Hedley, P. E., Guarino, L., et al. (2009). Does biotechnology have a role in the promotion of underutilised crops? *Food Policy, 34*(4), 319–328. https://doi.org/10.1016/J.FOODPOL.2009.02.003

Dien, B. S., Mitchell, R. B., Bowman, M. J., et al. (2018). Bioconversion of pelletized big bluestem, switchgrass, and low-diversity grass mixtures into sugars and bioethanol. *Frontiers in Energy Research, 6*, 129. https://doi.org/10.3389/fenrg.2018.00129

Diffenbaugh, N., Krupke, C. H., White, M. A., et al. (2008). Global warming presents new challenges for maize pest management. *Environmental Research Letters, 3*(4). https://doi.org/10.1088/1748-9326/3/4/044007

Dong, L., Si, T., Li, Y.-E., et al. (2021). The effect of conservation tillage in managing climate change in arid and semiarid areas—A case study in Northwest China. *Mitigation and Adaptation Strategies for Global Change, 26*(4), 17. https://doi.org/10.1007/s11027-021-09956-3

Dubey, P. K., Singh, G. S., & Abhilash, P. C. (2020). Adaptive agronomic practices for sustaining food production. In *Adaptive Agricultural Practices* (Springer briefs in environmental science). Springer. https://doi.org/10.1007/978-3-030-15519-3_2

Dulloo, M., Drucker, A., Gaisberger, H., et al. (2016). *Mainstreaming agrobiodiversity in sustainable food systems: Scientific foundations for an agrobiodiversity index – Summary* (pp. 22–25). Bioversity International.

Dwivedi, S. L., Ceccarelli, S., Blair, M. W., et al. (2016). Landrace germplasm for improving yield and abiotic stress adaptation. *Trends in Plant Science, 21*(1), 31–42. https://doi.org/10.1016/j.tplants.2015.10.012

Ebert, A. W. (2014). Potential of underutilized traditional vegetables and legume crops to contribute to food and nutritional security, income and more sustainable production systems. *Sustainability, 6*, 319–335. https://doi.org/10.3390/su6010319

Ehrlich, P. R., & Ehrlich, A. H. (1981). *Extinction: The causes and consequences of the disappearance of species* (pp. 72–98). Random House.

Ekanayake, S., Nair, B., Jansz, E. R., et al. (2003). Effect of processing on the protein nutritional value of Canavalia gladiate seeds. *Die Nahrung, 47*(4), 256–260.

Erjavec, E., Lovec, M., & Erjavec, K. (2015). From "Greening" to "Greenwash": the drivers and discourses of CAP 2020 reform. In: J.F. Swinnen (Ed.), *The Political Economy of the 2014-2020 Common Agricultural Policy: An Imperfect Storm*. Rowman & Littlefield, London and CEPS, Brussels, ISBN: 978-1-78348-484-3, p. 215–244.

Eshel, G., Shepon, A., Makov, T., et al. (2014). Land, irrigation water, greenhouse gas, and reactive nitrogen burdens of meat, eggs, and dairy production in the United States. *Proceedings of the National Academy of Sciences of the United States of America, 111*(33), 11996–12001. https://doi.org/10.1073/pnas.1402183111

EURISCO. (2022). European search catalogue for plant genetic resources. Available at: https://eurisco.ipk-gatersleben.de/. Accessed 5 Feb 2022.

European Commission. (2021). The common agricultural policy at a glance: The common agricultural policy supports farmers and ensures Europe's food security. https://ec.europa.eu/info/food-farming-fisheries/key-policies/common-agricultural-policy/cap-glance_en. Accessed 31 Jan 2022.

Eyzaguirre, P., Padulosi, S., & Hodgkin, T. (1999). IPGRI's strategy for neglected and underutilised species and the human dimension of agrobiodiversity. In S. Padulosi (Ed.), *Priority setting for underutilised and neglected plant species of the Mediterranean region. Report of the IPGRI conference, 9–11. February 1998, ICARDA, Aleppo. Syria*. International Plant Genetic Resources Institute.

Fahey, J. W. (2005). Moringa oleifera: A review of the medical evidence for its nutritional, therapeutic, and prophylactic properties. Part 1. *Phytochemistry, 47*, 123–157.

Fanzo, J., Haddad, L., Schneider, K. R., et al. (2021). Viewpoint: Rigorous monitoring is necessary to guide food system transformation in the countdown to the 2030 global goals. *Food Policy, 104*, 102163. https://doi.org/10.1016/J.FOODPOL.2021.102163

FAO. (1999). Agricultural biodiversity. In *Multifunctional Character of Agriculture and Land Conference*, Maastricht, 12–17 Sept 1999.

FAO. (2001). International treaty on plant genetic resource for food and agriculture, Rome.

FAO. (2010a). Proceedings of the International Scientific Symposium Biodiversity and Sustainable diets united against hunger. 3–5 Nov 2010. Available in: https://www.fao.org/3/i3004e/i3004e00.pdf

FAO (2010b). *The second report on the state of the world's plant genetic resources for food and agriculture* (pp. 183–201). Food and Agriculture Organization of the United Nations, Rome.

FAO. (2018a). Global livestock environmental assessment model. http://www.fao.org/fileadmin/user_upload/gleam/docs/GLEAM_2.0_Model_description.pdf

FAO. (2018b). The future of Food and Agriculture: Alternative pathways to 2050 | Knowledge for policy. https://knowledge4policy.ec.europa.eu/publication/future-food-agriculture-alternative-pathways-2050_en

FAO. (2019a). The state of the world's biodiversity for food and agriculture. In J. Bélanger & D. Pilling (Eds.), *FAO commission on genetic resources for food and agriculture assessments*. 572 pp.

FAO (2019b). *Mountain agriculture: Opportunities for harnessing zero hunger in Asia*. Food and Agriculture Organization of the United Nations, Bangkok.

FAO. (2021). The state of food and agriculture 2021. https://doi.org/10.4060/CB4476EN.

FAOSTAT, Food and Agriculture Organization of the United Nations. Available at: https://www.fao.org/faostat/en/#data/QCL. Accessed 25 Jan 2022.

Ferchichi, S., Hessini, K., Dell Aversana, E., et al. (2018). *Hordeum vulgare* and *Hordeum maritimum* respond to extended salinity stress displaying different temporal accumulation pattern of metabolites. *Functional Plant Biology, 45*(11), 1096–1109.

Finell, M. (2003). *The use of reed canary grass (*Phalaris arundinacea*) as a short fibre raw material for the pulp and paper industry* (Dissertation). Swedish University of Agricultural Sciences.

Foley, J. A., Ramankutty, N., Brauman, K. A., et al. (2011). Solutions for a cultivated planet. *Nature, 478*(7369), 337–342.

Foyer, C. H., Lam, H. M., Nguyen, H. T., et al. (2016). Neglecting legumes has compromised human health and sustainable food production. *Nature Plants, 2*, 16112.

Fraj, M. B., Al-Dakheel, A. J., McCann, I. R., et al. (2013). Selection of high yielding and stable safflower (*Carthamus tinctorius* L.) genotypes under salinity stress. *Agricultural Science Research Journal, 3*, 273–283.

Fraj, M. B., Al-Dakheel, A. J., McCann, I. R., et al. (2014). Selection of salt-tolerant triticale (x *Triticum secale* Wittmack) genotypes for grain and forage end-uses. *American-Eurasian Journal of Agricultural and Environmental, 14*, 445–454.

Freitas, R. M., Dombroski, J. L., Freitas, F. C., et al. (2019). Water use of cowpea under deficit irrigation and cultivation systems in semi-arid region. *Revista Brasileira de Engenharia Agrícola e Ambiental, 23*(4), 271–276.

Galmés, J., Conesa, M. À., Ochogavía, J. M., Perdomo, J. A., Francis, D. M., Ribas-Carbó, M., Savé, R., Flexas, J., Medrano, H., & Cifre, J. (2011). Physiological and morphological adaptations in relation to water use efficiency in Mediterranean accessions of Solanum lycopersicum. *Plant, Cell & Environment, 34*, 245–260. https://doi.org/10.1111/j.1365-3040.2010.02239.x

Gatsios, A., Ntatsi, G., Celi, L., et al. (2019). Nitrogen nutrition optimization in organic greenhouse tomato through the use of legume plants as green manure or intercrops. *Agronomy, 9*, 766. https://doi.org/10.3390/agronomy9110766

Gatsios, A., Ntatsi, G., Yfantopoulos, D., et al. (2021). Effects of different organic soil amendments on nitrogen nutrition and yield of organic greenhouse tomato crop. *Nitrogen, 2*, 347–358. https://doi.org/10.3390/nitrogen2030024

Gatsios, A., Ntatsi, G., Celi, L., et al. (2021a). Impact of legumes as a pre-crop on nitrogen nutrition and yield in organic greenhouse tomato. *Plants, 10*, 468. https://doi.org/10.3390/plants10030468

Gatsios, A., Ntatsi, G., Celi, L., et al. (2021b). Legume-based mobile green manure can increase soil nitrogen availability and yield of organic greenhouse tomatoes. *Plants, 10*, 2419. https://doi.org/10.3390/plants10112419

GENESYS. (2022). The global gateway to genetic resources. Available online at: https://www.genesys-pgr.org. Accessed 5 Feb 2022.

Ghafoor, I., Habib-ur-Rahman, M., Ali, M., et al. (2021). Slow-release nitrogen fertilizers enhance growth, yield, NUE in wheat crop and reduce nitrogen losses under an arid environment. *Environmental Science and Pollution Research, 28*, 43528–43543. https://doi.org/10.1007/s11356-021-13700-4

Grass, S. (2004). Utilisation of grass for production of fibres, protein, and energy. In *Biomass and agriculture. Sustainability, markets and policies* (pp. 169–177). OECD.

Gregory, P. J., Mayes, S., Hui, C. H., et al. (2019). Crops For the Future (CFF): An overview of research efforts in the adoption of underutilised species. *Planta, 250*, 979–988. https://doi.org/10.1007/s00425-019-03179-2

Gruère, G. P., Giuliani, A., & Smale, M. (2008). Marketing underutilized plant species for the poor: A conceptual framework. In *Agrobiodiversity Conservation and Economic Development* (pp. 86–105). Routledge, London, UK.
Guida, G., Sellami, M. H., Mistretta, C., et al. (2017). Agronomical, physiological and fruit quality responses of two italian long-storage tomato landraces under rain-fed and full irrigation conditions. *Agricultural Water Management, 180*, 126–135.
Habiyaremye, C., Barth, V., Highet, K., et al. (2017). Phenotypic responses of twenty diverse proso millet (*Panicum miliaceum* L.) accessions to irrigation. *Sustainability, 9*, 389.
Hadavi, E., & Ghazijahani, N. (2018). Closed and semi-closed systems in agriculture. In E. Lichtfouse (Ed.), *Sustainable agriculture reviews 33*. Springer. https://doi.org/10.1007/978-3-319-99076-7_10
Haddad, L., Hawkes, C., Webb, P. et al. (2016). A new global research agenda for food. *Nature 540*, 30–32. https://doi.org/10.1038/540030a
Hadebe, S., Modi, A. T., & Mabhaudhi, T. (2017). Drought tolerance and water use of cereal crops: A focus on sorghum as a food security crop in Sub-Saharan Africa. *Journal of Agronomy and Crop Science, 203*, 177–191.
Havlin, J. L. (2020). Soil: Fertility and nutrient management. In *Landscape and land capacity* (pp. 251–265). CRC Press, Boca Raton, Florida, USA
Hawes, C., Haughton, A. J., Bohan, D. A., et al. (2009). Functional approaches for assessing plant and invertebrate abundance patterns in arable systems. *Basic and Applied Ecology, 10*, 34–47.
Hawes, C., Begg, G. S., Iannetta, P., et al. (2016). A whole-systems approach for assessing measures to improve arable ecosystem sustainability. *Ecosystem Health and Sustainability, 2*, e01252.
Hermansen, J., Jørgensen, U., Lærke, P. E., et al. (2017). Green biomass - protein production through bio-refining. DCA Report.
Holm, L., & Møhl, M. (2000). The role of meat in everyday food culture: An analysis of an interview study in Copenhagen. *Appetite, 34*(3), 277–283. https://doi.org/10.1006/APPE.2000.0324
Huang, J., Xu, C., Ridoutt, B. G., et al. (2017). Nitrogen and phosphorus losses and eutrophication potential associated with fertilizer application to cropland in China. *Journal of Cleaner Production, 159*, 171–179.
Hunter, M. C., Smith, R. G., Schipanski, M. E., et al. (2017). Agriculture in 2050: Recalibrating targets for sustainable intensification. *Bioscience, 67*(4), 386–391. https://doi.org/10.1093/BIOSCI/BIX010
Hunter, D., Borelli, T., Beltrame, D. M. O., et al. (2019). The potential of neglected and underutilized species for improving diets and nutrition. *Planta, 250*(3), 709–729. https://doi.org/10.1007/s00425-019-03169-4
Huppmann, D., Rogelj, J., Kriegler, E., Krey, V., & Riahi, K. (2018). A new scenario resource for integrated 1.5 °C research. In Nature Climate Change (Vol. 8, Issue 12, pp. 1027–1030). Nature Publishing Group. https://doi.org/10.1038/s41558-018-0317-4
Ibeanu, V. N., Edeh, C. G., & Ani, P. N. (2020). Evidence-Based Strategy for Prevention of Hidden Hunger among Adolescents in a Suburb of Nigeria. *BMC Public Health, 20*(1), 1–10. https://doi.org/10.1186/s12889-020-09729-8
Imathiu, S. (2021). Neglected and underutilized cultivated crops with respect to indigenous African leafy vegetables for food and nutrition security. *Journal of Food Security, 9*(3), 115–125.
International assessment of agricultural knowledge, science and technology for development. (2009). *Agriculture at a crossroads: A global report*. Island Press.
IPCC. (2019). Climate change and land. An IPCC Special Report on climate change, desertification, land degradation, sustainable land management, food security, and greenhouse gas fluxes in terrestrial ecosystems. www.ipcc.ch
Jahn, S. A. A., Musnad, H. A., & Burgstaller, H. (1986). The tree that purifies water: Cultivating multipurpose Moringaceae in the Sudan. *Unasylva, 38*, 23–28.

Jamnadass, R., Mumm, R. H., Hale, I., et al. (2020). Enhancing African orphan crops with genomics. *Nature Genetics, 52*, 356–360. https://doi.org/10.1038/s41588-020-0601-x

Jansen, M., Guariguata, M. R., Raneri, J. E., et al. (2020). Food for thought: The underutilized potential of tropical tree-sourced foods for 21st century sustainable food systems. *People and Nature, 2*(4), 1006–1020.

Jungers, J. M., Fargione, J. E., Sheaffer, C. C., et al. (2013). Energy potential of biomass from conservation grasslands in Minnesota, USA. *PLoS One, 8*(4), e61209. https://doi.org/10.1371/journal.pone.0061209

Kamel, S. H., Zaki, Z. M., & Kassim, J. (2018). The effectiveness of *Psophocarpus tetragonolobus's* seed as turbidity removal. *International Journal of Engineering and Technology, 7*, 144–146.

Kamenya, S. N., Mikwa, E. O., Song, B., & Odeny, D. A. (2021). Genetics and breeding for climate change in orphan crops. *Theoretical and Applied Genetics, 134*, 1787–1815. https://doi.org/10.1007/s00122-020-03755-1

Kanatas, P. (2020). Mini-review: The role of crop rotation, intercropping, sowing dates and increased crop density towards a sustainable crop and weed management in arable crops. *Agraarteadus, 31*(1), 22–27. https://doi.org/10.15159/jas.20.11

Karavidas, I., Ntatsi, G., Vougeleka, V., et al. (2022). Agronomic practices to increase the yield and quality of common bean (Phaseolus vulgaris L.): A systematic review. *Agronomy, 12*, 271. https://doi.org/10.3390/agronomy12020271

Karkanis, A., Ntatsi, G., Lepse, L., et al. (2018). Faba bean cultivation – Revealing novel managing practices for more sustainable and competitive European cropping systems. *Frontiers in Plant Science, 9*, 1115. https://doi.org/10.3389/fpls.2018.01115

Karkanis, A., Polyzos, N., Kompocholi, M., et al. (2022). Rock samphire, a candidate crop for saline agriculture: Cropping practices, chemical composition and health effects. *Applied Sciences, 12*, 737. https://doi.org/10.3390/app12020737

Karunaratne, A. S., Walker, S., & Azam-Ali, S. N. (2015). Assessing the productivity and resource-use efficiency of underutilised crops: Towards an integrative system. *Agricultural Water Management, 147*, 129–134.

Katschnig, D., Broekman, R. A., & Rozema, J. (2013). Salt tolerance in the halophyte *Salicornia dolichostachya* Moss: Growth, morphology and physiology. *Environmental and Experimental Botany, 92*, 32–42.

Keatinge, J. D. H., Waliyar, F., Jamnadass, R. H., et al. (2010). Re-learning old lessons for the future of food-by bread alone no longer: Diversifying diets with fruit and vegetables. *Crop Science, 50*, 51–62.

Khoury, C. K., Bjorkman, A. D., Dempewolf, H., et al. (2014). Increasing homogeneity in global food supplies and the implications for food security. *Proceedings of the National Academy of Sciences of the United States of America, 111*, 4001–4006. https://doi.org/10.1073/pnas.1313490111

King, C., Richardson, M., McEniry, J., et al. (2013). Potential use of fibrous grass silage press-cake to minimise shrinkage cracking in low-strength building materials. *Biosystems Engineering, 115*, 203–210.

Koellner, T., De Baan, L., Beck, T., et al. (2013). UNEP-SETAC guideline on global land use impact assessment on biodiversity and ecosystem services in LCA. *The International Journal of Life Cycle Assessment, 18*, 1188–1202.

Koidou, M., Mountousis, I., Dotas, V., et al. (2019). Temporal variations of herbage production and nutritive value of three grasslands at different elevation zones regarding grazing needs and welfare of ruminants. *Archives Animal Breeding, 62*, 215–226.

Kuo, C. G., Chen, H. M., & Sun, H. C. (1992). Membrane thermostability and heat tolerance of vegetable leaves. In *Adaptation of food crops to temperature and water stress* (pp. 160–168). AVRDC - The World Vegetable Center.

Karunaratne, A. S., Walker, S., & Azam-Ali, S. N. (2015). Assessing the productivity and resource-use efficiency of underutilised crops: Towards an integrative system. *Agricultural Water Management, 147*, 129–134.

Lakew, B., Eglinton, J., Henry, R. J., Baum, M., Grando, S., & Ceccarelli, S. (2011). The potential contribution of wild barley (Hordeum vulgare ssp. spontaneum) germplasm to drought tolerance of cultivated barley (H. vulgare ssp. vulgare). *Field Crops Research, 120*(1), 161–168.

Lazcano, C., Zhu-Barker, X., & Decock, C. (2021). Effects of organic fertilizers on the soil microorganisms responsible for N_2O emissions: A review. *Microorganisms, 9*(5), 983. https://doi.org/10.3390/microorganisms9050983

Lewandowski, I., Clifton-Brown, J. C., Andersson, B., et al. (2003). Environment and harvest time affects the combustion qualities of Miscanthus genotypes. *Agronomy Journal, 95*, 1274–1280.

Li, X., & Siddique, K. H. M. (2020). Future smart food: Harnessing the potential of neglected and underutilized species for zero hunger. *Maternal and Child Nutrition, 16*(S3), 1–22. https://doi.org/10.1111/mcn.13008

Li, J., Wang, Y. K., Guo, Z., et al. (2020a). Effects of conservation tillage on soil physicochemical properties and crop yield in an arid loess plateau, China. *Scientific Reports, 10*(1), 4716. https://doi.org/10.1038/s41598-020-61650-7

Li, X., Yadav, R., & Siddique, K. H. M. (2020b). Neglected and underutilized crop species: The key to improving dietary diversity and fighting hunger and malnutrition in Asia and the Pacific. *Frontiers in Nutrition, 7*, 593711. https://doi.org/10.3389/fnut.2020.593711

Liakos, K. G., Busato, P., Moshou, D., et al. (2018). Machine learning in agriculture: A review. *Sensors (Switzerland), 18*(8). https://doi.org/10.3390/s18082674

Liang, J., Chen, X., Guo, P., et al. (2021). Grafting improves nitrogen-use efficiency by regulating the nitrogen uptake and metabolism under low-nitrate conditions in cucumber. *Scientia Horticulturae, 289*, 110454. https://doi.org/10.1016/j.scienta.2021.110454

Lienhardt, T., Black, K., Saget, S., et al. (2019a). Just the tonic! Legume biorefining for alcohol has the potential to reduce Europe's protein deficit and mitigate climate change. *Environment International, 130*, 104870. https://doi.org/10.1016/j.envint.2019.05.064

Lienhardt, T., Black, K., Saget, S., et al. (2019b). Data for life cycle assessment of legume biorefining for alcohol. *Data in Brief, 25*. https://doi.org/10.1016/j.dib.2019.104242

Linnemann, A. R., & Azam-Ali, S. (1993). Bambara groundnut (Vigna subterranea). In J. T. William (Ed.), *Pulses and vegetables* (pp. 13–57). Chapman and Hall – CRC Press, Boca Raton, Florida, USA.

Lowe, N. M. (2021). The global challenge of hidden hunger: Perspectives from the field. *Proceedings of the Nutrition Society, 80*(3), 283–289. https://doi.org/10.1017/S0029665121000902

Mabhaudhi, T., O'Reilly, P., Walker, S., et al. (2016). Opportunities for underutilised crops in Southern Africa's post–2015 development agenda. *Sustainability, 8*(4), 302.

Mabhaudhi, T., Chimonyo, V. G. P., Hlahla, S., et al. (2019). Prospects of orphan crops in climate change. *Planta, 250*(3). https://doi.org/10.1007/s00425-019-03129-y

Mafakheri, M., & Kordrostami, M. (2020). Newly revealed promising gene pools of neglected brassica species to improve stress-tolerant crops. In M. Hasanuzzaman (Ed.), *The plant family Brassicaceae*. Springer. https://doi.org/10.1007/978-981-15-6345-4_4

Mammadov, J., Buyyarapu, R., Guttikonda, S. K., et al. (2018). Wild relatives of maize, rice, cotton, and soybean: Treasure troves for tolerance to biotic and abiotic stresses. *Frontiers in Plant Science, 9*. https://doi.org/10.3389/fpls.2018.00886

Mandizvo, T., Odindo, A. O., & Mashilo, J. (2021). Citron watermelon potential to improve crop diversification and reduce negative impacts of climate change. *Sustainability, 13*(4), 2269. https://doi.org/10.3390/su13042269

Massawe, F., Mayes, S., & Cheng, A. (2016). Crop diversity: An unexploited treasure trove for food security. *Trends in Plant Science, 21*, 365–368. https://doi.org/10.1016/j.tplants.2016.02.006

Masson-Delmotte, V., Zhai, P., Pörtner, H.-O., et al. (2019). Global warming of 1.5°C An IPCC Special Report on the impacts of global warming of 1.5°C above pre-industrial levels and related global greenhouse gas emission pathways, in the context of strengthening the global response to the threat of climate change, sustainable development, and efforts to eradicate poverty Summary for Policymakers Edited by Science Officer Science Assistant Graphics Officer Working Group I Technical Support Unit. https://report.ipcc.ch/sr15/pdf/sr15_spm_final.pdf

Matthews, N. E., Stamford, L., & Shapira, P. (2019). Aligning sustainability assessment with responsible research and innovation: Towards a framework for Constructive Sustainability Assessment. *Sustainable Production and Consumption, 20*, 58–73. https://doi.org/10.1016/j.spc.2019.05.002

Mayes, S., Massawe, F. J., Alderson, P. G., et al. (2012). The potential for underutilized crops to improve security of food production. *Journal of Experimental Botany, 63*(3), 1075–1079. https://doi.org/10.1093/JXB/ERR396

Mecha, E., Natalello, S., Carbas, B., et al. (2021). Disclosing the nutritional quality diversity of Portuguese common beans—The missing link for their effective use on protein quality breeding programmes. *Agronomy, 11*, 221. https://doi.org/10.3390/agronomy11020221

Mejias, J. H., Salazar, F., Pérez-Amaro, L., et al. (2021). Nanofertilizers: A cutting-edge approach to increase nitrogen use efficiency in grasslands. *Frontiers in Environmental Science, 9*, 635114. https://doi.org/10.3389/fenvs.2021.635114

Meldrum G., Sthapit S., Rojas W., et al. (2016) Methodology mapping for resilient production systems: Approaches and results from surveys in Bolivia, India, and Nepal. Proceedings of the 3rd International Conference on Neglected and Underutilized Species: For a Food-Secure Africa, Accra, Ghana, 25–27 September 2013, pp. 108–112.

Mentis, M. (2020). Environmental rehabilitation of damaged land. *Forest Ecosystems, 7*(1). https://doi.org/10.1186/s40663-020-00233-4

Millennium Ecosystem Assessment (MEA). (2005). *Ecosystems and human well-being: Synthesis*. Island Press, Washington, DC, 137p. https://www.millenniumassessment.org/documents/document.356.aspx.pdf

Mintel. (2021). Global consumer trends 2022. Available in: https://downloads.mintel.com/private/r3Mts/files/889468/

Mládek, J., Hejcman, M., Hejduk, S., et al. (2011). Community seasonal development enables late defoliation without loss of forage quality in semi-natural grasslands. *Folia Geobotanica, 46*, 17–34.

Mndzebele, B., Ncube, B., Fessehazion, M. K., et al. (2020). Effects of cowpea-amaranth intercropping and fertiliser application on soil phosphatase activities, available soil phosphorus, and crop growth response. *Agronomy, 10*(1), 79. https://doi.org/10.3390/agronomy10010079

Morris, J. B. (2009). Morphological and reproductive characterization in hyacinth bean, *Lablab purpureus* (L.) sweet germplasm with clinically proven nutraceutical and pharmaceutical traits for use as a medicinal food. *Journal of Dietary Supplements, 6*(3), 263–279.

Mos, M., Banks, S. W., Nowakowski, D. J., et al. (2013). Impact of *Miscanthus x giganteus* senescence times on fast pyrolysis bio-oil quality. *Bioresource Technology, 129*, 335–342.

Mueller, N., Gerber, J., Johnston, M., et al. (2012). Closing yield gaps through nutrient and water management. *Nature, 490*, 254–257. https://doi.org/10.1038/nature11420

Murevanhema, Y. Y., & Jideani, V. A. (2013). Potential of bambara groundnut (*Vigna subterranea* (L.) verdc) milk as a probiotic beverage-a review. *Critical Reviews in Food Science and Nutrition, 53*(9), 954–967.

Murthy, H. N., & Bapat, V. A. (2020). Importance of underutilized fruits and nuts. In H. N. Murthy & V. A. Bapat (Eds.), *Bioactive compounds in underutilized fruits and nuts* (pp. 3–19). Springer Nature. https://doi.org/10.1007/978-3-030-30182-8

Naeem, M., Khan, M. M., & Morris, J. (2009). Agrobotanical attributes, nitrogen-fixation, enzyme activities and nutraceuticals of hyacinth bean (*Lablab purpureus* L.): A bio-functional medicinal legume. *American Journal of Plant Physiology, 4*, 58–69.

Nandal, U., & Bhardwaj, R. L. (2014). The role of underutilized fruits in nutritional and economic security of tribals: A review. *Critical Reviews in Food Science and Nutrition, 54*, 880–890. https://doi.org/10.1080/10408398.2011.616638

Nc'Nean Distillery. (2021). Sustainability report. https://ncnean.com/pages/sustainability. Accessed 31 Jan 2022.

Ndjiondjop, M. N., Manneh, B., Cissoko, M., et al. (2010). Drought resistance in an interspecific backcross population of rice (*Oryza* spp.) derived from the cross WAB56-104 (*O. sativa*) x CG14 (*O. glaberrima*). *Plant Science, 179*, 364–373.

Nganga, S. (2014). African leafy vegetables and household wellbeing in Kenya: A disaggregation by gender. *Current Research Journal of Social Sciences, 6*(4), 82–94.

Nielsen, D. C., & Vigil, M. F. (2017). Water use and environmental parameters influence proso millet yield. *Field Crops Research, 212*, 34–44.

Ntatsi, G., Karkanis, A., Yfantopoulos, D., et al. (2018a). Evaluation of the field performance, nitrogen fixation efficiency and competitive ability of pea landraces grown under organic and conventional farming systems. *Archives of Agronomy and Soil Science, 65*, 294–307. https://doi.org/10.1080/03650340.2018.1501155

Ntatsi, G., Karkanis, A., Yfantopoulos, D., et al. (2018b). Impact of variety and farming practices on growth, yield, weed flora and symbiotic nitrogen fixation in faba bean cultivated for fresh seed production. *Acta Agriculturae Scandinavica, Section B — Soil & Plant Science, 68*, 619–630. https://doi.org/10.1080/09064710.2018.1452286

O'Keeffe, S., Schulte, R., Sanders, J., et al. (2011). I. Technical assessment for first generation green biorefinery (GBR) using mass and energy balances: Scenarios for an Irish GBR blueprint. *Biomass and Bioenergy, 35*, 4712–4723.

Obasi, M., & Ezedinma, F. (1991). Floral biology of Kerstings groundnut (*Kerstingiella geocarpa* harms). *Ghana Journal of Agricultural Science, 24–27*, 145–149.

Oelke, E. A., Oplinger, E. S., Teynor, T. M., et al. (1992). Safflower. Alternative crops field manual. Universities of Wisconsin Cooperative Extension Service, University of Minnesota Center for Alternative Plant and Animal Products & Minnesota Extension Service. http://www.hort.purdue.edu/newcrop/afcm/safflower.html

Onyekachi, O. G., Boniface, O. O., Gemlack, N. F., et al. (2019). The effect of climate change on abiotic plant stress: A review. In A. De Oliveira (Ed.), *Abiotic and biotic stress in plants*. 13pp. Available at: https://www.intechopen.com/chapters/64791. Accessed 27 Jan 2022.

Oyelakin, O. S., & Olaniyi, O. W. (2019). Effects of intercropping maize (*Zea mays* L.) with Jack beans (*Canavalia ensiformis* L.) at different spacing and weeding regimes on crops productivity. *International Journal of Agricultural and Biosystems Engineering, 13*(3), 61–65.

Padulosi, S., Eyzaquirre, P., & Hodgkin, T. (1999). Challenges and strategies in promoting conservation and use of neglected and underutilized crop species. In J. Janick (Ed.), *Perspectives on new crops and new uses* (p. 140). ASHS Press.

Padulosi, S., Hodgkin, T., Williams, J., & Haq, N. (2002). 30 underutilized crops: trends, challenges and opportunities in the 21st century. In: *Managing plant genetic diversity* (eds J.M.M. Engels, V. Ramanatha Rao, A.H.D. Brown and M.T. Jackson) 323–487

Padulosi, S., Heywood, V., Hunter, D., et al. (2011). Underutilized species and climate change: Current status and outlook. In S. S. Yadav, R. J. Redden, J. L. Hatfield, H. Lotze-Campen, & A. E. Hall (Eds.), *Crop adaptation to climate change*. John Wiley & Sons, Ltd.: Blackwell Publishing Ltd.

Padulosi, S., Thompson, J., & Rudebjer, P. (2013). *Fighting poverty, hunger and malnutrition with neglected and underutilised species (NUS): Needs, challenges and the way forward.* Bioversity International 60 p. https://cgspace.cgiar.org/handle/10568/68927

Palou, E., López-Malo, A., Barbosa-Canovas, G. V., et al. (2020). High-pressure preservation of foods. In *Handbook of food preservation* (pp. 843–872). CRC Press. https://doi.org/10.1201/9780429091483-54

Parihar, A. K., Basandrai, A. K., Kushwaha, K. P., et al. (2018). Targeting test environments and rust-resistant genotypes in lentils (*Lens culinaris*) by using heritability-adjusted biplot analysis. *Crop & Pasture Science, 69*, 1113–1125.

Parvin, S., Uddin, S., Bourgault, M., et al. (2019a). Effect of heat wave on N2 fixation and N remobilization of lentil (*Lens culinaris* MEDIK) grown under Free Air CO_2 Enrichment in a Mediterranean-type environment. *Plant Biology, 22*(S1), 123–132.

Parvin, S., Uddin, S., Tausz-Posch, S., et al. (2019b). Elevated CO_2 improves yield and N2 fixation but not grain N concentration of faba bean (*Vicia faba* L.) subjected to terminal drought. *Environmental and Experimental Botany, 165*, 161–173. https://doi.org/10.1016/j.envexpbot.2019.06.003

Petrusán, J. I., Rawel, H., & Huschek, G. (2016). Protein-rich vegetal sources and trends in human nutrition: A review. *Current Topics in Peptide and Protein Research, 17*, 1–19.

Piñeiro, V., Arias, J., Dürr, J., et al. (2020). A scoping review on incentives for adoption of sustainable agricultural practices and their outcomes. *Nature Sustainability, 3*(10), 809–820. https://doi.org/10.1038/s41893-020-00617-y

Ponisio, L. C., Gonigle, L. K. M., Mace, K. C., et al. (2015). Diversification practices reduce organic to conventional yield gap. *Proceedings of the Royal Society B Biological Sciences, 282*, 20141396. https://doi.org/10.1098/rspb.2014.1396

Poore, J., & Nemecek, T. (2018). Reducing food's environmental impacts through producers and consumers. *Science, 360*(6392), 987–992. https://doi.org/10.1126/science.aaq0216

Popkin, B. M., Corvalan, C., & Grummer-Strawn, L. M. (2020). Dynamics of the double burden of malnutrition and the changing nutrition reality. *The Lancet, 395*(10217), 65–74. https://doi.org/10.1016/S0140-6736(19)32497-3

Popoola, J. O., Ojuederie, O. B., Omonhinmin, C. A., et al. (2019). Neglected and underutilized legume crops: Improvement and future prospects. In F. Shah, Z. Khan, & A. Iqbal (Eds.), *Recent advances in grain crops research*. IntechOpen. https://doi.org/10.5772/intechopen.87069

Popp, J., Pető, K., & Nagy, J. (2013). Pesticide productivity and food security. A review. *Agronomy for Sustainable Development, 33*(1), 243–255. https://doi.org/10.1007/s13593-012-0105-x

Profeta, A. (2019). The impact of health claims in different product categories. *Journal of International Food & Agribusiness Marketing, 32*(2), 123–140. https://doi.org/10.1080/08974438.2019.1599753

Profeta, A., Balling, R., & Enneking, U. (2006). Geschützte HerkunftsangabenStatus Quo und Entwicklung der Nutzung der Verordnung (EG) 510/06 › GJAE - German Journal of Agricultural Economics - DE. *German Journal of Agricultural Economics, 55*(8), 353–358. https://www.gjae-online.de/de/articles/geschutzte-herkunftsangabenstatus-quo-und-entwicklung-der-nutzung-der-verordnung-eg-510-06/

Profeta, A., Enneking, U., & Balling, R. (2007). Wahrnehmung von regionalen Lebensmittelspezialitäten in Deutschland – Eine deutschlandweite, repräsentative Konsumentenbefragung. *Berichte Über Landwirtschaft: Zeitschrift Für Agrarpolitik Und Landwirtschaft, 85*(2), 238–251.

Quesada, N., Iannetta, P. P. M., White, P. J., et al. (2019). What evidence exists on the effectiveness of the techniques and management approaches used to improve the productivity of field grown tomatoes under conditions of water-, nitrogen- and/or phosphorus-deficit? A Systematic Map Protocol. *Environmental Evidence, 8*, 26. https://doi.org/10.1186/s13750-019-0172-4

Rai, K. K., Pandey, N., Meena, R. P., et al. (2021). Biotechnological strategies for enhancing heavy metal tolerance in neglected and underutilized legume crops: A comprehensive review. *Ecotoxicology and Environmental Safety, 208*, 111750. https://doi.org/10.1016/j.ecoenv.2020.111750

Rao, N. K., & Shahid, M. (2016). Neglected and underutilized crops for sustainable agricultural production in marginal areas. In R. A. Hall & P. Rudebjer (Eds.), *Proceeding of the 3rd international conference on neglected and under-utilized crops for food secure Africa, 25–27 September, Accra, Ghana* (pp. 130–139). Bioversity International.

Rao, N. K., Shahid, M., Al-Shankiti, A., & Elouafi, I. (2014). Neglected and underutilized species for food and income security in marginal environments. *Acta Horticulturae, 1051*, 91–104.

Ravelombola, W., Shi, A., & Weng, Y. (2018). Association analysis of salt tolerance in cowpea (*Vigna unguiculata* (L.) Walp) at germination and seedling stages. *Theoretical and Applied Genetics, 131*, 79. https://doi.org/10.1007/s00122-017-2987-0

Razi, K., Bae, D.-W., & Muneer, S. (2021). Target-based physiological modulations and chloroplast proteome reveals a drought resilient rootstock in okra (*Abelmoschus esculentus*) genotypes. *International Journal of Molecular Sciences, 22*(23), 12996. https://doi.org/10.3390/ijms222312996

Reif, T. M., Zikeli, S., Rieps, A. M., et al. (2021). Reviving a neglected crop: A case study on lentil (*Lens culinaris* Medikus subsp. culinaris) cultivation in Germany. *Sustainability, 13*(1), 133. https://doi.org/10.3390/SU13010133

Rivero, R. M., Mittler, R., Blumwald, E., et al. (2022). Developing climate-resilient crops: Improving plant tolerance to stress combination. *The Plant Journal, 109*, 373–389. https://doi.org/10.1111/tpj.15483

Rocchi, L., Boggia, A., & Paolotti, L. (2020). Sustainable agricultural systems: A bibliometrics analysis of ecological modernization approach. *Sustainability (Switzerland), 12*(22), 1–16. https://doi.org/10.3390/su12229635

Rogelj, J. D., Shindell, K., Jiang, S., et al. (2018). Mitigation pathways compatible with 1.5°C in the context of sustainable development. In V. Masson-Delmotte, P. Zhai, H.-O. Pörtner, D. Roberts, J. Skea, P. R. Shukla, A. Pirani, W. Moufouma-Okia, C. Péan, R. Pidcock, S. Connors, J. B. R. Matthews, Y. Chen, X. Zhou, M. I. Gomis, E. Lonnoy, T. Maycock, M. Tignor, & T. Waterfield (Eds.), *Global warming of 1.5°C. An IPCC Special Report on the impacts of global warming of 1.5°C above pre-industrial levels and related global greenhouse gas emission pathways, in the context of strengthening the global response to the threat of climate change, sustainable development, and efforts to eradicate poverty*. In Press. https://www.ipcc.ch/site/assets/uploads/sites/2/2019/02/SR15_Chapter2_Low_Res.pdf

Ronga, D., Caradonia, F., Vitti, A., et al. (2021). Agronomic comparisons of heirloom and modern processing tomato genotypes cultivated in organic and conventional farming systems. *Agronomy, 11*, 349. https://doi.org/10.3390/agronomy11020349

Rouphael, Y., & Colla, G. (2020a). Editorial: Biostimulants in agriculture. *Frontiers in Plant Science, 1*, 40. https://doi.org/10.3389/fpls.2020.00040

Rouphael, Y., & Colla, G. (2020b). Toward a sustainable agriculture through plant biostimulants: From experimental data to practical applications. *Agronomy, 10*, 1461. https://doi.org/10.3390/agronomy10101461

Rubiales, D., & Flores, F. (2020). Adaptation of one-flowered vetch (*Vicia articulata* Hornem.) to Mediterranean rain fed conditions. *Agronomy, 10*(3), 383. https://doi.org/10.3390/agronomy10030383

Rubiales, D., Fondevilla, S., Chen, W., et al. (2015). Achievements and challenges in legume breeding for pest and disease resistance. *Critical Reviews in Plant Science, 34*, 195–236. https://doi.org/10.1080/07352689.2014.898445

Rubiales, D., Emeran, A. A., & Flores, F. (2020). Adaptation of grass pea (*Lathyrus sativus*) to Mediterranean environments. *Agronomy, 10*(9), 1295; doi:10.3390/agronomy10091295.

Rubiales, D., Annicchiarico, P., Vaz Patto, M. C., & Julier, B. (2021). Legume breeding for the agroecological transition of global agri-food systems: A European perspective. *Frontiers in Plant Science, 12*, 782574. https://doi.org/10.3389/fpls.2021.782574

Russel, S. J., & Norvig, P. (2021). *Artificial intelligence: A modern approach* (4th ed.). Hoboken Person. ISBN 9780134610993.

Sabatino, L., Iapichino, G., Consentino, B. B., et al. (2020). Rootstock and arbuscular mycorrhiza combinatorial effects on eggplant crop performance and fruit quality under greenhouse conditions. *Agronomy, 10*, 693. https://doi.org/10.3390/agronomy10050693

Saini, S., Saxena, S., Samtiya, M., et al. (2021). Potential of underutilized millets as nutri-cereal: An overview. *Journal of Food Science and Technology, 58*, 4465–4477. https://doi.org/10.1007/s13197-021-04985-x

Sardana, V., Mahajan, G., Jabran, K., et al. (2017). Role of competition in managing weeds: An introduction to the special issue. *Crop Protection, 95*, 1–7.

Savvas, D., & Gruda, N. (2018). Application of soilless culture technologies in the modern greenhouse industry - A review. *European Journal of Horticultural Science, **83***, 280–293. https://doi.org/10.17660/eJHS.2018/83.5.5

Scarano, A., Semeraro, T., Chieppa, M., et al. (2021). Neglected and underutilised plant species (NUS) from the Apulia Region worthy of being rescued and re-included in daily diet. *Horticulturae, 7*(7), 177. https://doi.org/10.3390/horticulturae7070177

Schlegel, P., Wyss, U., Arrigo, Y., et al. (2016). Mineral concentrations of fresh herbage from mixed grassland as influenced by botanical composition, harvest time and growth stage. *Animal Feed Science and Technology, 219*, 226–233.

Scottish Government. (2021). Cereal and oilseed rape harvest: Final. https://www.gov.scot/publications/cereal-oilseed-rape-harvest-2021-final-estimates/. Accessed 31 Jan 2022.

Scottish Whisky Association. (2021). Facts and figures. https://www.scotch-whisky.org.uk/insights/facts-figures/. Accessed 31 Jan 2022.

Shah, F., & Wu, W. (2019). Soil and crop management strategies to ensure higher crop productivity within sustainable environments. *Sustainability, 11*, 1485. https://doi.org/10.3390/su11051485

Shahzad, M., Hussain, M., Jabran, K., et al. (2021). The impact of different crop rotations by weed management strategies' interactions on weed infestation and productivity of wheat (*Triticum aestivum* L.). *Agronomy, 11*, 2088. https://doi.org/10.3390/agronomy11102088

Shannon, M. C., & Grieve, C. M. (1999). Tolerance of vegetable crops to salinity. *Scientia Horticulturae, 78*, 5–36.

Shayanowako, A., Morrissey, O., Tanzi, A. S., et al. (2021). African leafy vegetables for improved human nutrition and food system resilience in Southern Africa: A scoping review. *Sustainability, 13*, 2896.

Siddique, K. H. M., Li, X., & Gruber, K. (2021). Rediscovering Asia's forgotten crops to fight chronic and hidden hunger. *Nature Plants, 7*, 116–122. https://doi.org/10.1038/s41477-021-00850-z

Siracusa, L., Avola, G., Patanè, C., et al. (2013). Re-evaluation of traditional Mediterranean foods. The local landraces of 'Cipolla di Giarratana' (*Allium cepa* L.) and long-storage tomato (*Lycopersicon esculentum* L.): Quality traits and polyphenol content. *Journal of the Science of Food and Agriculture, 93*(14), 3512–3519.

Skendžić, S., Zovko, M., Živković, I. P., et al. (2021). The impact of climate change on agricultural insect pests. *Insects, 12*, 440. https://doi.org/10.3390/insects12050440

Slabbert, R., & van den Heever, E. (2006). Selection of traditional crops for improved drought tolerance in leafy amaranth: Moving towards sustainable food supply. In *International conference on indigenous vegetables and legumes prospectus for fighting poverty, hunger malnutrition* (Vol. 752, pp. 281–286). ISHS.

Smetana, S., Mathys, A., Knoch, A., et al. (2015). Meat alternatives: Life cycle assessment of most known meat substitutes. *The International Journal of Life Cycle Assessment, 20*(9), 1254–1267. https://doi.org/10.1007/s11367-015-0931-6

Smetana, S., Spykman, R., & Heinz, V. (2021). Environmental aspects of insect mass production. *Journal of Insects as Food and Feed, 7*(5), 553–571. https://doi.org/10.3920/JIFF2020.0116

Soares, J., Deuchande, T., Valente, L., et al. (2019). Growth and nutritional responses of bean and soybean genotypes to elevated CO2 in a controlled environment. *Plants, 8*, 465. https://doi.org/10.3390/plants8110465

Souza, L. A., López Andrade, S. A., Ribeiro Souza, S. C., et al. (2013). Evaluation of mycorrhizal influence on the development and phytoremediation potential of Canavalia Gladiata in Pb-contaminated soils. *International Journal of Phytoremediation, 15*, 465–476.

Springmann, M., Clark, M. A., Mason-D'Croz, D., et al. (2018). Options for keeping the food system within environmental limits. *Nature, 562*, 519–525. https://doi.org/10.1038/s41586-018-0594-0

Stagnari, F., Maggio, A., Galieni, A., & Pisante, M. (2017). Multiple benefits of legumes for agriculture sustainability: an overview. *Chemical and Biological Technologies in Agriculture, 4*(1), 1–13.

Stamp, P., Messmer, R., & Walter, A. (2012). Competitive underutilized crops will depend on the state funding of breeding programmes: An opinion on the example of Europe. *Plant Breeding, 131*(4), 461–464. https://doi.org/10.1111/j.1439-0523.2012.01990.x

Steffen, W., Richardson, K., Rockström, J., et al. (2015). Planetary boundaries: Guiding human development on a changing planet. *Science, 347*(6223), 1259855.

Steinfeld, H., Gerber, P., Wassenaar, T., et al. (2006). *Livestock's long shadow*. FAO of the UN.

Stošić, M., Ivezić, V., & Tadić, V. (2021). Tillage systems as a function of greenhouse gas (GHG) emission and fuel consumption mitigation. *Environmental Science and Pollution Research, 28*(13), 16492–16503. https://doi.org/10.1007/s11356-020-12211-y

Tello, A., Aganovic, K., Parniakov, O., et al. (2021). Product development and environmental impact of an insect-based milk alternative. *Future Foods, 4*. https://doi.org/10.1016/J.FUFO.2021.100080

Teuteberg, H. J. (2006). Der Verzehr von Nahrungsmitteln in Deutschland pro Kopf und Jahr seit Beginn der Industrialisierung (1850–1975). https://doi.org/10.4232/1.8252.

Thanh, P. T., Sripichitt, P., Chanprame, S., et al. (2006). Transfer of drought resistant character from wild rice (*Oryza meridionalis and Oryza nivara*) to cultivated rice (*Oryza sativa* L.) by backcrossing and immature embryo culture. *Kasetsart Journal of Social Sciences, 40*, 582–594.

Tonn, B., Thumm, U., & Claupein, W. (2010). Semi-natural grassland biomass for combustion: Influence of botanical composition, harvest date and site conditions on fuel composition. *Grass and Forage Science, 65*, 383–397.

Tran, F., Holland, J. E., Quesada, N., et al. (2021). What evidence exists on the effectiveness of the techniques and management approaches used to improve the productivity of field grown tomatoes under conditions of water-, nitrogen- and/or phosphorus-deficit? A systematic map. *Environmental Evidence, 10*, 14. https://doi.org/10.1186/s13750-021-00229-9

Traoré, H., Barro, A., Yonli, D., et al. (2020). Water conservation methods and cropping systems for increased productivity and economic resilience in Burkina Faso. *Water (Switzerland), 12*(4), 976.

Uhl, P., & Brühl, C. A. (2019). The impact of pesticides on flower-visiting insects: A review with regard to European risk assessment. *Environmental Toxicology and Chemistry, 38*(11), 2355–2370.

von Grebmer, K., Saltzman, A., Birol, E., et al. (2014). *Global hunger index: The challenge of hidden hunger*. Welthungerhilfe, International Food Policy Research Institute, and Concern Worldwide.

Van Delden, S. H., SharathKumar, M., Butturini, M., et al. (2021). Current status and future challenges in implementing and upscaling vertical farming systems. *Nature Food, 2*, 944–956. https://doi.org/10.1038/s43016-021-00402-w

Van Oosterom, E. J., O'Leary, G. J., Carberry, P. S., et al. (2002). Simulating growth, development, and yield of tillering pearl millet. III. Biomass accumulation and partitioning. *Field Crops Research, 79*(2–3), 85–106.

Vaz Patto, M. C., & Rubiales, D. (2014). Lathyrus diversity: Available resources with relevance to crop improvement. *Annals of Botany, 113*(6), 895–908. https://doi.org/10.1093/aob/mcu024

Vaz Patto, M. C., Amarowicz, R., Aryee, A., et al. (2015). Achievements and challenges in improving the nutritional quality of food legumes. *Critical Reviews in Plant Sciences, 34*, 105–143. https://doi.org/10.1080/07352689.2014.897907

Vermeulen, S., Campbell, B. M., & Ingram, J. S. I. (2012). Climate change and food systems. *Annual Review of Environment and Resources, 37*, 195–222. https://doi.org/10.1146/annurev-environ-020411-130608

Vernooy, R. (2021). *Action research on neglected and underutilised species. A methodological guide*. The Alliance of Bioversity International and CIAT.

Vidigal, P., Duarte, B., Cavaco, A. R., et al. (2018). Preliminary diversity assessment of an undervalued tropical bean (*Lablab purpureus* (L.) Sweet) through fatty acid profiling. *Plant Physiology and Biochemistry, 132*, 508–514.

Vijayalakshmi, D., Geetha, K., Gowda, J., et al. (2010). Empowerment of women farmers through value addition on minor millets genetic resources in Karnataka. *Indian Journal of Plant Genetic Resources, 23*(1), 132–135.

Villegas-Fernández, Á. M., Sillero, J. C., Emeran, A. A., et al. (2011). Multiple-disease resistance in Vicia faba: Multi-environment field testing for identification of combined resistance to rust and chocolate spot. *Field Crops Research, 124*, 59–65.

Wang, S. T., & Ebert, A. W. (2013). Breeding of leafy amaranth for adaptation to climate change. In R. Holmer, G. Linwattana, P. Nath, & J. D. H. Keatinge (Eds.), *High value vegetables in Southeast Asia: Production, supply and demand; proceedings of the SEAVEG 2012 regional symposium* (pp. 36–43). AVRDC - The World Vegetable Center.

Waramit, N., Moore, K. J., & Fales, S. L. (2012). Forage quality of native warm-season grasses in response to nitrogen fertilization and harvest date. *Animal Feed Science and Technology, 174*, 46–59.

Weinberger, K. (2003). Impact analysis of Mungbean research in South and Southeast Asia. Final report of GTZ Project. Shanhua, Taiwan. https://assets.publishing.service.gov.uk/media/57a08b1ced915d3cfd000b38/DFID_impact_case_study_Mungbean_FINAL_1_.pdf

Weinberger, K., & Swai, I. (2006). Consumption of traditional vegetables in Central and North-Eastern Tanzania. *Ecology of Food and Nutrition, 45*, 87–103.

Wilker, J., Navabi, A., Rajcan, I., et al. (2019). Agronomic performance and nitrogen fixation of heirloom and conventional dry bean varieties under low-nitrogen field conditions. *Frontiers in Plant Science, 10*, 952. https://doi.org/10.3389/fpls.2019.00952

Will, M. (2008). *Promoting value chains of neglected and underutilized species for pro-poor growth and biodiversity conservation: Guidelines and good practices*. Bioversity International; Global Facilitation Unit for Underutilized Species. isbn:978-92-9043-775-8. Available on-line: https://cgspace.cgiar.org/handle/10568/104850

Willett, W., Rockström, J., Loken, B., et al. (2019). Food in the Anthropocene: The EAT-lancet commission on healthy diets from sustainable food systems. *Lancet (London, England), 393*(10170), 447–492. https://doi.org/10.1016/S0140-6736(18)31788-4

Wilson, D. M., Dalluge, D. L., Rover, M., et al. (2013). Crop management impacts biofuel quality: Influence of switchgrass harvest time on yield, nitrogen and ash of fast pyrolysis products. *Bioenergy Research, 6*, 103–113.

Wirsig, A., Profeta, A., Häring, A., et al. (n.d.). Branding of rural regions and autochthon agricultural products linked to their terroir Section 2. Agricultural Economics and Rural Sociology Branding of rural regions and autochthon agricultural products linked to their terroir.

Wirsig, A., Profeta, A., Häring, A., & Cerjak, M. (2011). Branding of rural regions and autochthon agricultural products linked to their terroir. Proceedings. 43rd Proceedings. 46th Croatian and 6th Croatian and 3rd International Symposium on Agriculture. Opatija. Croatia (301-305) Agriculture. Opatija. Croatia (301- XXX), January, 301–305.

Wright, D., Neupane, S., Heidecker, T., et al. (2021). Understanding photothermal interactions will help expand production range and increase genetic diversity of lentil (*Lens culinaris* Medik.). *Plants, People and Planet, 3*, 171–181. https://doi.org/10.1002/ppp3.10158

Yachi, S., & Loreau, M. (1999). Biodiversity and ecosystem productivity in a fluctuating environment: The insurance hypothesis. *Proceedings of the National Academy of Sciences of the United States of America, 96*, 1463–1468. https://doi.org/10.1073/pnas.96.4.1463

Zhai, Z., Martínez, J. F., Beltran, V., et al. (2020). Decision support systems for agriculture 4.0: Survey and challenges. *Computers and Electronics in Agriculture, 170*. https://doi.org/10.1016/j.compag.2020.105256

Zhao, J., Zeng, Z., He, X., et al. (2015). Effects of monoculture and mixed culture of grass and legume forage species on soil microbial community structure under different levels of nitrogen fertilization. *European Journal of Soil Biology, 68*, 61–68. https://doi.org/10.1016/j.ejsobi.2015.03.008

Chapter 10
Alternative Proteins for Food and Feed

Stefanie Verstringe, Robin Vandercruyssen, Hannes Carmans, Alexandru Vasile Rusu, Geert Bruggeman, and Monica Trif

Abstract The protein supply is most critical, both for human consumption and animal feed. Integration of a variety of new or alternative protein sources from both terrestrial and aquatic origin into new and existing processes or products needs to be explored to develop and ensure more sustainable, resilient supply chains, featuring high consumer acceptability by a clean labeling approach and attractive market opportunities. The present chapter is dealing with new alternative protein and traditional protein sources from the terrestrial origin for food and feed such as insects, plant-based (legumes and grasses), and by-products of crops.

Keywords Sustainable · Protein · Plant-based · By-product · Insect

1 Introduction

With the possibility of the human population reaching 9 billion in 2050, many are looking for new and more sustainable food resources. The increasing population is increasing the global demand for meat and animal products. The need for food should even increase by 70%. Proteins are an essential component in the human and animal diet, and to meet their increasing demand, alternative food and feed resources are needed (Stiles, 2017; World Resources Institute, 2019).

The Food and Agriculture Organization of the United Nations (FAO) characterizes a sustainable diet as: "a diet that has hardly any negative impact on the

S. Verstringe (✉) · R. Vandercruyssen · H. Carmans · G. Bruggeman
Nutrition Sciences N.V., Drongen, Belgium
e-mail: stefanie.verstringe@nusciencegroup.com

A. V. Rusu (✉)
Biozoon Food Innovations GmbH, Strategic Research Department, Bremerhaven, Germany
e-mail: rusu@biozoon.de

M. Trif (✉)
Centre for Innovative Process Engineering GmbH, Food Research Department, Stuhr, Germany

C. M. Galanakis (ed.), *Biodiversity, Functional Ecosystems and Sustainable Food Production*, https://doi.org/10.1007/978-3-031-07434-9_10

environment, and that makes a positive contribution to food security and healthy life for present and future generations' returns" (Food and Agriculture Organization of the United Nations (Hrsg.), 2017).

The Bioeconomy Council has already dealt extensively with sustainability in the production of human food and animal feed of increasing global demand (especially among the growing middle class in emerging countries). In developing countries, the often inadequate protein supply is a relevant and urgent problem.

Innovations within the existing system can contribute to further optimization in the manufacture and consumption of agricultural products. In the future, however, novel and possibly more efficiently produced protein sources can also serve to ensure a more sustainable protein supply. Using traditional protein sources such as plant-based and fish more efficiently can also help (Napiórkowska-Krzebietke, 2017).

In the future, it should not only be possible to produce proteins more cheaply but also to reduce the environmental impact that their production currently causes. The first products should be on the market at the latest in 2025.

A healthy and sustainable supply of the world's population with proteins from alternative sources.

1.1 The Importance of Proteins in the Diet

Proteins, building blocks of human and animal tissues, necessary for the growth and regeneration of tissues, are made up of chains of up to 20 different amino acids, the exact sequence of which (resulting from the DNA sequence) determines the structure and thus the function of the protein (Alberts et al., 2002). It is estimated that over 100,000 different proteins are found in the human body. In addition, almost each of the 20 amino acids also has particular other functions in the metabolism. Of the 20 amino acids required for protein synthesis, among them nine essential amino acids (l-valine, l-leucine, l-isoleucine, l-lysine, l-threonine, l-methionine, l-histidine, l-phenylalanine, and l-tryptophan) cannot be synthesized in animals and humans. They must be ingested from food (Trif et al., 2020). The value of a protein for the nutrition of humans and animals is determined by the content or pattern of these essential amino acids (Trif et al., 2016; Rusu, Alvarez Penedo, Schwarze, & Trif, 2020b).

In the western world, in particular, many people ingest these proteins through meat or fish. This consumer behavior is already causing significant problems for us, the climate, and nature, and of course, the animals.

Comparing the health effects of meat and plant-based meat substitute products is pretty complicated, as these depend heavily on the (animal/plant) species, the degree of processing, and the preparation of the products. For example, the German Nutrition Society recommends low-fat meat and lean muscle meat. One advantage of plant-based foods over meat products is that they contain almost no cholesterol (Trif et al., 2019). In general, not heavily processed soy and gluten products have a low fat and high nutrient and protein content (Huber & Keller, 2017).

According to Protein Digestibility Corrected Amino Acid Score (PDCAAS), used to show protein utilization in the human body, the soy proteins are rated with the highest value of 1.0, same as milk and egg proteins (Biesalski et al., 2011). Wheat protein has a protein quality of 0.4 (Biesalski et al., 2011), and therefore protein utilization can be increased by combining legumes with cereals.

1.2 Novel Approaches to the Production of Protein-Rich Food and Animal Feed

Numerous publicly funded and commercial initiatives deal with the quantitative and qualitative adequate protein supply for the growing world population. One focus is on better supply in developing countries and a reduction and partial substitution of the consumption of animal products in the industrialized countries. Growing awareness of the negative consequences of animal-based food production and the emergence of new, biotechnological processes for protein production has also led to increased investor interest and, particularly in the USA, Israel, France, and the Netherlands, to considerable capital expenditure.

The new approaches aim at (a) *the development of new products based on traditional protein sources* and (b) *the utilization of new protein sources.*

Since the new foods should be not only healthy and sustainable but also tasty, another focus is on improving the taste, aroma, and structure of developed products (Rusu, Randriambelonoro, Perrin, et al., 2020c). Meat substitute products based on processed vegetable proteins have become increasingly important in the last few decades. These include already commercial products such as seitan (wheat protein), soy meat/tempeh (soybeans), and Quorn (fermented mycelium), likewise products based on lupines.

1.2.1 Development of New Products Based on Traditional Protein Sources

Soybeans and wheat are used directly for human nutrition in the form of vegetable meat substitutes.

The most common protein sources used in feed are soybean meal (SBM) and, to a lesser extent, fish meal, mostly used in aquaculture. The reason soybean is currently utilized as a protein source is that it contains a higher protein content (440–500 g kg^{-1}) than other protein-rich crops. Also, it has an excellent amino acid composition too, meaning that the essential amino acids are present in a reasonable amount.

To produce plant-based meat substitute products such as tofu, tempeh, soy meat, and seitan, agricultural land is required for soy and wheat cultivation. Only around 6% of the soybeans cultivated worldwide are used for direct human consumption, while around 70–75% is used as animal feed (IDH and IUCN NL, 2019). The grain is also primarily fed to animals. For example, in Germany, around 40% of the harvested wheat ends up in the feeding trough.

The cultivated soybean (*Glycine max*) is considered to have evolved from *Glycine soja*, which was initially a landrace. With the origin of soybean being in China, soybean was introduced to Europe in the seventeenth century. At the beginning of the nineteenth century, it was introduced in the USA, where after its use has rapidly spread out. Nowadays, soybean is the most critical cultivated legume crop worldwide, and around 352.6 million tons are produced, according to the Food and Agriculture Organization (FAO). It is cultivated globally on 123.6 million hectares, with an average yield of 3.1 t/ ha. South-America, with Brazil, Argentina, and Chile, is the biggest producer before the USA and Asia, with China and India.

Europe's production of soybean is low compared to other oilseeds; thus, they depend on imports from the leading producers, like South-America (de Camargo et al., 2019). Currently, around 93% of the soy demand in Europe is used for animal feed (European Commission, 2018). If there is no substitution, and if there are additional meat substitute products, the demand for soy also increases.

In animal husbandry, plants are used as animal feed, so the rate of conversion from plant calories to animal calories is high, which is also known as calorie loss. A fattening pig consumes an average of 250 kg of animal feed within 5–6 months until it has reached the slaughter weight of approx. 95 kg; the edible proportion of a whole animal is around 62% (Heinrich-Böll-Stiftung, 2014). In the case of a plant-based diet, a smaller amount of agricultural products is therefore required to feed the same amount of people.

SBM, as the by-product of the extraction of soybean, is estimated to account for two-thirds of the world's soybean output. Soybean can be processed mechanically or by solvent extraction to form different products. SBM, with a crude protein content around 490–500 g kg^{-1}, is one of those products. If the solvent extraction method is used, the oil content in the meal is lower than 20 g kg^{-1}. In mechanically extracted meals, the oil content exceeds 30 g kg^{-1}. SBM contains high amounts of the amino acid lysine, tryptophan, threonine, and isoleucine but low amounts of methionine. Thus supplementation of this sulfur amino acid is essential, especially in pig food. Another vital factor in feed is the digestibility of amino acids, which is very high in SBM. A disadvantage of SBM is the insufficient amount of B vitamins and phosphorus, of which 60–70% is bound to phytic acid. This causes a problem in the availability of phosphorus and other minerals for monogastric animals like pigs. SBM also contains oligosaccharides such as stachyose and raffinose. But the monogastric animals cannot digest them due to a lack of specific enzymes. On top of that, antinutritional factors such as lectin and trypsin inhibitors, abundantly present in the legume family, also pose problems for non-ruminants. Heat treatment is necessary to inactivate all of the above-mentioned factors. If not adequately treated, nutrient utilization can decrease, and the growth rate can be depressed (Florou-Paneri et al., 2014).

Fish meal is made by drying and grinding carcasses of fish. Two types of fish are used: fish specifically caught for the production of fish meal (the most common) and those coming from the by-products of the human food fishery waste. Fish consume a significant portion of fish oil (81%) and fish meal (more than 63%) in aquaculture. Although not very sustainable, the fish meal provides an excellent source of high-quality protein, and its nutritional features promote fast animal growth and increased yields of eggs, milk, meat, and farmed fish (Dórea, 2006).

Fish oil and fish meal are, next to SBM, an essential supplement in animal feed. The meal can enhance the docosahexaenoic acid (DHA) and ω-3 polyunsaturated fatty acid content of animal products. Moreover, they stimulate the production of unsaturated fatty acids such as DHA, eicosapentaenoic acid, and conjugated linoleic acid in the milk of dairy cows. However, there are concerns that persistent toxic substances bioaccumulated in fish or fish meal could end up in the human diet via the transfer of animal products. These substances, mostly organohalogen pollutants (OHP), are produced by industry and end up in aquatic environments. Mercury, a pollutant known to bioaccumulate in fish, is associated with the protein fraction, and OHPs are associated with the lipid fraction (Dórea, 2006). This would mean that concentrating fish protein and oil equals concentrating mercury and OHPs, respectively. Because of the bioaccumulative features of some of these substances, there are ongoing efforts to reduce the human health risks associated with fish consumption (Dórea, 2006). For this reason, the use of fish meal is becoming more and more limited, especially in aquaculture (Riddick, 2014).

At present, the animal industry is based on the utilization of SBM and fish meal as primary sources of proteins, despite their high price. Feed represents the high cost in animal production partially due to the competition between ingredients for humans and monogastric animals. Fish meal use and soybean import, mostly from South America, are becoming a controversial topic in terms of the environmental impact of its production. Studies have already shown that an SBM based diet has a high environmental impact per kg pig production (de Camargo et al., 2019) indeed. Another concern is the cultivation of soybean, which is inevitably linked with deforestation of Latin American rainforests (Nepstad et al., 2006).

With climate change being another challenge, many feed industries are looking for ways to reduce their carbon footprint. Another concern is the massive dependency on importing protein crops that makes the E.U. livestock sector vulnerable to price volatility and trade distortion (de Boer et al., 2014).

1.2.2 Utilization of New Protein Sources

Nowadays, companies and research teams are actively searching for safe and cost-effective alternatives in order to reduce their reliance on soybean import, for example. Non-conventional feedstuffs derived from biofuels, plants, fungi, algae, and non-chordate phyla could offer a solution and be used as an alternative (Florou-Paneri et al., 2014).

For the new protein-rich products, scientists are focusing on using not only plants, mushrooms, but by-products and leftovers from food production, for example, vegetables, beans, chickpeas, and quinoa, and also microbial biomass from edible mushrooms (Rusu, Schwarze, & Alvarez Penedo, 2020d), waste, bread crust or yeast residues and seedlings from beer production. The researchers have already begun with the chemical-nutritional characterization of the original raw materials (Dietrich et al., 2016).

Non-chordate animals like insects are increasing in popularity for human consumption, not only in animal nutrition. Insects could also contribute to more sustainable protein production in Europe but are still not widely accepted as food and so far (with a few exceptions) have not been approved for consumption in the E.U. yet. The feed the insects feed can also have an impact on human health. Insects that are collected in agriculturally used habitats and have fed on the plants there are, in some cases, high in pesticide loads (Rumpold & Schlüter, 2013; Rumpold et al., 2017). Overall, there is still a need for research on the health effects of insect consumption. Another problem with insect products right now is the price but will change in the future; as soon as it will be produced industrially, prices go down.

2 Plant-Based Alternatives as Traditional Protein Sources

2.1 Legumes and Grasses Cultivated

2.1.1 Introduction

The preferred crop utilized for food and feed remains soybean (*Glycine max* (*L.*) *Merr.*). However, soybean could be cultivated by a more sustainable means than it is generally cultivated today. Southern states of the United States of America (Illinois, Iowa, Minnesota, etc.) have a soybean production, which is increasing over the years, according to Shahbandeh (2020). These states dispose of large areas of arable land. Thus unsustainable practices such as deforestation are not employed to grow soybean.

Large-scale production of this crop in other climates and continents would be more sustainable as the large transportation distances, linked with the carbon footprint, would be reduced (de Boer et al., 2014). Since soybean was introduced to sub-Saharan Africa in the twentieth century, the cultivation area and production of it has increased exponentially in the past four decades. An important side note is that the soybean yield is amongst the lowest in the world due to lousy crop managing circumstances and limited resources. Many research teams and institutions are currently conducting soybean improvement methods to increase the yield to scale up soybean as a significant crop (Khojely et al., 2018).

Some countries in Europe (e.g., Italy) produce soybean already on a small scale. The unfavorable climate in more northern European countries prevents the cultivation of this crop, but with the climate becoming hotter and dryer in Europe, the cultivation of this crop gains more and more interest. This would make enterprises in this region less dependable on soybean import from South-America. According to their website, the institute ILVO in Belgium is currently conducting a breeding program to obtain a soybean variety that can grow in more moderate climates.

As described before, the reason soybean is mainly used as a protein source is that it contains a higher protein content (440–500 g kg^{-1}) and has a better amino acid composition in comparison with other crops. However, there is a range of other plants that could be cultivated in Europe that are promising to at least partially replace the soybean, wheat, and fish meal diet. Legumes and grasses are among those promising candidates (Lindberg et al., 2016).

Various crops with high protein content in the legume family of *Fabaceae* could be a plant-based alternative to soybean. The legume family has many advantages in agriculture since it can assimilate nitrogen from the air in symbiosis with *Rhizobia* species. Moreover, these plants are known to enrich the soil with water and nutrients and, more importantly, nitrogen, ensuring good yields for the next crop. This reduces the need for fertilizers, simultaneously reducing the emission of greenhouse gases (derived from the use and production of fertilizers). Crop rotation with legumes can assist in the control of diseases, weeds, and pests as they are not susceptible to the same pests. This could decrease the use of pesticides on arable land and increase the diversity of cropping systems. On top of that, the deep roots of grain legumes contribute to carbon sequestration and the slow release of nitrogen for the following crop (Lindberg et al., 2016).

Despite these positive effects, the cultivation area of grain legumes has declined over the last decades, from 5.8 M ha in 1961 to 1.8 M ha in 2013 (Zander et al., 2016). This decline could be explained by several political and economic factors. A political factor is the 1992 Blair House agreement, which allowed duty-free imports of oilseed and protein crops. Another explanation is the economic and yield advantage of cultivation of cereals, such as maize and wheat, over legume crops, making it more attractive to step over to cereal crops (Zander et al., 2016; Lindberg et al., 2016).

2.1.2 Candidate Crops

Nitrogen-fixating crops that can be cultivated in moderate climates are the common bean (*Phaseolus vulgaris*), faba bean (*Vicia faba*), pea (*Pisum sativum*), lupin (*Lupinus spp.*), alfalfa (*Medico sativa*), and red and white clover (*Trifolium pratense* and *Trefolium repens* respectively). Grain legumes (also called pulses), such as faba and common beans, peas, and lupins, could entirely or partially replace traditional protein sources for monogastric animals (Jezierny et al., 2010).

Low-alkaloid varieties are most suitable for monogastric nutrition, differing from the white-type lupins grown as a fodder crop for ruminants. There has been a resurgence of interest in growing lupins in Europe, as their nutritional profile is comparable with soybean. Australia, being the largest producer of lupins, has considerable experience feeding this crop to livestock. The dried seeds of the grain legumes, rich in protein, are used in compound feed mixtures to obtain the optimal nutritional composition for meat production (Lindberg et al., 2016).

Alfalfa and clover species are legumes, often found in grazing pastures for cattle, and their vegetative parts would also offer an alternative protein source for ruminants. Red clover seems to be the most abundant species when it comes to yield per ha. Furthermore, it is rich in lysine, methionine, and vitamin E. Ryegrass and other grass species can be used in animal feed despite their lower crude protein (C.P.) content. A significant hindrance to exploit these forage crops for cattle is the ability to cause bloat due to an accumulation of gasses in the rumen and reticulum.

The escape of fermentation gasses is prevented, which could potentially be lethal. To utilize these crops in feedstuffs, the protein needs to be extracted. One way to increase the C.P. content of the forages is via fractionation of the vegetative part of the plant into pulp, juice, and green protein (which contains the highest C.P. content). The available digestible juice fraction can be fed to monogastric animals as high-value protein feed. The remaining fiber fraction can be fed to ruminants or used for biogas as it contains more muscular bound proteins (Lindberg et al., 2016). Innovative biorefinery processes are currently used to extract proteins out of forage crops optimally.

2.1.3 Nutritive Value

Table 10.1 gives an overview of the chemical composition of grain legumes as compared to SBM. As can be seen in the table, the C.P. content of grain legumes is varying and generally lower than SBM. Among the grain legume alternatives, lupins have the highest C.P. content. However, not only the C.P. content but also the amino acid profile is vital as they build the proteins in the body.

Monogastric animals have a different digestive system than ruminants and require a balanced feed with an optimal amino acid profile. Ruminants, on the other hand, are provided with microbial protein and amino acids through the symbiosis with rumen microbiota. This microbiota can convert less digestible components

Table 10.1 Chemical composition (% in dry matter (D.M.)) of faba bean, common bean, lupine, and peas compared to SBM

Constituent	Faba bean	Common bean	Lupine	Pea	Soybean meal
Crude protein	30.1	24.8	35.2	24.6	51.6
Crude fiber	8.7	5.2	14.7	6.0	6.8
NDF[a]	15.9	20.0	23.9	12.9	10.5
ADF[a]	10.7	7.6	19.1	7.2	5.7
Ether extract	1.3	1.7	7.7	1.2	2.2
Ash	4.2	4.7	3.8	3.5	7.3
Calcium g/kg DM[a]	1.5	2.5	2.8		3.9
Phosphorus g/kg DM[a]	5.5	4.9	3.8		7.1
Lysine g/16 g N	6.1	6.5	4.6	7.0	6.1
Cysteine g/16 g N	1.2	1.1	1.5	1.4	1.5
Methionine g/16 g N	0.7	1.1	0.7	0.9	1.4

Values compiled by Sauvant et al. (2004), Jezierny et al. (2010), and Kragbaek (2014)
[a]Values obtained from feedipedia.com

such as cellulose and hemicellulose present in the grass, clover, and hay (Lindberg et al., 2016).

The proportion of lysine, an essential amino acid, is higher in peas, faba beans, and common beans than the lysine proportion in lupines. Therefore, supplementation of lysine in a lupine based diet is recommended. The sulfur-containing amino acids, cysteine, and methionine are underrepresented, and supplementation is strongly advised to prevent growth problems in pigs (Jezierny et al., 2010).

Another limiting factor is the digestibility of the pulses. Neutral detergent fiber (NDF) and acid detergent fiber (ADF) are parameters that represent the structural components of the plants and are generally related to indigestibility. These parameters are preferred to be as low as possible as they provide bulk and low energy.

Although having antimicrobial effects, the presence of secondary plant metabolites or antinutritional factors (ANF's) can hurt the performance and health of the animal and is the reason for the restricted use of legumes in the feed. They can be divided into two categories: a heat-labile and a heat-stable group. The first group, which includes protease inhibitors and lectins, is sensitive to high temperatures and can be deactivated via heat treatment. The heat-stable category includes tannins, alkaloids, saponins, and pyrimidine glycosides. But these compounds are stable under high temperatures. These ANF's need to be removed by, e.g., plant breeding or physical treatments such as dehulling and soaking since many of the ANF's are located in the hulls of the pulses. Germination is a biological treatment and has been proved to be adequate to remove secondary plant metabolites (Jezierny et al., 2010).

Besides the antinutritional factors, enzymes such as α-galactosidase, present in these plants, may lead to excessive fermentation and diarrhea in ruminants. Together with high levels of non-starch-polysaccharides, they can have a negative mark on the digestibility or metabolism. This and the fact that the plant biomass contains high levels of fiber, with lower digestibility, has restricted the use of these legumes in the feed (Jezierny et al., 2010; Osmane et al., 2017). Mechanical removal of the hulls, which contain all tannins, can increase the protein content avoiding the familiar tannin-protein interactions (Rusu, Alvarez Penedo, Bethke, et al., 2020e; Rusu, Schwarze, Bethke, et al., 2020f).

Trials suggest lupins can be used in pig diets from between 10 and 30 percent depending on the age and stage of the animals (Jezierny et al., 2010).

2.2 *By-Products of Crops as New Protein Sources*

2.2.1 Oilseed Meals and Dried Distiller Grains with Solubles

In recent years, many countries in Europe are utilizing oilseed meals to reduce the reliance on imported soybean meal. Sunflower (*Helianthus annuus L.*) and rapeseed (*Brassica napus L.*) are the two major oilseed crops produced in Europe. Oilseeds contain high concentrations of energy and good content of protein and fiber.

Oil is extracted from the oilseeds and used to produce biofuel or ethanol, and after pressing, the remaining cake is used in compound feed. This cake or meal is a by-product of oil extraction and is often used as an energy and protein source, especially in the diets of lactating cows. The chemical composition of the cake depends on the extraction method used. Solvent extraction and mechanical extraction are two main processing methods used to remove the oil. Solvent extraction results in a lower fat content in the meals than mechanical extraction (Rusu, Alvarez Penedo, Engelhardt, & Schwarze, 2020a). Ruminal bacteria are high degraders of the oilseed protein, and therefore it is desirable to treat the meals to reduce their degradability (Bernard, 2011). Dehulling of the seed can reduce the fiber content in the meals, yielding a higher protein content and a heat treatment to inactivate many ANF's such as protease inhibitors and lectins (Florou-Paneri et al., 2014).

Sunflowers are primarily grown because of the value of the oil for human consumption as it contains a lot of polyunsaturated fatty acids (PUFA's), which have potential beneficial health effects. After processing and dehulling, the protein content of over 40% can be obtained in meals. Sunflower is included in animal diets up to 20% when prices are favorable. As it is grown in the summer, the prices usually increase when out of season. Its use is restricted in diets for young pigs due to the low digestibility of the protein and the presence of phenolic compounds (Bernard, 2011; Kay, 2014).

Rapeseed is an oilseed crop that can grow in colder climates and is extensively produced in Europe and North-America. In North-America, it is called canola, and its meal has a slightly different chemical composition than the rapeseed meal. Rapeseed meal or canola meal is mainly grown to produce animal feed, vegetable oil, or biodiesel. In contrast to sunflower meal, rapeseed meal is valuable due to its ability to be grown as a winter crop. This is beneficial to the soil as it limits nitrogen run-off and provides good coverage during the winter (Florou-Paneri et al., 2014). Rapeseed and canola proteins have been used as a feed ingredient for a broad range of animal species, including poultry, pig, cattle, and fish (including salmon, trout, tilapia, and prawns) (van der Spiegel et al., 2013).

Sunflower meal and rapeseed meal are attractive alternatives since they have high contents of high-value protein, only slightly lower than soybean (Florou-Paneri et al., 2014). Their production was growing and was estimated to be 58.4 million tons in 2010–2011. They are often cheaper than SBM, with the cost price of rapeseed meal and sunflower meal being about half of the cost price of SBM, according to de Boer et al. (2014). Next to sunflower meal and rapeseed meal, also crops like flax, copra, cotton, wheat, and maize have interesting features as a protein source in animal feed.

Cottonseed meal could be used as an alternative protein source in diets for adult ruminants. Its use is limited in monogastric animals since it has high fiber content and variable protein content. The whole cottonseed with the hull contains high energy and high fiber content, which is shown to be effective in maintaining rumen function for dairy cows. To augment the digestibility of nutrients, different processing methods can be applied, such as mechanical delinting, pelleting, and extrusion (Bernard, 2011). Cottonseed meal has high levels of gossypol, which is produced by the plant as a defense mechanism against insect pests. Gossypol can be toxic for animals, especially pigs, and according to Florou-Paneri et al. (2014), even lethal in some cases. However, low-level gossypol varieties are available thanks to plant breeding and look promising for the future.

Copra meal is primarily produced in Asia, where it is a low-cost ingredient. It is obtained as a by-product after oil extraction. Protein levels of 20–25% can be obtained, and it has a high oil content. However, a low essential amino acid composition and high fiber content limit its inclusion in the feed (Kay, 2014; Mutsvangwa et al., 2016).

Flaxseed meal, a by-product of linseed oil, contains next to good protein content, essential ω-3 fatty acids such as α-linolenic acid. This can be associated with several health benefits; however, mixed results in trials with pigs have limited its use in feed for pigs (Florou-Paneri et al., 2014).

Dried distillers grains with solubles (DDGS) can reach protein levels up to 40% and are commonly used as a protein source in animal feed. Its quality varies depending on the type of feedstock and the processing steps being used. It is typically derived from corn or other grain crops such as wheat, triticale, or barley as a co-product of ethanol and biodiesel production. With the growth of the bioethanol industry, resulting in increased production of co-products, DDGS has excellent potential to be implemented in feed for dairy cows, pigs, and even ewes. Nowadays, it is increasingly used in feed for dairy cows in replacement of SBM (Hastad et al., 2005; Kleinschmit et al., 2007; Gaillard et al., 2017; Alshdaifat & Obeidat, 2019).

2.2.2 Nutritive Value

Table 10.2 shows the chemical composition of sunflower meal, rapeseed meal, DDGS, and soybean meal. From the table, it is clear that the fiber content in sunflower, rapeseed, and cottonseed meal is higher than the fiber content of SBM. The content of C.P. and crude fiber (C.F.) varies with the processing method; generally, solvent-extracted and dehulled meals contain higher protein content than meals obtained by other extraction methods. The complexity of the extraction processes is, therefore, linked with the market price.

Rapeseed meal is a good source of high-value protein and S-amino acids, with cysteine levels higher than SBM. Its application, however, is limited due to the presence of ANF's and relatively low protein and energy content (Lindberg et al., 2016). Glucosinolates, erucic acids, phytic acids, and phenolic compounds are known as tannins, and sinapines are considered antinutritional ingredients and are known to reduce animal performance. Glucosinolates in high amounts can inhibit liver and thyroid gland metabolism, which can dramatically depress the performance of the pig. Phenolic compounds such as tannins can bind proteins or other nutrients present in the diet, making them less available to the digestive system of monogastric animals (Florou-Paneri et al., 2014). High levels of phytate phosphorus tend to decrease the digestibility of protein and essential mineral elements, especially for monogastric animals as phytic acid cannot be digested in the gut. This causes plant

Table 10.2 Chemical composition (% in D.M.) of sunflower meal, rapeseed meal, DDGS, and SBM

CONSTITUENT	Sunflower meal	Rapeseed meal	DDGS[b]	Soybean meal
Crude protein	24–44	35	33.9	44–48
Crude fiber	12–32	10–15		3.0–7.0
NDF	38.7[a]	31.6[a]	23.9	10.5[a]
ADF	26.6[a]	20.7[a]	11.2	5.7[a]
Ether extract	1.8[a]	2.4[a]	6.7	2.2[a]
Calcium g/kg DM	0.46[a]	0.82[a]		0.39[a]
Phosphorus g/kg DM	1.3[a]	1.2[a]		0.71[a]
Lysine	1.34	2.06	0.79	3.0
Cysteine	0.57	0.72	0.66	0.72
Methionine	0.76	0.71	0.49	0.67
Threonine	1.34	1.87	1.06	1.36

Average values compiled from de Boer et al. (2014)
[a]Values obtained from feedipedia.com
[b]Values obtained from Gaillard et al. (2017); chemical composition of DDGS based on triticale (25% D.M.), wheat (55% D.M.), and barley (20% D.M.)

feed ingredients containing phytate-bound phosphorus to be ignored in feed formulation (She et al., 2015). In some DDGS derived from wheat, however, natural phytase is released and may hydrolyze some of the phytates. The addition of phytase, steeping, or fermentation of feed ingredients is other ways to get rid of phytate and can improve the digestibility of phosphorus. Some ANF's, such as lectins and protease inhibitors, are heat-labile and can be deactivated by feed processing. On the other hand, heat-stable ANF's such as alkaloids, condensed tannins, and pyrimidine glycosides will be insensitive to feed processing (Lindberg et al., 2016).

Compared to rapeseed meal, canola meal has lower glucosinolates and a much higher feeding value than rapeseed meal. For this reason, canola meal is preferred over rapeseed meal and sunflower meal to replace SBM in pig nutrition partially, and many studies have shown that the intake of canola meal has no detrimental effects on pig performance or carcass characteristics (Florou-Paneri et al., 2014). Rapeseed meal is commonly used in aquaculture as part of fish or shrimp compound feed since it has been shown that the meal can improve the performance of the aquatic animals. Also, the use of canola/rapeseed meal has no significant adverse or negative effect on the weight, size, and carcass quality of fish and shrimp (Enami, 2011).

A significant benefit of sunflower meal is the low content of antinutritional factors in comparison with rapeseed meal. For this reason, sunflower meal does not require heating or other treatment before being fed to pigs. Sunflower meal contains low quantities of some phenolic compounds such as chlorogenic acid and tannins and generally has no detrimental effects on animals. This also explains why sunflower meal has the highest values for digestibility of crude protein. Florou-Paneri et al. (2014) states that the average digestibility of the amino acids is high (89%) and comparable to that of SBM (90.6%).

Furthermore, sunflower meal is a valuable source of calcium, phosphorus, and B vitamins. This, in contrast with SBM, which is a common source of B vitamins. Additionally, the presence of polyunsaturated fatty acids (PUFA's) is another benefit since this can result in a higher content of PUFA's in muscle and fat of animals. PUFA's can contribute to the prevention of several diseases (cardiovascular disease, arthritis, etc.). Sunflower meal has a higher content in sulfur amino acids than SBM, especially methionine. It is considered to be a safe feed for pigs and other animals. However, its high fiber content, relatively low amounts of lysine, and deficiencies in other amino acids are limiting its application in the feed.

Other oilseed meals such as cottonseed meal, peanut meal as well as DDGS can be used in feed. The chemical composition of DDGS depends on the type of feedstock being used as well as on the processing method. In Table 10.2, the chemical composition of a mix of DDGS is given. DDGS holds a fair amount of C.P. and has a lower C.F. content than other meals, which makes it promising to substitute SBM in feed partially. The proteins in DDGS are highly digestible and moderately resistant to ruminal degradation, thus ideal to replace SBM in dairy cow diets partially. Gaillard et al. (2017) found that substituting SBM and canola cake with grain-based DDGS up to 13.5% of dietary D.M. did not affect the feed intake nor milk production. An unfavorable amino acid profile prevents higher inclusion levels in feed.

3 Insects as a Novel Alternative Protein Source

3.1 Introduction

Insects are invertebrates with an exoskeleton. It is estimated to be around 2000 edible species of insects in the world (Stiles, 2017; Jongema, 2017), and being consumed in more than 100 countries around the world.

Compared to their chordate counterparts, they are incredibly efficient in converting organic matter to proteins (Halloran et al., 2016). Insertion of insects into future food and feed systems shows excellent potential and could enhance food and feed security on a larger scale. Rearing insects could be an environmentally friendly and sustainable solution as insects grow and reproduce quickly. Insects can be fed waste biomass as a substrate, offering a solution to the growing waste problem nowadays (Makkar et al., 2014). Value from materials such as manure, which would otherwise be disposed of, can be recovered through biological reprocessing.

A significant benefit of rearing insects is that it does not require much land as mass farming of insects requires between 50–90% less land than conventional agriculture per kilogram of protein. More arable land is thus available for the cultivation of conventional crops. Insect utilization in feed and food could reduce greenhouse gas emissions from the livestock industry by 50% by 2050. Another advantage is that insect rearing can be done no matter which climate. With access to natural light, a suitable substrate, and oxygen, year-round breeding can be maintained in a greenhouse. This allows facilitation to a more controllable and predictable manufacturing system (Makkar et al., 2014; Stiles, 2017).

Because of the high lipid levels in insect and insect meal, isolation of oil and preparation of biodiesel would be impressive for the biofuel industry. Oil from the black soldier fly larvae, housefly maggot meal, mealworm, and silkworm could be isolated and used to prepare biodiesel. The rest of the defatted meal, being rich in C.P., could find a place as an invaluable protein-rich resource in the feed industry (Makkar et al., 2014).

The human body can, on average, produce more body protein from 100 grams of animal protein than from 100 grams of vegetable protein. The clear advantage of insects is that the proteins in the dried beetles are highly concentrated. For consumers' orientations without meat or fish as much as possible, insects are a valuable alternative source of protein. The high protein content of insects makes you so full that you can hardly eat more than 20 percent of the insect in a dish. Conversely, this means that around 50 grams of mealworms, for example, provide a full serving of protein, and in comparison, it doesn't cost that much more than good meat.

In a relatively new field of research, few studies have been carried out on the connection between the consumption of insects and individual diseases, such as colon cancer, obesity, or cardiovascular disease. However, the reasons are often animal fats, i.e., saturated fatty acids, as well as intensive frying and salting (WCRF & American Institute for Cancer Research, 2018). Insects have a high proportion of monounsaturated and polyunsaturated fatty acids, making them the right candidate for reducing the risk of disease (Fiebelkorn, 2017).

Further aspects are relevant to health effects. The potential allergic reactions that the consumption of insects, especially mealworms, grasshoppers, and crickets, can trigger in people with crustacean and house dust mite allergies must be taken into account (Ribeiro et al., 2018). This effect could also occur in people who work on insect farms; therefore, adequate occupational safety must be guaranteed (Dobermann et al., 2017).

Potentially pathogenic microorganisms from the intestinal flora of insects can usually be reduced by simple processing steps, such as thorough washing and heating, so that the microbial risk of insects, depending on the preparation, is comparable to that of other animal proteins. Contaminants such as heavy metals, dioxins, and polychlorinated biphenyls, which can arise through rearing, insect food, and packaging, do not pose a higher risk than with other animal products if the insects are rearing and processed correctly (Dobermann et al., 2017).

3.2 Candidate Insects

There are many species of insects, and several have the potential for application in feed products. Larvae from the black soldier fly (BSF) (*Hermetia illucens*, H.I.), mealworms (common: *Tenebrio molitor*, T.M.), and the common housefly (*Musca domestica*, M.D.) are the most suitable as an alternative animal feed source, especially in fish, poultry and pig (Sogari et al., 2019).

The black soldier fly is an insect from the *Stratiomyidae* family. The larvae of the black soldier fly are particularly interesting in the feed industry as the adults do not rely on the reserves stored from the larval stages (Makkar et al., 2014). The larvae can feed quickly and on a wide range of decaying organic waste material, such as rotting fruits and vegetables, coffee bean pulp, distiller grains, fish rests, and animal manure. Rearing insects to a dense population could be an efficient way to get rid of organic wastes by converting them into biomass, rich in protein and lipids. This biomass could have numerous industrial applications, e.g., alternative animal feedstuff, biodiesel, and chitin production (van Huis et al., 2013). Black soldier fly larvae are used live, chopped, or dried and ground. The potential production of BSF larvae has a significant advantage over other fly species since they are not a vector of diseases (van Huis et al., 2013). The capability of surviving in extreme conditions of drought, food shortage, and oxygen deficiencies is another benefit. However, like most other insects, the black soldier fly requires somewhat of a warm environment in order to biodegrade its substrate. In temperate regions, this can be an energy-consuming process (Makkar et al., 2014).

Mealworms are the larvae of the two species *Alphitobius diaperinus* (lesser mealworm) and *Tenebrio molitor* (yellow mealworm) from the *Tenebrionidae* family. They are already produced industrially as feed for pets and zoo animals since they are easy to breed and feed. They are mostly sold alive but are also fed dried, in powder form, or canned (Makkar et al., 2014). Mealworms are omnivorous and can consume various kinds of the plant as well as animal materials such as meat and feathers. They are mostly fed on cereal bran or flour, and supplements rich in proteins also form a part of their diet. To obtain a good yield, the substrate on which mealworms feed needs to contain moisture, which fresh fruits and vegetables can provide (Makkar et al., 2014).

The housefly is the most common fly species and is worldwide seen as a pest and a carrier of diseases. Therefore, the inclusion of maggot meal in feed raises concerns about potential contaminations as the maggot can carry over bacteria or fungi present in the substrate. Housefly maggots can feed on a large variety of substrates and are therefore useful to turn wastes into valuable biomass rich in protein and oil. Warm temperatures and moisture are ideal conditions, and to obtain a large population, only a small amount of substate is required; e.g., 1500 maggots can feed on 450 g of fresh manure. The most common type of substrate to feed on is poultry manure. Rotten fruit and animal offal, such as cattle blood and cattle and rumen content, could also be used. Several harvesting methods allow the larvae to be collected, after which they are washed, killed, and then dried and milled (Makkar et al., 2014).

Silkworms (*Bombyx mori*) are other potential candidate insects to be reared as an alternative resource of protein in feed. Currently, they are raised for their production of silk, which is derived from the cocoons when the silkworm enters the pupae phase. The spent silkworm is often discarded as waste material. However, it may be used in the oil industry and can yield various products such as paints, soaps, pharmaceuticals, plastics, and oils. Chitin, a long-chain polymer of N-acetylglucosamine, is another product that could be extracted. Its high protein content allows silkworm pupae meal to be used as livestock feed. Due to its high water content, spent pupae spoil rapidly. Drying and grounding could solve this issue, and defatted silkworm pupae meal could have an increased amount of protein (Makkar et al., 2014).

Locusts, grasshoppers, and crickets are insects that are also edible and contain a higher content of C.P. than previously mentioned insects. There are about 80 different species, and humans consume them in Asia, South-America, and Africa. Locusts are known as pests and cause massive damage to crops as they swarm, mostly in Africa. A locust swarm can consist out of 10 billion insects and has a considerable amount of biomass. Harvesting them is a means to control them and may help to reduce the application of chemical pesticides. In addition, they provide an extra source of nutrition (Makkar et al., 2014). In Thailand, the outbreak of a locust species has led to a campaign to promote the consumption of this locust, which is now being farmed (van Huis et al., 2013).

3.3 Nutritive Value

Like other animal foods, insects have besides the high protein content with a high biological value, a high content of unsaturated fatty acids, as is that of fiber in the form of chitin and essential minerals such as iron, calcium, and zinc, the vitamins riboflavin, pantothenic acid and biotin (Payne et al., 2016). Calcium is found in higher concentrations in vertebrates (Finke & Oonincx, 2017).

Table 10.3 compares the average values of chemical constituents of the insect mentioned above meals with those of fish meal and soymeal. The CP contents of insects are of the same order as in SBM; some are slightly lower, varying in range from 42% to 63% (Makkar et al., 2014). In comparison with SBM and fish meal, the lipid levels in insects are relatively high. Insect meals from the BSF, housefly, mealworm, and silkworm are also defatted, as they contain oil contents as high as 36%, and this increases their C.P. content further (C.P. contents of defatted meals are

Table 10.3 Average chemical composition (% in D.M.) of insect meals relative to those of fish meal and soybean meal

Constituent	Black soldier fly larvae	Housefly maggot meal	Meal-worm	Silkworm pupae meal	Fish meal	Soybean meal
Crude protein	42.1 (56.9)	50.4 (62.1)	52.8 (82.6)	60.7 (81.7)	70.6	51.8
Crude fibre	7.0[a]	5.7[a]	–	3.9[a]	–	6.8[b]
Ash	20.6[a]	10.1[a]	3.1[a]	5.8[a]	–	7.3[b]
Lipid	26.0	18.9	36.1	25.7	9.9	2.0
Calcium	7.56	0.47	0.27	0.38	4.34	0.39
Phosphorus	0.90	1.60	0.78	0.60	2.79	0.69

Values in parentheses are calculated values of the defatted meals (Makkar et al., 2014)
[a]Arango Gutierrez et al. (2004), Newton et al. (1977) and St-Hilaire et al. (2007a, b)
[b]Lindberg et al. (2016)

given in parentheses in Table 10.3). The defatted C.P. contents, especially from mealworm and silkworm, are higher than those of the conventional protein sources (Makkar et al., 2014).

Black soldier fly larvae are rich in both protein and fat. Cattle manure, fish offal, and oil-rich food waste give the highest percentages. The composition of fatty acids varies with the diet as well. The larvae are rich in calcium and phosphorus and incredibly rich in lysine (6–8% of the C.P.). The D.M. content of fresh larvae is relatively high, making them less costly to dehydrate (Makkar et al., 2014). Stiles (2017) reported that *H. illucens* larvae contain the right amount of lauric acid, a medium-chain fatty acid that has recognized antibacterial properties, and oils derived from the larvae could therefore be used in several applications (Stiles, 2017). Several species of insects have antibiotic properties as they produce bioactive antimicrobial or antifungal peptides or polypeptides. This is probably indispensable for insects to deal with food materials that are decaying and potentially harmful. This could potentially pave the way for the extraction of substances that could overcome antibiotic-resistant infection (Stiles, 2017).

Next to the high amount of CP (47–60%), essential amino acids, and fat (31–43%), mealworms contain about 60% water. This causes them to be relatively low in ash. The composition of the mealworm is highly variable since it depends on the substrate being used to feed the mealworm (Makkar et al., 2014; Stiles, 2017).

Housefly maggots are rich in digestible proteins, essential amino acids, fats, and micronutrients, making them ideal as a component of animal feed. Housefly larvae meal can be obtained after processing and has a protein content in the region of 37.5–63.1%. Therefore, it has nutritional content comparable to high-quality fish meal. Housefly larvae also contain high levels of essential amino acids, such as methionine and lysine, whereas plant-based sources of protein are often low in these critical compounds (Stiles, 2017).

Silkworm pupae meal is protein-rich with a high nutritional value. The CP content of a defatted meal can reach values higher than 82% and is found to be rich in cysteine, methionine, and lysine. However, only 73% of the C.P. content was 'true' protein. The presence of chitin, a component containing nitrogen, could explain this notability. The relatively undigestible proteins can be degraded during rumen fermentation, thus, would be most suitable to be fed to ruminants. Fat extraction of silkworm meal has been appointed since the high oil content could restrict its use (Makkar et al., 2014).

Lysine is the most important essential amino acid in food and feed, and other amino acid requirements are often expressed relative to the requirement of lysine. The lysine content in most insects seems to be sufficient for fish, poultry, and shrimp, but maybe limiting for pigs, while the content of arginine and sulfur-containing amino acids (cysteine and methionine) may be limiting for poultry. Supplementation is recommended in this case (Makkar et al., 2014).

Palatability is an essential factor for the inclusion of ingredients in food and feed products. Feeding studies conducted so far have confirmed the excellent palatability of diets containing insect meals. Most of these studies were first conducted on fish and poultry. For ruminants, the presence of high lipid levels could intervene with fiber digestion and rumen fermentation. For this reason, defatted insect meals would be more suitable for ruminants. However, the remaining meal of insects such as black soldier fly larvae contains a relatively high ash content and can decrease the intake of insect meal. High inclusion levels should be avoided as they could possibly cause other adverse effects in monogastric animals (Makkar et al., 2014).

In poultry, the inclusion of BSF larvae meal has been found to support growth. Higher feed conversion efficiency was observed in chicks fed a diet containing dried BSF larvae. Chicks fed with the BSF diet consumed only 93% as much feed, gaining weight at a rate of 96% instead of chicks fed with a control diet containing SBM (Newton et al., 2005).

Insects could also be used in aquaculture as studies on feeding black soldier fly to fish showed that the aroma and texture of the fish did not change. The processing of these larvae could increase the nutrient availability for fish (Makkar et al., 2014). Salmon fed on a diet of up to 50% insect meal showed no adverse effect on fish performance, indicating that about half of fish meal currently used in salmon aquaculture could be replaced (Stiles, 2017). However, in some cases, reduced fish performance has been observed. Further trials and economic analyses are needed to provide more information on, e.g., substrates types and processing methods affecting the utilization of the larvae by fish (Makkar et al., 2014).

The meal beetle (*Tenebrio molitor*) – better known as mealworm due to its appearance in the larval stage – would therefore be particularly suitable as food. But the industrial production of edible crickets is also very promising because it is possible to build on a wealth of experience from "smaller breeding farms."

3.4 Legislation

The existing laws in Europe are unclear on several issues regarding the use of novel protein sources in food and feed products, especially the case for insects. Despite their great potential in animal diets as a (partial) substitute for SBM and fish meal, the use of insects in feed for pigs and poultry is prohibited under regulation EC999/2001 (Stiles, 2017). This regulation poses that it is not possible to produce animal feed with animal products such as insects. As a consequence, proteins derived from insects are today not allowed for use in pig or poultry feed (https://ipiff.org/insects-eu-legislation/). This is in line with the report of the Codex Alimentarius Commission (CAC). They reported in 2010 that the food safety of edible insects had not been studied sufficiently (van der Spiegel et al., 2013). However, the recent regulation No 2017/893 authorized the use of insect proteins originating from certain insect species in feed for aquaculture animals. It could pave the way for new feed markets for insect producers. The European Commission services are currently exploring possibilities for authorizing insect proteins in poultry feed.

The withhold of insect feed for pigs and poultry is caused by concerns with regards to the safety of the implementation in feed products. van der Spiegel et al. (2013) stated that insects have to undergo a series of tests to be evaluated as safe.

Therefore, studies on insects need to identify safety issues before insects can enter the European market. Hazards of both chemical or microbiological origin can be introduced and can be harmful to animals and humans. The substrate used for the rearing of insects mostly consists out of waste materials and can be contaminated with antibiotics, pesticides, pathogens, heavy metals, and toxins, and mycotoxins. The insects themselves can contain allergens and other contaminants. These substances can accumulate quickly during rearing and can increase their concentrations. These safety hazards depend on several conditions, such as the insect species, the substrate used, the environment, and the production methods. However, the European Food Standards Agency (EFSA) has recently published a report, indicating limited risk from the introduction of insects into the food chain that was fed on food or feed grade materials (Stiles, 2017).

Furthermore, legislative texts define general principles and standards regarding food and feed safety. Consequently, insects may only be fed with dairy, eggs, or materials of vegetal origin, and the feeding of animals with slaughterhouse or rendering derived products such as manure and catering waste is prohibited. Under the so-called 'E.U. Animal health law,' insect producers are obliged to keep their animals in good health in order to prevent the spreading of diseases.

Nowadays, some laws in the E.U. legislation acknowledge the specificities of insect production activities. First of all, European insect producers must conform to E.U. environmental legislation. In this legislation, regulation EU1143/2014 restricts the farming of some insects'species in order to protect the environments or ecosystems in case of an accidental release of farmed insects. This is done in the form of a list establishing the 'invasive alien species.' For the production of insect meal, approval is conditional on fulfilling a specific processing method, as described in the E.U.' animal by-products legislation'. Regulation No 142/2011 allows production if one of the five described methods is used or if it can be demonstrated that the final product does not exceed certain levels of some well-defined bacterial agents.

Until the end of 2017, the sale of whole insects in Germany was still in a gray area. Selling ground insects or insect parts was prohibited. Within the E.U., only officially to enjoy insects in Belgium, the Netherlands, and, since 2017, also in Switzerland.

On 01.01.2018, a new E.U. regulation came into force that officially allows insects as food. An official application must now be submitted to the E.U. for each insect species. This usually decides on the application within a few months. If an insect species has been approved, both whole animals and parts of them can be sold. The fact that parts of insects can also be sold is interesting as consumers can get used to products with insect meals much more quickly than to whole insects.

In spring 2019, applications for various crickets, locusts, mealworms, and so-called buffalo worms had already been submitted. For products containing insects that were brought onto the market in Germany before the Novel Food Regulation (E.U.) 2015/2283 came into force, a transitional regulation applies. These may continue to be marketed if the required application for approval has been made by the beginning of 2019. The Novel Food Regulation not only regulates the approval of edible insects but also applies to other "new" foods.

4 Future Perspectives

4.1 *Candidate Crops*

The analysis of the environmental impact of plant-based substitute products is intended to help identify differences to conventional products and between the plant-based substitute products themselves. Where appropriate, results on tofu are also used concerning soybean cultivation.

More studies testing the nutritional value and digestibility of leguminous crops are needed in order to provide give better insights into the implementation of legumes in food and feed. In order to grow suitable grain legumes, more cultivars need to be developed and selected under a range of different cultivation conditions. Plant breeding is specialized in doing that (Lindberg et al., 2016). Considerable progress in plant breeding offers perspectives and has contributed to the commercial release of cultivars with nutritive value.

Lower contents of secondary plant metabolites have been obtained, resulting, e.g., in the development of zero-tannin faba bean cultivars (Duc et al., 1999). Next to eliminating ANF's, plant breeding also offers opportunities for growing legumes with a better amino acid composition and digestibility and higher-high-quality protein levels. However, because of the complex relationships of plant characteristics, a new cultivar is seen as a long term process. The rise of new gene-editing techniques (e.g., CRISPR Cas) is a fast and effective manner and could help develop new cultivars (Jezierny et al., 2010).

Next to improving legume crops, an increase in the production of legumes is needed to provide more stable market prices. This has shown to be a bottleneck as monocropping trends (in cereals) in response to market signals have weakened the position of legumes, especially in Europe. To become more appealing to farmers, several states of the E.U. have launched a couple of policies supporting the production of legumes. This can be done in the form of subsidies for the cultivation of legumes or import restrictions of soybean. Restricting policies that affect the markets of competing crops, such as the biofuel industries, can influence the profitability of these crops. However, non-market outputs such as the diversification of cropping systems, pest control, emission reduction, and biodiversity enhancement do momentarily not outweigh the economic advantage of other crops with a higher monetary return (Zander et al., 2016).

In the meantime, the E.U. has acknowledged the importation problem, and the European Innovation Partnership (EPI) Agri has launched a process to reduce the E.U.'s dependency on soy import for the feed industry. Experts from 20 E.U. countries are currently looking for alternatives to increase protein crop production in the E.U. (de Visser et al., 2014). According to E.U. regulation 1829/2003/E.C., genetically modified plants must be labeled, which means that the seitan and tofu or soy-based products sold in Germany, for example, are made from non-genetically modified wheat or soy and mainly from Europe or Canada (Stiftung Waren-test, 2016). In the production of conventional meat, on the other hand, genetically modified soy from South America is mainly used for animal feed (WWF International, 2014).

4.2 By-Products of Crops

Contrary to leguminous crops, the cultivation of oilseed crops has expanded over the last decades. Lindberg et al. (2016) reported that the rapeseed meal production in the E.U. has risen from 6 million tons up to 14.2 million tons over 10 years. This doubling is caused by the growing demand for renewable feedstocks as a substitute for petroleum-derived products. In combination with the conventional sources becoming more expensive, this offers exciting opportunities for the inclusion of by-products in the animal compound feed (Zanetti et al., 2013; Alshdaifat & Obeidat, 2019).

Rapeseed meal and sunflower meal are extensively produced in Europe, and Lindberg et al. (2016) stated that they compose about 30% of the vegetable protein-rich feed used in the European Union. However, some (nutritional) characteristics of oilseed crops have limited their use in feed compounds, and therefore they are only partly substituted for soymeal. The presence of ANF's, and the low digestibility of oilseed meals, are currently limiting the use of by-products in animal feed, particularly for monogastric animals such as pigs. High amounts of crude fiber and phytate are causing the indigestibility. Processing and extraction methods are used to solve this problem and to increase the protein content. Breeding cultivars with lower levels of fiber and AFN's has already led to crops with a more favorable composition about the amino acid profile and digestibility. However, other difficulties, such as the variable quality, mycotoxin presence, and protein, and amino acid digestibility still need to be overcome (Florou-Paneri et al., 2014).

Many oilseed crops could be cultivated in Europe, and new species with different characteristics and chemical compositions can be grown in Europe and have the potential to be used in diets (Zanetti et al., 2013). The most significant incentive to switch over to the use of DDGS or oilseed meals is the cost-effectiveness since the price of DDGS and oilseed meals are generally lower than the conventional protein sources. For this reason, oilseed meals and DDGS offer excellent alternatives to replace SBM in animal feed. Another essential factor for the implementation of diets is the reduction of the footprint. De Boer et al. (2014) performed a Life Cycle Assessment and estimated that the carbon footprint slightly increases when sunflower meal, as well as DDGS, are used to replace SBM partially. On the part of sunflower meal, fiber removal was taken into account to obtain a product with a high C.P. content. This caused the increase as fiber removal requires additional energy and equipment. The increase of carbon footprint on the part of DDGS is due to the drying of the product. However, a lot of assumptions were made in this study by de Boer et al., and more in-depth research is needed to prove if the use of oilseed meals and DDGS can be sustainable.

Nowadays, the most significant limitation is finding a stable protein source derived from oilseed crops. The crops producing the meals in Europe are often season-bound, causing availability problems and fluctuating market prices (Kay, 2014; Mutsvangwa et al., 2016). Higher amounts of sunflower and canola could be used in pig diets, provided that price and availability were favorable. As with DDGS, further investigation of their inclusion in pig diets combined with other technologies could allow for increased inclusion rates (Kay, 2014).

4.3 Insects

Insects could be a solution to the protein problem due to their nutritional value, low level of greenhouse gas emissions, and the little amount of water required to produce insects compared to common crops (Riddick, 2014). The CP content of insects depends on the species and is varying within the species too. It is about of the same order or higher than SBM and lower than fish meal. A study from Mexico (Ramos-Elorduy et al., 1997) reports that of 78 species of edible insects, the protein digestibility varied between 76% and 98%.

Along with the content of essential amino acids above 40%, the FAO considers this food of high nutritional quality (Ramos-Elorduy et al., 1997). Thus, the exploitation of insect species for applications in the food and feed industries as well as pharmaceutical and biodiesel industries seems promising since insects contain high-fat content, and some are known to have specific antimicrobial substances (Stiles, 2017).

So far, trials, where pig and poultry were fed on insect meal resulted in no significant differences in animal performance and weight gain than groups fed on current commercial diets. Furthermore, a significant increase in favorable gastrointestinal tract microorganisms such as lactobacilli was observed for piglets fed on insect meal. These findings suggest that insects could have great potential as a protein source in feed for monogastric livestock such as pigs and poultry. In addition, insects would form part of a more natural diet for chickens than the conventional feedstuffs (Stiles, 2017). Studies where black soldier fly larvae are included in diets for poultry, pig, and fish suggest that the conventional feed resources (SBM and fish meal) could be replaced by at least 25% by insect meal in diets of livestock and fish species. However, its inclusion level in diets still needs to be optimized, especially with regard to the levels of deficient amino acid supplementation. Trials on the inclusion of insect meals in diets of pigs and especially ruminants, are scarce due to the limited availability of insect meals. So far, small production systems of insects are used in researches. Therefore, rearing of insects on a larger scale is necessary in order to conduct more in-depth studies on nutrient utilization and growth in pigs and ruminants, as these require a substantial amount of insect meal (Makkar et al., 2014).

Despite their extraordinary potential as a (partial) substitute for SBM and fish meal, insects are not generally used yet in animal feed because of a lack of clear legislation (especially in Europe), cost-effectiveness, and information about the possible production systems. The general opinion of the public about insect use in feed poses another concern as, for most people, direct consumption of insect-based food remains an issue. However, Stiles (2017) reported that most people would be willing to eat animals fed on insect-based diets.

In order to use insects in feed, the market price of insects needs to decrease. It is reported by de Boer et al. (2014) that the cost price of mealworms can be about 50 times higher than the cost price of SBM. The sustainability of its use is another concern since a massive increase in carbon footprint was estimated when dried mealworms replaced SBM. This is caused by the elevated energy use for the warming and drying of the insects. The warming is needed for optimal growth as insects are generally poikilothermic; the drying of insects allows them to obtain a higher D.M. content. Larger scale insect production systems are needed to mitigate energy use and establish cost-effective insects or insect meals of a well-defined quality (de Boer et al., 2014; Makkar et al., 2014).

Other challenges to overcome regarding the use of insects in food and feed are the safety issues, including allergies in animals and humans, chemical and microbial contaminants (Domingues et al., 2020). At present, there is a lack of information and detailed analytical data, which makes it challenging to use as animal feed sources (Lindberg et al., 2016). Further investigation is needed to attack all the challenges and will hopefully lead to more perspective towards their use in the animal feed industry.

The new E.U. regulation will gradually result in the legalization of the sale of insects for food applications. According to a study by the Federal Institute for Risk Research (BfR), the market potential is there, and it is excellent: 10% of all German citizens would be willing to eat insects on a regular basis, and 30% would try insects at least once (Epp, 2016).

Possible transmission of diseases, poisons, or intolerances is the risk from the perspective of the population through insects as a source of food.

Ingredients and nutritional values of various types must be documented, and storage, transport, and preparation options must also be formulated. While it has been shown that the transmission of zoonotic infections to humans is much lower than that of livestock, more research is needed into the harmlessness of health when consuming more massive amounts and over a more extended period, especially if the insects are to be fed with food waste, for example. There are microbiological risks here, as insects can also be carriers of diseases. Therefore, before insects can be produced on an industrial scale, suitable technologies must be developed, and not only efficient and cost-effective breeding, but also ensuring safe and hygienic conditions in all stages of breeding.

5 Conclusions

Due to an increasingly critical examination remain the questions (1) which alternatives to conventional but also ecological production are conceivable and (2) which are possible alternative sources of protein to animal proteins are available. "Proteins of the future" is the theme for change of the current food system in the direction of sustainability. This leaves the increased public attention, growing economic activity, and the more intensive socio-political discourse.FundingThis work was supported from the Bio-Based Industries Joint Undertaking under the European Union's Horizon 2020 research and innovation program, under grant agreement No. 887259 (ALEHOOP), under grant agreement No. 720720 (FUNGUSCHAIN), and from the European Union's Horizon 2020 research and innovation program under grant agreement No. 862704 (NextGenProteins).

References

Alberts, B., Johnson, A., Lewis, J., et al. (2002). *Molecular biology of the cell* (4th ed.). Garland Science. The Shape and Structure of Proteins.

Alshdaifat, S. N., & Obeidat, B. S. (2019). The impact of feeding corn dried distillers grains with solubles on milk yield and composition in lactating Awassi ewes and digestibility and N partitioning in Awassi ewe lambs. *Italian Journal of Animal Science, 18*, 522–529.

Arango Gutierrez, G. P., Vergara Ruiz, R. A., & Mejia Velez, H. (2004). Compositional, microbiological and protein digestibility analysis of larval meal of *Hermetia illucens* (*Diptera:Stratiomyidae*) at Angelopolis-Antioquia, Colombia. *Revista Facultad Nacional de Agronomia Medellin, 57*, 2491–2500.

Bernard, J. K. (2011). *Feed ingredients/feed concentrates: Oilseed and oilseed meals. Encyclopedia of Dairy Sciences* (2nd ed., pp. 349–355).

Biesalski, H.-K., Grimm, P. & Nowitzki-Grimm, S. (2011). *Tasche-natlas der Ernährung* (5, überarb. und erw. Aufl.). : Thieme.

de Boer H.C., van Krimpen M.M., Blonk H. & Tyszler M. (2014). Replacement of soybean meal in compound feed by European protein sources. Livestock research Report Wageningen.

de Camargo, A. C., Favero, B. T., Morzelle, M. C., Franchin, M., Alvarez-Parrilla, E., de la Rosa, L. A., Geraldi, M. V., Junior, M. R. M., Shahidi, F., & Schwember, A. R. (2019). Is chickpea a potential substitute for soybean? Phenolic bioactives and potential health benefits. *International Journal of Molecular Sciences, 20*(11), 2644.

de Visser, C. L. M., Schreuder, R., & Stoddard, F. (2014). The E.U.'s dependency on soya bean import for the animal feed industry and potential for E.U. produced alternatives. *Protein Sources in Animal Feed, 21*, D407.

Dietrich, T., Del Carmen Villaran Velasco, M., Echeverría, P. J., Pop, B., & Rusu, A. (2016). Crop and plant biomass as valuable material for BBB. Alternatives for valorization of green wastes. In *Biotransformation of agricultural waste and by-products: The food, feed, fibre, fuel (4F) economy*.

Dobermann, D., Swift, J. A., & Field, L. M. (2017). Opportunities and hurdles of edible insects for food and feed. *Nutrition Bulletin, 42*(4), 293–308.

Domingues, C. H. F., Borges, J. A. R., Ruviaro, C. F., Guidolin, D. G. F., & Carrijo, J. R. M. (2020). Understanding the factors influencing consumer willingness to accept the use of insects to feed poultry, cattle, pigs and fish in Brazil. *PLoS One, 15*.

Dórea, J. G. (2006). Fish meal in animal feed and human exposure to persistent bioaccumulative and toxic substances. *Journal of Food Protection, 69*, 2777–2785.

Duc, G., Marget, P., Esnault, R., Le Guen, J., & Bastianelli, D. (1999). Genetic variability for feeding value of faba bean seeds (*Vicia faba*): Comparative chemical composition of isogenics involving zero-tannin and zero-vicine genes. *Journal of Agricultural Science, 133*, 185–196.

Enami, H. R. (2011). A review of using canola/rapeseed meal in aquaculture feeding. *Journal of Fisheries and Aquatic Science, 6*, 22–36.

Epp, A. (2016). *Sind essbare Insekten als Lebensmittel aus Sicht der Verbraucher sicher?* BfR-Symposium Insekten als Lebens- und Futtermittel.

European Commission. (2018). *Report from the Commission to the Council and the European Parliament*. On the development of plant proteins in the European Union (Europäische Kommission, Hrsg.). Brussel.

Fiebelkorn, F. (2017). Entomophagie. Insekten als Nahrungsmittel der Zukunft. *Biologie in unserer Zeit, 2*(47), 104–110.

Finke, M. D., & Oonincx, D. G. A. B. (2017). Nutrient content of insects. In A. van Huis & J. K. Tomberlin (Eds.), *Insects as food and feed. From production to consumption* (pp. 291–316). Wageningen Academic Publishers.

Florou-Paneri, P., Christaki, E., Giannenas, I., Bonos, E., Skoufos, I., Tsinas, A., Tzora, A., & Peng, J. (2014). Alternative protein sources to soybean meal in pig diets. *Journal of Food, Agriculture & Environment, 12*, 655–660.

Food and Agriculture Organization of the United Nations (Hrsg.). (2017). The future of food and agriculture. Trends and challenges.

Gaillard, C., Sørensen, M. T., Vestergaard, M., Weisbjerg, M. R., Basar, A., Larsen, M. K., Martinussen, H., Kidmose, U., & Sehested, J. (2017). Effect of substituting soybean meal and canola cake with grain-based dried distillers grains with solubles as a protein source on feed intake, milk production, and milk quality in dairy cows. *Journal of Dairy Science, 100*, 7980–7989.

Halloran, A., Roos, N., Eilenberg, J., Cerutti, A., & Bruun, S. (2016). Life cycle assessment of edible insects for food protein: A review. *Agronomy for Sustainable Development, 36*(4), 57.

Heinrich-Böll-Stiftung. (2014). Fleischatlas. Extra: Abfall und Verschwendung. Heinrich-Böll-Stiftung e.V., Berlin.

Hastad, C. W., Nelssen, J. L., Goodband, R. D., Tokach, M. D., Dritz, S. S., DeRouchey, J. M., & Frantz, N. Z. (2005). Effect of dried distillers grains with solubles on feed preference in growing pigs. *Journal of Animal Science, 83*, 73.

Huber, J. & Keller, M. (2017). *Ernährungsphysiologische Bewer-tung von konventionell und ökologisch erzeugten vegetarischen und veganen Fleisch- und Wurstalternativen.* S: Studie im Auftrag der Albert Schweitzer Stiftung für unsere Mitwelt, Berlin.

IDH and IUCN NL. (2019). *European Soy Monitor. Insights on the European supply chain and the use of responsible and deforestation-free soy in 2017.* https://www.idhsustainabletrade.com/uploaded/2019/04/European-Soy-Monitor.pdf.

Jezierny, D., Mosenthi, R., & Bauer, E. (2010). The use of grain legumes as a protein source in pig nutrition: A review. *Animal Feed Science and Technology, 157*, 111–128.

Jongema, Y. (Wageningen University & Research, Hrsg.). (2017). *List of edible insects of the world.*

Kay Z. (2014). *Nine protein alternatives for pig feeds.* https://www.wattagnet.com/articles/19229-nine-protein-alternatives-for-pig-feeds

Khojely, D. M., Ibrahim, S. E., Sapey, E., & Han, T. (2018). History, current status, and prospects of soybean production and research in sub-Saharan Africa. *The Crop Journal, 6*, 226–235.

Kleinschmit, D. H., Schingoethe, D. J., Hippen, A. R., & Kalscheur, K. F. (2007). Dried distillers grains plus Solubles with corn silage or alfalfa hay as the primary forage source in dairy cow diets. *Journal of Dairy Science, 90*, 5587–5599.

Kragbaek D. J. V. (2014). *Protein yields and amino acid characterization of juice and pulp from white clover, red clover, lucerne and ryegrass.* Aarhus University, Aarhus School of Engineering, Master Thesis.

Lindberg J.E., Lindberg G., Teräs J., Poulsen G., Solberg S.Ø., Tybirk K., Przedrzymirska J., Sapota G.P., Olsen M.L., Karlson H., Jóhannsson R., Smárason B.Ö., Gylling M., Knudsen M.T., Dorca-Preda T., Hermansen J.E., Kruklite Z. & Berzina I. (2016). Nordic alternative protein potentials.

Makkar, H. P. S., Tran, G., Heuzé, V., & Ankers, P. (2014). State-of-the-art on use of insects as animal feed. *Animal Feed Science and Technology, 197*, 1–33.

Mutsvangwa, T., Kiran, D., & Abeysekara, S. (2016). Effects of feeding canola meal or wheat dried distillers grains with solubles as a major protein source in low- or high-crude protein diets on ruminal fermentation, omasal flow, and production in cows. *Journal of Dairy Science, 99*, 1216–1227.

Napiórkowska-Krzebietke, A. (2017). Phytoplankton as a basic nutritional source in diets of fish. *Journal of Elementology, 22*, 831–841.

Nepstad, D. C., Stickler, C., & Almeida, O. T. (2006). Globalization of the amazon soy and beef industries: Opportunities for conservation. *Conservation Biology, 20*, 1595–1603.

Newton, G. L., Booram, C. V., Barker, R. W., & Hale, O. M. (1977). Dried Hermetia illucens larvae meal as a supplement for swine. *Journal of Animal Science, 44*, 395–400.

Newton L., Watson D.W., Dove R., Sheppard C. & Burtle G. (2005). Using the black soldier fly, *Hermetia illucens*, as a value-added tool for the management of swine manure. Animal and Poultry Waste Management Center, North Carolina State University, Raleigh, NC 17.

Osmane, B., Konosonoka, I. H., Trupa, A., & Proskina, L. (2017). Peas and beans as a protein feed for dairy cows. *Agronomy Research, 15*, 2026–2038.

Payne, C. L. R., Scarborough, P., Rayner, M., & Nonaka, K. (2016). Are edible insects more or less 'healthy' than commonly consumed meats? A comparison using two nutrient profiling models devel-oped to combat over- and undernutrition. *European Journal of Clinical Nutrition, 70*(3), 285–291.

Ramos-Elorduy, J., Moreno, J. M. P., Prado, E. E., Perez, M. A., Otero, J. L., & de Guevara, O. L. (1997). Nutritional value of edible insects from the state of Oaxaca, Mexico. *Journal of Food Composition and Analysis, 10*, 142–157.

Ribeiro, J. C., Cunha, L. M., Sousa-Pinto, B., & Fonseca, J. (2018). Allergic risks of consuming edible insects: a systematic review. *Molecular Nutrition and Food Research, 62*, 1–12. https://doi.org/10.1002/mnfr.201700030.

Riddick, E. W. (2014). Insect protein as a partial replacement for fishmeal in the diets of juvenile fish and crustaceans. *Mass Production of Beneficial Organisms*, 565–582.

Rumpold, B. A., & Schlüter, O. K. (2013). Nutritional composition and safety aspects of edible insects. *Molecular Nutrition & Food Research, 57*(5), 802–823.

Rumpold, B. A., Bußler, S., Jäger, H., & Schlüter, O. K. (2017). Insect processing. In A. van Huis & J. K. Tomberlin (Eds.), *Insects as food and feed. From production to consumption* (pp. 319–342). The Netherlands, Wageningen Academic Publishers.

Rusu, A., Alvarez Penedo, B., Engelhardt, S., & Schwarze, A.-K. (2020a). EXCORNSEED EU project: Separation, fractionation and isolation of biologically active natural substances from corn oil and other side streams to be used in food, specialty chemicals and cosmetic markets. *Bulletin UASVM Food Science and Technology, 77*(1).

Rusu, A., Alvarez Penedo, B., Schwarze, A.-K., & Trif, M. (2020b). The impact of Candida spp. on physiological alterations in gut microbiota. In T. Askun & J. S. M. Tondolo (Eds.), *Candida elimination diet therapy. Candidiasis*. IntechOpen. ISBN: 978-1-83881-112-9.

Rusu, A., Randriambelonoro, M., Perrin, C., et al. (2020c). Aspects influencing food intake and approaches towards personalising nutrition in the elderly. *Journal of Population Ageing, 13*, 239–256.

Rusu, A., Schwarze, A.-K., & Alvarez Penedo, B. (2020d). FUNGUSCHAIN EU project: Extracting value from the agricultural offcuts of commercial mushroom farming. *Bulletin UASVM Food Science and Technology, 77*(1).

Rusu, A., Alvarez Penedo, B., Bethke, M., Schwarze, A.-K., & Trif, M. (2020e). Smart technologies for personalized nutrition and consumer rngagement (Stance4health Eu H2020 – funded project). *Bulletin UASVM Food Science and Technology, 77*(1).

Rusu, A. V., Schwarze, A.-K., Bethke, M., Penedo, B. A., & Trif, M. (2020f). Tannin encapsulation for personalized product applications. *Mater. Proc., 2*, 5.

Sauvant D., Perez J.-M. & Tran G. (2004). *Table de composition et de valeur nutritive des matières premières destinées aux animaux d'élevage.*

Shahbandeh M. (2020). *Leading 10 soybean producing U.S. states 2019*. https://www.statista.com/statistics/192076/top-10-soybean-producing-us-states/

She, Y., Su, Y., Liu, L., Huang, C., Li, J., Li, P., Li, D., & Piao, X. (2015). Effects of microbial phytase on coefficient of standardized total tract digestibility of phosphorus in growing pigs fed corn and corn co-products, wheat and wheat co-products and oilseed meals. *Animal Feed Science and Technology, 208*, 132–144.

Sogari, G., Amato, M., Biasato, I., Chiesa, S., & Gasco, L. (2019). The potential role of insects as feed: A multi-perspective review. *Animals, 9*(4), 119.

St-Hilaire, S., Cranfill, K., McGuire, M. A., Mosley, E. E., Tomberlin, J. K., Newton, L., Sealey, W., Sheppard, C., & Irving, S. (2007a). Fish offal recycling by the black soldier fly produces a foodstuff high in omega-3 fatty acids. *Journal of the World Aquaculture Society, 38*, 309–313.

St-Hilaire, S., Sheppard, C., Tomberlin, J. K., Irving, S., Newton, L., McGuire, M. A., Mosley, E. E., Hardy, R. W., & Sealey, W. (2007b). Fly prepupae as a feedstuff for rainbow trout, Oncorhynchus mykiss. *Journal of the World Aquaculture Society, 38*, 59–67.

Stiftung Warentest. (2016). *Vegetarische Schnitzel & Co* (Vol. 10, pp. 20–29). Stiftung Warentest.

Stiles W. (2017). Potential sources of protein for animal feed: Insects.

Trif, M., Muresan, L., & Bethke, M. (2016). Personalised nutritional powder for elderly developed in OPTIFEL European project. *Bulletin UASVM Food Science and Technology, 73*(2).

Trif, M., Vodnar, D. C., Mitrea, L., Rusu, A. V., & Socol, C. T. (2019). Design and development of oleoresins rich in carotenoids coated microbeads. *Coatings, 9*(4), 235.

Trif, M., Braicu, C., & Rusu, A. (2020). Antiviral and immunostimulant activities of protein value chains from residual biomass. In D. M. Martirosyan (Ed.), *Functional foods and viral diseases* (Vol. 8, 1st ed.). Food Science Publisher. ISBN-13: 979-8675209637.

van der Spiegel, M., Noordam, M. Y., & van der Fels-Klerx, H. J. (2013). Safety of novel protein sources (insects, microalgae, seaweed, duckweed, and rapeseed) and legislative aspects for their application in food and feed production. *Comprehensive Reviews in Food Science and Food Safety, 12*, 662–678.

van Huis A., Van Itterbeeck J., Klunder H., Mertens E., Halloran A., Muir G. & Vantomme P. (2013). *Edible insects: Future prospects for food and feed security*. FAO Forestry Paper 171.

World Cancer Research Fund; American Institute for Cancer Research. (2018). *Continuous update project expert report 2018: Recommendations and public health and policy implications* (World Cancer Research Fund (WCRF) & American Institute for Cancer Research, Hrsg.)

World Resources Institute. (2019). *Creating a sustainable food fu-ture: A menu of solutions to feed nearly 10 billion people by 2050*.

WWF International (Ed.). (2014). *The growth of soy: Impacts and solutions*. WWF International, Hrsg..

Zander, P., Amjath-Babu, T. S., Preissel, S., Reckling, M., Bues, A., Schläfke, N., Kuhlman, T., Bachinger, J., Uthes, S., Stoddard, F., et al. (2016). Grain legume decline and potential recovery in European agriculture: A review. *Agronomy for Sustainable Development, 36*, 26.

Zanetti, F., Monti, A., & Berti, M. T. (2013). Challenges and opportunities for new industrial oilseed crops in EU-27: A review. *Industrial Crops and Products, 50*, 580–595.

Index

C. M. Galanakis (ed.), *Biodiversity, Functional Ecosystems and Sustainable Food Production*, https://doi.org/10.1007/978-3-031-07434-9

The manufacturer's authorised representative in the EU is Springer Nature Customer Service Centre GmbH, Europaplatz 3, 69115 Heidelberg, Germany. If you have any concerns regarding our products, please contact ProductSafety@springernature.com

Printed and bound by CPI Group (UK) Ltd, Croydon, CR0 4YY
28/09/2026
02224091-0016